Eocene Biodiversity

Unusual Occurrences and Rarely Sampled Habitats

TOPICS IN GEOBIOLOGY

Series Editors: F. G. Stehli, DOSECC, Inc., Gainesville, Florida
Douglas S. Jones, University of Florida, Gainesville, Florida

Current volumes in this series

Volume 4 THE GREAT AMERICAN BIOTIC INTERCHANGE
Edited by Francis G. Stehli and S. David Webb

Volume 5 MAGNETITE BIOMINERALIZATION AND MAGNETORECEPTION IN ORGANISMS
A New Biomagnetism
Edited by Joseph L. Kirschvink, Douglas S. Jones, and Bruce J. McFadden

Volume 6 *NAUTILUS*
The Biology and Paleobiology of a Living Fossil
Edited by W. Bruce Saunders and Neil H. Landman

Volume 7 HETEROCHRONY IN EVOLUTION
A Multidisciplinary Approach
Edited by Michael L. McKinney

Volume 8 GALÁPAGOS MARINE INVERTEBRATES
Taxonomy, Biogeography, and Evolution in Darwin's Islands
Edited by Matthew J. James

Volume 9 TAPHONOMY
Releasing the Data Locked in the Fossil Record
Edited by Peter A. Allison and Derek E. G. Briggs

Volume 10 ORIGIN AND EARLY EVOLUTION OF THE METAZOA
Edited by Jere H. Lipps and Philip W. Signor

Volume 11 ORGANIC GEOCHEMISTRY
Principles and Applications
Edited by Michael H. Engel and Stephen A. Macko

Volume 12 THE TERTIARY RECORD OF RODENTS IN NORTH AMERICA
William Korth

Volume 13 AMMONOID PALEOBIOGRAPHY
Edited by Neil H. Landman, Kazushige Tanabe, and Richard Arnold Davis

Volume 14 NEOGENE PALEONTOLOGY OF THE MANONGA VALLEY, TANZANIA
A Window into the Evolutionary History of East Africa
Edited by Terry Harrison

Volume 15 ENVIRONMENTAL MICROPALEONTOLOGY
The Application of Microfossils to Environmental Geology
Edited by Ronald E. Martin

Volume 16 PALEOBIOGEOGRAPHY
Bruce S. Lieberman

Volume 17 THE HISTORY AND SEDIMENTOLOGY OF ANCIENT REEF SYSTEMS
Edited by George D. Stanley Jr.

Volume 18 EOCENE BIODIVERSITY
Unusual Occurrences and Rarely Sampled Habitats
Edited by Gregg F. Gunnell

A Continuation Order Plan is available for this series. A continuation order will bring delivery of each new volume immediately upon publication. Volumes are billed only upon actual shipment. For further information please contact the publisher.

Eocene Biodiversity

Unusual Occurrences and Rarely Sampled Habitats

Edited by

Gregg F. Gunnell
Museum of Paleontology
Ann Arbor, Michigan

Kluwer Academic / Plenum Publishers
New York, Boston, Dordrecht, London, Moscow

Library of Congress Cataloging-in-Publication Data

Eocene biodiversity: unusual occurrences and rarely sampled habitats/edited by Gregg F. Gunnell.
p. cm. — (Topics in geobiology; v. 18)
Includes bibliographical references and index.
ISBN 0-306-46528-0
1. Vertebrates, Fossil. 2. Paleontology—Eocene. I. Gunnell, Gregg F. II. Series.

QE841 .E63 2001
566—dc21

00-052739

ISBN: 0-306-46528-0

233 Spring Street, New York, N.Y. 10013

http://www.wkap.nl/

10 9 8 7 6 5 4 3 2 1

A C.I.P. record for this book is available from the Library of Congress.

Printed in the United States of America

Contributors

John P. Alexander Division of Paleontology, American Museum of Natural History, New York, NY 10024-5192

Muhammad Arif Geological Survey of Pakistan, Quetta, Pakistan

Ana Maria Báez Departamento de Geología, Universidad de Buenos Aires, Buenos Aires, Argentina

William S. Bartels Department of Geological Seiences, Albion College, Albion, MI 49224

K. Christopher Beard Department of Vertebrate Paleontology, Carnegie Museum of Natural History, Pittsburgh, PA 15213-4080

Jonathan I. Bloch Department of Geological Sciences and Museum of Paleontology, University of Michigan, Ann Arbor, MI 48109

Gabriel J. Bowen Earth Sciences Department, University of California, Santa Cruz, Santa Cruz, CA 95064

Emily S. Bray University of Colorado Museum, Boulder, CO 80309-0315

Benjamin J. Burger Division of Paleontology, American Museum of Natural History, New York, NY 10024-5192

Robert Chandler Department of Biology, Georgia State College and University, Milledgeville, GA 31061

William C. Clyde Department of Earth Sciences, University of New Hampshire, Durham, NH 03824-3589

Jean-Yves Crochet Laboratoire de Paléontologie, Institut des Sciences de l'Evolution, CC64, Université Montpellier II, 34095 Montpellier Cédex 5, France

Mary R. Dawson Department of Vertebrate Paleontology, Carnegie Museum of Natural History, Pittsburgh, PA 15213-4080

Emmett Evanoff University of Colorado Museum, Boulder, CO 80309-0315

Jens Lorenz Franzen Forschungsinstitut Senckenberg, D-60325 Frankfurt am Main 1, Germany

Philip D. Gingerich Museum of Paleontology, University of Michigan, Ann Arbor, MI 48109-1079

Lance Grande Department of Geology, Field Museum of Natural History, Chicago, IL 60605-2496

Gregg F. Gunnell Museum of Paleontology, University of Michigan, Ann Arbor, MI 48109-1079

Munir ul-Haq Geological Survey of Pakistan, Quetta, Pakistan

Terry Harrison Department of Anthropology, New York University, New York, NY 10003

Jean-Louis Hartenberger Laboratoire de Paléontologie, Institut des Sciences de l'Evolution, CC64, Université Montpellier II, 34095 Montpellier Cédex 5, France

Bonnie Fine Jacobs Department of Geological Sciences, Southern Methodist University, Dallas, TX 75275-0395

Intizar Hussain Khan Geological Survey of Pakistan, Quetta, Pakistan

Kenneth R. Ludwig Berkeley Geochronology Center, Berkeley, CA 94709

Bernard Marandat Laboratoire de Paléontologie, Institut des Sciences de l'Evolution, CC64, Université Montpellier II, 34095 Montpellier Cédex 5, France

Claude Martinez ORSTOM Géofluides, Bassins, Eaux, CC58, Université Montpellier II, 34095 Montpellier Cédex 5, France

Elizabeth McGee Department of Biological Sciences, San José State University, San José, CA 95192-0100

Charles P. Msuya Department of Anatomy, Muhimbili University College of Health Sciences, Dar es Salaam, Tanzania

Roland Mundil Berkeley Geochronology Center, Berkeley, CA 94709

Paul C. Murphey University of Colorado Museum, Boulder, CO 80309-0315

Alison M. Murray Redpath Museum, McGill University, Montreal, QC H3A 2K6, Canada

D. Tab Rasmussen Department of Anthropology, Washington University, St. Louis, MO 63130

Kenneth D. Rose Department of Cell Biology and Anatomy, Johns Hopkins University School of Medicine, Baltimore, MD 21205

Bernard Sigé Centre de Paléontologie stratigraphique et Paléoécologie, Université Lyon I, 69622, Villeurbanne, France

Mary T. Silcox Department of Anthropology, Penn State University, University Park, PA 16802, mts 1ø@psu.edu

Gerhard Storch Forschungsinstitut Senckenberg, D-60325 Frankfurt am Main 1, Germany

Meg L. Thornton Department of Anthropology, Washington University, St. Louis, MO 63130

Lisa L. Torick University of Colorado Museum, Boulder, CO 80309-0315

James W. Westgate Department of Geology, Lamar University, Beaumont, TX 77710

Thomas E. Williamson New Mexico Museum of Natural History and Science, Albuquerque, NM 87104-1375

Preface

Biodiversity, as reflected by the incredible number of species living on earth and in the enormous complexity of their interactions, is a topic much in the news today. The consequences of human exploitation of dwindling planetary resources and the results of these actions on overall species diversity and survival are issues of great concern as governments and organizations grapple with policies that directly affect all organisms on earth. As efforts increase to understand the impact of human activity on biodiversity, the past has become an important resource for assessing pre-human global biodiversity and the consequences of naturally occurring perturbations on species diversity and longevity. The Eocene, the second Epoch of the Cenozoic Era ("The Age of Mammals") spanning the time period between 34 and 55 million years ago, has long been recognized as an important and interesting time period for several reasons. First, there is a rich Eocene fossil record from nearly every continent on earth. Many hundreds of thousands of fossils have been recovered from Eocene deposits in North America, Europe, Asia, and South America, spanning nearly the entire length of the Eocene. Africa is poorly represented except in the very latest part and Eocene rocks from Australia and Antarctica are just beginning to be examined.

Second, the Eocene was a time of great changes in global climates. The beginning of the Eocene was marked by an increase in global temperatures followed by a slow decrease through the early Eocene. Mean annual temperatures rose again towards the end of the early Eocene reaching their Cenozoic maximum into and through the middle Eocene and producing wide-spread equable climates, even at relatively high latitudes. The beginning of the late Eocene was marked by a dramatic drop in global temperatures. Because of the large number of fossil sites available for the Eocene, the impact of world-wide climatic change on extinction, origination, species distribution and diversity, and community structure can be examined in great detail.

Third, the beginning of the Eocene is marked by the initial appearance in the fossil record of many modern orders of mammals. The earliest representatives of primates (humans, apes, monkeys, prosimians), perissodactyls (odd-toed ungulates such as horses, tapirs, and rhinoceros), and artiodactyls (even-toed ungulates including sheep, goats, cattle, deer, among many others) all first appear at the beginning of the early Eocene. All of these orders, along with several others (notably rodents and carnivores) undergo broad diversification through the Eocene. Evolutionary processes and patterns that ultimately lead to modern members of these orders can be studied in detail from Eocene fossil assemblages.

The impetus for this volume grew out of my own 25 years of field work in the Eocene of North America and Asia. Initially, this work was designed to document and study the diversification of modern mammalian groups and was quite successful and satisfying. However, as field and laboratory work continued, there began to develop a suspicion that not all of the Eocene story was being told. It became apparent that most fossil samples, especially those from the American West, were derived from similar preservational circumstances and similar depositional settings. A program was initiated to look for other potential sources of fossil samples, either from non-traditional lithologies or from geographic areas that were not typically sampled. As this program of research grew it began to demonstrate that different lithologies and different geographic areas told different stories from those that had been developed based on more typical faunal assemblages.

This book was conceived as an introduction to non-traditional Eocene fossil samples, and as a place to document and discuss features of these fossil assemblages that are rare or unique. Chapters range from those that document special preservational aspects to those that come from rarely represented habitats. Specialized preservational circumstances include: lagerstätten where complete organisms are preserved, often with soft-body outlines (Chapters 1, 2, 7, 8); mass-death assemblages where a single taxon is preserved in large numbers (Chapters 13–14); and unique depositional settings where rare or unique taxa are preserved (Chapters 4–6, 11–12, and 15). Rarely sampled habitats include under-represented geographic areas (Chapters 3, 9–11) or different physiographic regions like upland areas (Chapter 16). Rare habitats can be an important aspect of unique depositional settings as well (see especially Chapters 4–5).

Authors of the various chapters present the unique and rare aspects of the fossil samples they discuss and then go beyond this to point out aspects that would not have been possible to document from more typical sarnples. These include such things as unique morphological forms and adaptations, unusual distributions or diversity patterns, and previously poorly understood habitat utilization patterns. This book demonstrates that, despite being one of the best sampled intervals of the Cenozoic, there is still much to be learned about the organisms that lived during the Eocene.

As with any work of this sort, many people have been involved with its production and it gives me great pleasure to acknowledge their contributions. First, and foremost, I thank all of the authors who participated in this volume. It was their hard work and dedication that made this book possible. Many of them served as reviewers of each others papers and many helped with ideas and discussion during the course of production. In addition, other colleagues gave freely of their valuable and limited time to help in research, review, illustration, photography, and discussion. I thank Ellen R. Miller, Dale A. Winkler, Mark D. Uhen, William J. Sanders, Kathleen M. Muldoon, Robyn J. Burnham,

Michael Gottfried, David Fox, Gerald, R. Smith, M. Elizabeth Strasser, Catherine Badgley, and Bonnie Miljour for their help during the production of this volume. I thank John P. Alexander who co-chaired and co-organized the symposium at the 1998 Society of Vertebrate Paleontology meetings in Snow Bird, Utah upon which this book is based. I thank my colleagues at Plenum. Amelia McNamara, Executive Editor, Kenneth Howell, Editor, and Arne Olsen, Manuel Alls, Jr. Editorial Assistants for their help, guidance, and patience. A special thanks goes to Terry Harrison for encouraging me to pursue this project and for guidance along the way. Finally, I thank my long-standing colleague and friend, William S. Bartels, for 25 years of field work, patience, discussion, support, and good times.

Gregg F. Gunnell.

Contents

Chapter 1 • An Updated Review of the Fish Faunas From the Green River Formation, the World's Most Productive Freshwater Lagerstätten

Lance Grande

Chapter 2 • Paleontological Investigations at the Eocene Locality of Mahenge in North-Central Tanzania, East Africa

Terry Harrison, Charles P. Msuya, Alison M. Murray, Bonnie Fine Jacobs, Ana Maria Báez, Roland Mundil, and Kenneth R. Ludwig

Chapter 3 • Early Wasatchian Mammals From the Gulf Coastal Plain of Mississippi: Biostratigraphic and Paleobiogeographic Implications

K. Christopher Beard and Mary R. Dawson

Chapter 5 • Unusual Vertebrate Microfaunas From the Willwood Formation, Early Eocene of the Bighorn Basin, Wyoming

Mary T. Silcox and Kenneth D. Rose

Chapter 6 • Stratigraphy and Taphonomy of Grizzly Buttes, Bridger Formation, and the Middle Eocene of Wyoming

John P. Alexander and Benjamin J. Burger

Chapter 7 • Taphonomic Analysis of the Messel Formation (Germany)

Jens Lorenz Franzen

Chapter 8 • Paleobiological Implications of the Messel Mammalian Assemblage

Gerhard Storch

Chapter 9 • The Eocene Mammalian Fauna of Chambi (Tunisia) in Its Geological Context

Jean-Louis Hartenberger, Jean-Yves Crochet, Claude Martinez, Bernard Marandat, and Bernard Sigé

Chapter 10 • Gandhera Quarry, A Unique Mammalian Faunal Assemblage From the Early Eocene of Baluchistan (Pakistan)

Philip D. Gingerich, Muhammad Arif, Intizar Hussain Khan, Munir ul-Haq, Jonathan I. Bloch, William C. Clyde, and Gregg F. Gunnell

Chapter 11 • Paleoecology and Biostratigraphy of Marginal Marine Gulf Coast Eocene Vertebrate Localities

James W. Westgate

Chapter 12 • Taphonomic Interpretation of Gnat-Out-of-Hell, an Early Uintan Small Mammal Locality in the Uinta Formation, Utah

Meg L. Thornton and D. Tab Rasmussen

Chapter 13 • A Mass Death Accumulation of *Coryphodon anthracoideus* (Mammalia: Pantodonta) at Roehler's *Coryphodon* Catastrophe Quarry (Lower Eocene, Wasatch Formation), Washakie Basin, Wyoming

Elizabeth McGee

Chapter 14 • *Meniscotherium* Mass-Death Assemblages

Thomas E. Williamson

Chapter 15 • Taphonomy, Fauna, and Depositional Environment of the *Omomys* Quarry, an Unusual Accumulation From the Bridger Formation (Middle Eocene) of Southwestern Wyoming (USA)

Paul C. Murphey, Lisa L. Torick, Emily S. Bray, Robert Chandler, and Emmett Evanoff

Chapter 16 • Basin Margins, Biodiversity, Evolutionary Innovation, and the Origin of New Taxa

Gregg F. Gunnell and William S. Bartels

Chapter 1

An Updated Review of the Fish Faunas From the Green River Formation, the World's Most Productive Freshwater Lagerstätten

LANCE GRANDE

LANCE GRANDE • Department of Geology, Field Museum of Natural History, Chicago, IL 60605-2496

Eocene Biodiversity: Unusual Occurrences and Rarely Sampled Habitats, edited by Gregg F. Gunnell, Kluwer Academic/Plenum Publishers, New York, 2001.

1. Introduction

The Green River Formation has provided what is clearly the richest, most diverse sample of Early Tertiary freshwater vertebrate faunas in the world (see numerous illustrations in Grande, 1984, for example). In addition to a wide array of non-fish fossils, the formation provides an unparalleled sample of the Early Tertiary fish fauna of North America, with hundreds of thousands of beautifully preserved skeletons belonging to at least 16 different families. The composition, relationships, and importance of the Green River Formation fish faunas have become increasingly evident as detailed studies on these faunas continue. An earlier review by Grande (1980, 1984) has been a commonly used identification guide to the fishes of the Green River Formation. It was written over 20 years ago when there were far fewer specimens, fewer detailed descriptions and almost no phylogenetic studies of Green River fish species together with other closely related fossil and living species. In light of progress since that earlier review, I believe that the time is right for a new synthesis. This paper provides an updated taxonomic assessment of the Green River fishes, summarizes recent works on them, and points out some remaining questions concerning various Green River taxa.

Institutional abbreviations used in this paper are as follows. **AMNH**, American Museum of Natural History, New York, New York; **BM**, Bell Museum, University of Minnesota, Minneapolis, Minnesota; **BMNH**, The Natural History Museum, London, England; **BYU**, Brigham Young University, Provo, Utah; **FMNH**, Department of Geology, Field Museum of Natural History, Chicago, Illinois; **KUVP**, University of Kansas Museum of Natural History, Lawrence, Kansas; **SMMP** Department of Paleontology, Science Museum of Minnesota, St. Paul, Minnesota; **UMA**, University of Massachusetts, Amherst; **USNM**, National Museum of Natural History, Washington, D.C.; **UW**, Geological Museum, University of Wyoming, Laramie, Wyoming; **YPM**, Peabody Museum of Natural History, Yale University, New Haven, Connecticut (includes Princeton collection). Because both fossil and living taxa are mentioned in this text, all exclusively fossil (i.e., extinct) taxa are preceded by a dagger symbol (†). Most size ranges given here are based on collections at FMNH, AMNH and BMNH except where noted otherwise.

2. The Green River System as a Paleogene Great Lakes Complex

The Green River Formation was deposited by one of the world's most extensive and long-lived Great Lakes systems. Parts of this system appear to have

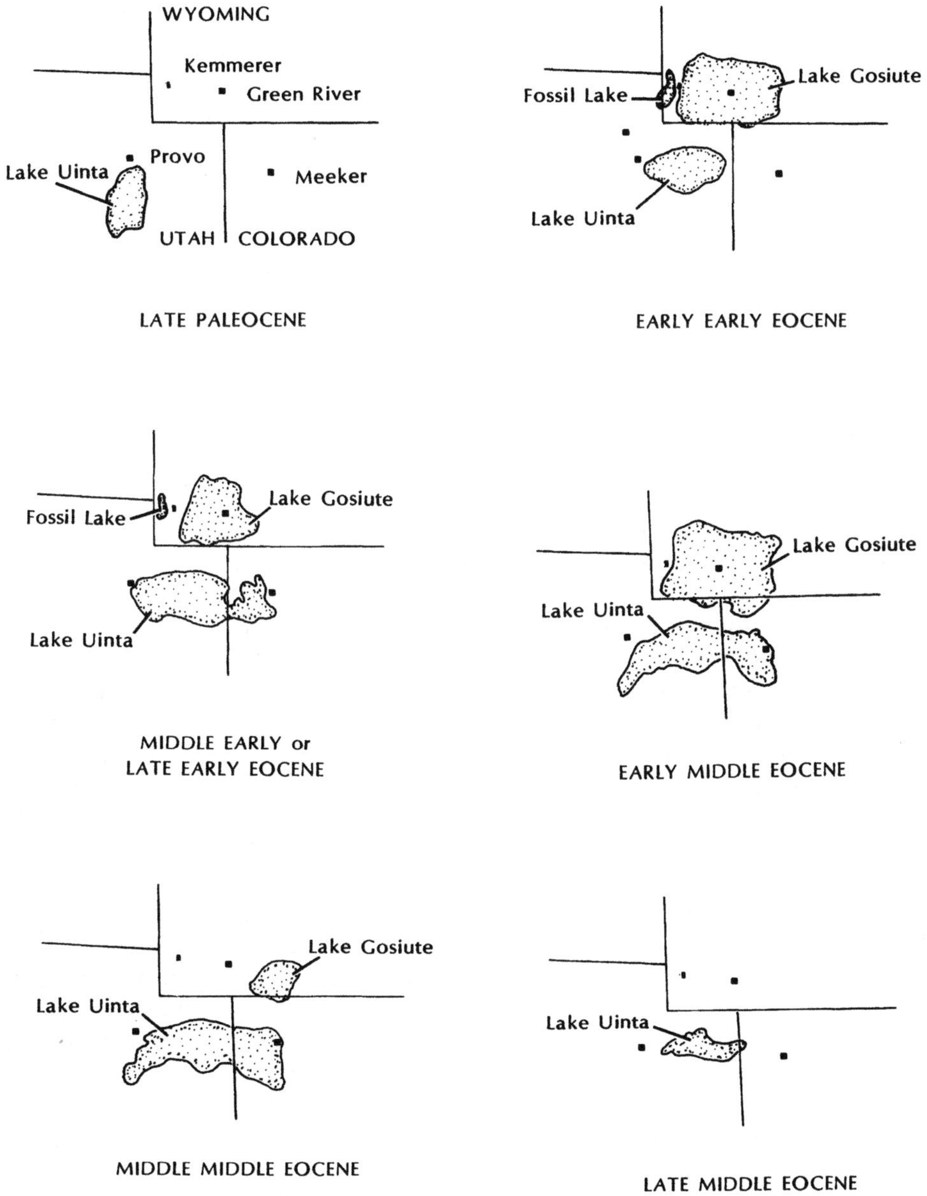

FIGURE 1. Maps of the Green River lake system during six points in time. After Schaeffer and Mangus (1965), McGrew and Casilliano (1975), and Grande (1984, 1989).

persisted from late Paleocene through late Eocene time (e.g., Figure 1), a longevity exceeding that of any extant lake system (Grande, 1994:15). The lake system is known to have included three lakes: Lake Uinta, Lake Gosiute, and Fossil Lake. Each lake had its own distinctive aquatic communities, environmental parameters, and duration (e.g., Grande, 1984).

Lake Uinta existed in what is now parts of Utah and Colorado, from late Paleocene through late Eocene time (Schaeffer and Mangus, 1965; McGrew and Casilliano, 1975; Grande, 1984:figures I.1, I.2; Figure 1). Although it appears to have been the longest lived of the three lakes and one of the two largest, its known fish fauna appears to have been relatively depauperate both in terms of numbers of fishes and taxonomic diversity (Grande, 1984:8, 9, 171–173). This may be a reflection of harsh ecological factors in the lake (e.g., Baer, 1969); but it may also be simply an artifact of poor preservation and collecting bias. The collecting bias is extreme, with the known sample size from Lake Uinta deposits being several orders of magnitude smaller than the enormous sample sizes known from Fossil Lake and Lake Gosiute deposits. Probably fewer than 1000 fishes have been excavated from Lake Uinta deposits. In contrast, sample sizes of tens of thousands to hundreds of thousands of fishes have been excavated from Fossil and Gosiute Lake deposits (Grande, 1984, 1999; Grande and Buchheim, 1994). Even with such large sample sizes from the Wyoming lakes, the known biodiversity there still continues to increase (Grande, 1999).

Lake Gosiute existed mostly in what is now southwestern Wyoming, although during its maximum extent, it reached into small parts of what is now northern Utah and Colorado (Figure 1; Buchheim and Surdam, 1981). Lake Gosiute sediments are known to have been deposited at least from early early Eocene through middle middle Eocene time (Schaeffer and Mangus, 1965; McGrew and Casilliano, 1975; Grande, 1984:figures I.1, I.2; Figure 1). The fish fauna of Lake Gosiute is abundant and diverse and is distinctive among the three lakes in that it contains abundant catfishes (†Hypsidoridae and Ictaluridae) and suckers (Catostomidae). The appearance of abundant catfishes and suckers is important because the groups are nearly unknown in the other two Green River lakes (particularly Fossil Lake), and Catostomidae and Ictaluridae make up an important part of the modern freshwater fish fauna of North America.

Fossil Lake existed in the extreme southwestern corner of Wyoming, west of Lake Gosiute, and at its maximum extent it may have reached into northern Utah (Schaeffer and Mangus, 1965; McGrew and Casilliano, 1975; Grande, 1984:figures I.1, I.2). Although it appears to have been the shortest lived and smallest of the three lakes (Figure 1), it has provided the largest and most diverse sample of fishes (e.g., Table 1). This is due mainly to the intense commercial quarrying activity in the area over the last 30 to 40 years. The fish fauna of Fossil Lake is distinctive among the three lakes in having extremely abundant percoids (belonging to at least four species). Percoids are extremely rare to absent in the known Lake Gosiute and Lake Uinta deposits. The prolific Fossil Lake deposits are also distinctive in their lack of suckers and catfishes, two groups abundantly represented in the Lake Gosiute deposits.

2.1. Abundance and Preservation of Fishes and Other Fossils

Within the Green River Formation, amateur, commercial and scientific excavation of fossils has concentrated mainly within the late early and early middle Eocene horizons of western Wyoming (hereafter termed EGRW for brevity). The EGRW deposits of Fossil Lake are from the late early Eocene Fossil Butte Member, and the EGRW deposits of Lake Gosiute are from the early middle Eocene Laney Member. These beds are enormously productive, containing numerous beautifully preserved fossil vertebrates (mostly fishes) that can be easily collected in large quantities using modern excavation techniques. The EGRW localities have provided the largest, most diverse samples of fossilized freshwater vertebrate communities anywhere on earth. The best collection, by far, of EGRW fishes in a public repository is at FMNH. Important collections also exist at UW, AMNH, BMNH, USNM, YPM and SMMP. Nearly every North American natural history museum has at least a small collection of these fishes.

In terms of productivity, diversity and fine skeletal preservation, the EGRW localities are unsurpassed by any other lagerstätten in the world, and are much more productive and diverse than any lagerstätten of freshwater origin. These lacustrine deposits have been mined for fossils for about 150 years, and have produced well over a million nearly complete to complete vertebrate skeletons. The deposits have produced an even greater number of non-vertebrate fossils (e.g., Grande, 1984), but this paper will focus only on the vertebrates. The most intensive commercial excavation of Green River fossils began in the 1960s and has generally continued to expand since then. Fishes make up the vast majority of vertebrate fossils in the formation. Grande and Buchheim (1994:34) conservatively estimated that during 1965 through 1990 approximately half a million fossil fish skeletons were excavated from the EGRW deposits of Fossil Lake, and that a similar number was excavated from the EGRW deposits of Lake Gosiute, representing an unparalleled sample size of approximately one million freshwater fossil vertebrate skeletons collected over 25 years. The fishes are not only abundant, they are also well preserved. Most specimens are complete skeletons, and extensive growth series exist for numerous species (McGrew, 1975; Grande, 1984, 1989, 1994; Grande and Buchheim, 1994).

3. An Updated Look at the Taxonomic Diversity of the Green River Fish Faunas

Most, if not all fish species of the Green River Formation can be found in the EGRW horizons. Paleontological evidence strongly suggests that the EGRW environment was freshwater subtropical during deposition (Grande, 1994). Saline phases were evidently present in parts of the Green River lakes at times (Buchheim and Surdam, 1981; Buchheim, 1994), but not within the EGRW vertebrate Lagerstätten horizons (Grande, 1994). The diverse freshwater fauna must have retreated via connecting streams to other freshwater sanctuaries during periods of elevated salinity in the lake. The valid species of Green River fishes are all briefly reviewed below, and are summarized in Table 1. The relative phylogenetic relationships among these species, as currently hypothesized, are illustrated in Figure 2.

Table 1. Updated taxonomic list of the Eocene fish fauna of the Green River Formation. For distribution column, F = Fossil Lake, G = Lake Gosiute, and U = Lake Uinta. The genus †"*Priscacara*" as currently used for species from the Green River Formation is probably not monophyletic (see discussion of Green River percoids in text on pp. 31).

Family	Genus	Species	Known distribution within Green River complex
Dasyatidae?	†*Heliobatis* Marsh, 1877	†*H. radians* Marsh, 1877	F
Dasyatidae?	†n. gen.	†n. sp.	F
Polyodontidae	†*Crossopholis* Cope, 1883	†*C. magnicaudatus* Cope, 1883	F
Lepisosteidae	*Lepisosteus*? Lacépède, 1803	†*L.*? *cuneatus* (Cope, 1885a)	F?, G, U
Lepisosteidae	"*Atractosteus*" Rafinesque, 1820	†"*A. simplex*" (Leidy, 1873a)	F, G?
Lepisosteidae	"*Atractosteus*" Rafinesque, 1820	†"*A. atrox*" (Leidy, 1873a)	F
Amiidae	*Amia* Linnaeus, 1766	†*A. pattersoni* Grande and Bemis, 1998	F, G
Amiidae	†*Cyclurus* Agassiz, 1844	†*C. gurleyi* (Romer and Fryxell, 1928)	F
Hiodontidae	†*Eohiodon* Cavender, 1966	†*E. falcatus* Grande, 1979	F
Osteoglossidae	†*Phareodus* Leidy, 1873b	†*P. encaustus* (Cope, 1871)	F, G
Osteoglossidae	†*Phareodus* Leidy, 1873b	†*P. testis* (Cope, 1877)	F
†Paraclupeidae	†*Diplomystus* Cope, 1877	†*D. dentatus* Cope, 1877	F, G, U
Clupeidae	†*Knightia* Jordan, 1907	†*K. eocaena* Jordan, 1907	F, G, U
Clupeidae	†*Knightia* Jordan, 1907	†*K. alta* (Leidy, 1873b)	F, G, U

3.1. Dasyatidae?

There are two stingray species known from the Green River Formation. Both are from the late early Eocene Fossil Butte Member deposits (Figure 3). One is †*Heliobatis radians* Marsh, 1877 (junior synonyms listed in Grande, 1984:24). This species reaches close to a meter in total axial length, but is also known

Family	Genus	Species	Known distribution within Green River complex
Clupeidae	†*Gosiutichthys* Grande, 1982b	†*G. parvus* Grande, 1982b	G
Gonorynchidae	†*Notogoneus* Cope, 1885b	†*N. osculus* Cope, 1885b	F
Catostomidae	†*Amyzon* Cope, 1872	†*A. gosiutensis* Grande et al., 1982	G
†Hypsidoridae	†*Hypsidoris* Lundberg and Case, 1970	†*H. farsonensis* Lundberg and Case, 1970	G
Ictaluridae	†*Astephus* Cope, 1873	†*A. antiquus* (Leidy, 1873a)	G, U, F*
Esocidae	*Esox* Linnaeus, 1758	†*E. kronneri* Grande, 1999	F*
Percopsidae	†*Amphiplaga* Cope, 1877	†*A. brachyptera* Cope, 1877	F
Percopsidae	†*Erismatopterus* Cope, 1871	†*E. levatus* (Cope, 1870)	G, U
†Asineopidae	†*Asineops* Cope, 1870	†*A. squamifrons* Cope, 1870	F, G, U
Percoidei *incertae sedis*	†*Mioplosus* Cope, 1877	†*M. labracoides* Cope, 1877	F, U
Percoidei *incertae sedis*	†*Priscacara* Cope, 1877	†*P. serrata* Cope, 1877	F, U
Percoidei *incertae sedis*	†"*Priscacara*" Cope, 1877	†"*P.*" *liops* Cope, 1877	F
Percoidei *incertae sedis*	†"*Priscacara*" Cope, 1877	†"*P.*" *hypsacantha* Cope, 1886	F

*Known by only one specimen from Fossil Lake.

from very young juvenile specimens (e.g., 85 mm specimen in Grande and Buchheim, 1994:Figure 8A). The other stingray species is an undescribed genus and species (description in progress by Carvalho, Grande and Maisey). This species is known from numerous young juveniles (e.g., Grande, 1984:figures II.6a and II.6b) up to large adults over 600 mm in total length (e.g., Figure 3B). The phylogenetic relationships of the Green River stingrays to other fossil and living stingrays are still unclear and in need of study. Although the Green River

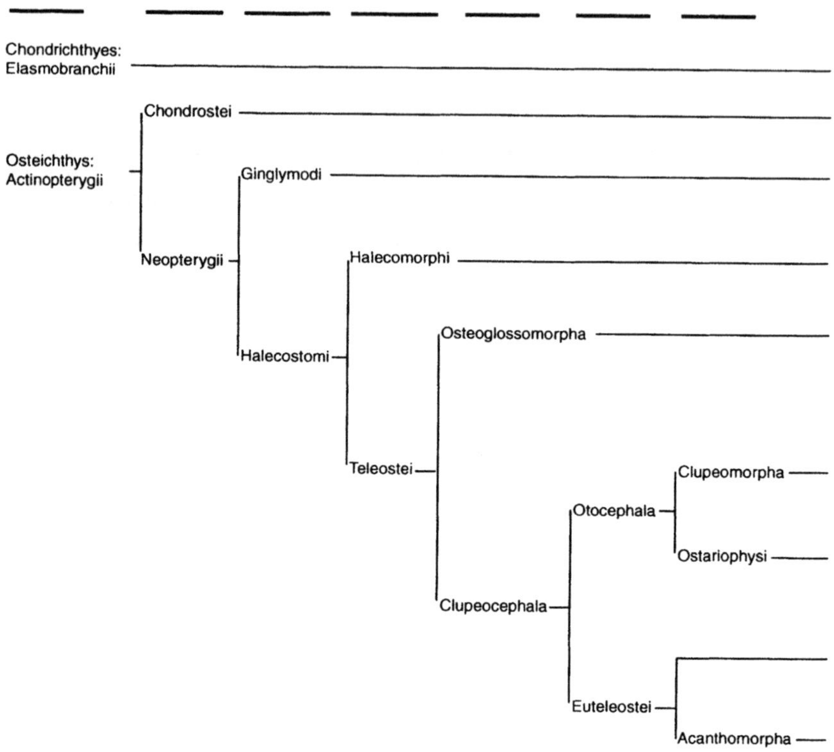

FIGURE 2. Branching diagram showing relative interrelationships among Green River fish taxa as far as known. Supraordinal actinopterygian relationships based on Grande *et al.*, in preparation. Columns of supraordinal taxa do not necessarily show taxa of equivalent rank. This is not a phylogeny of fishes (e.g., Elopomorpha is omitted because there are no elopomorphs in the Green River Formation). This diagram is meant only to show relative phylogenetic relationships among Green River Formation species. "Dasyatidae" is questionable as used here (see text).

stingrays have previously been included in the family Dasyatidae (e.g., Grande, 1984), some recent authors (e.g., Nishida, 1990; Miyake *et al.*, 1992) have divided the former Dasyatidae into more than one family. It is currently unclear whether the Green River rays belong to Dasyatidae or Urolophidae within these recent classifications. A comprehensive phylogenetic review of fossil and living batoids is needed to better identify the relationships of the Green River and other stingrays.

Stingrays are a relatively rare part of the known fish fauna, and are so far not reported from Lakes Gosiute or Uinta. In Fossil Lake, they are more common in the near-shore deposits (e.g., "F-2" of Grande and Buchheim, 1994)

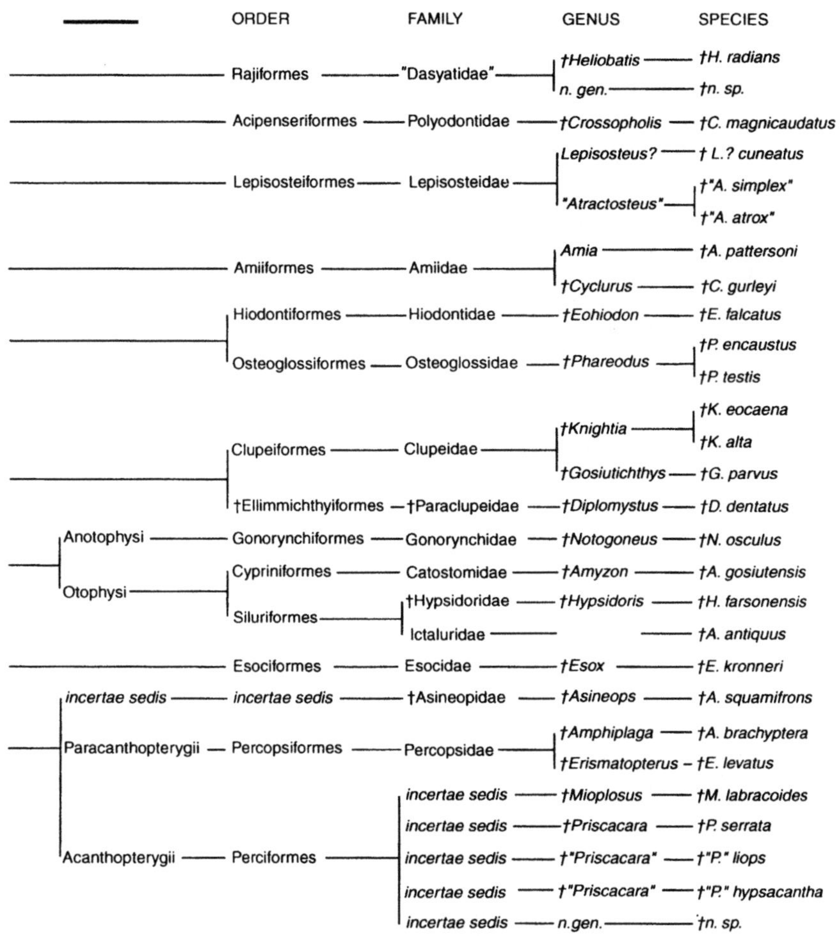

than in the mid-lake deposits. Together both stingray species make up less than 0.1% of the fish fauna in the quarry when they are most abundant (Grande, 1984:167–169).

3.2. Polyodontidae

There is one paddlefish in the Green River Formation: †*Crossopholis magnicaudatus* Cope, 1883 (Figure 4). This species is known only from Fossil Lake, where it is known by specimens from 260 mm total length to 1480 mm total length (Grande and Bemis, 1991). Within the Fossil Lake localities †*Crossopholis* is most common in the near-shore deposits (Grande and Buchheim, 1994), and in beds several meters above the F-1 beds of Grande and Buchheim (1994). †*Crossopholis* was recently redescribed in detail along with other members of the family by Grande and Bemis (1991, 1996) and Grande et al. (in Press).

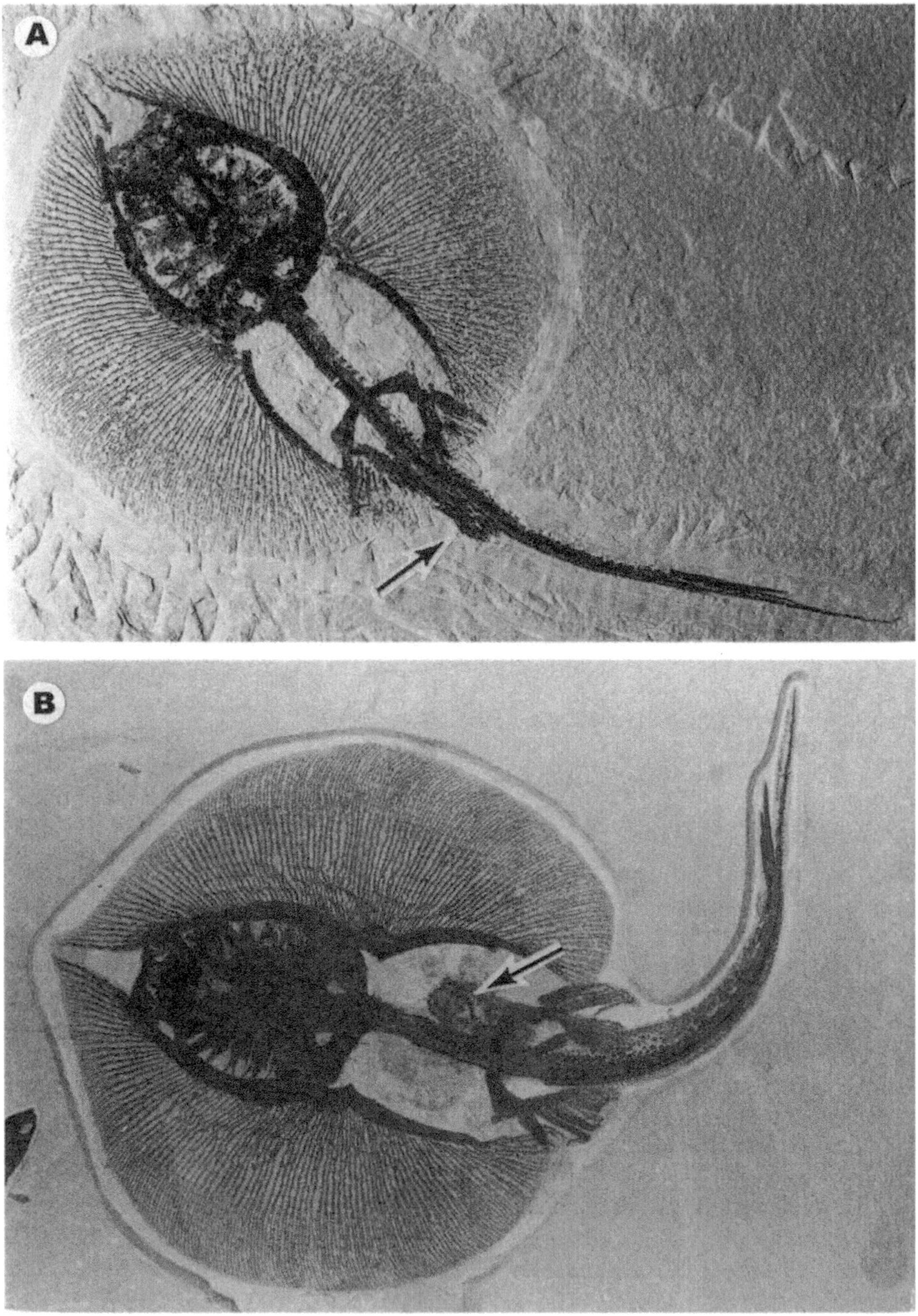

FIGURE 3. Two stingray species ("Dasyatidae") from the Green River Formation. **A**) †*Heliobatis radians* Marsh, 1877 (SMMP 83.2.4; 380 mm total axial length). Male specimen (arrow points to claspers). **B**) An undescribed genus and species (FMNH PF15166; 628 mm total axial length). Female specimen (arrow points to embryos preserved in pelvic region).

FIGURE 4. The Eocene paddlefish (Polyodontidae), †*Crossopholis magnicaudatus* Cope, 1883 (FMNH PF11897; 384 mm total length), from the Green River Formation. Photographic negative flopped for print so fish faces left.

It was found to be the sister taxon to the extant Mississippi River drainage paddlefish, *Polyodon*. †*Crossopholis* and *Polyodon* together make up a group that is the sister taxon to the living Chinese paddlefish, *Psephurus*. †*Crossopholis* is one of the rarest described fishes from Fossil Lake (Grande, 1984:167–169). Nevertheless, because of the enormous number of fishes that have been excavated from the Fossil Lake deposits over the last twenty years, dozens of †*Crossopholis* skeletons are now known.

3.3. Lepisosteidae

There are several well-preserved gar species known from the Green River Formation (e.g., Figure 5), but their anatomy and relationships are still in need of much study. Grande (1980, 1984) recognized three different gar species in the Green River Formation, including two typical long-snouted gars from EGRW and one peculiar, extremely short-snouted form from Lake Uinta deposits; but this preliminary key was based on only a very small sample size. The Green River gars (including a much larger sample size), along with all other fossil and living gars, are currently the focus of a comprehensive phylogenetic study by Grande and Bemis (in progress). One of the anticipated results of that study will be an improved understanding of the Green River gar taxonomy.

The generic assignments of the Green River gars are also in need of further study. The monophyly (and validity) of the genus *Atractosteus* is controversial (Wiley and Schultze, 1984; Gottfried and Krause, 1998:276–277). The long-snouted Green River gars (Figure 5A, B) have been included in *Lepisosteus* by some authors (e.g., Grande, 1980, 1984) but in *Atractosteus* by others (e.g., Wiley, 1976). In addition, the short-snouted Green River gar, †"*Lepisosteus*" *cuneatus* (Cope, 1885a) has never been described in detail. Preliminary work on many new well-preserved specimens (in progress) indicates that this and one or two other undescribed species may belong to a widespread group of primitively short-headed, blunt-toothed forms lacking several derived features

FIGURE 5. Three gar species (Lepisosteidae) from the Green River Formation. For explanation of quotation marks, see text. **A)** †"*Atractosteus simplex*" (Leidy, 1873a) (FMNH PF12597; 674 mm total length). This is one of two long-snouted species of gar known from the Green River Formation. Photographic negative flopped for print so fish faces left. **B)** †"*Atractosteus atrox*" (Leidy, 1873a) (FMNH PF14931; 735 mm total length). This is the second of two long-snouted species of gar known from the Green River Formation. It is the largest known species in the Green River deposits, reaching at least two meters in length. **C)** †"*Lepisosteus*" *cuneatus* (Cope, 1885a) (BMNH P.61229; 350 mm total length). This very short-snouted gar species may belong in a new genus (Grande and Bemis, in progress).

of all living and most fossil gar species. These short-headed forms may necessitate the naming of a new gar genus.

The two documented species of long-snouted gars from the Green River Formation have been referred to as †"*Atractosteus simplex*" (Leidy, 1873a) and †"*A. atrox*" (Leidy, 1873a) by some authors (e.g., Wiley, 1976; Wiley and Schultze, 1984). Other authors have referred to these taxa as †"*Lepisosteus simplex*" Leidy, 1873a, and †"*L. atrox*" Leidy, 1873a (e.g., Eastman, 1900; Hay, 1902; Grande, 1980, 1984). As with the genus names, the commonly used species names for the long-snouted Green River gars are problematic (thus the quotation marks). The problems result from the names "*simplex*" and "*atrox*" being tied to fragmentary holotypes that may make those names *nomen dubium* (*sensu* Grande and Bemis, 1998:18–19). Although the questionable validity of the names for these and other species of gars is currently under study (Grande and Bemis, in progress), there are at least two separate species of long-snouted gars in the Green River Formation. The "*atrox*" type can be differentiated from the "*simplex*" type by its characteristic ornamentation on the skull bones and anterior scales, in having a greater number of scales and caudal fin rays (e.g., Figure 5A and 5B), and in having more elongate premaxillae (Grande, 1984:Figure II.18). The "*atrox*" type is the largest known fish species in the Green River Formation, with one articulated UW specimen measuring over two meters (seven feet) in total length, and another disarticulated specimen of similar size in the FMNH collection. The "*simplex*" type is known to reach a meter in total length. †"*Atractosteus simplex*" and †"*A. atrox*" occur mainly in Fossil Lake deposits, where they are relatively rare (less than 0.05% of the fish fauna). The long-snouted gars are much rarer in Lake Gosiute and Lake Uinta deposits, where they are so far known by only a few individuals not identified to species (e.g., a juvenile illustrated in Grande, 1984:Figure II.14).

The described short-snouted Green River gar (Figure 5C) has so far been considered by all recent authors (e.g., Grande, 1980, 1984; Wiley, 1976) to belong in *Lepisosteus*. The short-snouted Green River gar was first described from Lake Uinta deposits (†*Lepisosteus cuneatus*) where it is now known by over 100 individuals (e.g., Grande, 1984:figures II.16, II.17). The Lake Uinta specimens rarely exceed 350 mm in total length. At first the short-snouted gars were thought to be absent from Fossil Lake deposits (e.g., Grande, 1980, 1984), but more recently a number of short-snouted specimens have been discovered from those deposits as well (Grande and Buchheim, 1994; Grande and Bemis, in progress), including some large individuals (e.g., one UMA specimen 940 mm in total length). It has not yet been determined whether the short-snouted species from Fossil Lake is conspecific with †*Lepisosteus cuneatus* or if it represents a new species. Within Fossil Lake, the short-snouted variety appears to be the rarest of the gar species and is known by only a few specimens, whereas in Lake Uinta the short-snouted †*L. cuneatus* is the most common gar species and is known by over 100 specimens (e.g., mass-mortality blocks deposited at FMNH, USNM, SMMP, BYU, and AMNH, and a collection of individuals at KUVP). One short-snouted gar specimen has also been reported from Lake Gosiute deposits (Grande, 1984:41). It was in a private collection, and its current status is unknown. Young juvenile gar specimens from the Green River Formation (e.g., Grande and Buchheim, 1994:Figure 8C) are exceedingly rare.

3.4. Amiidae

There are two valid bowfin species in the Green River Formation (Figure 6), each belonging to a different genus of the subfamily Amiinae (*sensu* Grande and Bemis, 1998). Although the articulated skeletons of Green River amiids have earlier been assigned to other species and/or genera, detailed study of the family Amiidae (Grande and Bemis, 1998) indicates that the two valid species are †*Cyclurus gurleyi* (Romer and Fryxell, 1928), and †*Amia pattersoni* Grande and Bemis, 1998. For a comprehensive phylogenetic review and description of this family (including the two Green River species), see Grande and Bemis (1998).

FIGURE 6. Two bowfin species (Amiidae) and genera from the Green River Formation. **A)** †*Amia pattersoni* Grande and Bemis, 1998 (Holotype, FMNH 14091; 1092 mm total length). **B)** †*Cyclurus gurleyi* (Romer and Fryxell, 1928) (Holotype, FMNH UC2201; 683 mm total length).

Table 2. Some distinguishing features between the two valid species of Amiidae from the Green River Formation, based on the study sample in Grande and Bemis (1998). For more detailed explanation and comparison see that volume.

Distinguishing features	**†*Cyclurus gurleyi***	**†*Amia pattersoni***
Total vertebral centra	70-79 (n = 5)	83-87 (n = 8)
Total preural caudal centra	63-72 (n = 5)	75-79 (n = 8)
Number of teeth on adult premaxilla	10-12* (n = 7)	6-8 (n = 11)
Body shape of adult	shorter/deeper, football-shaped (Grande and Bemis, 1998:fig. 158; table 62)	more elongate (Grande and Bemis, 1998:fig. 120; table 42)
Anterior coronoid teeth	short and blunt (Grande and Bemis, 1998:fig. 134)	elongate and sharply pointed (Grande and Bemis, 1998:fig. 134)
Gular shape	relatively shorter and wider (Grande and Bemis, 1998:fig. 136)	relatively longer and narrower (Grande and Bemis, 1998:fig. 136)

*One juvenile specimen not included here had 9 (Grande and Bemis, 1998:table 64).

Both amiid species occur in Fossil Lake and Lake Gosiute deposits, although most known specimens of both species are from Fossil Lake deposits. Even so, in the Fossil Lake deposits amiids are relatively rare, making up far less than 0.1% of the fish fauna (Grande, 1984:167–169). Amiid material known from Lake Uinta deposits is even rarer than from the other two Green River lakes and consists only of fragments unassignable to species.

Previous studies of amiids from the Green River Formation such as Boreske (1974) and Grande (1980, 1984) were based on much smaller study samples of fossil and living material than the study of Grande and Bemis (1998). The Grande and Bemis study clearly showed that †*Amia fragosa* and †*Amia uintaensis* are not valid identifications for the Green River species, and that the two Green River species do not belong in the same genus. The identification key in Grande (1980, 1984) used to distinguish between the two Green River species was based on a very small sample size, and should be superseded by the distinguishing features listed in Grande and Bemis (1998). Some of the main distinguishing features are given here in Table 2. †*Cyclurus gurleyi* is known to reach about 789 mm total length, and †*Amia pattersoni* exceeds 1400 mm in total length (Grande and Bemis, 1998). Young juvenile amiid specimens are extremely rare in the Green River Formation.

3.5. Hiodontidae

There is one species representing the gold-eye and moon-eye family in the Green River Formation, †*Eohiodon falcatus* Grande, 1979 (Figure 7). This species occurs in Fossil Lake and is unknown from the other two lakes. It is a relatively rare component of the known Fossil Lake fish fauna (making up far less than 0.1% of the known fish fauna; Grande, 1984:167–169) and within Fossil Lake, nearly all known specimens are from the near-shore deposits ("F-2" deposits of Grande and Buchheim, 1994). It has been suggested (Li *et al.*, 1997) that the Green River species, †*E. falcatus*, may be a junior synonym of the Eocene Washington State and Canadian species, †*E. woodruffi* Wilson, 1978, based on similar meristic and measurement data for the two nominal species. Detailed morphological comparisons of the skull morphology between these two nominal species are needed to better resolve the validity or synonymy of the species. This species is known to reach 200 mm in total length, and several young juveniles are also known (Grande and Buchheim, 1994; Li *et al.*, 1997).

3.6. Osteoglossidae

There is one osteoglossid genus with two valid species in the Green River Formation (Figure 8): †*Phareodus encaustus* (Cope, 1871) and †*Phareodus testis* (Cope, 1877). †*Phareodus encaustus* reaches a total length of about 650 mm, but †*P. testis* is known up to only 350 mm total length (Li *et al.*, 1997). Both species occur in Fossil Lake, where together they make up less than six

FIGURE 7. The Eocene mooneye (Hiodontidae), †*Eohiodon falcatus* Grande, 1979 (FMNH PF12516; 197 mm total length), from the Green River Formation.

FIGURE 8. Two osteoglossid species (Osteoglossidae) from the Green River Formation. **A)** †*Phareodus encaustus* (Cope, 1871) (UMA F10155; 570 mm total length). **B)** †*Phareodus testis* (Cope, 1877) (FMNH PF12411; 177 mm total length).

percent of the fish fauna (Grande, 1984:tables II.8–II.10). Synonymies of other nominal species of †*Phareodus* are listed in Grande (1980, 1984:9–74). These synonymies were verified in Li *et al.* (1997). Osteoglossids are much scarcer in Lake Gosiute and Lake Uinta sediments than in the Fossil Lake deposits. The genus †*Phareodus* also occurs in Australia and Europe, and the Australian species is more closely related to the Green River species than the European species (Li *et al.*, 1997).

3.7. †Paraclupeidae

The genus †*Diplomystus* was determined to be a non-clupeiform clupeomorph (i.e., within Clupeomorpha but outside Clupeiformes) by Grande (1982a, 1985) following previous work by Patterson (1967), who discovered that certain fossil clupeomorphs lacked some derived features of living clupeomorphs. †*Diplomystus* was revised in Grande (1982a, 1985) and placed in the new family †Ellimmichthyidae and order †Ellimmichthyiformes. In a later study (Chang and Grande, 1997) the name †Paraclupeidae Chang and Chou (1977) was found to have priority over the name †Ellimmichthyidae Grande (1982a). Grande (1982a) was unaware of the name proposed by Chang and Chou (1977) for the †*Diplomystus* family group.

†*Diplomystus dentatus* Cope, 1877 (Figure 9), from the Green River Formation, is the type species of the genus †*Diplomystus*. It is the only valid species of the genus in the formation (junior synonyms listed in Grande, 1980, 1984:91). It is also the youngest known occurrence of the Early Cretaceous through Middle Eocene order, †Ellimmichthyiformes. †*Diplomystus dentatus* is relatively common in the Fossil Lake deposits, making up 28% to 40% of the fish community in some quarries (Grande, 1980, 1984:tables II.8–II.10). It is relatively scarcer in the Lake Gosiute and Lake Uinta deposits, where it makes up less than one percent of the fish communities where surveyed (e.g., Grande, 1980, 1984:tables II.11, II.12). This species reaches its largest known size in Fossil Lake deposits, where it can reach 650 mm total length. Larval specimens of this species from 18 to 22 mm total length are very common in the mid-lake deposits of Fossil Lake (Grande and Buchheim, 1994).

Another species of †*Diplomystus*, very closely related to the Green River species, occurs in Eocene freshwater deposits of China (Zhang *et al.*, 1985; Grande, 1985; Chang and Grande, 1997). The Chinese species was first

FIGURE 9. A primitive clupeomorph fish of the family †Paraclupeidae, †*Diplomystus dentatus* Cope, 1877 (FMNH PF11935; 531 mm total length), from the Green River Formation. Photographic negative flopped for print so fish faces left.

described, but unnamed, in Chang and Chow (1978). It was also referred to as "†*Diplomystus* n. sp. A" in Grande (1985). It was named †*D. Shengliensis* in Zhang *et al.* (1985).

3.8. Clupeidae

The herring family is represented in the Green River Formation by at least three species in two genera (Figure 10). This was clearly the dominant family of fishes in the Green River lake system for most of the lakes' known duration, particularly in Fossil Lake and Lake Gosiute. The genus †*Knightia* (subfamily Pellonulinae) is currently thought to contain two valid species in the Green River Formation (Grande, 1982b, 1984): †*K. eocaena* Jordan, 1907 (by far the more common of the two), and †*K. alta* (Leidy, 1873b), a deeper-bodied form (see Grande, 1982b, 1984). These two species are separated only by body depth, which is somewhat arbitrary because the body depth of one species grades into that of the other (Grande, 1984). †*Knightia* ranges from extremely deep-bodied (e.g., Grande, 1984:Figure II.39) to extremely slim (e.g., Grande, 1984:Figure II.37), with all grades of intermediates. Such an extreme range of morphometric variation indicates that there are probably at least two species of †*Knightia* in the Green River Formation, but other diagnostic characters congruent with body depth ranges are needed to make species identification less arbitrary. Growth series for both nominal species can be found in Fossil Lake deposits (e.g., Grande, 1982b:figures 1, 5).

The Green River species of †*Knightia* are in need of further systematic/morphological study. A look at the comparative osteology of living clupeids indicates that an unambiguous key to fossil clupeids such as †*Knightia* may be extremely difficult to construct. There are about 200 species of extant clupeids (Nelson, 1994) and over 120 nominal species of fossil clupeids (Grande, 1985), and the skeletal morphologies of many different species are very similar or indistinguishable from each other. And even when osteological differences are known for living clupeid species, they often consist of structural or numerical differences of gill rakers or other fine details not visible in most fossil specimens. Consequently, an unambiguous, detailed understanding of †*Knightia*'s species diversity is not yet available, and may always be somewhat arbitrary.

In Fossil Lake, †*Knightia* makes up from 30% to over 99% of the fish community in the surveyed quarries (e.g., Grande, 1980, 1984:tables II.8–II.10), and over 99% in certain horizons several meters above the so-called "F-1" beds of Grande and Buchheim (1994). †*Knightia eocaena* also occurs in mass mortality horizons of up to hundreds of individuals per square meter, indicating there were vast schools of these fishes (e.g., Grande, 1984:Figure II.41; Schaeffer and Mangus, 1965:20, 21). †*Knightia* is also extremely common in some Lake Gosiute quarries (e.g., over 75% of the fish fauna at the Farson locality reported in Grande, 1980, 1984:170–171). In other Lake Gosiute quarries, the dominant fish is another clupeid, †*Gosiutichthys* (discussed below). Collections at

FIGURE 10. Three species of the herring family (Clupeidae) from the Green River Formation. **A)** †*Knightia eocaena* Jordan, 1907 (FMNH PF12389; 81 mm total length). Subfamily: Pellonulinae (*sensu* Grande, 1982b). **B)** †*Knightia alta* (Leidy, 1873b) (FMNH PF9671; 124 mm total length). Subfamily: Pellonulinae (*sensu* Grande, 1982b). **C)** †*Gosiutichthys parvus* Grande, 1982b (AMNH 10480; 46 mm total length). Subfamily: Clupeinae (*sensu* Grande, 1982b).

AMNH indicate that †*Knightia* was also a proportionately dominant taxon in at least some Lake Uinta deposits. †*Knightia* reached its largest known size in the Fossil Lake deposits, where it sometimes exceeds 200 mm (FMNH PF12033, 206 mm total length). Juvenile specimens of †*Knightia* are also common in the Fossil Lake deposits. Species of †*Knightia* also occur in Paleocene freshwater deposits of Montana (Grande, 1982b) and Eocene freshwater deposits of China (Chang and Chow, 1978; Grande, 1985; Zhang *et al.*, 1985).

The genus †*Gosiutichthys* (subfamily Clupeinae) is represented in the Green River Formation by one species, †*Gosiutichthys parvus* Grande, 1982 (Figure 10C). This species is so far reported only from Lake Gosiute, where it is extremely common. In some of the Lake Gosiute quarry sites it makes up over 99.9% of the fish fauna (Grande, 1982b:18) and occurs in mass mortalities with densities up to thousands per square meter. This species is much smaller than †*Knightia*. It has a maximum known total length of 80 mm and rarely exceeds 40 mm. Although this species was first noticed by the author as a specimen mounted in a belt buckle, and first described in 1982, this species is now known by tens if not hundreds of thousands of specimens, most of which have been commercially collected, probably from pits around 18-Mile Canyon, near Green River, Wyoming. So far, †*Gosiutichthys* has not been reported from Fossil Lake or Lake Uinta.

3.9. Gonorynchidae

The family Gonorynchidae is represented in the Green River Formation by one species: †*Notogoneus osculus* Cope 1885b (Figure 11). The Green River species is the type species for the genus, which ranges from Late Cretaceous to late Oligocene in age, and is known from North America, Australia, Europe and The Ukraine (Grande and Grande, 1999). †*Notogoneus osculus* is a relatively scarce component of the fish fauna in Fossil Lake sediments and

FIGURE 11. The gonorynchiform species (Gonorynchidae), †*Notogoneus osculus* Cope, 1885b (FMNH PF11968; 527 mm total length), from the Green River Formation.

FIGURE 12. The sucker (Catostomidae), †*Amyzon gosiutensis* Grande *et al.*, 1982 (AMNH 10460; 255 mm total length), from the Laney Member Lake Gosiute deposits of the Green River Formation. Catostomids, like the catfish families †Hypsidoridae and Ictaluridae, are well known from Lake Gosiute deposits but conspicuously absent from Fossil Lake deposits.

evidently was not a near-shore species. In the Fossil Lake deposits, †*Notogoneus* is known from the mid-lake F-1 deposits, but generally absent from the near-shore F-2 and Warfield Springs deposits (Grande and Buchheim, 1994:44). In the F-1 quarries where it is most abundant, it makes up less than 3% of the fish community (e.g., Grande, 1980, 1984:table II.8). This species is now known by hundreds of nearly complete skeletons between 200 and 1000 mm in total length because of the intensive excavation activity within the "F-1" Fossil Lake deposits. Specimens less than 200 mm are scarce. †*Notogoneus* is as yet unknown from Lake Gosiute and Lake Uinta deposits.

3.10. Catostomidae

The sucker family, Catostomidae, is known from only one of the three Green River Formation lake deposits: the Laney Member, Lake Gosiute deposits. The family is represented in these deposits by †*Amyzon gosiutensis* Grande *et al.* 1982 (Figure 12), a species that is occasionally found together with the ictalurid catfish, †*Astephus antiquus* (e.g., Grande, 1980, 1984:Figure II.59) or the percopsiform, †*Erismatopterus*, and frequently with the clupeid, †*Gosiutichthys*. †*Amyzon* is a relatively rare member of the fish fauna and found mostly within the †*Gosiutichthys* beds near 18-Mile canyon and near the Fontanelle Reservoir, where it is now known by well over 100 nearly complete skeletons in museum and private collections. Specimens in the FMNH collection range from 70 to 240 mm total length, which is the size range known to me. Very young juveniles have not been reported. †*Amyzon gosiutensis* appears to be extremely similar to †*Amyzon aggregatum* Wilson, 1977, from the Eocene of British Columbia. Based on meristic and simple morphometric comparisons between the two species, Bruner (1991) suggested that they may be conspecific. Before such a synonymy is fully accepted, a more detailed comparison of the skull morphology of the Wyoming and the British Columbian forms should be made.

The apparent absence of suckers (Catostomidae) from the fish-rich Fossil Lake deposits is worth noting (e.g., Grande and Buchheim, 1994:45) because of their abundance in the Lake Gosiute deposits. This inter-lake faunal difference is discussed in SUMMARY.

3.11. †Hypsidoridae

The catfish †*Hypsidoris farsonensis* Lundberg and Case, 1970 (Figure 13) was originally described (Lundberg and Case, 1970) as belonging in the extant North American family Ictaluridae. Grande (1987), in a more detailed description based on much better-preserved and -prepared material, showed that †*Hypsidoris* did not (as first described) have the diagnostic features of Ictaluridae. He also showed that †*Hypsidoris* was sister group to a group containing over 30 nominal families of catfishes. Consequently this genus was placed in its own family †Hypsidoridae and superfamily †Hypsidoroidea. Later studies (Grande and Lundberg, 1988; Arratia, 1992; Grande and dePinna, 1998) corroborated this classification. A second species of the genus, †*Hypsidoris oregonensis* from the Eocene of Oregon, was later added to the family (Grande and dePinna, 1998), and the family currently includes only these two species.

Within the Green River Formation, †*Hypsidoris* is currently known only from the "Farson beds" of the Laney Member Lake Gosiute deposits, and is unknown from Fossil Lake or Lake Uinta. It makes up far less than 0.1% of the fish fauna in the Farson beds, and is known today by at least fifty specimens in museum (FMNH and AMNH) and private collections. Known size range of this species is 37 to 281 mm total length (Grande, 1987). Juveniles are extremely rare for this species. For further discussion see Grande (1987) and Grande and De Pinna (1998).

3.12. Ictaluridae

There is one valid species of ictalurid catfish from the Green River Formation, †*Astephus antiquus* (Leidy, 1873a) (Figure 14). Grande and Lundberg (1988) provide the most recent revision, detailed description and phylogenetic assessment of this species. †*Astephus* is known by well over 200 individuals from Lake Gosiute and Lake Uinta, but only a single incomplete individual has ever been reported from Fossil Lake deposits. Unfortunately, this specimen was in a private collection and is now lost. In Lake Gosiute deposits, where it is most abundant, †*Astephus* makes up from 0% to 0.2% of the fish fauna in the quarries indirectly surveyed (e.g., Grande, 1980, 1984:table II.11). At the Farson locality ("G-4" of Grande, 1980, 1984) †*Astephus* occurs together with another

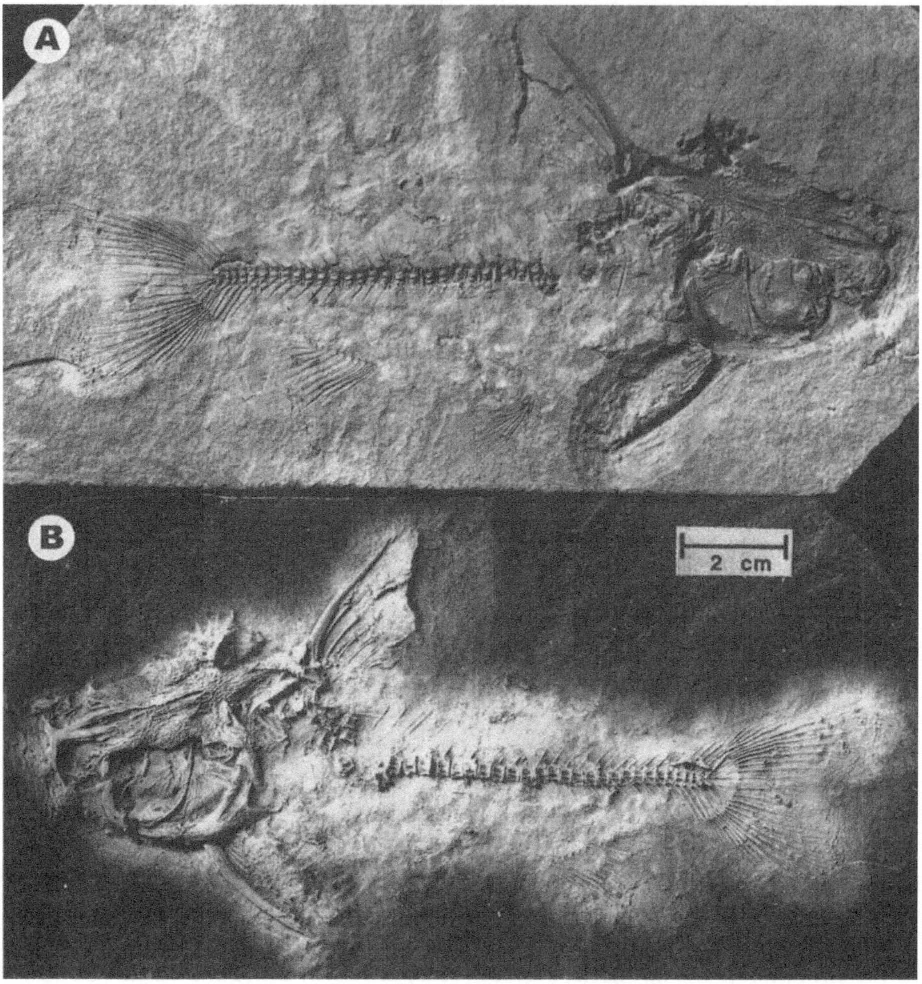

FIGURE 13. The extinct catfish family †Hypsidoridae is represented in the Green River Formation by †*Hypsidoris farsonensis* Lundberg and Case, 1970. This species is known only from the Laney Member "Farson Beds" (see text) where they are preserved as detailed negative impressions in matrix with most or all bone material missing. In order to study the skeletal detail, latex rubber peels are used. **A**) Original specimen showing negative impression in matrix (FMNH PF10356; 171 mm total length). **B**) Ammonium chloride coated black latex peel of **A**.

catfish, †*Hypsidoris farsonensis* (see Grande, 1987; Grande and Lundberg, 1988; and above). The size range of †*A. antiquus* at FMNH and BMNH is 43–323 mm total length (Grande and Lundberg, 1988), but specimens up to 600 mm exist in private collections. †*Astephus* is a phylogenetically primitive member of the family Ictaluridae (Lundberg, 1975; Grande and Lundberg, 1988).

FIGURE 14. The North American catfish family Ictaluridae is represented in the Green River Formation by †*Astephus antiquus* (Leidy, 1873a). This specimen is BMNH P.61233; 190 mm total length. Ictalurids are virtually unknown from Fossil Lake deposits, but known by hundreds of specimens from Lake Gosiute deposits. Note preservation of skin and barbels.

3.13. Esocidae

The order Esociformes, and family Esocidae (the pike and pickerel family) was unknown in the Green River Formation until very recently. Its late discovery (Grande, 1998) is remarkable because well over 1 million fish skeletons had been excavated from the Green River Formation over the last 100 years before the first esociform specimen was discovered. The Green River species, †*Esox kronneri* Grande 1999 (Figure 15), appears to be a pickerel. It is known

FIGURE 15. The pike and pickerel family, Esocidae, is represented in the Green River Formation by the pickerel †*Esox kronneri* Grande, 1999 (Holotype FMNH PF14918; 118 mm total length). This is the only esociform specimen known from the Green River Formation. Photographic negative flopped for print so fish faces left.

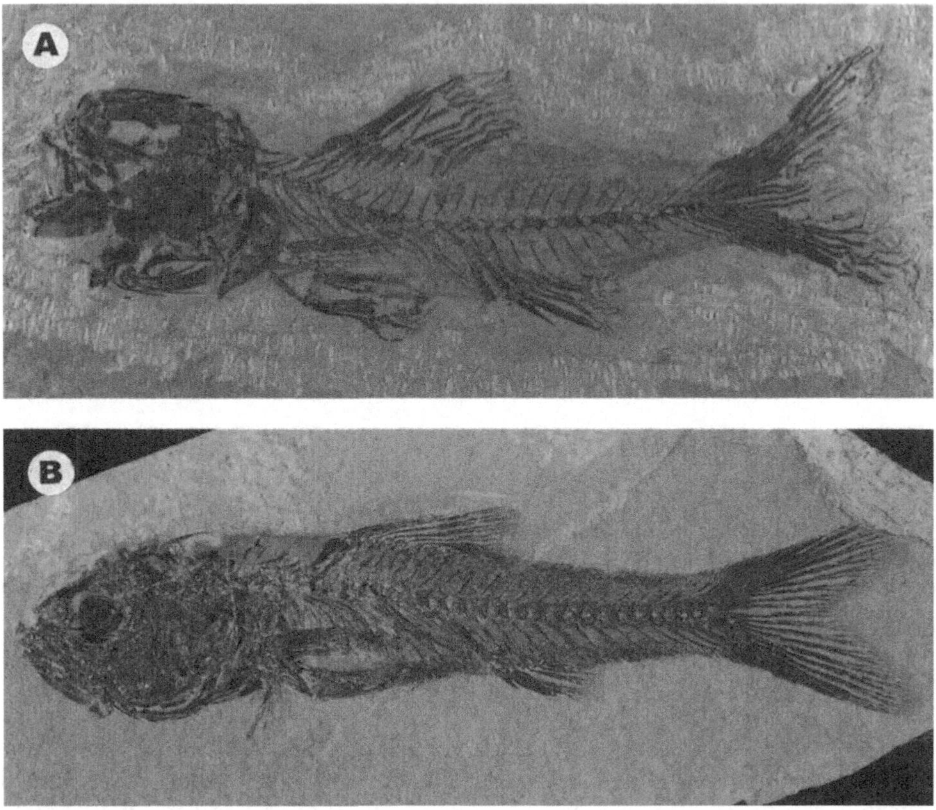

FIGURE 16. Two species of the trout-perch family, Percopsidae, from the Green River Formation. **A)** †*Amphiplaga brachyptera* Cope, 1877 (FMNH PF9888; 61 mm total length). Photographic negative flopped on print so fish faces left. **B)** †*Erismatopterus levatus* (Cope, 1870) (FMNH PF10326; 73 mm total length).

by only a single and well-preserved specimen from Fossil Lake (118 mm total length), making it the rarest of the fish species known from the Green River Formation. For further details, see Grande (1999).

3.14. Percopsidae

There are two species of the trout-perch family in the Green River Formation (Figure 16): †*Amphiplaga brachyptera* Cope 1877 and †*Erismatopterus levatus* (Cope, 1870). Like other percopsids, these are small species. †*Amphiplaga* has a maximum known size of about 150 mm total length, and †*Erismatopterus* has a maximum known size of 100 mm. †*Amphiplaga brachyptera* is known primarily from Fossil Lake sediments, and †*Erismatopterus brachyptera* is known primarily from Lake Gosiute and Lake Uinta deposits (e.g., Grande, 1980, 1984:table II.12). In the Fossil Lake quarries,

†*Amphiplaga* is extremely rare in the mid-lake quarries (i.e., the F-1 quarries *sensu* Grande and Buchheim, 1994) and much more abundant in the near-shore F-2 quarries (*sensu* Grande and Buchheim, 1994). There are only a few known specimens from the mid-lake F-1 quarries, but hundreds, perhaps thousands have been excavated from the near-shore F-2 quarries (where they make up about 1% of the fish fauna in a sample documented in Grande, 1980, 1984:table II.10). The Lake Gosiute and Lake Uinta percopsid, †*Erismatopterus*, is rare in most localities, but in certain Lake Gosiute deposits it is occasionally found in mass-mortalities of up to 100 per square meter (e.g., Grande, 1984:Figure II.74). The two Green River percopsids were described in detail in Rosen and Patterson (1969), and the placement of these two species in Percopsidae was verified more recently by Patterson and Rosen (1989).

3.15. †Asineopidae

There is one valid species of this extinct family present in the Green River Formation: †*Asineops squamifrons* Cope, 1870 (Figure 17). It is reported from deposits of all three of the Green River Formation lakes, but it is rare in the Fossil Lake and Lake Uinta deposits. In the Lake Gosiute quarries that have been surveyed (e.g., Grande, 1980, 1984:table II.11) it makes up far less than 1% of the fish fauna, and it is much rarer in the known Fossil Lake deposits (e.g., Grande, 1980, 1984:tables II.8–II.11) and Lake Uinta deposits. Hundreds of specimens are known from Lake Gosiute deposits in museum and private collections, but only about one or two dozen specimens are known from the

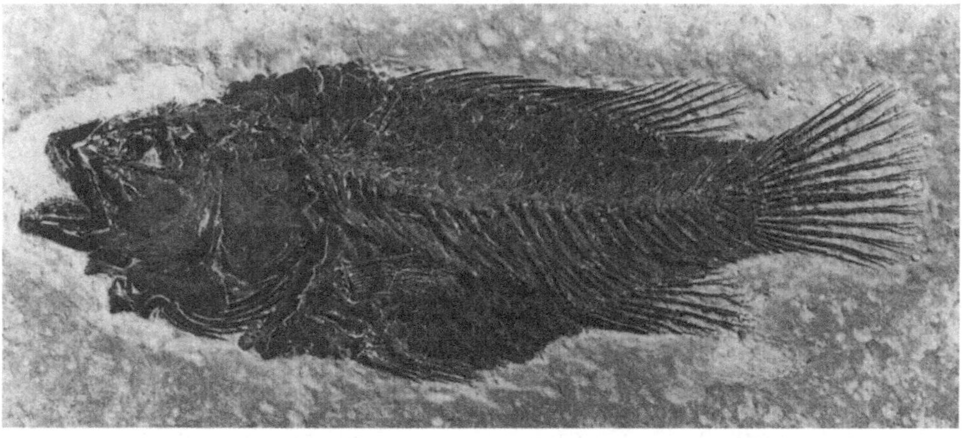

FIGURE 17. The enigmatic acanthomorph, †*Asineops squamifrons* (Cope, 1870) (FMNH PF13033; 102 mm total length), from the Green River Formation. This species is usually included in the family †Asineopidae, which still remains Acanthomorpha *incertae sedis*. Photographic negative flopped on print so fish faces left.

FIGURE 18. †*Mioplosus labracoides* Cope, 1877, one of the numerous percoids of the Fossil Lake deposits of the Green River Formation. This specimen is FMNH PF10332; 225 mm total length. The phylogenetic relationships of this species are still poorly known; consequently family assignment is still tenuous (see text). Photographic negative flopped on print so fish faces left.

Fossil Lake F-1 and F-2 deposits of Grande and Buchheim (1994). One locality within Fossil Basin (not F-1 or F-2) contains a mass mortality of small †*Asineops* (Buchheim, pers. comm., 1998). Although this species is much rarer in the Fossil Lake F-1 and F-2 deposits than in Lake Gosiute deposits, the specimens known from F-1 and F-2 are usually much larger than those from Lake Gosiute (e.g., 3 specimens at FMNH exceed 30 cm in length, a length not known to be attained by Lake Gosiute specimens). Maximum known length for this species is about 335 mm total length in Fossil Lake deposits, and about 170 mm total length in Lake Gosiute deposits.

†*Asineops squamifrons* was first described by Cope (1870, 1885) as having an affinity to the pirate-perch family, Aphredoderidae. Later, with more detailed study (e.g., Rosen and Patterson, 1969), this species was found to lack the diagnostic characters of that family. †*Asineops* currently remains in its own family, †Asineopidae, whose relationships within Acanthomorpha are currently unknown (Figure 2; Patterson, 1993:639). Thus it is not yet clearly assignable to order. †*Asineops squamifrons* was described in detail in Rosen and Patterson (1969).

3.16. Percoids (Family Relationships Uncertain)

There are numerous percoid species described from the Fossil Lake deposits of the Green River Formation including at least four valid described and one undescribed species (Figures 18, 19, 20). The described species are usually grouped together in two genera: †*Mioplosus* and †*Priscacara*. Percoids are conspicuously absent from most of the Lake Gosiute and Lake Uinta deposits. Although Buchheim and Surdam (1981) report †*Priscacara* as part of the Lake Gosiute fauna, the only example they provide (Buchheim and Surdam, 1981:Figure 14–9B) is not from Lake Gosiute deposits, but from Fossil Lake deposits (personal observation of their illustrated specimen at UW). †*Asineops* specimens from Lake Gosiute deposits are sometimes misidentified as †*Priscacara* specimens in museum collections and by collectors in the field. I

FIGURE 19. More percoids from the Green River Formation. These are species of †*Priscacara* and †"*Priscacara*" (see text for explanation of quotation marks). The family relationships of these species are uncertain (see text). They occur almost exclusively in the Fossil Lake deposits of the Fossil Butte Member. **A)** †*Priscacara serrata* Cope, 1877 (FMNH PF10296; 230 mm total length). This is the type species of the genus. **B)** †"*Priscacara*" *liops* Cope, 1877 (BM uncataloged; 130 mm total length).

FIGURE 20. More percoids from the Green River Formation. Like the species in Figures 18 and 19, the family relationships of these species are uncertain (see text). Both of these species are known only from the Fossil Lake deposits within the Green River Formation. **A)** †"*Priscacara*" *hypsacantha* Cope, 1886, a valid percoid species in the form genus †"*Priscacara*" that should eventually be placed into a different genus. This specimen is the holotype (AMNH 2453; 61 mm total length). **B)** "New percoid genus A" (description in progress). Specimen is FMNH PF14941; 85 mm total length. Photographic negative flopped on print so fish faces left.

have not personally seen any specimens of †*Priscacara* or †*Mioplosus* from Lake Gosiute deposits, so I suspect that if such specimens exist, they are extremely rare.

†*Mioplosus*, according to Grande (1980, 1984), contains one valid species: †*M. labracoides* Cope, 1877 (Figure 18). This species is known to reach a total length of at least 420 mm (Grande, 1984:Figure II.82b). Several subjective junior synonyms for this species are also listed in Grande (1980, 1984:142). †*Mioplosus* is a common fish in the Fossil Lake deposits, known by thousands of specimens in museums and private collections. In surveyed quarries of Fossil

Lake, it makes up two to over four percent of the fish fauna (Grande, 1980, 1984:tables II.8–II.10). There is even a mass-mortality horizon of small †*Mioplosus* together with numerous snails (*Viviparus* sp.) several meters above the F-1 beds of Fossil Lake. †*Mioplosus* has not been reported from Lake Gosiute sediments, and it is extremely rare in Lake Uinta deposits. †*Mioplosus* has often been placed in the family Percidae by various authors (e.g., Cope, 1885a; Woodward, 1901; Berg, 1940; Romer, 1966; Grande, 1980, 1984) but more recently it has been placed in Percichthyidae (Cavender, 1986:720). The interrelationships of living percoids need to be better resolved before the precise relationships of †*Mioplosus* can be adequately understood.

†*Priscacara* is currently used as a non-monophyletic "form-genus" with regard to Green River taxa, and it is sometimes included in its own family †Priscacaridae (Jordan, 1923; Berg, 1940; Grande, 1980, 1984). By "form-genus" I mean that the taxon has been used for all of the percoids from the Green River Formation that have a sub-oval body outline. In that sense, the name is used here in quotation marks when it refers to other than the type species. The monophyly of this genus, as it is currently used, is unsupported by empirical data and somewhat doubtful. This is largely because the interrelationships of the thousands of living percoid species are very poorly understood, and because sub-oval body shapes occur in many different percoid families. There are at least three valid described species in the Green River Lake deposits that are usually included in †*Priscacara* (Figure 19, 20A). These are †*P. serrata* Cope, 1877 (the type species); †"*P.*" *liops* Cope, 1877; and †"*P.*" *hypsacantha* Cope, 1886. †*Priscacara serrata* (Figure 19A) has heavy molariform pharyngeal teeth, thick fin spines, and reaches 400 mm in length. †"*Priscacara*" *liops* is a smaller form rarely exceeding 150 mm in length, lacking the heavy molariform pharyngeal teeth, and possessing thinner fin spines and more numerous median fin rays (Figure 19B). It is sometimes placed in the genus †*Cockerellities* Jordan (1923), and a case could be made to maintain it in that genus unless unambiguous character information can be found to tie †"*P.*" *liops* to †*P. serrata*. The third valid species, †"*Priscacara*" *hypsacantha* Cope, 1886 (Figure 20A), is a small species with a maximum total length of about 70 mm. This species should probably be assigned to a new genus for two reasons. First, there is no evidence suggesting that it is more closely related to †*Priscacara serrata* than to other percoid taxa. Second, there are some significant morphological differences between this species and †*P. serrata* (e.g., note the strongly spinate posterior edge of the preopercle in Figure 20A). New names and hypotheses of relationship for this species will not be made here, but will be addressed in a forthcoming project. Most or all other nominal species of †"*Priscacara*" from the Green River Formation (†*P. clivosa* Cope, 1878; †*P. pealei* Cope, 1878; †*P. cypha* Cope, 1877; †*P. dartonae* Eastman, 1917; †*P. oxyprion* Cope, 1878; and †*P. testudinaria* Cope, 1885a; types listed in Grande, 1980, 1984:150) are subjective junior synonyms of the three valid species discussed above).

†*Priscacara serrata* and †"*P.*" *liops* are common in the mid-lake "F-1" quarries (*sensu* Grande and Buchheim, 1994) of Fossil Lake, where together these two species make up over 20% of the fish fauna in the surveyed quarries (Grande, 1980, 1984:tables II.8, II.9). In the near-shore Fossil Lake quarries of

F-2, these two percoids (particularly †"*P.*" *liops*) are not as common, and they make up less than 3% of the fish fauna in surveyed quarries (Grande, 1980, 1984:table II.10). †"*Priscacara*" *hypsacantha* is the rarest described percoid known from the Green River Formation, and it is known only from Fossil Lake deposits. It is known by only a few specimens. Of the three †*Priscacara* or †"*Priscacara*" species listed above, †"*P.*" *liops* is by far the most common. It even occurs in mass-mortality horizons of up to several individuals per square meter within at least two lamina of the F-1 unit of Fossil Lake (*sensu* Grande and Buchheim, 1994).

In Lake Gosiute and Lake Uinta deposits, †"priscacarids" and other percoids are extremely rare or absent. Grande (1980, 1984:table II.12) reported that †"*Priscacara*" was extremely rare but present in Lake Gosiute deposits based on reports published elsewhere by other authors. Since then I have checked these specimens and found them all to be †*Asineops*. In addition, I have subsequently been unable to personally establish whether †"*Priscacara*" is present *at all* in the Laney Member Lake Gosiute sediments. †"*Priscacara*" is known from Lake Uinta sediments by only a few specimens described by Cope (1885a) as †"*Priscacara*" *testudinaria* (a species possibly synonymous with †*Priscacara serrata*).

The phylogenetic relationships of the †*Priscacara* and †"*Priscacara*" species are yet unclear. They have been included in Cichlidae by Woodward (1901: = his "Chromidae", 554), Pellegrin (1904), Haseman (1912), Berg (1940:476) and others. Cope (1877, 1885a) suggested an affinity with the Pomacentridae for these fishes. More recently, Cavender (1986:720) placed these species in his interpretation of Percichthyidae. As with †*Mioplosus*, unambiguous phylogenetic placements of †"priscacarids" await better phylogenetic resolution among the living percoid species.

A probable fifth species of percoid known from the Fossil Lake deposits (Figure 20B) is an undescribed taxon referred to in Grande (1980, 1984:159–160) as "new percoid genus A". It is similar to (and possibly conspecific with) †"*Priscacara*" *hypsacantha*, in being small (less than 100 mm total length) and having long spines on the posterior and ventral margins of the preopercle. It appears to differ from †"*P.*" *hypsacantha* in having a more elongate body shape and possibly other characters currently under study. This small percoid is a relatively rare species, making up less than 0.1% of all percoids found in surveyed quarries. The highest known concentration of this species is in the near-shore F-2 deposits, and it is nearly absent from the mid-lake F-1 deposits. There are over 30 specimens of this species in the collection at FMNH. Paul Buchheim (pers. comm., 1999) indicated that he has discovered a new Fossil Lake locality that may have a mass mortality horizon of this species.

4. Summary of Inter-Lake Faunal Comparisons

In addition to providing anatomical and phylogenetic information, the huge Eocene sample sizes from Fossil Lake and Lake Gosiute allow us to make some interesting faunal comparisons. Clupeomorphs (†Paraclupeidae and particularly Clupeidae) were, by far, the dominant fishes (proportionately) in both lakes. In a sample of 6667 fishes from three Fossil Lake quarries, clupeomorphs make up from 67% to 70% of the fishes found there (Grande, 1980, 1984:167–169); and in Lake Gosiute quarries, clupeomorphs make up from over 75% to over 99.9% of the fishes found (Grande, 1982b, 1984:171). The most striking differences between the Fossil Lake and the Lake Gosiute fish faunas are: A) the apparent absence of suckers and near absence of catfishes known from Fossil Lake deposits (vs. their abundance in Lake Gosiute deposits), and B) the apparent absence to extreme rarity of percoids from known Lake Gosiute deposits (vs. their extreme abundance in Fossil Lake deposits). As reluctant as I am to use absence data based on the fossil record, the large sample sizes (i.e., hundreds of thousands of fish skeletons and dozens of different sample sites known from each Eocene EGRW Lake) make the absence of these taxa from these localities an intriguing contrast. Even if a few specimens were to show up in the future, the contrast in species proportions would remain robust. Suckers, catfishes and percoids are all important members of the modern freshwater fish fauna of North America. The faunal differences between the Fossil Lake and Lake Gosiute EGRW deposits could be attributed either to paleoecological factors (i.e., differences in environmental conditions between the sample sites of the two lakes) and/or the slight differences in age between the main sample sites of Fossil Lake and of Lake Gosiute (late early Eocene in the former and early middle Eocene in the latter; Grande, 1994). Other faunal contrasts between Eocene Fossil Lake and Lake Gosiute include the apparent absence in Lake Gosiute of hiodontids, gonorynchids, stingrays and polyodontids; although the distribution of these families within the various Fossil Lake localities is spotty.

5. Comment on the Biodiversity of the Green River Fish Fauna

This review is conservative in its estimate of species-level diversity in that it synonymizes many species that cannot be diagnosed by unambiguous characters. Most of these synonymies were made previously in Grande (1980, 1984), where larger sample sizes allowed better control for intraspecific variation than was available for the original species descriptions of the late 19th and early 20thcenturies.

Among fossil freshwater fish localities, the known taxonomic diversity of the Green River fish assemblage exceeds that of any other known teleost-dominated locality in the world. But the situation appears to be quite different when species-level diversity comparisons are made to diverse extant faunas. For example, the known species diversity of the Green River Lake Complex appears much smaller than that of extant long-lived lake complexes such as the Great Lakes of eastern Africa. But before we take this type of comparison too literally, three factors must be considered. First, the known family-level diversity of fishes from the Green River lake system (which could increase further with future discoveries) is relatively high. Lake Gosiute had at least 10 families present, and Fossil Lake had 14. This compares favorably with 9 to 14 families in each of the extant Great Lakes of eastern Africa (Lowe-McConnell, 1975:278–279), suggesting that the species diversity of the Green River lakes could have been much higher than currently understood through examination of the fossil record. Second, species diversity (and possibly generic diversity) of Green River fishes may be grossly underrepresented in the fossil lakes, particularly among clupeids and percoids, because of incomplete preservation. Fossils are never as well preserved as living taxa. Even in exceptionally well-preserved localities such as the Green River Formation, usually only the hard parts (i.e., ossified or calcified parts) of the anatomy are preserved. In extant fish localities other features such as color patterns or soft-part anatomy (not preserved in the fossils) can be used to diagnose separate species. Third, being able to distinguish separate biological populations within all the individual ecological zones of an extant lake (at least potentially) can often make it retrospectively easier to recognize diagnostic features (including hard-part features) unique to distinct species. This is not possible in the Green River Lakes, because most of the lakes' deposits are not preserved (they have either been eroded away or were never preserved in the first place), and we do not observe the community in ecological time. The existing fossil deposits are intermittent in both space (geographically) and time (stratigraphically). In short, we will never be able to interpret the scope of species diversity that existed in the Green River lakes in the same way we can interpret species diversity in an extant lake. Comparison of species diversity between fossil and extant lake systems is probably not practical or historically meaningful, in general. For biodiversity comparisons between fossil and living fish faunas, comparisons at the family level should be more paleoecologically accurate than at the species level, because family-level differentiation will normally include more obvious skeletal characteristics that are likely to be preserved in fossils.

6. Future Studies

There has been much progress during the last thirty years in better understanding the anatomy, identity, evolutionary relationships and community composition of fishes from the Green River Formation. Progress will continue to be made (e.g., with the gar species, stingray species, percoid species); and updates on this extremely important fauna will be forthcoming. The Green River Formation lagerstätten still have much to tell us about the early development of the modern North American fish fauna. In fact, it is clear that the fossil fish fauna of the Green River Formation should be included in any comprehensive study of the origin and evolutionary relationships of the modern North American freshwater fish fauna (Grande, 1999:272). To better understand the relationships of many of the Green River fish species, we need a better understanding of the comparative osteology and phylogenetic relationships of many extant groups. The precise taxonomic identities of Green River species are often discovered only when they are included in comprehensive phylogenetic studies of fossil and living taxa together.

ACKNOWLEDGMENTS. For reading and commenting on an earlier draft of this paper I thank Gregg Gunnell and an anonymous reviewer. For loan of material I thank John Maisey (AMNH), Bruce Erickson (SMMP), Peter Forey (BMNH) and Rick and Gale Hebdon (Thayne, Wyoming). I also thank Elaine Zeiger for word processing assistance. Some of the information in this paper was the result of National Science Foundation grants DEB9119561 (to Grande and Bemis) and DEB9220938 (to Bemis and Grande).

References

Agassiz, L., 1833–1844, Recherches sur les Poissons Fossiles. 5 volumes plus supplement. Published by the author, printed by Petitpierre: Neuchâtel.

Arratia, G., 1992, Development and variation of the suspensorium of primitive catfishes (Teleostei: Ostariophysi) and their phylogenetic relationships, *Bonner Zoologische Monographien* **32**:1–149.

Baer, J. L., 1969, Paleoecology of cyclic sediments of the lower Green River Formation, central Utah, *Brigham Young University Geological Studies* **16**:3–95.

Berg, L. S., 1940, Classification of fishes both Recent and fossil, *Trudy Zoologicheskogo Instituta, Leningrad* **5**:87–517. [1965 edition reprinted by the Thai National Document Center, Bangkok, with English translation, 304 pp.]

Boreske, J. R., 1974, A review of the North American fossil amiid fishes, *Bulletin of the Museum of Comparative Zoology, Harvard University* **146(1)**:1–87.

Bruner, J. C., 1991, Comments on the genus *Amyzon* (Family Catostomidae), *Journal of Paleontology* **65**:678–686.

Buchheim, H. P., 1994, Eocene Fossil Lake: a history of fluctuating salinity, in: *Sedimentology and geochemistry of modern and ancient saline lakes* (R. Renaut, and W. Last, eds.), Society for Sedimentary Geology, Special Publication **50**:239–247.

Buchheim, H. P., and Surdam, R. R., 1981, Paleoenvironments and fossil fishes of the Laney Member, Green River Formation, Wyoming, in: *Communities of the Past* (J. Gray, A. J. Boucot, and W. B. N. Berry, eds.), Hutchinson, Ross, Stroudsburg, Pennsylvania, pp. 415–452.

Cavender, T. M., 1966, Systematic position of the North American Eocene fish, "*Leuciscus*" *rosei* Hussakof, *Copeia* **1966**:311–320.

Cavender, T. M., 1986, Review of the fossil history of North American freshwater fishes, in: *The Zoogeography of North American Freshwater Fishes* (C. H. Hocutt and E. O. Wiley, eds.), John Wiley and Sons, New York, pp. 699–724.

Chang, M.-M., and Chou, C.-C., 1977, On late Mesozoic fossil fishes from Zhejiang province, China, *Memoirs, Institute of Vertebrate Paleontology and Paleoanthropology. Academia Sinica*, **12**:1–60. [in Chinese with English summary].

Chang, M.-M., and Chou, C.-C., 1978, On the fossil fishes in Mesozoic and Cenozoic oil-bearing strata from east China and their sedimentary environment, *Vertebrata PalAsiatica*, **16(4)**:229–237. [In Chinese.]

Chang, M.-M., and Grande, L., 1997, Redescription of †*Paraclupea chetungensis*, an early clupeomorph from the Lower Cretaceous of southeastern China, *Fieldiana: Geology*, new series **(37)**:1–19.

Cope, E. D., 1870, Observations on the fishes of the Tertiary shales of Green River, Wyoming Territory, *American Philosophical Society Proceedings* **XI**:380–384.

Cope, E. D., 1871, On the fishes of the Tertiary shales of Green River, Wyoming Territory, in: *United States Geological and Geographical Survey of the Territories Annual Report*, Volume IV (F. V. Hayden, ed.), pp. 425–431.

Cope, E. D., 1872, On the Tertiary coal and fossils of Osino, Nevada, *Proceedings of the American Philosophical Society* **12**:478–481.

Cope, E. D., 1873, On the extinct Vertebrata of the Eocene of Wyoming observed by the expedition of 1872, with notes on the geology. Sixth annual report of the United States Geological Survey of the territories embracing portions of Montana, Idaho, Wyoming and Utah; being a report of progress of the explorations for the year 1872. 545–649 (F. V. Hayden, U. S. Geologist).

Cope, E. D., 1877, A contribution to the knowledge of the ichthyological fauna of the Green River Shales, *Bulletin, United States Geological and Geographical Survey* **3(art.34)**:807–819.

Cope, E. D., 1878, Descriptions of fishes from the Cretaceous and Tertiary deposits west of the Mississippi River, *Bulletin, United States Geological and Geographic Survey of the Territories* **IV**:67–77.

Cope, E. D., 1883, A new chondrostean from the Eocene, American Naturalist **17**:1152–1153.

Cope, E. D., 1885a, The Vertebrata of the Tertiary Formations of the West. Book 1. *United States Geological Survey of the Territories*, 1009 pp. [Although this publication bears two dates (1884 on p. i, 1883 on p. ix), it was not issued until 1885 (Hay, 1902).]

Cope, E. D., 1885b, Eocene paddle-fish and Gonorhynchidae, *American Naturalist* **19**:1090–1091.

Cope, E. D., 1886, On two new forms of polyodont and gonorhynchid fishes from the Eocene of the Rocky Mountains, *Memoirs of the National Academy of Sciences, U.S.A.* **3**:161–166.

Eastman, C. R., 1900, Fossil lepidosteids from the Green River shales of Wyoming, *Bulletin of the Museum of Comparative Zoology*, **36(3)**:67–75.

Eastman, C. R., 1917, Fossil fishes in the collection of the United States National Museum, *Proceedings of the U.S. National Museum* **52**:235–304.

Gottfried, M. D., and Krause, D. W., 1998, First record of gars (Lepisosteidae, Actinopterygii) on Madagascar: Late Cretaceous remains from the Mahajunga Basin, *Journal of Vertebrate Paleontology*, **18(2)**:275–279.

Grande, L., 1979, *Eohiodon falcatus*, a new species of hiodontid (Pisces) from the late Early Eocene Green River Formation of Wyoming, *Journal of Paleontology* **53(1)**:103–111.

Grande, L., 1980 (first edition), Paleontology of the Green River Formation, with a review of the fish fauna, *Geological Survey of Wyoming, Bulletin* **63**:1–333.

Grande, L., 1982a, A revision of the fossil genus *Diplomystus* with comments on the interrelationships of clupeomorph fishes, *American Museum Novitates* **2728**:1–34.

Grande, L., 1982b, A revision of the fossil genus *Knightia*, with a description of a new genus from the Green River Formation (Teleostei, Clupeidae), *American Museum Novitates* **2731**:1–22.

Grande, L., 1984 (second edition), Paleontology of the Green River Formation, with a review of the fish fauna, *Geological Survey of Wyoming, Bulletin* **63**:1–333.

Grande, L., 1985, Recent and fossil clupeomorph fishes with materials for revision of the subgroups of clupeoids, *Bulletin, American Museum of Natural History* **181**:231–372.

Grande, L., 1987, Redescription of †*Hypsidoris farsonensis* (Teleostei: Siluriformes) with a reassessment of its phylogenetic relationships, *Journal of Vertebrate Paleontology* **7(1)**:24–54.

Grande, L., 1989. The Eocene Green River lake system, Fossil Lake, and the history of the North American fish fauna, in: *Mesozoic/Cenozoic Vertebrate Paleontology: Classic Localities, Contemporary Approaches* (J. Flynn, ed.), 28th International Geological Congress fieldtrip guidebook T322, pp. 1–77. American Geophysical Union, pp. 18–28.

Grande, L., 1994, Studies of paleoenvironments and historical biogeography in the Fossil Butte and Laney Members of the Green River Formation, *Contributions to Geology, University of Wyoming* **30**:15–32.

Grande, L., 1998, The first pickerel (Teleostei: Esocidae) from the Green River Formation, and a review of the world's most productive freshwater lagerstätten, *Journal of Vertebrate Paleontology, Abstracts* **18**:47A.

Grande, L., 1999, The first *Esox* (Esocidae: Teleostei) from the Eocene Green River Formation, and a brief review of esocid fishes, *Journal of Vertebrate Paleontology* **19(2)**:271–292.

Grande, L., and Bemis, W., 1991, Osteology and phylogenetic relationships of fossil and Recent paddlefishes (Polyodontidae) with comments on the interrelationships of Acipenseriformes, *Society of Vertebrate Paleontology Memoir* **1**:i–viii, 1–121; supplement to *Journal of Vertebrate Paleontology* **11(1)**.

Grande, L., and Bemis, W., 1996, Interrelationships of Acipenseriformes, with comments on "Chondrostei", in: *Interrelationships of Fishes* (M. Stiassny, L. Parenti, and G. D. Johnson, eds.), Academic Press, San Diego, California, pp. 85–115.

Grande, L., and Bemis, W., 1998, A comprehensive phylogenetic study of amiid fishes (Amiidae) based on comparative skeletal anatomy. An empirical search for interconnected patterns of natural history, *Society of Vertebrate Paleontology Memoir* **4**:i–x, 1–690; supplement to *Journal of Vertebrate Paleontology* **18(1)**.

Grande, L., and Buchheim, H. P., 1994, Paleontological and sedimentological variation in Early Eocene Fossil Lake, *Contributions to Geology* **30**:33–56.

Grande, L., and de Pinna, M., 1998, Description of a second species of the catfish †*Hypsidoris* and a reevaluation of the genus and the Family †Hypsidoridae, *Journal of Vertebrate Paleontology* **18**:451–474.

Grande, L., and Grande, T., 1999, A new species of †*Notogoneus* (Teleostei: Gonorynchidae) from the Upper Cretaceous Two Medicine Formation of Montana, with a discussion of the Cretaceous record of freshwater fishes from North America, *Journal of Vertebrate Paleontology.* **19(4)**:612–622.

Grande, L., and Lundberg, J. G. L., 1988, Revision and redescription of the genus †*Astephus* (Siluriformes: Ictaluridae) with a discussion of its phylogenetic relationships, *Journal of Vertebrate Paleontology* **8**:139–171.

Grande, L., Eastman, J. T., and Cavender, T. M., 1982, *Amyzon gosiutensis*, a new catostomid fish from the Green River Formation, *Copeia* **(3)**:523–532.

Grande, L., Forey, P., and Bemis, W. E., In preparation, The origin and diversification of ray-finned fishes (Actinopterygii), the other half of Vertebrata.

Grande, L., Jin, F., Yabumoto, Y., and Bemis, W. [in press], †*Protopsephurus liui*, a well-preserved primitive paddlefish (Acipenseriformes: Polyodontidae) from the Early Cretaceous of China. *Journal of Vertebrate Paleontology.*

Haseman, J. D., 1912, The relationship of the genus *Priscacara*, *Bulletin of the American Museum of Natural History* **31**:97–101.

Hay, O. P., 1902, Bibliography and catalogue of the fossil vertebrata of North America, *Bulletin of the United States Geological Survey, Washington* **179**:1–868.

Jordan, D. S., 1907, The fossil fishes of California, with supplementary notes on other species of extinct fishes, *University of California Publications, Bulletin of the Department of Geology* **5(7)**:95–144.

Jordan, D. S., 1923, A classification of fishes including families and genera as far as known, *Stanford University Publications, University Series, Biological Sciences* **3**:77–243.

Lacépède, B. G. E., 1803, Histoire naturelle des poissons **5**. Paris.

Leidy, J., 1873a, Notice of remains of fishes in the Bridger Tertiary Formation of Wyoming, *Proceedings of the Academy of Natural Sciences, Philadelphia* **1873**:97–99.

Leidy, J., 1873b, Contributions to the extinct vertebrate fauna of the western territories, *Report of United States Geological Survey of the Territories* (pt.1):1–358 (F. V. Hayden, U. S. Geologist).

Li, G.-Q., Grande, L., and Wilson, M. V. H., 1997, The species of †*Phareodus* (Teleostei: Osteoglossidae) from the Eocene of North America and their phylogenetic relationships, *Journal of Vertebrate Paleontology* **17**:487–505.

Li, G.-Q., Wilson, M. V. H., and Grande, L., 1997, Review of †*Eohiodon* (Teleostei: Osteoglossomorpha) from western North America, with a phylogenetic reassessment of Hiodontidae, *Journal of Paleontology* **71**:1109–1124.

Linnaeus, C., 1758, *Systema Naturae.* Editio X. [Systema naturae per regna tria naturae, secundum classes, ordines, genera, species, cum characteribus, differentiis, synonymis, locis. Tomus I. Editio decima, reformata.] Holmiae. 824 pp.

Linnaeus, C., 1766, *Systema Naturae.* Editio Duodecima, Reformata. Impensis Direct Laurentii Salvii: Holmiae **1**:1–532.

Lowe-McConnell, R. H., 1975, *Fish Communities in Tropical Freshwaters.* Longman, London and New York.

Lundberg, J. G., 1975, The fossil catfishes of North America, *University of Michigan Museum of Paleontology, Claude W. Hibbard Memorial* **2(11)**:1–51.

Lundberg, J. G., and Case, G. R., 1970, A new catfish from the Eocene Green River Formation, Wyoming, *Journal of Paleontology* **44**:451–457.

Marsh, O. C., 1877, Notice of some new vertebrate fossils, *American Journal of Science* (III)**14**:249–256.

McGrew, P. O., 1975, Taphonomy of Eocene fish from Fossil Basin, Wyoming. *Fieldiana* **33**(14).

McGrew, P. O., and Casilliano, M., 1975, The geological history of Fossil Butte National Monument and Fossil Basin, *National Park Service Occasional Paper*, no. 3.

Miyake, T., McEachran, J. D., Walton, P. J., and Hall, B. K., 1992, Development and morphology of rostral cartilages in batoid fishes (Chondrichthyes: Batoidea), with comments on homology within vertebrates, *Biological Journal of the Linnean Society* **46**:259–298.

Nelson, J. S., 1994, *Fishes of the World*, 3rd ed., John Wiley and Sons, New York.

Nishida, K., 1990, Phylogeny of the suborder Myliobatidoidei, *Memoirs of the Faculty of Fisheries, Hokkaido University*, **37**:1–108.

Patterson, C., 1967, Are the teleosts a polyphyletic group? in: *Problèmes actuels de paléontologie (evolution des vertébrés).* Colloques Internationaux. Centre national de la Recherche Scientifique, no. 163, pp. 93–109.

Patterson, C., 1993, Osteichthyes: Teleostei, in: *The Fossil Record 2* (M. J. Benton, ed.), Chapman and Hall, London, pp. 621–656.

Patterson, C., and Rosen, D. E., 1989, The Paracanthopterygii revisited: Order and disorder, in: *Papers on the Systematics of Gadiform Fishes* (D. M. Cohen, ed.), Science Series, Natural History Museum of Los Angeles County, Science Series No. 32, pp. 5–36.

Pellegrin, J., 1904, Contribution à l'étude anatomique, biologique et taxonomique des poissons de la famille des cichlidés, *Mémoires de la Société Zoologique de France* **16**:41–402.

Rafinesque-Schmaltz, C. S., 1820, Ichthyologia Ohiensis, or natural history of the fishes inhabiting the river Ohio and its tributary streams, preceded by a physical description of the Ohio and its branches. Lexington, Kentucky.

Romer, A. S., 1966, *Vertebrate Paleontology*, 3rd ed., University of Chicago Press.

Romer, A. S., and Fryxell, F. M., 1928, *Paramiatus gurleyi*, a deep-bodied amiid fish from the Eocene of Wyoming, *American Journal of Science* **16**:519–527.

Rosen, D. E., and Patterson, C., 1969, The structure and relationships of the paracanthopterygian fishes, *American Museum of Natural History, Bulletin* **141**:357–474.

Schaeffer, B., and Mangus, M., 1965, Fossil lakes from the Eocene, *American Museum of Natural History* **74**:11–21.

Wiley, E. O., 1976, The phylogeny and biogeography of fossil and Recent gars (Actinopterygii: Lepisosteidae), *University of Kansas Museum of Natural History Miscellaneous Publication* **64**:1–111.

Wiley, E. O., and Schultze, H.-P. 1984, Family Lepisosteidae (gars) as living fossils, in: *Living Fossils* (N. Eldredge and S. Stanley, eds.), Springer Verlag, New York, pp. 160–165.

Wilson, M. V. H., 1977, Middle Eocene freshwater fishes from British Columbia, *Life Science Contributions, Royal Ontario Museum* **113**:1–61.

Wilson, M. V. H., 1978, *Eohiodon woodruffi* n. sp. (Teleostei: Hiodontidae) from the Middle Eocene Klondike Mt. Formation near Republic, Washington, *Canadian Journal of Earth Science* **15**:679–686.

Woodward, A. S., 1901, Catalogue of the fossil fishes in the British Museum, vol. 4: Containing the actinopterygian Teleostomi of the suborders Isospondyli (in part), Ostariophysi, Apodes, Percesoces, Hemibranchii, Acanthopterygii and Anacanthini. Taylor and Francis, London.

Zhang, M., Zhou, J., and Qin, D., 1985, Tertiary fish fauna from coastal region of Bahia Sea, *Institute of Vertebrate Paleontology and Paleoanthropology, Memoir*, **17**:1–136. [in Chinese with English summary]

Chapter 2

Paleontological Investigations at the Eocene Locality of Mahenge in North-Central Tanzania, East Africa

TERRY HARRISON, CHARLES P. MSUYA, ALISON M. MURRAY, BONNIE FINE JACOBS, ANA MARIA BÁEZ, ROLAND MUNDIL, and KENNETH R. LUDWIG

TERRY HARRISON • Department of Anthropology, New York University, New York, NY 10003. CHARLES P. MSUYA • Department of Anatomy, Muhimbili University College of Health Sciences, Dar es Salaam, Tanzania. ALISON M. MURRAY • Redpath Museum, McGill University, Montreal, QC, H3A 2K6, Canada. BONNIE FINE JACOBS • Department of Geological Sciences, Southern Methodist University, Dallas TX 75275-0395. ANA MARIA BÁEZ • Departamento de Geología, Universidad de Buenos Aires, Buenos Aires, Argentina. ROLAND MUNDIL and KENNETH R. LUDWIG • Berkeley Geochronology Center, Berkeley, CA 94709.

Eocene Biodiversity: Unusual Occurrences and Rarely Sampled Habitats, edited by Gregg F. Gunnell, Kluwer Academic/Plenum Publishers, New York, 2001.

1. Introduction

Since the early 1930s, a number of geologists prospecting for diamonds in the region of Singida in north-central Tanzania have reported the occurrence of superbly preserved fossils in lake beds overlying kimberlite pipes (Teale, 1931, 1932; Eades and Reeve, 1938; Williams, 1939; Greenwood, 1960; Mannard, 1962; Greenwood and Patterson, 1967). The sediments containing the fossils consist of shales and mudstones deposited in small crater lakes. These fossil localities have never been the subject of detailed paleontological investigation. As a consequence, in 1996 the senior author initiated excavations at the middle Eocene locality of Mahenge, the most productive fossil-bearing locality in the region (Greenwood, 1960; Mannard, 1962). The results of this expedition confirm the exceptional productivity and scientific importance of this site, and serve to highlight the paleontological potential of the region in general. A sizeable collection of fossil fishes, frogs, insects and plant remains was recovered. The fossils are remarkably well-preserved and, in the case of vertebrates, are generally represented by entire and articulated skeletons, occasionally preserving details of their soft anatomy (Greenwood, 1960; this report).

Because paleontological sites with strictly continental sediments of Paleogene age are almost entirely unknown in sub-Saharan Africa, Mahenge offers a unique opportunity to document the diversity, biogeography and evolutionary history of the flora and fauna of Africa during this time period. The fact that all of the vertebrates so far recovered from Mahenge are found nowhere else, serves to emphasize just how little we know about Eocene African faunas. Detailed systematic studies of the fauna and flora are still underway, but we present here a preliminary synthesis of our findings to date.

2. History of Paleontological Research

Exploration of the region began in 1930 with the report that C. J. T. Boshoff, a South African prospector, had discovered diamonds at Kiomboi on the Iramba Plateau. There was a major influx of prospectors and geologists during the early 1930s, but interest waned with the discovery that all of the kimberlite occurrences were either barren or unprofitable. Nevertheless, at least one European prospector, known only by his Kiswahili name of *Bwana chimba panya* (Mr. digging rat), remained active during the late 1930s and 1940s, until he succumbed from malaria. During this period, Reeve (with the Geological Survey of Tanganyika) apparently discovered a kimberlite occurrence in the vicinity

Table 1. List of localities in the Singida Region that have yielded fossil remains

Localities	Fossils Recovered	References
Blangetti	Silicified wood	Mannard, 1962: 240
Kinginya	Silicified wood	Mannard, 1962: 167-169
Magoba	Silicified wood	Mannard, 1962: 117, Figure 36a
Magwe ya Ghana	Silicified wood	Mannard, 1962: 48
Mahenge	Fishes, frogs, insects, fruits, leaves, & seeds	Mannard, 1962: 244; Greenwood, 1960; Greenwood and Patterson, 1967; this report
Maluga Kulamba	Silicified wood	Mannard, 1962: 205
Minusi?	Fishes	Greenwood and Patterson, 1967
Munyu (Mtawira)	Plant stem and leaf impressions & fragments of large bone	Teale, 1932; Mannard, 1962: 140, Figure 48a
Pyamahumbe	Fishes, leaves, plant stems & seeds	This report
Songeli (=Sungili)	Stems, leaves, seeds, & root casts	Teale, 1932; Mannard, 1962: 200; this report

of Mahenge (Williams, 1939), although no such occurrence is described by Eades and Reeve (1938) or identified on their geological map. From 1953 to 1959, Williamson Diamond Limited, based at the productive kimberlite pipe at Mwadui in Shinyanga Region, sponsored a major program of prospecting in the Singida Region. A systematic survey was conducted from 1957–1959 by George Mannard, a Canadian geologist (known locally as *dagandi* in Kinyiramba—the one who searches from place to place, Mannard, 1962).

The occurrence of fossil-bearing lake beds in the area was first noted by Du Toit (Teale, 1931, 1932), who recovered leaf impressions, stems, seeds and pieces of a large bone at Songeli and Mtawira. Mannard (1962) subsequently reported the occurrence of fossil leaves and silicified wood at six additional localities (see Table 1 for a complete list of paleontological sites in the region), and recovered a small collection of fossil fishes from Mahenge. The latter material was sent to Humphrey Greenwood for study (at the Natural History Museum, London), who later received additional fish specimens from the site of Minusi. Greenwood (1960) and Greenwood and Patterson (1967) referred these fish to *Palaeodenticeps tanganikae*, a new genus and species of clupeomorph belonging to the Denticipitidae; *Singida jacksonoides*, a new genus and species of osteoglossomorph belonging to its own extinct family, the

Singididae; and a genus of cichlid with affinities to *Haplochromis*, a taxon widespread in East Africa today.

In 1994, TH and CPM made a brief excursion to the sites of Sungili, Mtawira and Mahenge, and recovered fossil wood, leaves, root casts, and fishes. This prompted a full-scale expedition to Mahenge during the summer of 1996, in which over five hundred specimens were collected.

3. Research Area and Geological Context

The locality at Mahenge (4°47′38 S; 34°15′28 E) is situated 63 km west of the town of Singida in north-central Tanzania (Figures 1 and 2). The site is reached by taking the Singida-Usule road westwards as far as Sepuka, and then diverting along a small track to Mwaru. This latter village is usually identified on maps as Mahenge, and is the source of the name given to the fossil locality by Mannard (1962). However, the regional and district administrative officers in Singida, as well as the local inhabitants, refer to the present-day village as Mwaru. The locality is situated 0.4 km south of the village, and is reached by car via a forest trail that is used sporadically by trucks for transporting charcoal.

The Singida kimberlite field, with more than fifty kimberlite pipes and dykes, is located on a broad peneplain, just to the south of the Iramba Plateau

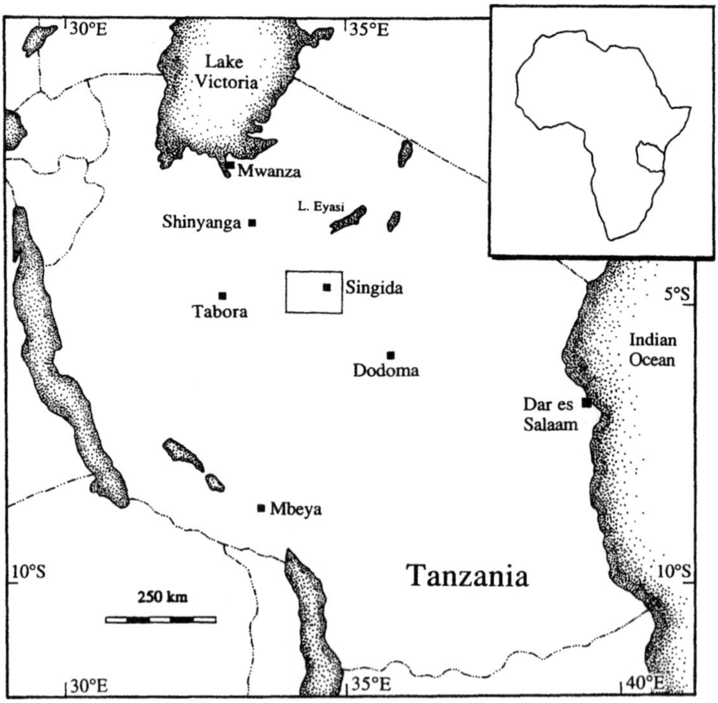

FIGURE 1. Map showing the location of the research area in north-central Tanzania. See Figure 2 for detail of inset.

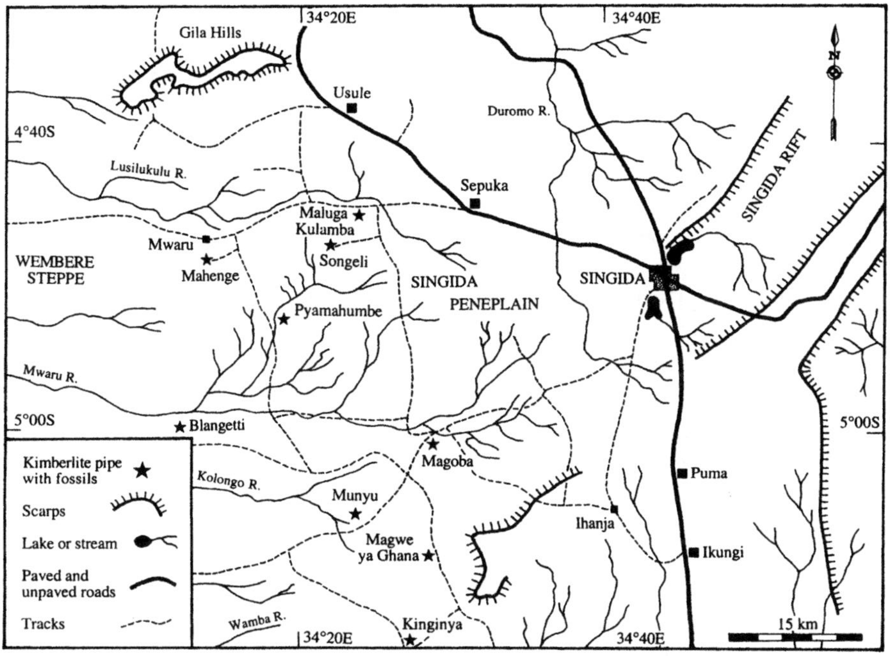

FIGURE 2. Map of research area showing the main geographical features and the location of known fossil localities, including Mahenge.

(Mannard, 1962). The latter represents a large tilted fault-block at the southwestern extremity of the Gregory Rift, with a maximum elevation of 1680 m above sea level (Teale, 1932; Eades, 1936). Steep fault-scarps border the Wembere depression and Eyasi trough to the north and west, and the Mpura-Durumo valley to the east. To the south, the plateau grades imperceptibly into the peneplain. The combination of mature peneplanation and long-term tectonic stability south of the Iramba Plateau has resulted in a topography of relatively uniform relief, with elevations ranging from 1100 m to 1500 m, and creating a gently undulating terrain. Outcrops of granite are common, forming small inselbergs or craggy tors that rise above the level of the peneplain. The most prominent granitic outcrop in the area is the Gila Hills, a table-topped plateau located about 20 km north of Mahenge (Figure 2).

The region is situated on the eastern margin of a vast tract of miombo woodland which stretches as far west as Lake Tanganyika. This type of woodland, dominated by several species of trees belonging to *Julbernardia*, *Isoberlinia*, and *Brachystegia*, is typical of the Central Plateau of Tanzania (Gillman, 1949; Kingdon, 1974; Lind and Morrison, 1974). It is a dry, deciduous woodland experiencing an annual precipitation of only 520 ± 260 mm, with almost no rain occurring during the lengthy dry season from June to November (Eades and Reeve, 1938). Owing to the irregular rainfall and the low relief of the terrain, ephemeral streams and rivers in the area are dry throughout most of the year. Dense vegetation covers much of the research area and this creates logistical problems for aerial reconnaissance and surface exploration.

The region is sparsely populated. The low and unreliable rainfall is an important factor; 800 mm is generally considered the minimum to support domestic livestock and intensive cultivation. In addition, tsetse flies (*Glossina* spp.), important vectors of bovine trypanosomiasis or nagana, have prevented people and their cattle from penetrating the densely wooded and bush-covered areas to the south (Harrison *et al.*, 1997). The Wanyiramba and Wanyituru, the predominant tribal groups in the area, are small-scale agriculturalists and cattle herders that generally live in cleared areas close to the margins of the woodland. As a consequence, few people, except itinerant woodcutters and charcoal burners, are encountered in the remotest regions of the research area.

The basement rocks in the region are part of the Dodoman system, formerly referred to as the Lower Basement Complex (Stockley, 1935, 1943; Grantham and Temperley, 1939; Teale and Oates, 1946; Quennell *et al.*, 1956). These rocks consist of a complex of pellitic schists, granitic gneisses and migmatites, ranging in composition from potassic granite to diorite. Dykes and veins of dolerite are common, and those of pegmatites, aplites, dolerite and quartz occur locally (Teale, 1932; Eades and Reeve, 1938; Mannard, 1962). An intrusive pegmatite dyke within the Dodoman has yielded a provisional Sr-Rb date of 3.25 ± 0.325 Ga (Quennell *et al.*, 1956).

The Nyanzian system, formerly known as the Upper Basement Complex, is inferred to be younger and superpositional to the Dodoman, but the nature of the contact between the two systems is unknown because of the lack of pertinent outcrops. The Nyanzian system consists primarily of a series of volcanic rocks, schists, and banded ironstones (Grace and Stockley, 1931; Eades, 1936; Eades and Reeve, 1938; Williams and Eades, 1939; Grantham *et al.*, 1945; Stockley, 1947; Quennell *et al.*, 1956; Borg, 1992; Walraven *et al.*, 1994). The most distinctive component of the Nyanzian is the Banded Ironstone Formation (BIF), with a U-Pb age of 2.7 Ga (Stockley, 1934; Borg, 1992; Walraven *et al.*, 1994), although outcrops in the Singida region are few (Eades, 1936; Eades and Reeve, 1938). Micas from a pegmatite segregation in the Nyanzian granite at Mwadui in Shinyanga Region have yielded K-Ar dates of 2,518 ± 42 Ma and 2,526 ± 42 Ma (Edwards and Howkins, 1966).

The Nyanzian was subject to several episodes of batholithic intrusion of granitoid rocks, with a widespread event of migmatization occurring at ~2.4–2.6 Ga (Old and Rex, 1971; Dodson *et al.*, 1975; Cahen *et al.*, 1984). The resulting granitoid shield forms a significant component of the Tanzanian craton (Kuntz, 1909; Eades, 1936; Wade, 1937; Stockley, 1948; Quennell *et al.*, 1956; Old and Rex, 1971; Dodson *et al.*, 1975; Cahen *et al.*, 1984; Walraven *et al.*, 1994). The Iramba Plateau and Singida peneplain straddle the divide between the Nyanzian and Dodoman belts respectively (Mannard, 1962).

During the Cretaceous and early Tertiary, the basement rocks were disrupted by the intrusion of a complex of kimberlite pipes (Williams, 1939; Mannard, 1962). The kimberlite occurrences lie on a broad NNW-SSE zone originating at the southern shore of Lake Victoria (Dawson, 1970; Mitchell, 1986).

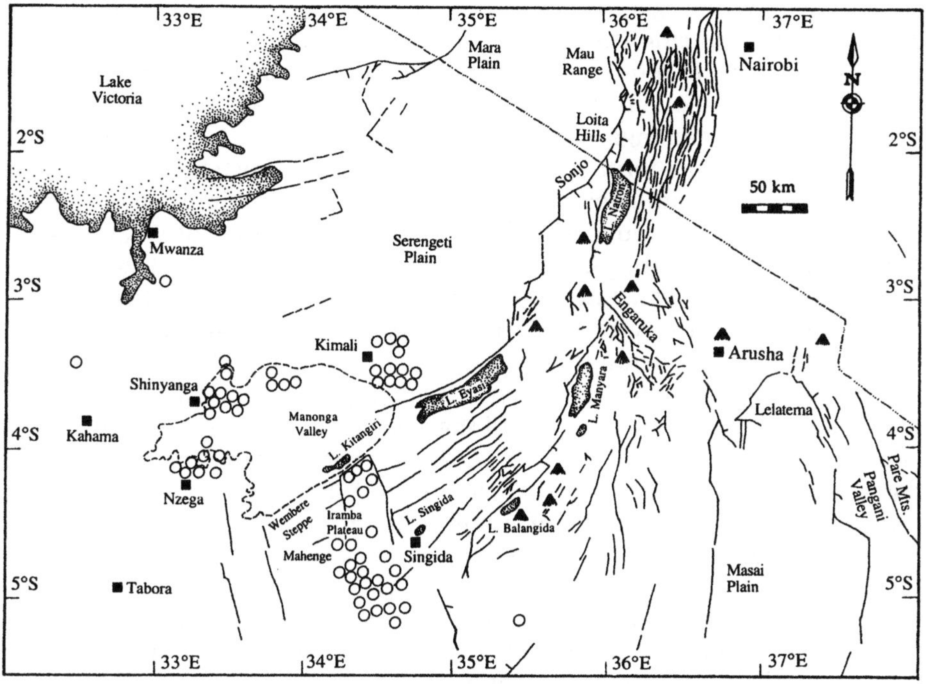

FIGURE 3. Map showing the distribution of kimberlite occurrences (open circles) in northern Tanzania in relation to the fault patterns, major volcanic centers, and the Manonga Valley (King, 1970; Harrison and Mbago, 1997).

This kimberlite province has been subdivided geographically into a number of separate fields (Edwards and Howkins, 1966; Mitchell, 1986; Figure 3). The largest of these is the Singida field, where more than fifty kimberlite bodies have so far been identified (Mannard, 1962). The Singida field appears to be part of a much larger kimberlite complex that also includes the Nzega, Shinyanga, Iramba and Kimali fields to the east and north (Edwards and Howkins, 1966; Mitchell, 1986). The present-day geographical separation of these fields may have been exaggerated, in part, by subsequent geological events. As can be seen in Figure 3, the kimberlite fields are located around the margins of the Manonga Valley and Wembere depression (Harrison & Mbago, 1997; Verniers, 1997). The thick series of late Miocene and Pliocene terrestrial and lacustrine sediments deposited in these basins may have obscured kimberlite outcrops in the central part of the province. However, individual fields are defined as geographically and genetically closely related groups of kimberlites of similar age, usually located within a 50 km diameter (Mitchell, 1986). The wide distribution (the main occurrences are spread over an area exceeding 200 km in diameter), different ages (at least for occurrences in the Singida, Shinyanga and Nzega fields—see dating below), and geochemistry of the kimberlites in the different clusters support the recognition of separate fields.

This sequence of geological events, comprising different episodes of kimberlite intrusion during the late Mesozoic and early Tertiary, warping of the basement complex to produce the Manonga Basin in the late Miocene and early Pliocene, and subsequent convergence of the Eyasi and Wembere grabens at the center of the Manonga Basin to form a triple junction, are not coincidental. It points to a long-term susceptibility to tectonic instability in this region of the craton, and indicates that the intrusion of the kimberlites is directly related to epeirogenic movements along fractures or lines of weakness in the basement at great depth, possibly coinciding with the contact zone between the Dodoman and Nyanzian belts (Williams, 1939; Dawson, 1970; Mitchell, 1986). Increased activity of the African superplume at ≈45 Ma could well be implicated in the intrusion of these kimberlites (Kerr, 1999; Haggerty, 1999).

Preliminary geological research at Mahenge suggests that the structure of the kimberlite pipe and the nature of the associated sediments conform closely to those described for similar occurrences in other parts of Africa (e.g., Edwards and Howkins, 1966; Hawthorne, 1975; Smith, 1986; Rayner *et al.*, 1991). Based on these data, and on studies of modern crater lakes, it is possible to reconstruct the general events that contributed to the formation of the crater and sedimentary sequence at Mahenge (Figure 4).

As kimberlites intruded through the basement rock and pierced the crustal layer the rising magma came into contact with permeating surface water to produce a violently explosive phreatomagmatic eruption (Lorenz, 1973; Scarth, 1994). As the conduit penetrated to the surface it shattered the surrounding country rock. Subsequent explosions further weakened the walls of the conduit, and these collapsed inwards to form a trumpet-shaped diatreme. Eventually, the throat of the diatreme became choked with coarse breccia composed of large blocks of shattered granite and lava, with massive, poorly-stratified pyroclastic kimberlite filling the interstices (Edwards and Howkins, 1966; Hawthorne, 1975). Close to the surface, the pyroclastic kimberlite is leached and oxidized to form a soft, friable, cream-colored “yellow ground” (Mannard, 1962). The edge of the crater is defined by a breccia collar, composed of fractured blocks of granite, that are exposed in the Luwala River, a small stream that crosses the center of the pipe from northeast to southwest. Based on test pits and from the location of the breccia collar, it is possible to reconstruct the crater as being roughly circular in outline, with a maximum diameter of 550 m (Figure 5).

The initial explosive events would have been short-lived, probably lasting no more than a few days, because of the lack of availability of water (Scarth, 1994). At the surface, the diatreme formed a crater bounded by steep-sided walls of exposed country rock. Minor eruptions later produce fine-grained pyroclastic kimberlites deposited as a series of well-stratified tuffs, up to 1 m in thickness, in and around the crater (Mannard, 1962; Hawthorne, 1975). These accumulated to form a low tuff cone, probably less than 10 m in height, around the rim of the crater. The primary tuffs typically contain pyrope (i.e., red magnesium garnet or “Cape ruby”), magnesian ilmenite, dark green diopside, olivine (generally altered to serpentine) and zircon (Teale, 1932; Hawthorne, 1975). The presence of this assemblage of minerals can be used as an aid to locating subsurface kimberlites.

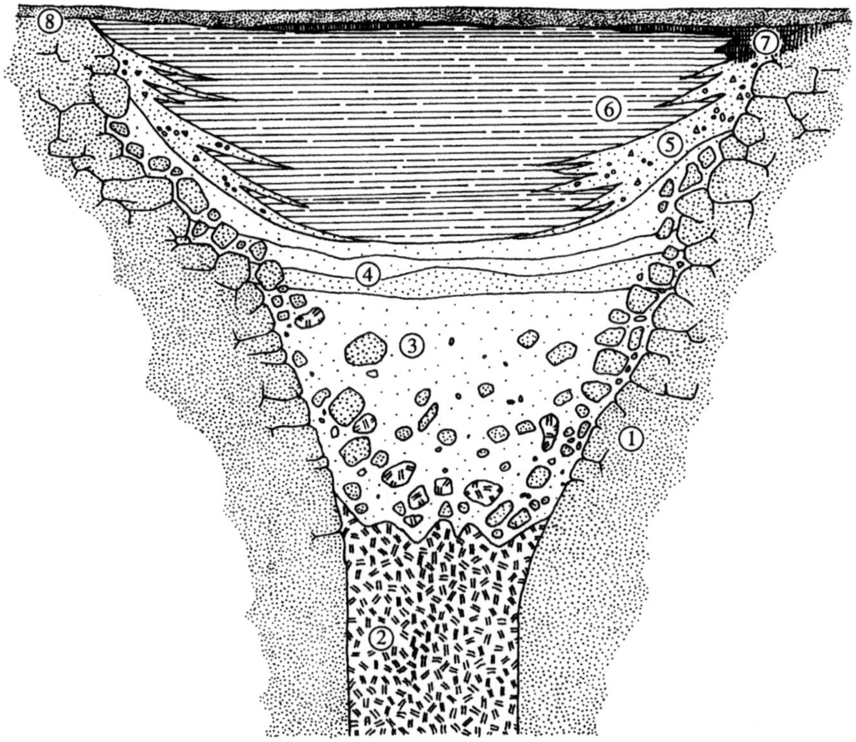

FIGURE 4. Schematic section (not to scale) through an idealized kimberlite pipe to illustrate the main geological features (after Mannard, 1962; Hawthorne, 1975; Smith, 1986). (1) shattered basement rock; (2) intrusive kimberlite—mostly "yellow ground" consisting of leached and oxidized primary kimberlite; rarely less oxidized "blue ground" and hardebank; (3) pyroclastic kimberlite and breccia—massive to crudely stratified pyroclastic kimberlite with shattered and permeated blocks of basement rock; (4) stratified kimberlite tuffs—alternating beds of coarse- to fine-grained tuffs; (5) sedimentary tuffaceous beds at the edge of the lake—silts, sands and poorly-sorted bouldery conglomerates with much granitic detritus; (6) sedimentary tuffaceous beds at the center of the lake—well-stratified microlaminated mudstones and shales containing well-preserved fossils; (7) silicified sedimentary beds and crater wall—alteration of sedimentary beds by chalcedonic silica and silicification of granite due to hydrothermal activity; (8) mbuga clay and loam—superficial layer comprising dark gray mbuga clay and sandy loams.

After cessation of volcanic activity, the crater filled with water from local run-off to produce a shallow crater lake or maar (Lorenz, 1973). The crater lake at Mahenge occupies a roughly circular area with a maximum diameter of 370m (Mannard, 1962; Harrison, unpublished data; Figure 5). The inner scree slope of the tuff cone would have been relatively steep, probably resulting in a restricted shoreline (Edwards and Howkins, 1966; Smith, 1986). Material eroding from the tuff cone was washed into the lake by means of slumping and heavy seasonal rains, and these formed a series of sedimentary tuffaceous beds (Mannard, 1962; Smith, 1986; see Figure 4). In the center of the lake, the sediments produced well-stratified microlaminated mudstones and shales that grade laterally into coarser siltstones and sandstones located closer to the

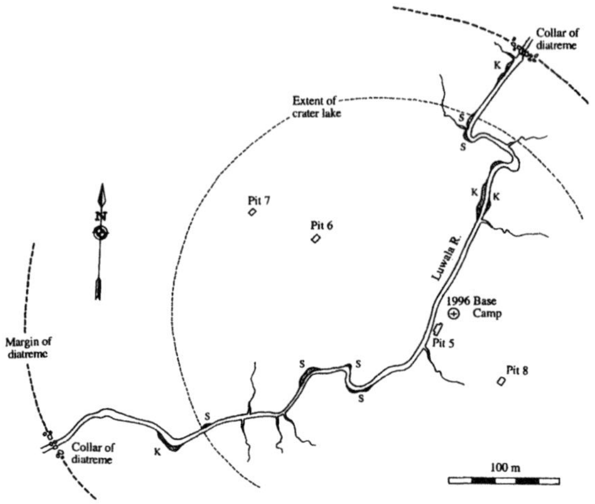

FIGURE 5. Map of Mahenge showing the location of the major outcrops of lake sediments (S) and kimberlites (K) in the Luwala River. Also shown are some of Mannard's pits and the position of the eroded collar of the diatreme. These natural exposures and test pits allow estimation of the shape and diameter of the diatreme and crater lake.

margin of the lake. These beds dip slightly (5°–20°) towards the center of the pipe, probably as a consequence of slumping (Williams, 1939; Harrison, unpublished data). The mudstones are calcareous, pale yellow-gray to buff-colored, and contain numerous well-preserved fossils.

The sedimentary tuffaceous beds and granites are silicified in places, probably as a result of intermittent hydrothermal activity. A thin crust of chalcedonic silica occurs around the margin of the crater, and a remnant of silicified shaly beds has been located along its southwestern edge (Williams, 1939; Mannard, 1962). A surface debris of brightly colored, angular chert chips and banded jaspers derived from the eroding surface of this layer is commonly associated with kimberlite diatremes in the Singida area (Eades and Reeve, 1938).

The lake sediments are covered by a superficial layer, up to 2 m thick, of dark gray to black mbuga clays and gray sandy loam (Figure 4). The general area is relatively flat, and there are no surface clues from the composition of the soil, the type of vegetation, or differences in topography to indicate the presence of an underlying kimberlite pipe (Mannard, 1962; Edwards and Howkins, 1966). The lake sediments were exposed primarily as a result of test pits dug by Mannard in the late 1950s. However, several outcrops of kimberlite and lake sediments containing fossils are exposed along the course of the Luwala River (Figure 5).

4. Stratigraphy

Excavations at Mahenge in 1996 have permitted the reconstruction of a detailed stratigraphic section of the fossiliferous sediments (Figure 6). The bottom of the section consists of a massive yellow-gray (5Y 8/1) clayey sandstone (of which only 0.6 m has been exposed). The sediments are relatively soft and friable, and exhibit only weak indications of horizontal bedding. No fossils have yet been recovered from this horizon. This bed apparently represents the result of a major influx of pyroclastic sediments towards the end of the eruptive phase that were deposited either as an ash-fall directly into the crater or as reworked sediments derived from the inner slopes of the tuff cone (Figure 6).

Overlying the sandstone is a series of fine-grained, well-stratified mudstones with an overall thickness of 1.59 m (Figure 6). These are predominantly yellow-gray in color (5Y 8/1 and 5Y 7/2), but occasionally pale orange (10YR 8/2). The sediments consist of an alternating series of well-consolidated, compact calcareous mudstones, with or without distinct laminations, and soft friable shales. The sequence can be subdivided into 16 cycles, each comprising a couplet of one or more consolidated mudstones and a friable shale (Figure 6). Each cycle has an average thickness of 10 cm (range = 4–17 cm). The cycles in the lower part of the sequence are slightly thicker on average (10.5 cm) than

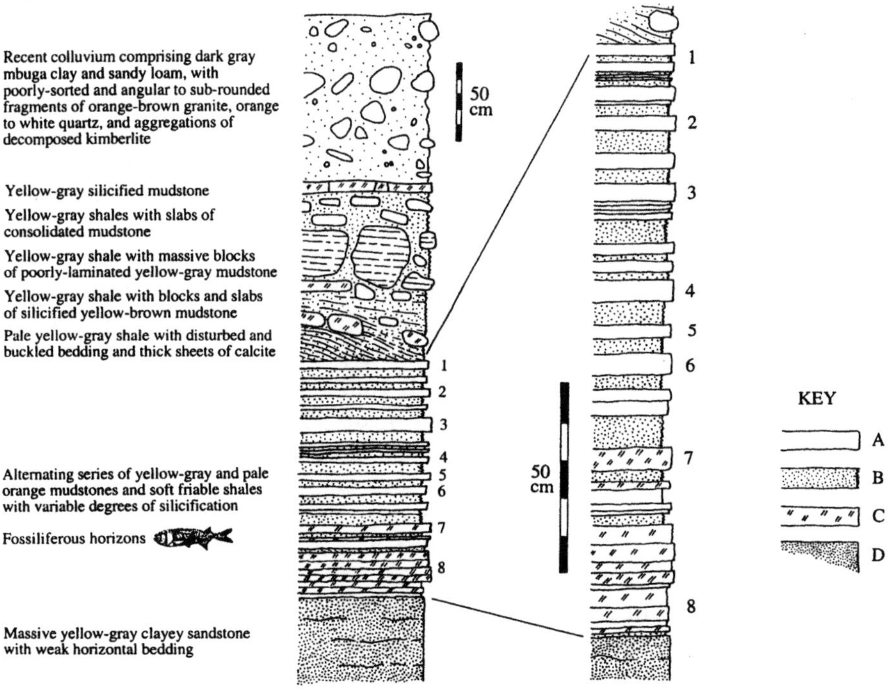

FIGURE 6. Stratigraphic section at Pit 5c at Mahenge. The numbers refer to marker horizons comprising compact and easily identifiable mudstones in the fossil-bearing beds. Fossils are found throughout the sequence from Level 1 through Level 8. Key to right-hand column: (A) compact mudstone with microlaminations strongly to weakly developed; (B) friable shales; (C) partially silicified mudstones; (D) clayey sandstones.

those from the upper part (9.4 cm), and the proportion of consolidated mudstones is also somewhat higher (70% of the section thickness compared to 53%). The mudstones have an average thickness of 3.5 cm, but can form prominent pavement-like layers up to 6 cm thick. Microscopic examination shows that these layers have fine subparallel laminations, with an average thickness of 2.9 mm, often separated by calcite (or more rarely carbonaceous) films. The precipitation of calcium carbonate presumably occurred initially at the sediment-water interface during the annual turnover of the water column (see below). If our estimate of sedimentation rate and compaction factors is reasonable[1], each lamination represents an average cycle of between 15 and 40 years. It is interesting to note that a similar phenomenon has been reported for modern-day African crater lakes, such as Barombi Mbo in Cameroon, where a 15 year periodicity in sedimentation is related to major flooding of large rivers (Giresse *et al.*, 1994).

In addition to calcite films between microlamination, there are often relatively thick sheets of calcite formed at the interbedding planes of the main sedimentary units. Smith (1986) has interpreted similar calcite layers at Stompoor, South Africa (late Cretaceous) as the result of periodic droughts that caused temporary shrinkage of the crater lake and precipitation of calcite on the lake floor. However, in the Mahenge sequence calcite is not restricted to the bedding planes, but often penetrates deeply into joints and cracks in the rock, implying that its precipitation occurred after consolidation and lithification of the sediments.

The thickest and most easily definable of the mudstones have been designated as marker beds (Figure 6). These were numbered Levels 1 through 8 from the top of the sequence down. The sediments between each marker bed were identified as sub-levels. Thus, Level 1 was separated from Level 2 by Sub-level 1. Fossils were recovered from all levels and sublevels, except for Sub-levels 4 and 6. However, fossils were not evenly distributed throughout the sequence; the most numerous and best-preserved specimens were recovered from the lowermost levels. For example, the percentage of fossil fishes obtained from different levels are: Level 1 to Sub-level 4, 18.5% (n = 43); Level 5 to Sub-level 6, 24.6% (n = 27); and Level 7 to Level 8, 56.9% (n = 215). However, since these represent units of uneven thickness, a more telling index of frequency is the number of fish specimens recovered per cm thickness of sediment: Level 1 to Sub-level 4 = 0.6; Level 5 to Sub-level 6 = 0.8; and Level 7 to Level 8 = 4.2. In addition, all of the insects and frogs, and most of the leaves (56%), were recovered from Level 7 through Level 8.

The mudstones show varying degrees of silicification, but this is most evident in the lower one-third of the sequence. This presumably relates to hydrothermal activity after the close of the main eruptive event. Silicification

[1]This estimate is based on a crater lake of 370 m diameter, with a diameter-depth ratio of 5–10 (giving an estimated depth of 37–74 m), a sediment accumulation rate of 0.7–2.0 mm per year, and a sediment compaction factor of 10 (Bradley, 1929; Smith, 1986; Scarth, 1994; Maley *et al.*, 1990, 1991; Giresse *et al.*, 1991; Cornen *et al.*, 1992).

makes the mudstones extremely hard and more difficult to fracture along bedding planes, resulting in a greater tendency for fractures to be conchoidal.

Above the fossiliferous beds is a layer, more than 1 m thick, consisting of slumped blocks and slabs of consolidated mudstones in a matrix of friable yellowish-gray shale (Figure 6). This layer presumably results from a single episode of slumping, produced by heavy seasonal flooding or local tectonic activity. The lowermost section (up to 30 cm thick) consists of pale yellow-gray shale. Bedding is distinct, but buckled and disturbed, and thick sheets of calcite occur throughout. Overlying this is a poorly-laminated and distorted series of yellowish-gray shales (up to 90 cm thick) with inclusions that grades upwards as follows: (1) small blocks and slabs of silicified yellow-brown mudstone without distinct laminations; (2) massive slump blocks, some more than 60 cm in diameter, of poorly-laminated yellow-gray mudstone; and (3) slabs of consolidated mudstones, up to 5 cm thick, contained in a matrix richly impregnated with calcite (Figure 6). The partial preservation of the original bedding structure implies that the transport distance of slumped material was relatively limited. Capping this layer is a well-consolidated, heavily silicified and poorly-laminated yellow-gray mudstone, 8 cm thick (Figure 6).

The uppermost layer consists of a recently formed colluvium (1.1 m thick) of dark gray mbuga clay and sandy loams containing numerous detrital fragments of orange-brown weathered granite, quartz, and aggregations of decomposed kimberlite (Figure 6). The larger clasts are poorly sorted, angular to sub-rounded pebbles and cobbles, some exceeding 20 cm in diameter.

5. Lake Paleoecology

Excellent comparative information is available on present-day crater lakes in tropical Africa (e.g., Trewavas *et al.*, 1972; Green, 1972; Green *et al.*, 1973; Melack, 1978; Beadle, 1981; Maley *et al.*, 1990, 1991; Giresse *et al.*, 1991, 1994; Cornen *et al.*, 1992), and this has proved especially helpful in reconstructing the paleoecology and taphonomic relationships at Mahenge. The remarkably complete preservation of the fossils, and the absence of any evidence of damage by benthic scavengers or bioturbation, is suggestive of thermal-chemical stratification of the water column. This indicates that the crater lake at Mahenge was relatively deep—greater than 18 m (Kling, 1988). The cooler, anoxic hypolimnion would have been hostile to most organisms (Beadle, 1981; Smith, 1986). Therefore, dead animals that settled to the bottom of the lake would have been rarely disturbed by scavengers.

The presence of varves in the sediments at Mahenge indicates that there were seasonal fluctuations in temperature that resulted in a periodic turnover of the water column (Bradley, 1948). Tropical lakes are generally relatively stable in terms of stratification, but they do experience seasonal oscillations in the depth of the thermocline, and even complete turnover in which mixing of the hypolimnion and epilimnion occurs (Beadle, 1981; Kling, 1987; Giresse *et al.*, 1991, 1994). Heavy cloud cover, lower ambient temperatures, increased influx of cooler water from run-off, and changing wind patterns during the rainy season all contribute to reducing the temperature of the epilimnion, and resulting in circulation of water from the hypolimnion (Talling, 1969; Livingstone and Melack, 1984; Kling, 1987). Turnover during cooler weather may result in an algal bloom or the circulation of toxic gases, such as carbon dioxide, methane or hydrogen sulphide, which can lead to catastrophic mass deaths of fish and other animals (Kling, 1987; Kling *et al.*, 1987; Livingstone and Melack, 1984; Elder and Smith, 1988). However, the density and distribution of the fish remains at Mahenge indicate an attritional death assemblage rather than one due to mass mortality. In sum, the lithological evidence suggests that the sediments at Mahenge were deposited in a relatively deep and stratified crater lake that experienced periodic turnover of its water column as a result of marked seasonal variation in climatic conditions.

6. Age of the Fossil Beds

There has been a good deal of speculation in the past about the age of the kimberlite occurrences in the Singida region. Teale (1932) suggested that sandstone fragments found as xenoliths in the Kiomboi pipe in the Iramba field are lithological reminiscent of Upper Karroo sediments; placing the eruptive phase later than the early Jurassic. Du Toit, who accompanied Teale on a reconnaissance to Singida in 1931, considered the kimberlites to be younger than the early Cretaceous, possibly of Tertiary age (Teale, 1932; Eades and Reeve, 1938; Williams, 1939). Eades and Reeve (1938) suggested that the lake beds may have been deposited during the main phase of regional peneplanation, which they speculated to have occurred in the Miocene.

Based on his preliminary assessment of the systematics of the fossil fishes from Mahenge, particularly the cichlids, Greenwood (1960) suggested that the sediments were probably middle to late Tertiary in age. However, the recovery of additional fossils led Greenwood and Patterson (1967) to re-evaluate the relationships of the fish fauna, and to favor a Paleogene (probably Oligocene) age.

Estimated ages of kimberlites in neighboring fields provide circumstantial evidence to support a late Cretaceous-early Tertiary eruptive phase for the Singida kimberlite field. The diamondiferous kimberlite pipe at Mwadui in Shinyanga Region (the largest such occurrence in Africa, with a surface area of 1.5 km^2), located about 150 km to the northwest of Mahenge, is overlain by a thick series of lake sediments, up to 380 m in depth (Edwards and Howkins, 1966). These have yielded fossil vertebrates and plant remains, as yet undescribed, as well as a microfloral assemblage that indicates an age of late Cretaceous or younger (Edwards and Howkins, 1966). A kimberlite near Nzega, 120 km to the west of Mahenge, has yielded U-Pb dates of 52.2 and 53.2 Ma, and fission track dates of 54.3 ± 14 Ma and 51.1 ± 3.8 Ma (Davis, 1977; Naeser and McCallum, 1977; Haggerty *et al.*, 1983). This places the intrusive event for the Nzega field in the early Eocene (= Ypresian).

During the 1996 expedition a single 508 mg hyacinth-colored zircon crystal was recovered from the stream bed at Mahenge. RM and KRL analyzed two

Table 2. Summary of U-Pb isotopic analyses of Mahenge zircon

Split #	ppm U	Th/U (1)	206Pb/204Pb (2)	206Pb/238U Age, Ma (3)	207Pb/206Pb Age, Ma (3)
1	16.1	0.31	278.5	45.96 ±0.31	79 ±80
2	15.7	0.29	170.7	45.79 ±0.20	78 ±110
Weighted mean				45.83 ±0.17 Ma	

(1) Calculated from radiogenic 208Pb/206Pb assuming Th/Pb - U/Pb concordance.

(2) Measured.

(3) Calculated assuming a common Pb with 206Pb/204Pb=18.6±0.4, 207Pb/204Pb=15.60±0.35, and corrected by +0.10 Ma for estimated deficiency of initial 230Th. Errors are 2-sigma.

milligram-sized splits of fragments from the crushed crystal, which yielded essentially identical and internally-concordant 206Pb/238U ages of 45.83 ± 0.17 Ma (Table 2). Because of the young age, low uranium-content (~16 ppm U) and internal concordance of the zircon, the date can be regarded as accurate. This indicates that intrusion of kimberlites in the Singida field occurred during the middle Eocene (= Lutetian), slightly later than those of the Nzega field, and probably also later than those of the Shinyanga field (e.g., Mwadui). It also confirms that the kimberlite occurrences in Tanzania are younger than those from southern Africa (which are mainly late Jurassic to Cretaceous in age, ranging from ~65–150 Ma) (Dawson, 1970; Hawthorne, 1975; Haggerty *et al.*, 1983; Mitchell, 1986).

An important point here is that the radiometric age determination provides a date for the intrusive event (i.e., the formation of the crater and eruption of the kimberlite), rather than the age of the fossiliferous sediments that subsequently filled the crater lake. However, studies of similar crater lakes in Europe and Africa show that sediments begin to accumulate relatively soon after initial formation of the crater (Lorenz, 1973; Smith, 1986; Rayner and McKay, 1986; Rayner, 1987; Giresse *et al.*, 1991; Cornen *et al.*, 1992). Although sedimentation rates and degree of compaction may vary, conservative estimates derived from data on modern maars indicate that the crater at Mahenge would have been filled in 0.2–1.0 myrs (see above). For example, McKay and Rayner (1986) estimate that the much larger crater at Orapa in Botswana (late Cretaceous), with at least 80 m of sediment, would have been completely filled in less than 0.5 myrs. We can confidently assume, therefore, that the fossils date to ~45–46 Ma. Equally important, if our estimates of sedimentation rates are correct, is that the deposits which yielded all of the fossils represents a very narrow time window of only 8000–22,700 years.

7. Renewed Fieldwork at Mahenge

Fossil vertebrates were originally discovered at Mahenge by George Mannard in the late 1950s as part of his intensive survey of the diamond-bearing potential of the Singida kimberlites (Mannard, 1962). During his investigations at Mahenge, Mannard excavated more than twenty test pits that penetrated the surface soil into the underlying kimberlite and basement rocks. The depth of Mannard's test pits in the region do not exceed 10 m in depth (Mannard, 1962), but it would appear that those at Mahenge were less than 5 m, and the majority of pits at the edge of the pipe were less than 2 m in depth. Mannard (1962: 243) provided a sketch map of the location of his pits and kimberlite outcrops, but he provided no further documentation or key to the individual pits.

In 1994 TH and CPM were able to relocate many of Mannard's pits, which remained unfilled except for some incidental back-filling of the excavated sediments. Individual pits and pit complexes were given reference numbers; those located within the boundary of the pipe (i.e., those that exposed lake sediments or kimberlite) were assigned an Arabic numeral, while those located outside the pipe (i.e., those that penetrated the granitic basement only) were assigned a Roman numeral. Pit complexes, consisting of several pits in close association, were given the same number, but were differentiated by a letter suffix. Mannard's sketch map is reproduced in Figure 7, with the addition of the pit reference numbers introduced in 1994.

The main pit complexes of Mannard (Pits 5 and 6) are located near the center of the crater, in close proximity to the Luwala River (Figures 5 and 7). The pit complex on the eastern bank of the Luwala, designated Pit 5, consists of three separate pits. Pit 5c is the deepest of these pits (~4 m deep). On the western side of the Luwala is a complex of pits that have been identified as Pit 6. Both of these pit complexes have yielded the remains of fossil vertebrates and plants. A few fossils were recovered from Mannard's spoil heaps at Pit 6 in 1994, but no further work was conducted in 1996.

The main focus of the 1996 season was on pit 5c. The recovery of more than forty fossil fish from Mannard's spoil heaps in 1994 suggested that this would be the most productive site to begin operations. Excavations involved clearing Mannard's original pit (which measured 6.0 m × 3.4 m) and enlarging it to form a small quarry (10.0 m × 5.7 m). This allowed compilation of a stratigraphic section and a more detailed assessment of the sedimentology (see above). In addition, the course of the Luwala River and the horizontal extent of the crater lake and diatreme were mapped (Figure 5).

Excavation of the quarry involved removal of the horizontal shales and mudstones one layer at a time. Although masonry saws were occasionally used to cut large slabs, in most cases the weight of the overlying sediments had caused the layers to crack into small irregular slabs that could easily be removed from the quarry by hand. After removal, fossils were recovered by splitting slabs along natural cleavage planes using fine masonry chisels. In general, only relatively complete specimens were retained and catalogued. All of the specimens discovered in 1994 and 1996 are housed in the National

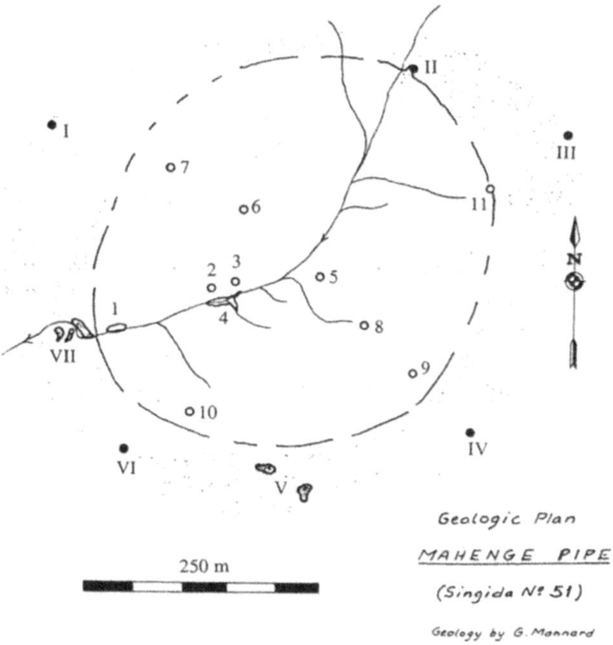

FIGURE 7. Sketch map adapted from Mannard (1962) to show the location of Mannard's test pits and their current pit designation numbers. See text for further details.

Museums of Tanzania, Dar es Salaam, while earlier collections are housed in the Natural History Museum, London.

8. Other Localities in the Singida Area

Of the 54 kimberlite pipes in the Singida field only ten have lacustrine sediments derived from a former crater lake (Mannard, 1962). The mean surface area of kimberlite pipes in the region is approximately 34,000 m^2, but most of the occurrences (78%) are smaller than this, with over half having an area of less than 15,000 m^2. The smallest pipes with crater lake sediments are Magoba, Munyu, and Kinginya; medium-sized pipes with surface areas of between 16,000 m^2 and 20,000 m^2. However, crater lake deposits are most commonly associated with large kimberlite bodies, and some significance can probably be attached to the fact that 6 of the 7 pipes (83%) with a surface area of over 100,000 m^2 are positively associated with lake sediments. These include Pyamahumbe (105,000 m^2), Mahenge (238,000 m^2) and Songeli (256,000 m^2)—all of which have sediments known to contain fossils (see Figure 2, Table 1).

There are good reasons for this association. Large kimberlite pipes are more likely to produce a depression of sufficient volume to accommodate a crater lake deep enough to allow the accumulation of fine-grained sediments and a stratified water column that is conducive to the preservation of relatively complete organisms. The data suggest that kimberlite pipes smaller than 15,000 m^2 are unlikely to be associated with crater lakes, while those between 15,000 m^2 and 100,000 m^2 may support small lakes that are generally too shallow for the preservation of fossils (the exception here is Munyu, which has well-preserved leaves in lake sediments overlying a pipe with an estimated area of just over 20,000 m^2). One can assume, therefore, that relatively few of the kimberlite pipes in the Singida region have the potential to yield fossils similar to those at Mahenge. Apart from the localities listed in Table 1 which have already yielded fossils, two further kimberlite occurrences with lacustrine sediments overlying relatively large pipes—Tabola (130,000 m^2) and Kolongo (140,000 m^2)—would probably merit further investigation.

In 1994 and 1996 team members made brief excursions from Mahenge to other kimberlite occurrences in the Singida area to explore their potential for further paleontological investigations. Visits were made to Songeli (= Sungili or Sunguli), Munyu (= southern "Mtawira"), Karusi ya Kasunga, Lusilukulu, Tabola (= Ntola), and Pyamahumbe (see Figure 2).

Songeli is of interest for several reasons. It is the largest diatreme in the Singida Region (with a maximum diameter of at least 760 m), and is the second largest in Tanzania after Mwadui. Kimberlites were first discovered in the area in 1931 (Teale, 1931, 1932; Eades and Reeve, 1938; Williams, 1939; Mannard, 1962). Teale (1931) reported the discovery of well-stratified lake beds overlying the pipe containing fossil stems, leaf impressions and seeds. In 1994 the Songeli stream was surveyed as far as its confluence with the Msungwe River and we mapped extensive outcrops of kimberlite in the gullies. No fossils were recovered, although we recorded fossilized wood and root casts *in situ* in the yellow ground and pyroclastic kimberlite. In addition, the river gravels and sands contain a coarse litter of brightly colored chert fragments derived from the silicified cap of the diatreme, as well as numerous Middle Stone Age handaxes and flakes made from the same material.

Several kimberlite occurrences have been recorded in the vicinity of Mtawira (Teale, 1932). These have been identified as Mtawira, Kolongo, Munyu and Munyu North pipes by Mannard (1962). Du Toit recovered impressions of plant remains and pieces of a large bone from Munyu (Teale, 1932), while Eales collected some well-preserved fossil leaves in a shale 6 m below the surface (Mannard, 1962). The Mtawira area is heavily vegetated and quite remote, being difficult to reach by car. Swarms of tsetse flies make prospecting burdensome. The former pits and spoil heaps of previous prospectors are now much overgrown, and little useful information can be gleaned from the exposures without opening further test pits.

At Karusi ya Kasunga, Lusilukulu, and Tabola, about 15 km to the northeast of Mahenge, numerous shallow pits were investigated, many of which were overgrown and partially filled. Local informants told us that some of the pits were dug prior to World War II—presumably the work of the South African prospector, Pienaar, who was known to have surveyed the area in the early

1930s. Outcrops of kimberlite were recorded, but there was no evidence of lake sediments.

The pipe at Pyamahumbe, with a maximum diameter of 360 m, is located on a flat, sparsely forested terrain. A small river transects the pipe and exposes limited outcrops of kimberlite and lacustrine sediments. The uppermost layer of sediments consists of a thin cherty bed, up to 30 cm thick, that exhibits strong warping and folding. This bed represents a series of heavily silicified mudstones, with individual layers up to 5 cm thick, that retain much of their original laminar structure. They are mostly gray-green with bright orange streaks and mottling. Large white crystalline and red-orange inclusions are possibly of organic origin. The beds are difficult to split mechanically because of their chemical alteration, but they do contain fossils—mostly stem and leaf fragments, seeds and/or fruits, and isolated fish scales and bones. Better material might be recovered with more extensive sampling. The cherty mudstone grades down into a friable light gray shale that rests unconformably on kimberlite. The lake sediments are covered by a 1.0–1.5 m thick layer of dark gray mbuga clay.

In addition, mention should be made of two localities, not visited by our expedition, that are potentially of some paleontological interest. Local informants who worked with Mannard in the late 1950s reported that large fossil bones were recovered from a site near the village of Masorsa. However, Mannard (1962) makes no mention of these discoveries, and since he was aware of the obvious significance of the fossil fish from Mahenge, it seems unlikely that he would not have documented such an occurrence. We are inclined, therefore, to discount the validity of this claim. The other locality, potentially much more important, is the site of Minusi. Greenwood and Patterson (1967) made reference to a specimen of *Singida jacksonoides* and two specimens of cichlids collected at this site by Mr. Philip J. King. The fossils are preserved in buff colored mudstones comparable to those at Mahenge. Unfortunately, the provenience and details of the discovery are unknown, except that the specimens were found by accident during construction of a house (Patterson, personal communication). Minusi cannot be found on maps and the name is unfamiliar to local inhabitants. It is possible that the name represents a corruption of Minughe or Minyughe—a river and village located about 30 km to the southeast of Mahenge. However, none of the pipes in the vicinity of Minughe are associated with lake sediments, except for Magoba, which has previously only yielded silicified wood (Mannard, 1962; Table 1).

9. The Mahenge Fauna and Flora

A detailed analysis of the fauna and flora from Mahenge is still underway, but we present here a preliminary account of our initial findings. Our study is based on more than six hundred specimens recovered from Mahenge in 1994 and 1996 (Table 3). The collection consists predominantly of fossil fishes

Table 3. Number of specimens recovered from sites in the Singida Region in 1994 and 1996

	Frogs	Fishes	Inverte-brates	Plants	Other[1]	Total
Mahenge 1996	4	466	3	74	17	564
Mahenge 1994	0	47	2	11	0	60
Pyamahumbe 1996	0	1	0	6	2	9
Total	4	514	5	91	19	633

[1] Includes coprolites and indeterminate specimens

(81.2%), while frogs (0.6%), insects and gastropods (0.8%), plants (14.4%) and indeterminate specimens (3.0%) make up the remainder.

9.1. Fishes

Only two species of fish have previously been described from Mahenge: *Palaeodenticeps tanganikae* and *Singida jacksonoides* (Greenwood, 1960; Greenwood and Patterson, 1967). In addition, Greenwood (1960) and Greenwood and Patterson (1967) also noted the occurrence of haplochromine cichlids, although preservation of the material was not adequate to assign them to a particular species. The ichthyofauna can now be expanded to include two further taxa—a catfish (Siluriformes) and a second species of osteoglossomorph.

Additional excellent examples of *Singida jacksonoides* have now been recovered. This species represents the most common fish at the site, and is found throughout the fossiliferous sequence. Specimens have a standard length (SL) from 20 mm to about 180 mm, with the majority being under 70 mm SL. Some specimens preserve impressions of scales.

Singida jacksonoides is the sole representative of an extinct family, the Singididae (Greenwood and Patterson, 1967). It is the only fossil osteoglossoid fish known from Africa, and its closest extant relatives (i.e., species belonging to the Osteoglossidae, Pantodontidae and Heterotididae) form relictual taxa in the Nile and Zairean basins. *Singida* is unique among osteoglossomorphs in lacking teeth, but otherwise its cranium is similar to that of other osteoglossoids, especially to the extant Australasian *Scleropages* and the South American *Osteoglossum* (Greenwood and Patterson, 1967). In a recent phylogenetic analysis, Li and Wilson (1996) concluded that *Singida* represents the sister-taxon to the extant *Scleropages* + *Osteoglossum* within the family Osteoglossidae. However, *Singida* is more primitive than all known osteoglossoids, including the Paleogene phareodontines, in retaining a hiodontoid-like caudal skeleton that retains two uroneurals (Greenwood and Patterson, 1967). It is possible, therefore, that it represents a stem osteoglossoid, a clade that had already diverged from other osteoglossomorphs by the early Cretaceous (Lundberg, 1993; Li and Wilson, 1996).

The distribution of extant osteoglossoids in southern tropical freshwater regions suggests that the initial diversification of the modern families occurred on Gondwana prior to the separation of South America and Africa in the late Cretaceous (Nelson, 1969; Taverne, 1979; Lundberg, 1993; but see Cracraft, 1974; Briggs, 1987 for alternative views). This biogeographical scenario is given additional support from the fossil record. The occurrence of the earliest representatives of the Heterotididae in the Cretaceous (i.e., *Chandlerichthys* from the early Cretaceous of North America, *Laeliichthys* from the late early Cretaceous of Brazil and *Paradercetis* from the late Cretaceous of West Africa; Da Silva Santos, 1985; Grande, 1986; Lundberg, 1993), and the wide geographical distribution of the extinct members of the Osteoglossidae in the early Tertiary, including the Phareodontinae, both point to a South American-African distribution of these families prior to the break up of western Gondwana (Lundberg, 1993). Regardless of whether *Singida* proves to be a stem osteoglossoid or the sister taxon to extant osteoglossids, the biogeographical evidence implies that *Singida* is part of a distinct lineage that originated before the close of the Mesozoic.

In addition to *Singida jacksonoides*, the recent collections from Mahenge have yielded what appears to be a second, previously undescribed, species of osteoglossomorph. This taxon occurs throughout the deposits. The specimens are all fairly large, about 150 mm SL. Scales are associated with most specimens. The new species is distinguishable from *Singida jacksonoides* by the presence of teeth. One of us (AMM) is currently working on a detailed description for publication.

Palaeodenticeps tanganikae is the only known fossil representative of the Denticipitidae, a family based on a single extant species, *Denticeps clupeoides*, confined today to the rivers of western Nigeria and Benin (Clausen, 1959; Greenwood, 1960, 1965; Gras, 1961). Apparently, *Denticeps* is a relictual member of what was formerly a more diverse and geographically widespread taxon. The retention of largely plesiomorphic traits in *Denticeps* and *Palaeodenticeps* suggests that the Denticipitidae is the primitive sister group of all other clupeomorphs (Greenwood, 1960, 1968; Lecointre and Nelson, 1996).

Palaeodenticeps is relatively rare at Mahenge, where it represents less than 2% of fish specimens. Interestingly, Greenwood (1960) examined eight specimens of *Palaeodenticeps tanganikae* collected by Mannard at Mahenge, while the much larger fossil fish collection obtained in 1996 included only two individuals. *Palaeodenticeps tanganikae* are small fish (about 25–28 mm SL), but since they are not much smaller than some of the *Singida jacksonoides* and cichlid specimens, it seems unlikely that their rarity is caused by sampling or taphonomic biases, and probably reflects an ecological distinction.

Cichlids are common and are found in all levels. All appear to be "haplochromines" with ctenoid scales. Preliminary comparisons suggest that there may be some diversity among the taxa represented, possibly with different communities being sampled at different horizons. Greenwood (1960) suggested that all of the specimens that he examined could be included in a single genus, with closest affinities to extant *Haplochromis*. The relatively small size, stout unicuspid teeth, and deep dentaries of the fossil cichlids suggest that they were feeding on phytoplankton, organic debris, and small aquatic insect larvae. The mid-Eocene age of the Mahenge cichlids makes them the earliest known members of the family, at least from Africa, and they should eventually provide important clues to understanding the evolutionary history and zoogeographic origins of the group (Van Couvering, 1982; Stiassny, 1991; Lundberg, 1993).

Three specimens of a catfish were recovered. These are all from the lowest levels, Sub-level 7 and Level 8. The absence of catfish above the bottommost levels may indicate a change in the faunal community structure over time. It is conceivable that these catfish, whose modern-day counterparts are air-breathers tolerant of deoxygenated and swampy conditions, were better able to exploit the lake when it was first formed (i.e., when the water column was at its deepest and sedimentation rates were relatively rapid). However, these horizons also correspond to the levels in which the greatest number of fossils are preserved, so the restricted stratigraphic distribution of the catfish could simply reflect a problem of sampling. The catfish and the osteoglossoids would have been the major predators of other fish and invertebrates in the lake.

Preservation of the fish specimens varies. Most of the *Singida jacksonoides* specimens, as noted by Greenwood and Patterson (1967), are preserved as natural molds, with some "cheesy" bone still adhering. The other osteoglossomorph, although found in the same layers as *S. jacksonoides*, has well-preserved, hard bones, often associated with scales. The cichlids vary in preservation, some being natural molds and others having bones and scales preserved, but the bone is quite fragile. The catfish specimens are all well-preserved, including the ornamentation of the cranial bones, but the matrix is extremely hard and difficult to remove. Greenwood (1960) reported isolated vertebrae, ribs and fin rays from Mahenge. In 1996 only a single isolated bone, a fragment of a preopercle, and portions of several vertebral columns belonging to unidentified taxa were recovered—almost all of the remaining specimens were entire. In addition, coprolites of fish are extremely common throughout the sequence.

The greatest diversity of the fish fauna is in the lowest levels of Mahenge. At least five different species have been identified, although it is likely that multiple species of cichlids are represented at each level. The general correlation between lake size and number of species of fish in modern-day tropical lakes (Barbour and Brown, 1974) indicates that the crater lake at Mahenge would have supported fewer than ten species. However, data on modern

African lake communities suggest that a slightly higher diversity might be possible. For example, the modern-day crater lake at Barombi Mbo, although considerably larger than the lake at Mahenge (being 2.5 km across), might provide a reasonably good model. The fauna comprises 17 species of fish, including eleven species of cichlids, a cyprinid, three species of cyprinodontids (found primarily in the inflow stream) and two species of clariid catfish (Trewavas *et al.*, 1972). Most species are relatively small (less than 160 mm SL), while the catfish, the main predators, are somewhat larger (up to 300 mm SL).

As noted by Greenwood (1960) and Greenwood and Patterson (1967) the closest living relatives of *Singida* and *Palaeodenticeps* are riverine species, rather than lacustrine, implying that the fossil species had different ecological and behavioral characteristics. It is conceivable, however, that *Palaeodenticeps*, being a relatively rare species at Mahenge, may have lived in the feeder streams or adjacent parts of the lake. In addition, studies of modern African fish faunas indicate that small crater lakes are readily colonized at times of heavy seasonal flooding, particularly by cichlids, and that rapid changes in behavior of the immigrants are a common theme (Trewavas *et al.*, 1972; Kingdon, 1989).

9.2. Anurans

Partially articulated skeletons of four individuals were recovered in the recent collections from Mahenge. They appear to belong to a single species of pipid frog, representing different developmental stages. They are relatively small in size, with a maximum head and body length of about 40 mm. Several of the individuals appear to preserve stomach contents in the abdominal region, including coarse sand grains, presumably derived from the margin of the crater lake (probably ingested accidentally along with its prey shortly before the time of death). SEM analysis of one of the specimens (by W. J. Sanders, University of Michigan) found no trace of any organic materials in the detritus. Similar materials have been recovered from the stomachs of extant African lake fishes that prey on near shore insect larvae (Trewavas *et al.*, 1972).

The species from Mahenge shares several derived character-states with Pipinae, a clade represented today in northeastern South American by *Pipa* and in sub-Saharan Africa by the Hymenochirini (i.e., *Pseudhymenochirus* and *Hymenochirus*). Preliminary examination of this material shows the presence of the following pipine synapomorphies: wedge-shaped skull in lateral profile, wide parasphenoid in the floor of the braincase, coracoids broadly expanded at their sternal ends, squamosal lacking a well-developed zygomatic ramus, vomer appears to be absent, the first two presacral vertebrae are fused, the sacrum is formed by at least three vertebrae, and the sacral diapophyses are greatly expanded. A close relationship with the Hymenochirini is suggested by the blunt anterior margin of the parasphenoid, the deep eustachian canals crossing the otic capsules in an anterolateral-posteromedial direction, and the ventrally curved postzygapophyses. However, the species from Mahenge appears to be more plesiomorphic than the extant hymenochirins in the presence of pterygoids bearing an anterior arm. Although some instances of hyperossification are evident, such as the fusion of the nasals and frontoparietal, extension of the frontoparietal over the otic capsules, and ossification of the planum antorbitale, there are no indications of the bizarre specializations seen in the new hymenochirin genus from the late Cretaceous of Niger (Báez and Rage, 1998; Báez *et al.*, in press).

This record confirms that pipid frogs, confined to Africa and South America at present, are components of the freshwater faunas of Western Gondwana. The combination of fossil occurrences (Haughton, 1931; Nevo, 1968; Estes, 1975, 1977; Duellman, 1993; Báez, 1996) and current hypotheses of pipoid phylogenetic relationships (Cannatella and Trueb, 1988a, b; Báez and Trueb, 1997) suggest that pipids were well-diversified prior to the complete separation of Africa and South America in the middle Cretaceous. Although the biology of extant African pipines (*sensu* Cannatella and Trueb, 1988a, b) is poorly known, they are fully aquatic and restricted to equatorial lowland forests, ranging from Nigeria and Cameroon to eastern Zaire. The occurrence of a pipine at Mahenge demonstrates that this clade had a more eastern distribution in the past.

9.3. Invertebrates

Only a few insects were recovered from Mahenge, including a beautifully preserved ant (Formicidae, Hymenoptera) and a partially preserved winged insect of indeterminate taxonomic status.

Mollusks are extremely rare. A single poorly-preserved and unidentifiable shell of a small gastropod was recovered from Mannard's spoil heap at Pit #5 in 1994. There are two possible explanations for this isolated occurrence. First, in a thermally stratified lake the hypolimnion is generally acidic and calcium carbonate shells tend to dissolve completely (Wuttke, 1992). Second, in modern African crater lakes, such as that at Barombi Mbo in Cameroon (Giresse *et al.*, 1991), gastropods are restricted to the feeder streams, and do not occur in the lake itself. If this is typical, rare finds of mollusks are likely to be aquatic gastropods derived from the in-flow streams or possibly shells of terrestrial gastropods washed into the lake from the same source as the terrestrial plant material.

9.4. Plants

Paleobotanical studies were conducted on an initial collection of 70 specimens from Mahenge, which consists almost entirely of leaf compressions/impressions, but also includes a few stem fragments. Although the collection is small, it adds a significant new dimension to our currently limited knowledge of African Paleogene environments. Mahenge plant fossils represent the first Eocene vegetative remains from equatorial Africa. They complement what little is known from West African palynological core samples (Van Hoeken-Klinkenberg, 1966; Salard-Cheboldaeff, 1979, 1981; Salard-Cheboldaeff and Dejax, 1991), fossil wood from mainly poorly-dated deposits of West and North Africa (Boureau *et al.*, 1983; Dupéron-Laudoueneix and Dupéron, 1995), and isolated fruit or seed specimens from Egypt and West Africa (Chandler, 1954; Chesters, 1955). Preliminary study of the Mahenge paleoflora also provides some valuable data that contribute to an improved understanding of the evolutionary history of East African floras.

A minimum of 20 species is represented among the leaf specimens. Five of these are in the Leguminosae family and have been classified as *Acacia* sp. nov. (subfamily Mimosoideae), cf. *Aphanocalyx* (subfamily Caesalpinioideae), cf. *Cynometra* (subfamily Caesalpinioideae), and two species of unknown affinity (Herendeen and Jacobs, submitted). The presence of *Acacia* at Mahenge marks the first unequivocal occurrence of this genus in Africa.

A complex of character patterns among extant species of *Aphanocalyx* and its relative, *Monopetalanthus*, create problems for systematic placement within this group. Further documentation (from additional Mahenge fossils) of character distribution in relatively primitive members of these groups may aid in understanding the phylogenetic relationships of extant relatives. The legume family is diverse in Africa today, but has a limited fossil record, especially in the Paleogene. Caesalpinioid and mimosoid legumes are recorded among Eocene pollen and wood specimens (Salard-Cheboldaeff, 1979; Gros, 1992; Dupéron-Laudoueneix and Dupéron, 1995), and a probable Caesalpinioid leaf is known from the Oligocene Jebel Qatrani Formation, Egypt (Bown *et al.*, 1982). However, the Mahenge site confirms that legumes were well-established and diverse in equatorial Africa at this time.

A qualitative study of leaf physiognomy among the 20 species provides a general indication of the paleoenvironmental setting. Of the 18 dicot leaf taxa for which measurements could be made, all are smaller than mesophyll (>20.25 to 45.00 cm2) size. In modern equatorial Africa, the proportion of species with mesophyll leaves in a plant community is positively correlated with both mean annual rainfall and wet months rainfall (months having more than 50 mm on average; Hall and Swaine, 1976; Jacobs, 1999). The predominance of small-leaved species at Mahenge is an indication that rainfall was less than that which would support forest vegetation today (generally >1100 mm/yr).

All but one of 18 dicot taxa have entire (smooth) margins. Today, across a broad latitudinal temperature gradient, the proportion of species with an entire margin is directly correlated with mean annual temperature (Bailey and Sinnott, 1915; Wolfe, 1995). This indicates that the Mahenge flora represents a hot or warm environment with markedly seasonal or relatively low (<1000 mm/yr) mean annual precipitation. This interpretation is compatible with the presence of *Acacia*, a large genus confined today to warm or hot regions where rainfall supports woodland or wooded grassland environments. More precise climate reconstruction, and perhaps a better understanding of the relatively open structure of this plant community, might be possible with an enlarged collection of specimens.

Taphonomic studies of stream-fed lacustrine depositional systems have shown that leaves are derived from both allochthonous upstream vegetation and local plant communities (Spicer, 1981; Ferguson, 1985; Greenwood, 1991). However, just over one-third (6 of 16) of the species for which leaf shape could be determined have leaves that are linear, narrowly oblong, or narrowly elliptic. Plants which grow in or immediately adjacent to fast-flowing streams often have narrow (stenophyllous) leaves (Richards, 1996). It is possible, therefore, that the narrow leaves derive from (possibly distant) streamside vegetation, although there is currently no sedimentological or geomorphological evidence of in-flowing streams at Mahenge. Nevertheless, the small size of the vast majority of the collection, including taxa without narrow leaves, supports the hypothesis that the surrounding vegetation was not forest, but a more open plant community indicative of a drier climate.

Much of the Paleogene palynological work is biostratigraphic, associated with oil exploration in coastal West and North Africa (e.g., Boltenhagen, 1965). However, Salard-Cheboldaeff and Dejax (1991) summarize the paleoenvironmental significance of West African Late Cretaceous and Tertiary palynological records. They document an overall increase in diversity of angiosperm taxa, especially since the middle Eocene, and attribute this to angiosperm diversification, the northward movement of Africa between the terminal Cretaceous and middle Eocene, and the increasing availability of moisture along coastal West Africa after the later Cretaceous split with South America. The marked increase in angiosperm richness after the middle Eocene is taken to represent the origin of wet lowland forest. On the basis of fossil wood, Boureau *et al.* (1983) reconstruct a tropical forest along the western and northern coasts of the continent during the middle Eocene, with a savanna environment in-between. However, as these reconstructions are based on fossil wood, there are not data to address the existence of grasses that would be associated with a savanna environment today. Axelrod and Raven (1978) suggested lowland rainforest had been established by the Paleocene, and extended northward from about 15°S, but this was before some of the West African pollen records had been published. The Mahenge paleoflora indicates the absence of wet forest at paleolatitude 15°S, at least in the vicinity of the site.

A global review of the origin of grass-dominated ecosystems indicates that grasses may have diversified during the middle Eocene in tropical Africa and South America (Jacobs *et al.*, 1999). Palynological assemblages consistently record the presence of *Monoporites annulatus*, a form genus that represents the grass family (Salard-Cheboldaeff, 1979; Germeraad *et al.*, 1968). However, paleobotanical evidence of the savanna biome is not common until at least the middle Miocene. The botanical affinities of Eocene tropical grasses and their ecological role(s) are unknown. The Mahenge leaf assemblage includes at least one probable grass blade fragment. Additional monocot fossils may shed light on the question of the role of grasses in Eocene tropical plant communities.

10. Discussion and Conclusions

When fully studied and described, the fossils from Mahenge will yield important information on the evolution, paleoecology and biogeography of the fauna and flora of East Africa during the early Tertiary. The site is of especially significance because it samples a geographical province and time period for which we have almost no fossil record. The following discussion briefly highlights the scientific importance of the site and identifies potential avenues of future research.

10.1. A Unique Temporal and Geographical Window

Paleogene sites with continental sediments are extremely rare in sub-Saharan Africa, so the evolutionary history and biogeography of the fauna and flora of the region during this time period is almost entirely undocumented. The fossils from Mahenge help to fill a major gap. Preliminary study of the fossil fishes and frogs provides a good indication of just how exotic this fauna is—all of the species are unique, being found nowhere else except for localities in the Singida region.

Only a limited exchange of terrestrial faunas was possible between Afro-Arabia and Eurasia throughout much of the Paleogene. At this time a shallow epicontinental sea joined the Atlantic to the Indian Ocean, and separated Africa from Eurasia. The isolation of Africa is reflected in high levels of endemism at early Tertiary sites in North Africa (Sigé *et al.*, 1990; Godinot and Mahboubi, 1992, 1994; Gheerbrant *et al.*, 1993), and the development of several distinctive African mammalian clades (i.e., Proboscidea, Hyracoidea, Tenrecidae, Macroscelidea, Chrysochloridae, and Tubulidentata) (Springer *et al.*, 1997, 1999; Liu and Miyamoto, 1999). However, Storch and Schaarschmidt (1992) have suggested that limited faunal interchange did occur between Africa and Eurasia during the early Eocene. The first appearances of certain mammals in central Europe may signal the arrival of taxa from Africa, possibly including the earliest artiodactyls, perissodactyls, and primates. Other vertebrate groups that arrive in Europe at this time include the Ceratophryinae, ziphodont mesosuchians and Phorusrhacidae, all of which are known to have South American distributions, and imply a broader connection with faunas of western Gondwana (Storch and Schaarschmidt, 1992).

Unfortunately, we know nothing of the diversity or distinctiveness of early Tertiary faunas in East Africa. In fact, the only sites with terrestrial mammals of Eocene age in sub-Saharan Africa are M'Bodione Dadere in Senegal (middle to late Eocene) and Malembe in Angola (late Eocene), both with small faunas sampling predominantly marine, near-shore facies, similar to those in North Africa (Sudre, 1979; Pickford, 1986). Although no mammals have been discovered at Mahenge, it is possible that with better sampling of the sediments such fossils will eventually be discovered (see Rose and Silcox, 1999). The recovery of fossil mammals from a terrestrial continental setting in East Africa would contribute significant new insights into the evolutionary history and biogeography of mammalian faunas in Africa.

10.2. Zoogeography

As mentioned above, the zoogeographic relationships of the Mahenge fauna clearly point to a strong association with western Gondwana. For example, of the fossil groups represented at Mahenge, the Osteoglossoidea, Cichlidae and Pipidae all have living species found today (although not exclusively) in both Africa and South America. The fossil record suggests that these taxa were widely distributed in western Gondwana prior to the final separation of Africa and South America in the late early Cretaceous (~95–100 Ma) (Pitman *et al.*, 1993). The distribution is best explained by a continental drift-based vicariance model (Lundberg, 1993). The Denticipitidae only has a relictual distribution in West Africa today, but its occurrence at Mahenge implies that the group may have been more widely distributed across tropical Africa during the early Tertiary. A more detailed assessment of the phylogenetic relationships of the Mahenge fish and frogs (and possibly the insects) will contribute to a better appreciation of the origins, zoogeographic distribution, and diversity of Gondwanan faunas.

On a more local level, an analysis of the ichthyofauna and amphibians, in conjunction with study of the regional geology and geomorphology, may provide helpful clues to reconstructing the early development of East African hydrological systems prior to the formation of the Rift Valley. Past connections with the Zaire Basin and West African river systems are indicated by the taxonomic relationships of the Mahenge fish (Greenwood and Patterson, 1967).

10.3. Paleoecology

The plant macrofossils contribute to an improved understanding of the systematics, evolutionary history and paleobiology of the East Africa flora, and also provide intriguing new data on the vegetation and paleoecology. A preliminary analysis of leaf physiognomy indicates that the vegetation at Mahenge was a predominantly dry, seasonal woodland, rather than humid tropical rainforest.

Given what we know about early Eocene climate and vegetation patterns on a worldwide scale, the paleoecological inference drawn from the Mahenge fossil plants is somewhat surprising. Several independent lines of evidence have demonstrated that the early Eocene was a period of global warming (e.g., Frakes and Kemp, 1973; Savin, 1977; Shackleton, 1986; Miller *et al.*, 1987; McGowran, 1990; Zachos *et al.*, 1993; Sloan and Rea, 1995), which had a significant impact on the taxonomic and ecological composition of floras at high latitudes (Frakes and Kemp, 1973; Wolfe, 1980; Hubbard and Boulter, 1983; Upchurch and Wolfe, 1987; Schaarschmidt, 1992). However, the effect, if any, of elevated global temperatures on the floras of equatorial Africa (Mahenge was probably located about 10°S of its present location during the Eocene) is unknown.

Placed temporally midway between the Latest Paleocene Thermal Maximum (LPTM) at ~58 and the Early Oligocene Glacial Maximum at ~35–36 Ma (Shackleton, 1986; Zachos *et al.*, 1993), one would anticipate that the climate at Mahenge would have been somewhat warmer than modern-day temperatures. However, oxygen isotopic data indicate a reduced latitudinal sea surface temperature gradient relative to the present, so temperatures may not have been very different in equatorial Africa from present-day conditions (Sloan and Rea, 1995). Oxygen isotope studies show that during the early Paleogene deep sea temperatures at high latitudes warmed from 8–12°C, while tropical ocean temperatures remained constant or were somewhat cooler than the present-day (Shackleton and Boersma, 1981, Zachos *et al.*, 1993). This is referred to as the "cool tropic paradox" (D'Hondt and Arthur, 1996).

It has been claimed that during the early Eocene eastern Africa would have supported humid subtropical to tropical rainforest (Axelrod and Raven, 1978; Parrish *et al.*, 1982; Parrish, 1987). However, Coetzee (1993) indicates that microfloral evidence shows that the present-day Congo basin was savanna woodland at that time, with subtropical woodlands extending into East Africa, while Salard-Cheboldaeff (1981) and Salard-Cheboldaeff and Dejax (1991) have shown the co-existence of dry and humid forests in West Africa. When integrated with data from Mahenge and from published reports of plant macrofossils from other Eocene sites in Africa (Adamson, 1931; Rennie, 1931; Van Hoeken-Klinkenberg, 1966; Koeniguer, 1971; Axelrod and Raven, 1978; Gros, 1992; Dupéron-Laudoueneix and Dupéron, 1995), the evidence indicates that equatorial Africa was dominated by a combination of dry seasonal woodlands and more humid forests. While high latitude regions experienced elevated temperatures and the establishment of subtropical and tropical floras, equatorial Africa had dry woodlands associated with cooler temperatures and markedly seasonal precipitation.

ACKNOWLEDGMENTS. Numerous colleagues and students have contributed to the success of this project. The following scientific personnel participated in the 1996 field season: Becky Dudzik, Christine Kiyembe, Terri Harrison, Terry Harrison, Avelin Malyango, Michael Mbago, Charles Msuya, Alison Murray, Chris Robinson, Leahanne Sarlo, Fred Szalay, and John Van Couvering. Logistical support was provided by Tobias Lubulila, Freddie Nsinge, the late Hamisi Salum, Hassani Athumani, Saidi Athumani Makagale, Abdala Athumani Makagale, Haji Athumani Selemani, Omari "Mavumbi" Imanegela, Mohammed Omari, Saidi Ramadhani, Mohammed Musa, John Joseph and Myassa Abubaka. John Van Couvering contributed his expertise to the interpretation of the sedimentology and stratigraphy. Richard K. J. Jumanne of Williamson Diamond Limited provided access to samples and gave a tour of the mine at Mwadui. Bill Sanders assisted us with casting and SEM analysis of the fossil frogs, and along with an anonymous reviewer he provided useful comments that helped to improve the content and clarity of the text. In addition, we thank the following individuals for their help and advice: John Alexander, Peter Andrews, Eric Baker, Margaret Collinson, Wendy Dirks, Peter Forey, the late P. Humphrey Greenwood, Judith Harris, Terri Harrison, Jonathan Kingdon, John Krigbaum, Jessica Manser, Malcolm McKenna, the late Colin Patterson, Andrew Ross, Bill Sanders, Leahanne Sarlo, Kathy Stewart, Fred Szalay, Simon Waane, Sally Young, and all of the villagers at Mwaru. We are especially grateful to Mzee Magwisha for the dawa ya simba. Permission to conduct research in Tanzania was granted by the Tanzania Commission for Science and Technology, the Unit of Antiquities, and the District and Regional Administrative Offices in Singida. This research was supported by grants from the National Geographic Society, the Boise Fund of Oxford University, and New York University Research Challenge Fund. Finally, we thank Gregg Gunnell for inviting us to contribute to this volume and to the symposium at the 1996 annual meetings of the Society of Vertebrate Paleontology.

This paper is dedicated to the memory of the late Hamisi Salum.

References

Adamson, R., 1931, Note on some petrified wood from Banke, Namaqualand, *Trans. R. Soc. S. Afr.* **19**:255–258.

Axelrod, D. I., and Raven, P. H., 1978, Late Cretaceous and Tertiary vegetation history of Africa, in: *Biogeography and Ecology of Southern Africa, Vol. 1* (M. J. A. Werger, ed.), Junk, The Hague, pp. 77–130.

Báez, A., 1996, The fossil record of the Pipidae, in: *The Biology of* Xenopus, (R. C. Tinsley and H. R. Kobel, eds.), Zoological Society of London, Clarendon Press, Oxford, pp. 329–347.

Báez, A., and Rage, J.-C., 1998, Pipid frogs from the Upper Cretaceous of In Beceten, Niger, *Palaeontology* **41**:669–691.

Báez, A., and Trueb, L., 1997, Redescription of the Paleogene *Shelania pascuali* from Patagonia and its bearing on the relationships of fossil and Recent pipoid frogs, *Sci. Pap., Nat. Hist. Mus. Univ. Kansas* **4**:1–41.

Báez, A., Trueb, L., and Calvo, J., 2000, The earliest known pipoid frog from South America: a new genus from the middle Cretaceous of Argentina, *J. Vert. Paleontol* **20**:490–500.

Bailey, I. W., and Sinnott, E. W., 1915, A botanical index of Cretaceous and Tertiary climates, *Science* **41**:831–834.

Barbour, C. D., and Brown, J. H., 1974, Fish species diversity in lakes, *Am. Nat.* **108**:423–489.

Beadle, L. C., 1981, *The Inland Waters of Tropical Africa: An Introduction to Tropical Limnology*, 2nd Edition, Longman, New York.

Boltenhagen, E., 1965, Introduction a la palynologie stratigraphique du bassin sédimentaire de l'Afrique équatoriale, *Ext. 1° Coll. Afr. Micropaléontol. (Dakar, 1963) Mem. B.R.G.M.* **32**:305–326.

Borg, G., 1992, New aspects of the lithostratigraphy and evolution of the Siga Hills, an Archaean granite-greenstone terrain in NW Tanzania, *Z. Angew. Geol.* **38**:89–93.

Boureau, E., Salard-Cheboldaeff, M., Koeniguer, J.-C., and Louvet, P., 1983, Evolution des flores et de la végétation Tertiaires en Afrique, au nord de l'Equateur, *Bothalia* **14**:355–367.

Bown, T. M., Kraus, M. J., Wing, S. L., Fleagle, J. G., Tiffney, B. H., Simons, E. L., and Vondra, C. F., 1982, The Fayum primate forest revisited, *J. hum. Evol.* **11**:603–632.

Bradley, W. H., 1929, The varves and climate of the Green River Epoch, *U.S. Geol. Survey, Professional Papers* **158-E**:87–110.

Bradley, W. H., 1948, Limnology and the Eocene lakes of the Rocky Mountain region, *Geol. Soc. Amer. Bull.* **59**:635–648.

Briggs, J. C., 1987, *Biogeography and Plate Tectonics*, Elsevier, Amsterdam.

Cahen, L., Snelling, N. J., Delhal, J., and Vail, J. R., 1984, *The Geochronology and Evolution of Africa*, Clarendon Press, Oxford.

Cannatella, D. C., and Trueb, L., 1988a, Evolution of pipoid frogs: Intergeneric relationships of the aquatic frog family Pipidae (Anura), *Zool. J. Linnean Soc.*, **94**:1–38.

Cannatella, D. C., and Trueb, L., 1988b, Evolution of pipoid frogs: Morphology and phylogenetic relationships of *Pseudhymenochirus*, *J. Herpetol.* **22**:439–456.

Chandler, M. E. J., 1954, Some upper Cretaceous and Eocene fruits from Egypt, *Bull. Br. Mus. (Nat. Hist.) Geol.* **2**:147–187.

Chesters, K. I. M., 1955, Some plant remains from the upper Cretaceous and Tertiary of West Africa, *J. Nat. Hist.* **8**:498–503.

Clausen, H. S., 1959, Denticipitidae, a new family of primitive isospondylous teleosts from West African fresh-water, *Medd. Dan. Naturn. Foren.* **121**:141–151.

Coetzee, J. A., 1993, African flora since the terminal Jurassic, in: *Biological Relationships between Africa and South America* (P. Goldblatt, ed.), Yale University Press, New Haven, pp. 37–61.

Cornen, G., Bandet, Y., Giresse, P., and Maley, J., 1992, The nature and chronostratigraphy of Quaternary pyroclastic accumulations from Lake Barombi Mbo (West-Cameroon), *J. Volcanol. Geotherm. Res.* **51**:357–374.

Cracraft, J., 1974, Continental drift and vertebrate distribution, *Ann. Rev. Ecol. Syst.* **5**:215–262.

Da Silva Santos, R., 1985, *Laeliichthys ancestralis*, novo gênero e espécie de osteoglossiformes do Aptiano da Formaçao Areado, Estado de Mina Gerais, Brasil, *VIII Congr. Brasil. Paleontol. 1983, MME-DNPM, Ser. Geol.* **27**, Paleontol./Estrat. **2**:151–167.

Davis, G. L., 1977, The ages and uranium contents of zircon from kimberlites and associated rocks, *Carnegie Inst. Washington, Yearbk.* **76**:631–635.

Dawson, J. B., 1970, The structural setting of African kimberlite magmatism, in: *African Magmatism and Tectonics* (T. N. Clifford and I. G. Gass, eds.), Hafner, Darien, CT, pp. 321–335.

D'Hondt, S., and Arthur, M. A., 1996, Late Cretaceous oceans and the cool tropic paradox, *Science* **271**:1838–1841.

Dodson, M. H., Gledhill, A. R., and Shackleton, R. M., 1975, Age differences between archaean cratons of eastern and southern Africa, *Nature* **254**:315–318.

Duellman, W. E., 1993, Amphibians in Africa and South America: Evolutionary history and ecological comparisons, in: *Biological Relationships between Africa and South America* (P. Goldblatt, ed.), Yale University Press, New Haven, pp. 156–199.

Dupéron-Laudoueneix, M., and Dupéron, J., 1995, Inventory of Mesozoic and Cenozoic woods from Equatorial and North Equatorial Africa, *Rev. Palaeobot. Palynol.* **84**:439–480.

Eades, N. W., 1936, The geology of the Iramba Plateau, *Short Paper Geol. Surv. Tanganyika* **15**:1–33.

Eades, N. W., and Reeve, W. H., 1938, Explanation of the geology of Degree Sheet No. 29 (Singida), *Bull. Geol. Surv. Tanganyika* **11**:1–59.

Edwards, C. B., and Howkins, J. B., 1966, Kimberlites in Tanganyika with special reference to the Mwadui occurrence, *Econ. Geol.* **61**:537–554.

Elder, R. L., and Smith, G. R., 1988, Fish taphonomy and environmental inference in paleolimnology, *Palaeogeogr., Palaeoclimat., Palaeoecol.* **62**:577–592.

Estes, R., 1975, *Xenopus* from the Paleocene of Brazil and its zoogeographic importance, *Nature* **254**:48–50.

Estes, R., 1977, Relationships of the South African fossil frog *Eoxenopoides reuningi* (Anura, Pipidae), *Ann. S. Afr. Mus.* **73**:49–80.

Ferguson, D. K., 1985, The origin of leaf-assemblages—new light on an old problem, *Rev. Palaeobot. Palynol.* **46**:117–188.

Frakes, L. A., and Kemp, E. M., 1973, Palaeogene continental positions and evolution of climate, in: *Implications of Continental Drift to the Earth Sciences, Vol. 1* (D. H. Tarling and S. K. Runcord, eds.), Academic Press, New York, pp. 539–559.

Germeraad, J. H., Hopping, C. A., and Muller, J., 1968, Palynology of Tertiary sediments from tropical areas, *Rev. Paleobot. Palynol.* **6**:189–348.

Gheerbrant, E., Capetta, H., Feist, M., Jaeger, J.-J., Sudre, J., Vianey-Liaud, M., and Sigé, B., 1993, La succession des faunes des vertébrés d'age paléocene supérieur et éocène inférieur dans le bassin d'Ouazazate, Maroc. Contexte géologique, portée biostratigraphique et paléogéographique, *Newsl. Stratigr.* **28**:33–58.

Gillman, C., 1949, A vegetation-types map of Tanganyika Territory, *Geogr. Rev.* **39**:7–37.

Giresse, P., Maley, J., and Brenac, P., 1994, Late Quaternary palaeoenvironments in the Lake Barombi Mbo (West Cameroon) deduced from pollen and carbon isotopes of organic matter, *Palaeogeogr., Palaeoclimat., Palaeoecol.* **107**:65–78.

Giresse, P., Maley, J., and Kelts, K., 1991, Sedimentation and palaeoenvironment in crater lake Barombi Mbo, Cameroon, during the last 25,000 years, *Sediment. Geol.* **71**:151–175.

Godinot, M., and Mahboubi, M., 1992, Earliest known simian primate found in Algeria, *Nature* **357**:324–326.

Godinot, M., and Mahboubi, M., 1994, Les petits primates Simiiformes de Glib Zegdou (Eocène, Algérie), *C. R. Acad. Sci. Paris,* Sér. II **319**:357–364.

Grace, C., and Stockley, G. M., 1931, Geology of part of the Usongo area, Tabora Province, Tanganyika Territory, *J. East Af. Nat. Hist. Soc.* **37**:185–192.

Grande, L., 1986, The first articulated freshwater teleost fish from the Cretaceous of North America, *Palaeontology* **29**:365–371.

Grantham, D. R., and Temperley, B. N., 1939, Preliminary report of the geology and gold occurrences of the Kahama region, *Short Pap. Geol. Surv. Tanganyika* **21**.

Grantham, D. R., Temperley, B. N., and McConnell, R. B., 1945, Explanation of the geology of Degree Sheet No. 17 (Kahama), *Bull. Geol. Surv. Tanganyika* **15**:1–35.

Gras, R., 1961, Contribution à l'étude des poissons du Bas-Dahomay. Description de quatre espèces nouvelles, *Bull. Mus. natl. Hist. Nat.* **32**:401–410.

Green, J., 1972, Ecological studies on crater lakes in West Cameroon. The zooplankton of Barombi Mbo, Mboandong, Lake Kotto and Lake Soden, *J. Zool., Lond.* **166**:283–301.

Green, J., Corbet, S. A., and Betney, E., 1973, Ecological studies on crater lakes in West Cameroon. The blood of endemic cichlids in Barombi Mbo in relation to stratification and their feeding habits, *J. Zool., Lond.* **170**:299–308.

Greenwood, D. R., 1991, The taphonomy of plant macrofossils, in: *The Processes of Fossilization* (S. K. Donovan, ed.), Belhaven Press, London, pp. 141–169.

Greenwood, P. H., 1960, Fossil denticipitid fishes from East Africa, *Bull. Br. Mus. (Nat. Hist.), Geol.* **5**:1–11.

Greenwood, P. H., 1965, The status of *Acanthothrissa* Gras, 1961 (Pisces, Clupeidae), *Ann. Mag. Nat. Hist.* **7**:337–338.

Greenwood, P. H., 1968, The osteology and relationships of the Denticipitidae, a family of clupeomorph fishes, *Bull. Br. Mus. (Nat. Hist.), Zool.* **16**:215–273.

Greenwood, P. H., and Patterson, C., 1967, A fossil osteoglossoid fish from Tanzania (E. Africa), *J. Linn. Soc. (Zool.)* **47**:211–223.

Gros, J. P., 1992, Les bois fossiles d'Éthiopie. Inventaire systématique, *Geobios* **25**:3–12.

Haggerty, S. E., 1999, A diamond trilogy: Superplumes, supercontinents, and supernovae, *Science* **285**:851–860.

Haggerty, S. E., Raber, E., and Naeser, C. W., 1983, Fission track dating of kimberlite zircons, *Earth Planet. Sci. Lett.* **63**:41–50.

Hall, J. B., and Swaine, M. D., 1976, Classification and ecology of closed-canopy forest in Ghana, *J. Ecol.* **64**:913–951.

Harrison, T., and Mbago, M. L., 1997, Introduction: Paleontological and geological research in the Manonga Valley, Tanzania, in: *Neogene Paleontology of the Manonga Valley, Tanzania: A Window into the Evolutionary History of East Africa* (T. Harrison, ed.), Topics in Geobiology, Vol. 14, Plenum Press, New York, pp. 1–32.

Harrison, T., Mbago, M. L., and Msuya, C. P., 1997, New archaeological sites in the Manonga Valley, north-central Tanzania, *Nyame Akuma* **47**:53–60.

Haughton, S., 1931, On a collection of fossil frogs from the clays at Banke, *Trans. R. Soc. S. Afr.* **19**:233–249.

Hawthorne, J. B., 1975, Model of a kimberlite pipe, *Phys. Chem. Earth* **9**:1–15.

Herendeen, P. S., and Jacobs, B. F., (2000), Fossil legumes from the middle Eocene (46.0 Ma) Mahenge flora, Singida, Tanzania, *Am. J. Botany* **87**:1358–1366.

Hubbard, R. N. L. B., and Boulter, M. C., 1983, Reconstruction of Palaeogene climate from palynological evidence, *Nature* **301**:147–150.

Jacobs, B. F., 1999, Estimation of rainfall variables from leaf characters in tropical Africa, *Palaeogoegr., Palaeoclimat., Paleoecol.* **145**:231–250.

Jacobs, B. F., Kingston, J. D., and Jacobs, L. L., 1999, Origin of grass-dominated ecosystems, *Ann. Missouri Bot. Garden* **86**:590–643.

Kerr, R. A., 1999, The great African plume emerges as a tectonic player, *Science* **285**:187–188.

King, B. C., 1970, Vulcanicity and rift tectonics in East Africa, in: *African Magmatism and Tectonics* (T. N. Clifford and I. G. Gass, eds.), Hafner, Darien, CT, pp. 263–283.

Kingdon, J., 1974, *East African Mammals, An Atlas of Evolution in Africa. Vol. I*, University of Chicago Press, Chicago.

Kingdon, J., 1989, *Island Africa: The Evolution of Africa's Rare Animals and Plants*, Princeton University Press, Princeton.

Kling, G. W., 1987, Seasonal mixing and catastrophic degassing in tropical lakes, Cameroon, West Africa, *Science* **237**:1022–1024.

Kling, G. W., 1988, Comparative transparency, depth of mixing, and stability of stratification in lakes of Cameroon, West Africa, *Limnol. Oceanogr.* **33**:27–40.

Kling, G. W., Clark, M. A., Comton, H. R., Devine, J. D., Evans, W. C., Humphrey, A. M., Koenigsberg, E. J., Lockwood, J. P., Tuttle, M. L., and Wagner, G. N., 1987, The August 1986 gas disaster at Lake Nyos, Cameroon, West Africa, *Science* **236**:169–175.

Koeniguer, J. C., 1971, Sur les bois fossiles du Paléocène de Sessao (Niger), *Rev. Palaeobot. Palynol.* **12**:303–323.

Kuntz, J., 1909, Beitrag zur Geologie der Hochländer Deutsch-Ostafrika, *Z. Prakt. Geol.* **17**:208.

LeCointre, G., and Nelson, G., 1996, Clupeomorpha, sister-group of Ostariophysi, in: *Interrelationships of Fishes* (M. L. J. Stiassny, L. R. Parenti, and G. D. Johnson, eds.), Academic Press, New York, pp. 193–207.

Li Guo-Qing, and Wilson, M. V. H., 1996, Phylogeny of Osteoglossomorpha, in: *Interrelationships of Fishes* (M. L. J. Stiassny, L. R. Parenti, and G. D. Johnson, eds.), Academic Press, New York, pp. 163–174.

Lind, E. M., and Morrison, M. E., 1974, *East African Vegetation*, Longman, London.

Lorenz, V., 1973, On the formation of maars, *Bull. Volcanol.* **37**:183–204.

Liu, F.-G. R., and Miyamoto, M. M., 1999, Phylogenetic assessment of molecular and morphological data for eutherian mammals, *Syst. Biol.* **48**:54–64.

Livingstone, D. A., and Melack, J. M., 1984, Some lakes of subsaharan Africa, in: *Lakes and Reservoirs* (F. B. Taub, ed.), Elsevier, Amsterdam, pp. 467–497.

Lundberg, J. G., 1993, African-South American freshwater fish clades and continental drift: Problems with a paradigm, in: *Biological Relationships between Africa and South America* (P. Goldblatt, ed.), Yale University Press, New Haven, pp. 156–199.

Maley, J., Livingstone, D. A., Giresse, P., Brenac, P., Kling, G., Stager, C., Thouveny, N., Kelts, K., Haag, M., Fournier, M., Bandet, Y., Williamson, D., and Zogning, A., 1991, West Cameroon Quaternary lacustrine deposits: preliminary results, *J. Afr. Earth Sci.* **12**:147–157.

Maley, J., Livingstone, D. A., Giresse, P., Thouveny, N., Brenac, P., Kelts, K., Kling, G. W., Stager, C., Haag, M., Fournier, M., Bandet, Y., Williamson, D., and Zogning, A., 1990, Lithostratigraphy, volcanism, paleomagnetism and palynology of Quaternary lacustrine deposits from Barombi Mbo (West Cameroon): preliminary results, *J. Volcanol. Geotherm. Res.* **42**:319–335.

Mannard, G. W., 1962, *The Geology of the Singida Kimberlite Pipes, Tanganyika*, Ph.D. Thesis, McGill University, Montreal.

McGowran, B., 1990, Fifty million years ago, *Am. Sci.* **78**:30–39.

McKay, I. J., and Rayner, R. J., 1986, Cretaceous insects from Orapa, Botswana, *S. Afr. J. Entomol.* **49**:7–17

Melack, J. M., 1978, Morphometric, physical and chemical features of the volcanic crater lakes of western Uganda, *Arch. Hydrobiol.* **84**:430–453.

Miller, K. G., Fairbanks, R. G., and Mountain, G. S., 1987, Tertiary oxygen isotope synthesis, sea level history, and continental margin erosion, *Paleoceanography* **2**:1–19.

Mitchell, R. H., 1986, *Kimberlite: Mineralogy, Geochemistry, and Petrology*, Plenum, New York.

Naeser, C. W., and McCallum, M. E., 1977, Fission-track dating of kimberlitic zircons, *Ext. Abstr., 2nd. Int. Kimberlite Conf.*, American Geophysical Union, Washington, D.C.

Nelson, G. J., 1969, Infraorbital bones and their bearing on the phylogeny and geography of the osteoglossomorph fishes, *Amer. Mus. Novit.* **2394**:1–37.

Nevo, E., 1968, Pipid frogs from the early Cretaceous of Israel and pipid evolution, *Bull. Mus. comp. Zool., Harvard* **136**:255–318.

Old, R. A. and Rex, D. C., 1971, Rubidium and strontium age determinations of some Precambrian granitic rocks, S. E. Uganda, *Geol. Mag.* **108**:353–360.

Parrish, J. T., 1987, Global palaeogeography and palaeoclimate of the Late Cretaceous and Early Tertiary, in: *The Origin of Angiosperms and Their Biological Consequences* (E. M. Friis, W. G. Chaloner, and P. R. Crane, eds.), Cambridge University Press, Cambridge, pp. 51–73.

Parrish, J. T., Ziegler, A. M., and Scotese, C. R., 1982, Rainfall patterns and the distribution of coals and evaporites in the Mesozoic and Cenozoic, *Palaeogeogr., Palaeoclimat., Palaeoecol.* **40**:67–101.

Pickford, M., 1986, Première découverte d'une faune mammalienne terrestre paléogène d'Afrique sub-saharienne, *C. R. Acad. Sc. Paris* **302**, Sér II:1205–1210.

Pitman, W. C., Cande, S., LaBrecque, J., and Pindell, J., 1993, Fragmentation of Gondwana: The separation of Africa and South America, in: *Biological Relationships between Africa and South America* (P. Goldblatt, ed.), Yale University Press, New Haven, pp. 15–34.

Quennell, A. M., McKinlay, A. C. M., and Aitken, W. G., 1956, *Summary of the Geology of Tanganyika. Part I: Introduction and Stratigraphy*, Government Printer, Dar es Salaam, Tanganyika.

Rayner, R. J., 1987, March flies from an African Cretaceous springtime, *Lethaia* **20**:123–127.

Rayner, R. J., and McKay, I. J., 1986, The treasure chest of the Orapa diamond mine, *Botswana Notes Rec.* **18**:55–61.

Rayner, R. J., Waters, S. B., McKay, I. J., Dobbs, P. N., and Shaw, A. L., 1991, The mid-Cretaceous palaeoenvironment of central Southern Africa (Orapa, Botswana), *Palaeogeogr., Palaeoclimat., Palaeoecol.* **88**:147–156.

Rennie, J., 1931, Note on fossil leaves from Banke clays, *Trans. R. Soc. S. Afr.* **19**:251–253.

Richards, P. W., 1996, *The Tropical Rain Forest*, Cambridge University Press, New York.

Rose, K. D., and Silcox, M. T., 1999, Primate evolution at the Society of Vertebrate Paleontology, *Evol. Anthropol.* **8**:5–6.

Salard-Cheboldaeff, M., 1979, Palynologie Maestrichtienne et Tertiaire du Cameroun. Etude qualitative et répartition verticale des principales espèces, *Rev. Palaeobot. Palynol.* **28**:365–387.

Salard-Cheboldaeff, M., 1981, Palynologie Maestrichtienne et Tertiaire du Cameroun. Resultats botaniques, *Rev. Palaeobot. Palynol.* **32**:401–439.

Salard-Cheboldaeff, M., and Dejax, J., 1991, Evidence of Cretaceous to Recent West African intertropical vegetation from continental sediment spore-pollen analysis, *J. Afr. Earth Sci.* **12**:353–361.

Savin, S. M., 1977, The history of the Earth's surface temperature during the past 100 million years, *Ann. Rev. Earth. Planet Sci.* **5**:319–356.

Scarth, A., 1994, *Volcanoes: An Introduction*, Texas A and M University Press, College Station, TX.

Schaarschmidt, F., 1992, The vegetation: fossil plants as witnesses of a warm climate, in: *Messel: An Insight into the History of Life and of the Earth* (S. Schaal and W. Ziegler, eds.), Clarendon Press, Oxford, pp. 29–52.

Shackleton, N. J., 1986, Paleogene stable isotope events, *Palaeogeogr., Palaeoclimat., Palaeoecol.* **57**:91–102.

Shackleton, N. J., and Boersma, A., 1981, The climate of the Eocene ocean, *J. Geol. Soc. Lond.* **138**:153–157.

Sigé, B., Jaeger, J.-J., Sudre, J., and Vianey-Liand, M., 1990, *Altiatlasius koulchii* n. gen. et sp., primate omomyidé du Paléocène supérieur du Maroc, et les origines des Euprimates, *Palaeontographica* **A214**:31–56.

Sloan, L. C., and Rea, D. K., 1995, Atmospheric carbon dioxide and early Eocene climate: A general circulation modeling sensitivity study, *Palaeogeogr., Palaeoclimat., Palaeoecol.* **119**:275–292.
Smith, R. M. H., 1986, Sedimentation and palaeoenvironments of Late Cretaceous crater-lake deposits in Bushmanland, South Africa, *Sedimentology* **33**:369–386.
Spicer, R. A., 1981, The sorting and deposition of allochthonous plant material in a modern environment at Silwood Lake, Silwood Park, Berkshire, England, *U.S. Geol. Survey, Professional Paper* **1143**:1–77.
Springer, M. S., Cleven, G. C., Madsen, O., de Jong, W. W., Waddell, V. G., Amrine, H. M., and Stanhope, M. J., 1997, Endemic African mammals shake the phylogenetic tree, *Nature* **388**:61–64.
Springer, M. S., Amrine, H. M., Burk, A., and Stanhope, M. J., 1999, Additional support for Afrotheria and Paenungulata, the performance of mitochondrial versus nuclear genes, and the impact of data partitions with heterogeneous base composition, *Syst. Biol.* **48**:65–75.
Stiassny, M. L. J., 1991, Phylogenetic intrarelationships of the family Cichlidae: An overview, in: *Cichlid Fishes: Behaviour, Ecology and Evolution*, (M. H. A. Keenleyside, ed.), Croom Helm, London, pp. 1–35.
Stockley, G. M., 1934, The Musoma Goldfield, *Ann. Rep. Geol. Surv. Tanganyika* **1933**:11–16.
Stockley, G. M., 1935, The Musoma Goldfield, *Ann. Rep. Geol. Surv. Tanganyika* **1934**:12.
Stockley, G. M., 1943, The pre-Karoo stratigraphy of Tanganyika, *Geol. Mag.* **80**:161–170.
Stockley, G. M., 1947, The geology of the country around Mwanza Gulf, *Short Pap. Geol. Surv. Tanganyika* **29**:1–21.
Stockley, G. M., 1948, The geology and mineral resources of Tanganyika Territory, *Bull. Geol. Surv. Tanganyika* **20**:1–37.
Storch, G., and Schaarschmidt, F., 1992, The Messel fauna and flora: a biogeographical puzzle, in: *Messel: An Insight into the History of Life and of the Earth*, (S. Schaal and W. Ziegler, eds.), Clarendon Press, Oxford, pp. 293–297.
Sudre, J., 1979, Nouveaux mammiferès Eocenes du Sahara Occidental, *Palaeovertebrata* **9**: 83–115.
Talling, J. F., 1969, The incidence of vertical mixing, and some biological and chemical consequences, in tropical African lakes, *Int. Ver. Theor. Angew. Limnol. Verh.* **17**:998–1012.
Taverne, L., 1979, Ostéologie, phylogénèse et systématique des Téléostéens fossiles et actuels du super-ordre de Osteoglossomorphes. Partie 3, Evolution des structures ostéologique et conclusions générales relatives à la phylogénèse et à la systématique du super-ordre, *Acad. Roy. Belgique, Mém. Cl. Sci.*, 8° Ser. 2, **43**:1–168.
Teale, E. O., 1931, Shinyanga diamond fields, *Short Paper Geol. Surv. Tanganyika* **9**:5–6.
Teale, E. O., 1932, The kimberlite and associated occurrences of the Iramba Plateau, *Short Paper Geol. Surv. Tanganyika* **10**:1–10.
Teale, E. O., and Oates, F., 1946, The mineral resources of Tanganyika Territory, *Bull. Geol. Surv. Tanganyika* **16**:3–5.
Trewavas, E., Green, J., and Corbet, S. A., 1972, Ecological studies on crater lakes in West Cameroon fishes of Barombi Mbo, *J. Zool., Lond.* **167**:41–95.
Upchurch, G. R., and Wolfe, J. A., 1987, Mid-Cretaceous to early Tertiary vegetation and climate: Evidence from fossil leaves and wood, in: *The Origin of Angiosperms and Their Biological Consequences* (E. M. Friis, W. G. Chaloner, and P. R. Crane, eds.), Cambridge University Press, Cambridge, pp. 75–106.
Van Couvering, J. A. H., 1982, Fossil cichlid fish of Africa, *Spec. Pap. Paleontol. Assoc.* **29**:1–103.
Van Hoeken-Klinkenberg, P. M. J., 1966, Maastrichtian, Paleocene and Eocene pollen and spores from Nigeria, *Leidse Geol. Mededel.* **38**:37–48.
Verniers, J., 1997, Detailed stratigraphy of the Neogene sediments at Tinde and other localities in the central Manonga basin, in: *Neogene Paleontology of the Manonga Valley, Tanzania: A Window into the Evolutionary History of East Africa* (T. Harrison, ed.), Topics in Geobiology, Vol. 14, Plenum Press, New York, pp. 33–65.
Wade, F. B., 1937, A stratigraphical classification and table of Tanganyika Territory, *Bull. Geol. Surv. Tanganyika* **9**:1–62.
Walraven, F., Pape, J., and Borg, G., 1994, Implications of Pb-isotopic compositions at the Geita gold deposit, Sukumaland Greenstone Belt, Tanzania, *J. Afr. Earth Sci.* **18**:111–121.

Williams, G. J., 1939, The Kimberlite province and associated diamond deposits of Tanganyika Territory, *Bull. Geol. Surv. Tanganyika* **12**:1–38.
Williams, G. J., and Eades, N. W., 1939, Explanation of the geology of Degree Sheet No. 18 (Shinyanga), *Bull. Geol. Surv. Tanganyika* **13**:5–20.
Wolfe, J. A., 1980, Tertiary climates and floristic relationships at high latitudes in the Northern Hemisphere, *Palaeogeogr., Palaeoclimat., Palaeoecol.* **30**:313–323.
Wolfe, J. A., 1995, Paleoclimate estimates from Tertiary leaf assemblages, *Ann. Rev. Earth Planet. Sci.* **23**:119–142.
Wuttke, M., 1992, Conservation—dissolution—transformation. On the behaviour of biogenic materials during fossilization, in: *Messel: An Insight into the History of Life and of the Earth* (S. Schaal and W. Ziegler, eds.), Clarendon Press, Oxford, pp. 265–275.
Zachos, J. C., Lohmann, K. C., Walker, J. C. G., and Wise, S. W., 1993, Abrupt climate change and transient climates during the Paleogene: A marine perspective, *J. Geol.* **101**:191–213.

Chapter 3

Early Wasatchian Mammals From the Gulf Coastal Plain of Mississippi: Biostratigraphic and Paleobiogeographic Implications

K. CHRISTOPHER BEARD and MARY R. DAWSON

1. Introduction

The Wasatchian North American Land Mammal Age (or NALMA) corresponds to an interval spanning approximately 55.5–50.5 million years before the present (Prothero, 1995; Clyde *et al.*, 1997). The Wasatchian fossil record is among the richest of any NALMA, being documented by abundant vertebrate remains throughout most of its duration (Krishtalka *et al.*, 1987; Gingerich, 1989; Bown *et al.*, 1994). This stratigraphically dense record of Wasatchian

K. CHRISTOPHER BEARD and MARY R. DAWSON • Department of Vertebrate Paleontology, Carnegie Museum of Natural History, Pittsburgh, PA 15213-4080.

Eocene Biodiversity: Unusual Occurrences and Rarely Sampled Habitats edited by Gregg F. Gunnell, Kluwer Academic/Plenum Publishers, New York, 2001.

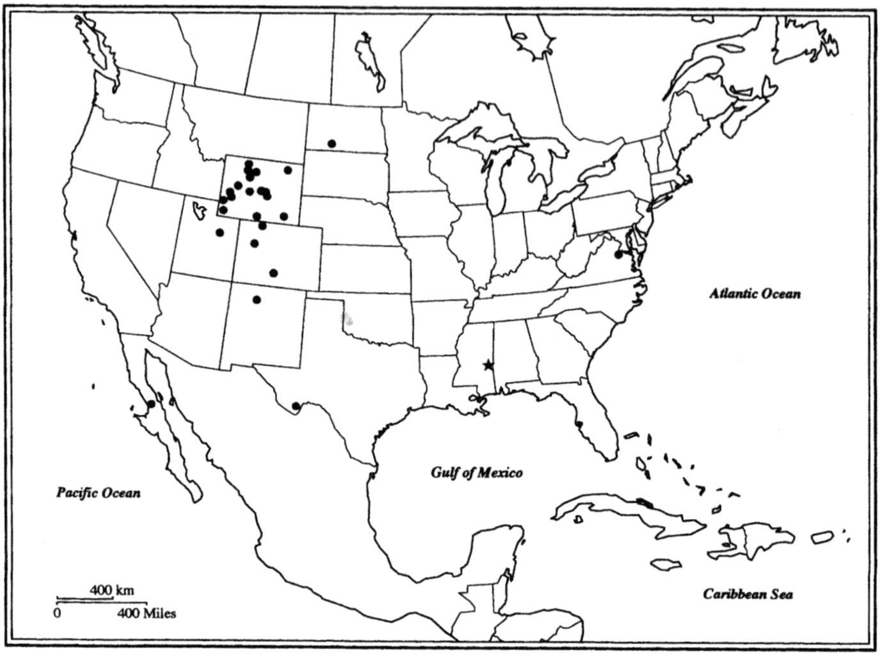

FIGURE 1. Map showing the geographic distribution of Wasatchian faunas from the United States and Mexico, designated by dots. The Red Hot local fauna is indicated by a star. Additional Wasatchian faunas occur on Ellesmere and Axel Heiberg islands in the Canadian Arctic archipelago, beyond the limits of this map.

vertebrates owes its existence to Laramide tectonics, which provided ideal depositional settings for the accumulation and rapid burial of vertebrate remains of this age along much of the Rocky Mountain Cordillera. As a result, however, the fossil record of Wasatchian vertebrates is strongly biased geographically. Abundant samples of Wasatchian vertebrates from intermontane basins of the Rocky Mountain region, particularly in Wyoming, stand in sharp contrast to the paucity of similar data from other parts of the North American craton (Figure 1). Vertebrate faunas of Wasatchian age are known from only a few areas outside of the Rocky Mountain region: the Big Bend region of west Texas (Wilson, 1967), Ellesmere and Axel Heiberg Islands in the Canadian Arctic archipelago (Dawson *et al.*, 1976, 1993; Estes and Hutchison, 1980; McKenna, 1980), Baja California, Mexico (Novacek *et al.*, 1991), Stafford County, Virginia (Rose, 1999) and Lauderdale County, Mississippi. Here, we provide the first substantial information regarding the Wasatchian mammals from the latter area.

The first discovery of Wasatchian mammals in Mississippi was made by Gerard R. Case and his colleagues in 1979, during the course of field work that focused on the early Cenozoic sharks of the Gulf Coastal Plain (Case, 1994a). Case's mammal specimens remained unknown to paleomammalogists for many years, but in April of 1990 a field party from the Carnegie Museum of Natural

History returned to Case's site to follow up on his original discovery. Several subsequent field seasons have led to the recovery of a moderately diverse mammalian fauna from a stratigraphic unit that Ingram (1991) named the "T4 sand," one of several sand lenses in the upper part of the Tuscahoma Formation in the vicinity of Meridian, Mississippi. Floral remains from this unit include fern rhizomes (Call *et al.*, 1993) and abundant pollen (Frederiksen, 1998). The vertebrate assemblage from the T4 sand includes sharks and rays (Case, 1994a), bony fishes (Case, 1994b), snakes (Holman *et al.*, 1991; Holman and Case, 1992), and fragmentary fossils of birds, lizards, and crocodilians in addition to the assemblage of mammals emphasized here. We designate this assemblage the "Red Hot local fauna," following the longstanding use of this name for the well-known stratigraphic section that outcrops there (Ingram, 1991; Dockery, 1998). In the immediate vicinity, Wasatchian mammals of somewhat younger age have also been recovered from the overlying Bashi Formation (Beard and Tabrum, 1991).

Recently, a small mammalian assemblage from the Williamsburg Formation of South Carolina was described by Schoch (1998) as possibly pertaining to the Wasatchian NALMA, although an earlier age assignment was preferred by that author. Because Wasatchian index taxa are absent from the South Carolina assemblage, we concur with Schoch (1998) that the Williamsburg Formation mammals most likely antedate the Wasatchian.

One of the most important events in the evolution of North American mammalian faunas occurred at the beginning of the Wasatchian NALMA, when artiodactyls, perissodactyls, primates and hyaenodontid creodonts first dispersed to that continent, apparently via the Bering land bridge from Asia (Beard, 1998; Beard and Dawson, 1999). Because of its unique stratigraphic context, the Red Hot local fauna further constrains the timing and correlation of this major episode of intercontinental dispersal. This assemblage also provides a rare window into the early Cenozoic mammals that once inhabited the southeastern part of the United States. As a result, earlier hypotheses regarding the paleobiogeographic relationships of this region can be reassessed, and new inferences can be drawn.

2. The Mammalian Component of the Red Hot Local Fauna

The Red Hot local fauna includes some 24 species of land mammals at present (Table 1), making it by far the most diverse mammalian assemblage known from the early Cenozoic of the southeastern United States. Spatial constraints prohibit us from providing detailed descriptions of the entire mammalian fauna here, but selected specimens are illustrated in Figures 2–3. In this section we discuss the taxonomic diversity and abundance of the mammalian component of the Red Hot local fauna, the possible taphonomic biases that led to the diversity and abundance patterns observed there, and some of the mammal taxa that are conspicuously absent from the Red Hot local fauna.

Table 1. Mammalian faunal composition, early Wasatchian Red Hot local fauna, uppermost Tuscahoma Formation, Lauderdale County, Mississippi

Taxon	TNS	MNI
Mimoperadectes labrus	1	1
Peratherium constans	2	1
Scenopagus sp.	4	2
Talpavoides dartoni	1	1
Diacocherus sp.	2	1
Erinaceomorpha, gen. et sp. nov.	7	3
Plagioctenodon sp.	4	1
cf. *Placentidens* sp.	3	2
Palaeosinopa lutreola	10	2
Apatemys chardini	2	2
Uintasoricinae, gen. et sp. nov.	20	4
Tetonius sp.	18	3
cf. *Anemorhysis* sp.	11	2
Ectocion parvus	2	1
Haplomylus speirianus	3	1
Hyracotherium sp., cf. *H. sandrae*	1	1
Diacodexis sp.	1	1
Coryphodon sp., cf. *C. eocaenus*	1	1
Prototomus sp., cf. *P. phobos*	1	1
Viverravus sp.	12	2
Miacis sp.	3	1
Microparamys sp.	5	2
Paramyinae, gen. et sp. undet.	9	2
Cylindrodontidae, gen. et sp. nov. (A)	26	5
Cylindrodontidae, gen. et sp. nov. (B)	20	3
Rodentia, gen. et sp. undet.	27	

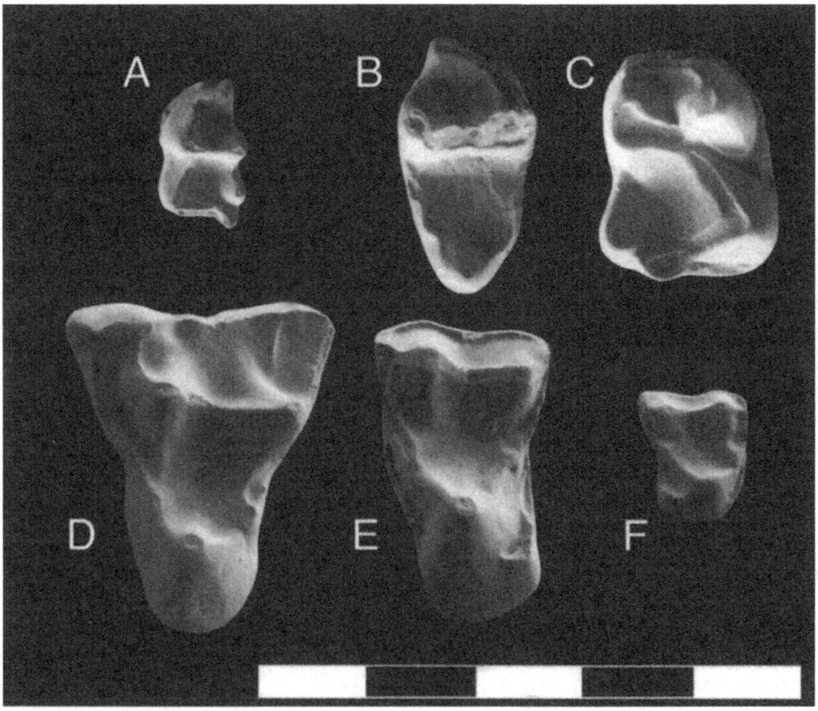

FIGURE 2. Selected fossil mammal specimens from the Red Hot local fauna, early Wasatchian of Mississippi. The Carnegie Museum of Natural History (acronym CM) is the permanent repository for all specimens illustrated here and in Figure 3. (A) *Peratherium constans*, left M_2 or M_3, CM 70400; (B) *Viverravus* sp., right M_2, CM 70453; (C) *Haplomylus speirianus*, right M_2, CM 70403; (D) *Mimoperadectes labrus*, left M^2, CM 67850; (E) *Tetonius* sp., left M^2, CM 70436; (F) Uintasoricinae, gen. et sp. nov., right M^1, CM 62696. Scale = 5 mm.

2.1. Mammalian Faunal Composition and Diversity

In terms of both the total number of specimens (TNS) and the minimum number of individuals (MNI) required to account for the observed TNS, the mammalian component of the Red Hot local fauna is dominated by rodents, lipotyphlans, and uintasoricine and omomyid primatomorphs (Figures 4–5). This pattern of relative abundance differs from that frequently encountered in roughly contemporaneous faunas from the Bighorn Basin of northwestern Wyoming, where all of these small mammal taxa are typically much less abundant (Rose, 1981b; Gingerich, 1989). Rather, the early Wasatchian faunas of the Bighorn Basin are dominated by mammals of medium to large body size, especially perissodactyls, artiodactyls, notharctid primates, and phenacodontid and hyopsodontid condylarths. Early Wasatchian abundance patterns more nearly approximating that shown by the Red Hot local fauna have been reported by McKenna (1960) from the Four Mile Creek area of northwestern Colorado. As is the case for the Mississippi assemblage, these Four Mile Creek

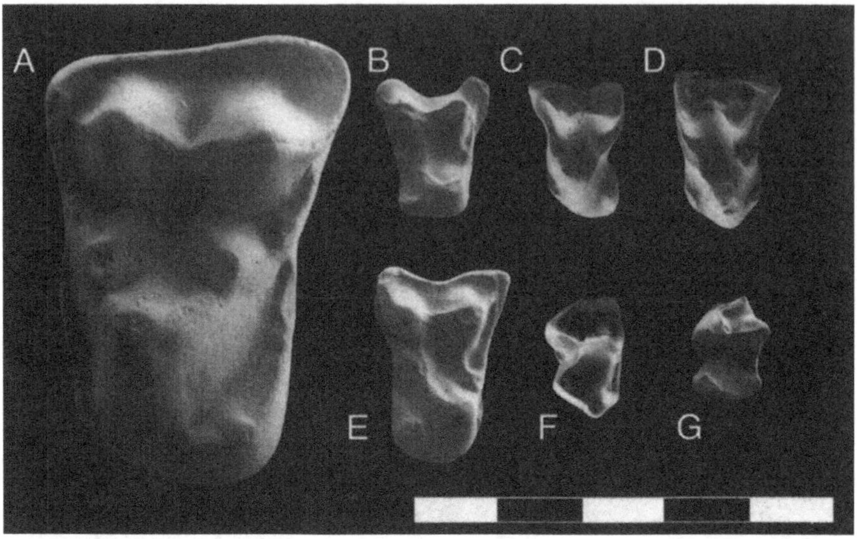

FIGURE 3. Selected fossil mammal specimens from the Red Hot local fauna, early Wasatchian of Mississippi. (A) *Palaeosinopa lutreola*, left M^1, CM 62681; (B) *Apatemys chardini*, right M^2, CM 70442; (C) *Plagioctenodon* sp., left M^2, CM 67848; (D) cf. *Placentidens* sp., right M^2, CM 70455; (E) Erinaceomorpha, gen. et sp. nov., right M^1, CM 67849; (F) *Scenopagus* sp., left M_2, CM 70444; (G) *Talpavoides dartoni*, left M_2, CM 70443. Scale = 5 mm.

faunas are often dominated by rodents. In contrast, rodents have proven to be much rarer components of contemporaneous mammalian faunas from the Bighorn Basin (Ivy, 1990). These markedly different patterns of abundance may be attributable, at least in part, to different taphonomic biases and modes of specimen recovery (see below).

Among the larger mammals represented in the Red Hot local fauna, the abundance of the pantolestid *Palaeosinopa* is noteworthy. Pantolestids are rarely such common elements of early Wasatchian faunas from the western United States. Possibly, the relatively high abundance of pantolestids in the Red Hot local fauna reflects their paleoecological preferences. Pantolestids are often reconstructed as piscivorous, otter-like forms based on their anatomy and fossilized gut contents (Matthew, 1909; Koenigswald, 1980; Pfretzschner, 1993). If so, the estuarine depositional setting from which the Red Hot local fauna was derived would presumably have been a highly suitable habitat for these animals.

2.2. Taphonomic Biases

Ingram (1991:11) characterized the T4 sand that yielded the Red Hot local fauna as follows: "It is jade-green and composed of fine- to very fine-grained quartz sand, fine-grained glauconite, and mica. It is cross-bedded with multiple, cross-cutting scours evidencing the cut-and-fill depositional characteristics associated with estuarine channel facies." Vertebrate fossils are particularly abundant at the base of the T4 sand, where they were concentrated by sediment winnowing as a lag deposit.

Mammals are represented in the Red Hot local fauna almost exclusively by isolated teeth. Only rarely are disarticulated mammalian postcranial elements recovered, and jaws bearing one or more teeth are unknown. Isolated teeth of mammals show variable levels of abrasion, suggesting different degrees of transport prior to deposition. Nevertheless, many mammalian specimens are tiny and delicate, yet virtually pristine (Figures 2–3), implying deposition relatively near their source.

As noted earlier, the mammalian component of the Red Hot local fauna differs from many Wasatchian assemblages from the western United States in its abundance of small taxa (Figures 4–5). In contrast, Gingerich (1989:67) noted that small mammals were rare components of Wa0 assemblages from the Bighorn Basin. To a large extent, this differential pattern of abundance of small mammals probably reflects the different methods used to recover these fossil

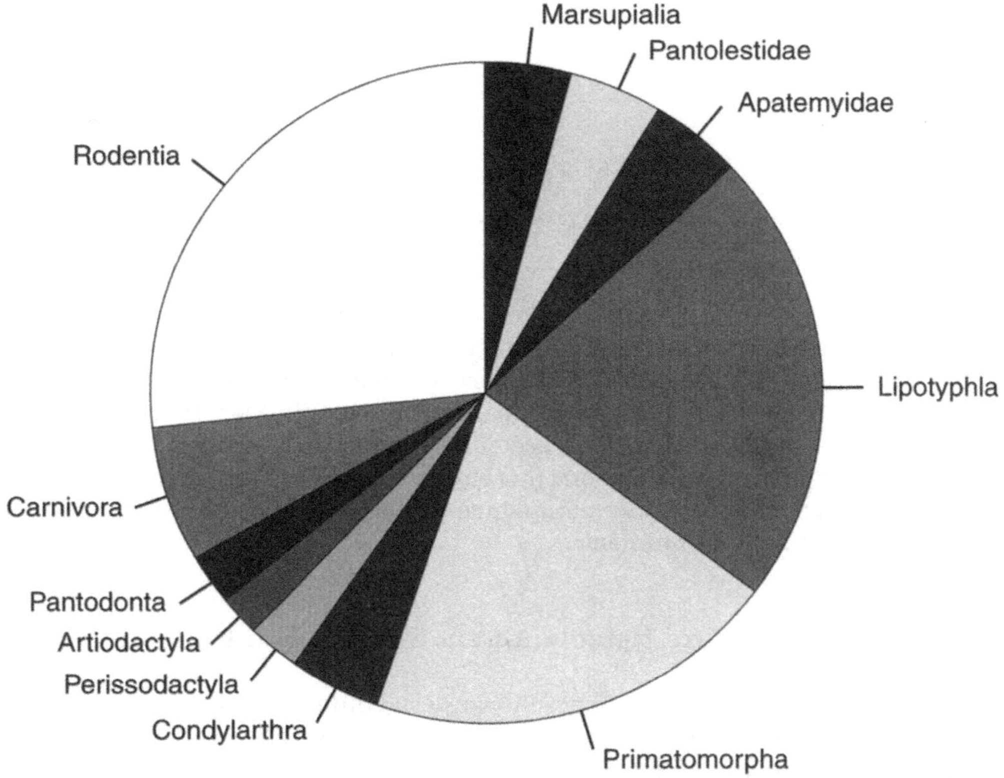

FIGURE 4. Relative abundance of mammalian higher taxa comprising the Red Hot local fauna, based on MNI data from Table 1.

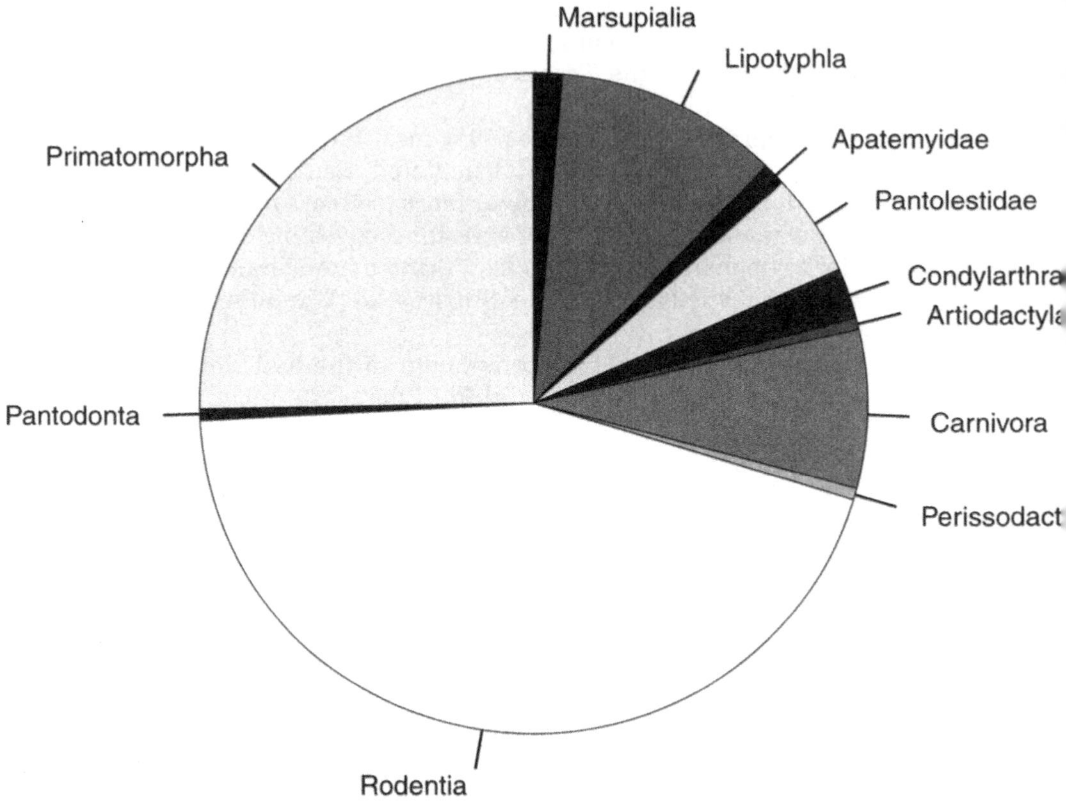

FIGURE 5. Relative abundance of mammalian higher taxa comprising the Red Hot local fauna, based on TNS data from Table 1.

assemblages. Faunas recovered primarily through surface-prospecting, such as the Wa0 assemblage described by Gingerich (1989), are widely acknowledged to underestimate the abundance of small mammals (Winkler, 1983; Wilf *et al.*, 1998). On the other hand, many Wasatchian mammals of medium to large body size are unrepresented (or underrepresented) in the Red Hot local fauna. Their absence or anomalously low abundance probably reflects size-sorting in an estuarine channel environment.

2.3. Mammalian Taxa Notably Absent from the Red Hot Local Fauna

Early Wasatchian faunas from the western United States are characterized by exceptionally high generic richness, caused by the influx of a large number of immigrant taxa coupled with a low level of extinction of endemic forms (Maas *et al.*, 1995). The moderate generic richness known for the Red Hot local fauna therefore suggests that only a fraction of the ancient mammalian fauna that once inhabited the region has been sampled to date. As noted in the preceding section, taphonomic biases have skewed the composition of the Red Hot local fauna, so that mammals of medium to large body size are poorly represented. Here, we note the absence of two mammalian taxa from the Red Hot local fauna that cannot be attributed to this type of taphonomic size-sorting.

Multituberculates are moderately diverse and reasonably abundant in early Wasatchian faunas from the Rocky Mountain region, especially at localities that have either been quarried or screen-washed (Krause, 1982, 1986). For example, Krause (1986:table 1) notes that multituberculates comprise roughly 8% of all mammalian individuals at several sites in the Wasatch Formation in the Four Mile Creek area of northwestern Colorado (McKenna, 1960). Despite the fact that the Red Hot local fauna was collected under optimal conditions for retrieving micromammals such as multituberculates, none have so far been recovered there.

Many early Wasatchian faunas of the western United States are dominated by the small hyopsodontid condylarth *Hyopsodus* (Rose, 1981b; Gingerich, 1989). This pattern of abundance has provided the basis for both detailed studies of evolutionary tempo and mode (Gingerich, 1974, 1976; Redline, 1997) and the distribution of *Hyopsodus* species across ancient floodplain environments (Bown and Beard, 1990). *Hyopsodus* also occurs in Sparnacian faunas of northwestern Europe (Hooker, 1979) and in Bumbanian faunas of eastern Asia (Russell and Zhai, 1987). Given its widespread distribution and frequently high abundance, the absence of *Hyopsodus* in the Red Hot local fauna is surprising.

3. Discussion

The Red Hot local fauna improves our understanding of Wasatchian mammals in the following three ways: (1) it constitutes our best point of reference for characterizing the mammalian fauna that inhabited the southeastern United States during that interval, as discussed above; (2) because of its unique stratigraphic context, the Red Hot local fauna permits more precise correlations between the early part of the Wasatchian NALMA and roughly contemporaneous mammal-bearing strata in northwestern Europe; (3) the Red Hot local fauna allows us to assess whether Wasatchian faunas of the southeastern United States show greater biogeographic affinities with western Europe or the North American continental interior.

3.1. Biostratigraphic Implications of the Red Hot Local Fauna

The beginning of the Wasatchian NALMA is distinguished by one of the most important episodes of mammalian faunal turnover in the North American Cenozoic record (Rose, 1981a; Gingerich, 1989; Maas *et al.*, 1995). Given that this turnover is widely attributed to intercontinental dispersal of higher-level mammalian taxa (Gingerich, 1989; Krause and Maas, 1990; Woodburne and Swisher, 1995; Beard, 1998; Beard and Dawson, 1999), knowledge of the exact timing of this event is important in constraining its etiology. Several studies of paleomagnetic stratigraphy have demonstrated that early Wasatchian faunas occur in rocks that were deposited during Chron C24r (Butler *et al.*, 1981; Clyde *et al.*, 1994; Tauxe *et al.*, 1994). More precise correlation of early Wasatchian faunas has been achieved on the basis of carbon isotope chemostratigraphy, which demonstrates that basal Wasatchian faunas correlate with a dramatic negative shift in $\delta^{13}C$ that is also observable in marine strata (Koch *et al.*, 1992). This chemostratigraphic correlation is consistent with magnetostratigraphic data, because the negative excursion in $\delta^{13}C$ occurs in the lower part of Chron C24r, at roughly 55.5 Ma in the timescale of Berggren *et al.* (1995).

Despite these advances in correlation of the early Wasatchian to global timescales, detailed understanding of the relative chronology of similar episodes of faunal turnover in North America (at the Clarkforkian-Wasatchian boundary) and Europe (at the Thanetian-Sparnacian or MP6–MP7 boundary) has remained elusive. Both of these turnover events occurred during the relatively long interval encompassed by Chron C24r (Berggren and Aubry, 1998). However, it has so far proven difficult to tie the earliest Sparnacian (or MP7) faunas of Europe unambiguously to the carbon isotope excursion (Hooker, 1998:437). Given the imprecision or ambiguity of these independent (i.e., nonbiological) methods of correlating between North American and European mammal faunas from this interval, it is unsurprising that no consensus exists regarding how such faunas compare chronologically. Based solely or primarily on biostratigraphic evidence from fossil mammals, various workers have argued that earliest Wasatchian and MP7 faunas in North America and Europe are essentially synchronous (e.g., Gingerich, 1989; Krause and Maas, 1990; Hooker, 1996; Lucas, 1998). Others have concluded from the same body of evidence that faunal turnover in Europe preceded that in North America, so that at least some MP7 faunas antedate the earliest Wasatchian (Gingerich and Rose, 1977; Rose, 1980, 1981a; Godinot, 1981, 1982, 1996; Godinot *et al.*, 1987; Hooker, 1998). These different correlations have important paleobiogeographic ramifications (Beard and Dawson, 1999).

Several of the taxa known from the Red Hot local fauna suggest that this assemblage correlates with early, if not earliest, Wasatchian faunas from the western United States. Significantly, two of the mammals that occur in the Red Hot local fauna, *Mimoperadectes labrus* and *Ectocion parvus*, are also known from earliest Wasatchian zone Wa0 in the Clark's Fork Basin of northwestern Wyoming (Gingerich, 1989). Gingerich (1989:51) noted that *Ectocion parvus* may approximate a guide fossil for zone Wa0, because "virtually all *E. parvus* known to date come from zone Wa0. . .". Other mammals known from the Red Hot local fauna also support correlation with the early Wasatchian. These include *Haplomylus speirianus*, which is restricted to early Wasatchian zones Wa1–3 in the Clark's Fork Basin (Gingerich, 1994), a small species of *Hyracotherium* possibly conspecific with *H. sandrae*, and a small, very primitive species of the omomyid primate *Tetonius*. None of the mammals known from the Red Hot local fauna conflicts with an early Wasatchian correlation for this assemblage.

For purposes of further constraining the chronology of faunal turnover across the Clarkforkian-Wasatchian boundary in North America vis-à-vis that which occurred in Europe across the MP6–MP7 boundary, the stratigraphic setting of the Red Hot local fauna is fortuitous indeed. Associated with the mammals and other vertebrates in the T4 sand are abundant dinoflagellates, which allow the early Wasatchian NALMA to be correlated directly with marine biochronologies for the first time. Moreover, the early Wasatchian NALMA can also be placed within the sequence stratigraphic framework developed by Baum and Vail (1988) for the North American Gulf Coastal Plain. Mammalian faunas pertaining to reference level MP7 in Europe, including the important Dormaal fauna from Belgium which is the "reference locality" for MP7 (Godinot, 1987), can also be placed within a sequence stratigraphic framework. As we discuss in greater detail below, sequence stratigraphy provides a novel and particularly compelling means of estimating the relative chronology of the Red Hot local fauna on the west side of the Atlantic and Dormaal and its correlatives on the east side of this oceanic basin.

3.1.1. Correlation with Marine Biochronologies

The estuarine setting in which the T4 sand was deposited allowed the remains of terrestrial vertebrates and pollen to become entombed with marine taxa, including sharks, bony fishes and dinoflagellate cysts. Dinoflagellates have proven to be useful in biostratigraphic zonation, particularly in estuarine facies such as that represented by the T4 sand (Powell, 1988; Powell *et al.*, 1996; De Coninck, 1999).

The dinoflagellate assemblage from the T4 sand includes *Apectodinium augustum*, *A. parvum*, and *A. homomorphum*, among other taxa (Lucy E. Edwards, written communication to S. L. Ingram, February 16, 1993). This dinoflagellate flora can be assigned to the *Apectodinium augustum* (Aau) interval biozone in the usage of Powell *et al.* (1996). De Coninck (1999) refers to the same biozone as his "Event 1." These dinoflagellate zones correlate with calcareous nannoplankton zone NP9 (Powell, 1988). The pollen records from the T4 sand reported by Frederiksen (1998) also imply a correlation with NP9.

3.1.2. Sequence Stratigraphic Framework

A rapid fall in sea level, corresponding to the TP2.3/TE1.1 sequence boundary of Baum and Vail (1988), occurred during the *Apectodinium augustum* interval biozone. On the eastern Gulf Coastal Plain, this drop in eustatic sea level resulted in an erosional unconformity between the Tuscahoma and Bashi formations. Strata belonging to the upper part of the Tuscahoma Formation were deposited during the latter part of sequence TP2.3, prior to the fall in eustatic sea level (Ingram, 1991). Locally, the T4 sand is the highest stratigraphic unit that can be assigned to sequence TP2.3. An unconformity immediately above the T4 sand corresponds to the Type 1 boundary between the TP2.3 and TE1.1 sequences. In contrast to most other Gulf Coastal Plain sections, incised valley fill deposits corresponding to the lowstand systems tract at the base of sequence TE1.1 are preserved in the Red Hot local section (Ingram, 1991). These strata, which consist of fossiliferous interbedded sands and mudstones, are assigned to the Bashi Formation, which also includes transgressive marine strata upsection. Abundant pollen and plant macrofossils occur in the mudstones belonging to the lower part of the Bashi Formation. Frederiksen (1998) noted the abundance of *Platycarya* pollen in these mudstones, an occurrence that is widely regarded to signify Eocene age.

3.1.3. Correlation between the Red Hot Local Fauna and Dormaal, Belgium

The diverse vertebrate fauna from the Dormaal Member of the Tienen Formation in Belgium has long been considered one of the oldest Sparnacian (MP7) mammal faunas in Europe (Godinot *et al.*, 1978; Hooker, 1996, 1998). In recent years, renewed collecting and detailed study of the Dormaal fauna has significantly expanded our understanding of this important assemblage (Smith and Smith, 1995; Smith *et al.*, 1996; Smith, 1997), and Steurbaut (1998) has succeeded in placing the Dormaal vertebrates within a rigorous sequence stratigraphic framework. Steurbaut's (1998) work is timely indeed, because it allows us to assess the relative chronology of Dormaal and the Red Hot local fauna independently of the mammals themselves. Given the persistent debate regarding the diachroneity or synchroneity of faunal turnover across the MP6-MP7 boundary in Europe with respect to similar turnover across the Clarkforkian-Wasatchian boundary in North America (see above), such an independent means of assessing chronology is critical.

The Dormaal vertebrate assemblage occurs in fluvial strata that Steurbaut (1998:figure 12) interprets as representing the lowstand systems tract of sequence TE1.1. If this interpretation is correct, the strata that yield the Dormaal vertebrate assemblage correlate with the lowermost beds of the Bashi Formation in the Red Hot local section, which likewise represent the lowstand systems tract of TE1.1 (Ingram, 1991). Therefore, assuming that the Dormaal fauna is not significantly reworked, the Red Hot local fauna from the upper part of sequence TP2.3 is older than the vertebrate assemblage from Dormaal.

What does this result imply regarding diachroneity versus synchroneity for mammalian faunal turnover across the MP6–MP7 and Clarkforkian-Wasatchian boundaries, respectively? Any definitive answer to this question depends on establishing the relative chronologies of the earliest MP7 faunas of Europe and the earliest Wasatchian faunas of North America. Although the Dormaal assemblage is often cited as being among the oldest, if not the oldest, of European MP7 faunas, the possibility remains that MP6–MP7 faunal turnover occurred prior to the accumulation of the Dormaal fauna. Likewise, although the Red Hot local fauna certainly appears to correlate with early Wasatchian faunas from the western United States, we do not yet know how closely it approximates the Clarkforkian-Wasatchian boundary. Despite these caveats, the hypothesis that faunal turnover in Europe preceded that in North America conflicts with our finding that the Red Hot local fauna antedates the Dormaal fauna. Rather, this result agrees with the paleobiogeographic model we recently proposed (Beard and Dawson, 1999). Under our model, the higher-level mammal taxa whose first appearance marks the beginning of the Wasatchian in North America and the beginning of zone MP7 in Europe are thought to have originated in Asia prior to dispersing across Beringia to North America. Once dispersal to North America was achieved, it would have been highly feasible for these same taxa to invade Europe by crossing a second high latitude land bridge that then connected Greenland with Scotland (McKenna, 1983; Knox, 1998). Based on this paleobiogeographic reconstruction, we predicted that the earliest Sparnacian mammals of Europe would be unlikely to antedate the earliest Wasatchian mammals of the United States (Beard and Dawson, 1999). By framing the debate over relative chronology in a sequence stratigraphic context for the first time, the Red Hot local fauna provides welcome corroboration for this biogeographic model.

3.2. Biogeographic Affinities of the Red Hot Local Fauna

For nearly a century paleontologists have known that the early Eocene mammalian faunas of western North America show strong affinities with those of western Europe (e.g., Osborn, 1910). Given this context, it is useful to assess whether the Red Hot local fauna shows greater biogeographic affinities with North American or European faunas of similar antiquity. Indeed, Schiebout (1979, 1981) predicted that early Eocene mammalian faunas from the southeastern United States would have strong European affinities. However, as noted by Krause and Maas (1990), Schiebout's prediction rested partly on a paleogeographic reconstruction for this interval that is now considered to be flawed. Specifically, Schiebout (1981:figure 1) believed that an embayment from the Gulf of Mexico formed an epicontinental seaway as far inland as southern Canada during the Paleocene. Such a paleogeographic setting might have segregated a distinct faunal province in the southeastern United States from contemporary faunas in the Rocky Mountain region. However, more recent paleogeographic studies reconstruct this epicontinental seaway (the Cannonball Sea) as a southward embayment of the Arctic Ocean that would have lacked any direct marine connection with the Gulf of Mexico (Marincovich *et al.*, 1990).

In order to assess the biogeographic affinities of the Red Hot local fauna, we calculated generic-level Faunal Resemblance Indices (FRIs) for the Mississippi assemblage with respect to composite faunal lists for the early Wasatchian of the Bighorn Basin, Wyoming, and the early Sparnacian of western Europe. FRIs quantify the degree of similarity between faunas by calculating the percentage of taxa (in this case, genera) shared by both faunas relative to the total number of taxa in the least diverse fauna. By this measure, the Red Hot local fauna clearly shows greater affinity with early Wasatchian faunas from the Bighorn Basin (FRI = 71) than it does with Sparnacian faunas from western Europe (FRI = 42).

Despite clear evidence for faunal contact between North America and Europe at this time, a number of mammalian genera appear to have been endemic to North America during the early Wasatchian. Of these endemic North American forms, the following genera occur in the Red Hot local fauna: *Mimoperadectes*, *Talpavoides*, *Scenopagus*, *Tetonius*, cf. *Anemorhysis*, *Ectocion*, *Haplomylus* and *Viverravus*. In addition, representatives of two higher-level taxa that were apparently restricted to North America during this interval are recorded in the Red Hot local fauna. These higher taxa include the Uintasoricinae (Primatomorpha:Microsyopidae) and the Cylindrodontidae (Rodentia), both of which are represented by new genera in the early Wasatchian assemblage from Mississippi. Undoubted uintasoricines are restricted to North America (Gunnell, 1989), although Russell *et al.* (1992) questionably referred *Avenius* from the early Eocene of France to this group. More recent work suggests that *Avenius* is not a member of the Uintasoricinae, although *Avenius* and its nearest relatives may be closely related to this endemic North American clade (Hooker *et al.*, 1999). Aside from the new records from Mississippi, the oldest definite representative of the Cylindrodontidae is the genus *Mysops*, first recorded from the early Bridgerian of western North America (Gazin, 1976). The late Lostcabinian rodent *Dawsonomys* has been referred questionably to Cylindrodontidae (Korth, 1984), but this reference cannot be strongly supported on the basis of the morphology of the single known specimen. Outside of North America, the oldest known cylindrodontids occur in middle Eocene (Irdinmanhan) faunas of Asia (Tong, 1997; Dashzeveg and Meng, 1998). Cylindrodontids have never been found in Europe. The phylogenetic position of the Cylindrodontidae is uncertain, although paramyine affinities have been suggested (Wilson, 1949).

Only one of the mammalian taxa currently known from the Red Hot local fauna, cf. *Placentidens* sp., points toward European, rather than North American, affinities for the Mississippi assemblage. However, this generic allocation is based on fragmentary specimens and must be considered tentative. Moreover, the Mississippi specimens also resemble other taxa of early Cenozoic "insectivores" from both North America and Europe, including *Wyonycteris*, *Ceutholestes*, and *Remiculus*. Given the ambiguity of the fossil material at hand, we attach little biogeographic significance to this record.

To summarize the available biogeographic data, fossil mammals of the Red Hot local fauna show clear and consistent biogeographic affinities with faunas of similar age from western North America. Current evidence suggests only moderate provincialism within North America during the early Wasatchian. Modest early Wasatchian provincialism is indicated, for example, by the precocious presence of cylindrodontid rodents in the Red Hot local fauna. Schiebout's (1979, 1981) prediction that early Eocene faunas of the southeastern United States would show strong European affinities is not supported by the composition of the Red Hot local fauna.

4. Summary

The early Wasatchian Red Hot local fauna from the uppermost Tuscahoma Formation, Lauderdale County, Mississippi, documents for the first time the presence of a moderately diverse early Cenozoic mammalian fauna in the southeastern United States. Small mammalian taxa such as rodents, insectivores and omomyid and uintasoricine primatomorphs dominate the fossil assemblage. Although larger mammalian taxa are poorly documented in the Red Hot local fauna, this low level of representation probably reflects the depositional setting and taphonomic history of the fossil assemblage rather than the composition of the mammalian fauna that then occupied the region. Correlation between the stratigraphic section that produced the Red Hot local fauna and contemporaneous strata in northwestern Europe is achieved on the basis of sequence stratigraphy and dinoflagellate biostratigraphy. These independent (i.e., non-mammalian) lines of evidence indicate that the early Wasatchian Red Hot local fauna antedates the diverse Dormaal mammal fauna from Belgium, which is widely regarded as one of the oldest Sparnacian (or MP7) mammal faunas known from Europe. This result agrees with the paleobiogeographic model of Beard and Dawson (1999), in which the dispersal of higher-level mammalian taxa characteristic of the Wasatchian/Sparnacian is reconstructed as proceeding from Asia to North America to Europe. The early Wasatchian Red Hot local fauna shows closer biogeographic affinities with roughly contemporaneous faunas from the western United States than it does with faunas of similar age from western Europe.

ACKNOWLEDGMENTS. For access to private land which allowed us to collect the Red Hot local fauna, we are grateful to Mr. Paul Broadhead and Mr. Charley Phillips of Mid-State Land and Timber Company, Meridian, Mississippi. We thank the numerous individuals who helped collect the fossil vertebrates comprising the Red Hot local fauna, both in the field and in the vertebrate paleontology preparation laboratory at Carnegie Museum of Natural History. Of these, we particularly want to acknowledge the efforts of Robert Fox, whose diligence in the field and in sorting screen-washed concentrate was without parallel. David T. Dockery, III and Stephen L. Ingram (Office of Geology, Mississippi Department of Environmental Quality) facilitated all aspects of our field work, and have been invaluable colleagues and enthusiastic supporters of this research. Alan Tabrum coordinated many aspects of field work and supervised the sorting of screen-washed concentrate. We gratefully acknowledge the financial support of the National Geographic Society (grant number 4299–90) and the O'Neil and Netting Funds, Carnegie Museum of Natural History.

References

Baum, G. R., and Vail, P., 1988, Sequence stratigraphic concepts applied to Paleogene outcrops, Gulf and Atlantic basins, in: *Sea-Level Changes–An Integrated Approach* (C. K. Wilgus, B. S. Hastings, C. G. St. C. Kendall, H. W. Posamentier, C. A. Ross, and J. C. Van Wagoner, eds.),

SEPM (Society for Sedimentary Geology) Special Publication No. 42, Tulsa, Oklahoma, pp. 309–327.

Beard, K. C., 1998, East of Eden: Asia as an important center of taxonomic origination in mammalian evolution, *Bull. Carnegie Mus. Nat. Hist.* **34**:5–39.

Beard, K. C., and Dawson, M. R., 1999, Intercontinental dispersal of Holarctic land mammals near the Paleocene/Eocene boundary: Paleogeographic, paleoclimatic and biostratigraphic implications, *Bull. Soc. Géol. France* **170**:697–706.

Beard, K. C., and Tabrum, A. R., 1991, The first early Eocene mammal from eastern North America: An omomyid primate from the Bashi Formation, Lauderdale County, Mississippi, *Mississippi Geol.* **11(2)**:1–6.

Berggren, W. A., and Aubry, M.-P., 1998, The Paleocene/Eocene Epoch/Series boundary: Chronostratigraphic framework and estimated geochronology, in: *Late Paleocene-Early Eocene Climatic and Biotic Events in the Marine and Terrestrial Records* (M.-P. Aubry, S. G. Lucas, and W. A. Berggren, eds.), Columbia University Press, New York, pp. 18–36.

Berggren, W. A., Kent, D. V., Swisher, C. C., III, and Aubry, M.-P., 1995, A revised Cenozoic geochronology and chronostratigraphy, in: *Geochronology, Time Scales and Global Stratigraphic Correlation* (W. A. Berggren, D. V. Kent, M.-P. Aubry, and J. Hardenbol, eds.), SEPM (Society for Sedimentary Geology) Special Publication No. 54, Tulsa, Oklahoma, pp. 129–212.

Bown, T. M., and Beard, K. C., 1990, Systematic lateral variation in the distribution of fossil mammals in alluvial paleosols, lower Eocene Willwood Formation, Wyoming, in: *Dawn of the Age of Mammals in the Northern Part of the Rocky Mountain Interior, North America* (T. M. Bown and K. D. Rose, eds.), Geological Society of America Special Paper 243, Boulder, Colorado, pp. 135–151.

Bown, T. M., Rose, K. D., Simons, E. L., and Wing, S. L., 1994, Distribution and stratigraphic correlation of upper Paleocene and lower Eocene fossil mammal and plant localities of the Fort Union, Willwood, and Tatman formations, southern Bighorn Basin, Wyoming, *U.S. Geol. Survey Prof. Pap.* **1540**:1–103.

Butler, R. F., Gingerich, P. D., and Lindsay, E. H., 1981, Magnetic polarity stratigraphy and biostratigraphy of Paleocene and lower Eocene continental deposits, Clark's Fork Basin, Wyoming, *J. Geol.* **89**:299–316.

Call, V. B., Manchester, S. R., and Dilcher, D. L., 1993, *Wetherellia* fruits and associated fossil plant remains from the Paleocene/Eocene Tuscahoma-Hatchetigbee interval, Meridian, Mississippi, *Mississippi Geol.* **14(1)**:10–18.

Case, G. R., 1994a, Fossil fish remains from the late Paleocene Tuscahoma and early Eocene Bashi formations of Meridian, Lauderdale County, Mississippi. Part I. Selachians, *Palaeontographica Abt. A* **230**:97–138.

Case, G. R., 1994b, Fossil fish remains from the late Paleocene Tuscahoma and early Eocene Bashi formations of Meridian, Lauderdale County, Mississippi. Part II. Teleosteans, *Palaeontographica Abt. A* **230**:139–153.

Clyde, W. C., Stamatakos, J., and Gingerich, P. D., 1994, Chronology of the Wasatchian Land-Mammal Age (early Eocene): Magnetostratigraphic results from the McCullough Peaks section, northern Bighorn Basin, Wyoming, *J. Geol.* **102**:367–377.

Clyde, W. C., Zonneveld, J.-P., Stamatakos, J., Gunnell, G. F., and Bartels, W. S., 1997, Magnetostratigraphy across the Wasatchian/Bridgerian NALMA boundary (early to middle Eocene) in the western Green River Basin, Wyoming, *J. Geol.* **105**:657–669.

Dashzeveg, D., and Meng, J., 1998, A new Eocene cylindrodont rodent (Mammalia, Rodentia) from the eastern Gobi of Mongolia, *Am. Mus. Novitates* **3253**:1–18.

Dawson, M. R., West, R. M., Langston, W., Jr., and Hutchison, J. H., 1976, Paleogene terrestrial vertebrates: Northernmost occurrence, Ellesmere Island, Canada, *Science* **192**:781–782.

Dawson, M. R., McKenna, M. C., Beard, K. C., and Hutchison, J. H., 1993, An early Eocene plagiomenid mammal from Ellesmere and Axel Heiberg Islands, Arctic Canada, *Kaupia* **3**:179–192.

De Coninck, J., 1999, Appearances of dinoflagellate species recorded in the Tienen Formation (Landen Group) and in the Kortrijk Formation (Ieper Group) in the Belgian Basin. Their relation to transgression phases in the southern part of the North Sea Basin, *Bull. Soc. Géol. France* **170**:77–84.

Dockery, D. T., III, 1998, Molluscan faunas across the Paleocene/Eocene Series boundary in the North American Gulf Coastal Plain, in: *Late Paleocene-Early Eocene Climatic and Biotic Events in the Marine and Terrestrial Records* (M.-P. Aubry, S. G. Lucas, and W. A. Berggren, eds.), Columbia University Press, New York, pp. 296–322.

Estes, R., and Hutchison, J. H., 1980, Eocene lower vertebrates from Ellesmere Island, Canadian Arctic archipelago, *Palaeogeogr. Palaeoclimatol. Palaeoecol.* **30**:325–347.

Frederiksen, N. O., 1998, Upper Paleocene and lowermost Eocene angiosperm pollen biostratigraphy of the eastern Gulf Coast and Virginia, *Micropaleontol.* **44**:45–68.

Gazin, C. L., 1976, Mammalian faunal zones of the Bridger Middle Eocene, *Smithsonian Contrib. Paleobiol.* **26**:1–25.

Gingerich, P. D., 1974, Stratigraphic record of early Eocene *Hyopsodus* and the geometry of mammalian phylogeny, *Nature* **248**:107–109.

Gingerich, P. D., 1976, Paleontology and phylogeny: Patterns of evolution at the species level in early Tertiary mammals, *Am. J. Sci.* **276**:1–28.

Gingerich, P. D., 1989, New earliest Wasatchian mammalian fauna from the Eocene of northwestern Wyoming: Composition and diversity in a rarely sampled high-floodplain assemblage, *Univ. Michigan Pap. Paleontol.* **28**:1–97.

Gingerich, P. D., 1994, New species of *Apheliscus*, *Haplomylus*, and *Hyopsodus* (Mammalia, Condylarthra) from the late Paleocene of southern Montana and early Eocene of northwestern Wyoming, *Contrib. Mus. Paleontol. Univ. Michigan* **29**:119–134.

Gingerich, P. D., and Rose, K. D., 1977, Preliminary report on the American Clark Fork mammal fauna, and its correlation with similar faunas in Europe and Asia, *Gêobios, Mém. spéc.* **1**:39–45.

Godinot, M., 1981, Les mammifères de Rians (Éocène inférieur, Provence), *Palaeovertebrata* **10**:43–126.

Godinot, M., 1982, Aspects nouveaux des échanges entre les faunes mammaliennes d'Europe et d'Amerique du Nord a la base de l'Éocène, *Géobios, Mém. spéc.* **6**:403–412.

Godinot, M., 1987, Mammalian reference levels MP1–10, *Münchner Geowiss. Abh.* (A) **10**:21–23.

Godinot, M., 1996, Le renouvellement des faunes de mammifères en Europe et en Amérique du Nord autour de la limite Paléocène-Eocène, *Strata* (Ser. 1) **8**:18–20.

Godinot, M., de Broin, F., Buffetaut, E., Rage, J.-C., and Russell, D., 1978, Dormaal: Une des plus anciennes faunes éocènes d'Europe, *C. R. Acad. Sci., Paris* (Sér. D) **287**:1273–1276.

Godinot, M., Crochet, J.-Y., Hartenberger, J.-L., Lange-Badre, B., Russell, D. E., and Sigé, B., 1987, Nouvelles données sur les mammifères de Palette (Eocène inférieur, Provence), *Münchner Geowiss. Abh.* (A) **10**:273–288.

Gunnell, G. F., 1989, Evolutionary history of Microsyopoidea (Mammalia, ?Primates) and the relationship between Plesiadapiformes and Primates, *Univ. Michigan Pap. Paleontol.* **27**:1–157.

Holman, J. A., and Case, G. R., 1992, A puzzling new snake (Reptilia: Serpentes) from the late Paleocene of Mississippi, *Ann. Carnegie Mus.* **61**:197–205.

Holman, J. A., Dockery, D. T., III, and Case, G. R., 1991, Paleogene snakes of Mississippi, *Mississippi Geol.* **11(1)**:1–12.

Hooker, J. J., 1979, Two new condylarths (Mammalia) from the early Eocene of southern England, *Bull. Br. Mus. Nat. Hist.* (Geol.) **32**:43–56.

Hooker, J. J., 1996, Mammalian biostratigraphy across the Paleocene-Eocene boundary in the Paris, London and Belgian basins, in: *Correlation of the Early Paleogene in Northwest Europe* (R. W. O. Knox, R. M. Corfield, and R. E. Dunay, eds.), Geological Society Special Publication No. 101, London, pp. 205–218.

Hooker, J. J., 1998, Mammalian faunal change across the Paleocene-Eocene transition in Europe, in: *Late Paleocene-Early Eocene Climatic and Biotic Events in the Marine and Terrestrial Records* (M.-P. Aubry, S. G. Lucas, and W. A. Berggren, eds.), Columbia University Press, New York, pp. 428–450.

Hooker, J. J., Russell, D. E., and Phélizon, A., 1999, A new family of Plesiadapiformes (Mammalia) from the Old World lower Paleogene, *Palaeontol.* **42**:377–407.

Ingram, S. L., 1991, The Tuscahoma-Bashi section at Meridian, Mississippi: First notice of low-stand deposits above the Paleocene-Eocene TP2/TE1 sequence boundary, *Mississippi Geol.* **11(4)**:9–14.

Ivy, L. D., 1990, Systematics of late Paleocene and early Eocene Rodentia (Mammalia) from the Clarks Fork Basin, Wyoming, *Contrib. Mus. Paleontol. Univ. Michigan* **28**:21–70.

Knox, R. W. O., 1998, The tectonic and volcanic history of the North Atlantic region during the Paleocene-Eocene transition: Implications for NW European and global biotic events, in: *Late Paleocene-Early Eocene Climatic and Biotic Events in the Marine and Terrestrial Records* (M.-P. Aubry, S. G. Lucas, and W. A. Berggren, eds.), Columbia University Press, New York, pp. 91–102.

Koch, P. L., Zachos, J. C., and Gingerich, P. D., 1992, Correlation between isotope records in marine and continental carbon reservoirs near the Palaeocene/Eocene boundary, *Nature* **358**:319–322.

Koenigswald, W. von, 1980, Das Skelett eines Pantolestiden (Proteutheria, Mamm.) aus dem mittleren Eozän von Messel bei Darmstadt, *Paläontol. Z.*, **54**:267–287.

Korth, W. W., 1984, Earliest Tertiary evolution and radiation of rodents in North America, *Bull. Carnegie Mus. Nat. Hist.* **24**:1–71.

Krause, D. W., 1982, Multituberculates from the Wasatchian land-mammal age, early Eocene, of western North America, *J. Paleontol.* **56**:271–294.

Krause, D. W., 1986, Competitive exclusion and taxonomic displacement in the fossil record: The case of rodents and multituberculates in North America, in: *Vertebrates, Phylogeny, and Philosophy* (K. M. Flanagan and J. A. Lillegraven, eds.), Contributions to Geology, University of Wyoming, Special Paper 3, Laramie, Wyoming, pp. 95–117.

Krause, D. W., and Maas, M. C., 1990, The biogeographic origins of late Paleocene-early Eocene mammalian immigrants to the western interior of North America, in: *Dawn of the Age of Mammals in the Northern Part of the Rocky Mountain Interior, North America* (T. M. Bown and K. D. Rose, eds.), Geological Society of America Special Paper 243, Boulder, Colorado, pp. 71–105.

Krishtalka, L., Stucky, R. K., West, R. M., McKenna, M. C., Black, C. C., Bown, T. M., Dawson, M. R., Golz, D. J., Flynn, J. J., Lillegraven, J. A., and Turnbull, W. D., 1987, Eocene (Wasatchian through Duchesnean) biochronology of North America, in: *Cenozoic Mammals of North America: Geochronology and Biostratigraphy* (M. O. Woodburne, ed.), University of California Press, Berkeley, pp. 77–117.

Lucas, S. G., 1998, Fossil mammals and the Paleocene/Eocene Series boundary in Europe, North America, and Asia, in: *Late Paleocene-Early Eocene Climatic and Biotic Events in the Marine and Terrestrial Records* (M.-P. Aubry, S. G. Lucas, and W. A. Berggren, eds.), Columbia University Press, New York, pp. 451–500.

Maas, M. C., Anthony, M. R. L., Gingerich, P. D., Gunnell, G. F., and Krause, D. W., 1995, Mammalian generic diversity and turnover in the late Paleocene and early Eocene of the Bighorn and Crazy Mountains Basins, Wyoming and Montana (USA), *Palaeogeogr. Palaeoclimatol. Palaeoecol.* **115**:181–207.

Marincovich, L., Jr., Brouwers, E. M., Hopkins, D. M., and McKenna, M. C., 1990, Late Mesozoic and Cenozoic paleogeographic and paleoclimatic history of the Arctic Ocean Basin, based on shallow-water marine faunas and terrestrial vertebrates, in: The Geology of North America, Volume L, The Arctic Ocean Region (A. Grantz, L. Johnson, and J. F. Sweeney, eds.), Geological Society of America, Boulder, Colorado, pp. 403–426.

Matthew, W. D., 1909, The Carnivora and Insectivora of the Bridger Basin, middle Eocene, *Mem. Am. Mus. Nat. Hist.* **9**:289–567.

McKenna, M. C., 1960, Fossil Mammalia from the early Wasatchian Four Mile fauna, Eocene of northwest Colorado, *Univ. California Publ. Geol. Sci.* **37**:1–130.

McKenna, M. C., 1980, Eocene paleolatitude, climate, and mammals of Ellesmere Island, *Palaeogeogr. Palaeoclimatol. Palaeoecol.* **30**:349–362.

McKenna, M. C., 1983, Holarctic land mass rearrangement, cosmic events, and Cenozoic terrestrial organisms, *Ann. Missouri Bot. Gard.* **70**:459–489.

Novacek, M. J., Ferrusquía-Villafranca, I., Flynn, J. J., Wyss, A. R., and Norell, M., 1991, Wasatchian (early Eocene) mammals and other vertebrates from Baja California, Mexico: the Lomas las Tetas de Cabra fauna, *Bull. Am. Mus. Nat. Hist.* **208**:1–88.

Osborn, H. F., 1910, *The Age of Mammals*, Macmillan, New York.
Pfretzschner, H.-U., 1993, Muscle reconstruction and aquatic locomotion in the middle Eocene *Buxolestes piscator* from Messel near Darmstadt, *Kaupia* **3**:75–87.
Powell, A. J., 1988, A modified dinoflagellate cyst biozonation for latest Palaeocene and earliest Eocene sediments from the central North Sea, *Rev. Palaeobot. Palynol.* **56:** 327–344.
Powell, A. J., Brinkhuis, H., and Bujak, J. P., 1996, Upper Paleocene-lower Eocene dinoflagellate cyst sequence biostratigraphy of southeast England, in: *Correlation of the Early Paleogene in Northwest Europe* (R. W. O. Knox, R. M. Corfield, and R. E. Dunay, eds.), Geological Society Special Publication No. 101, London, pp. 145–183.
Prothero, D. R., 1995, Geochronology and magnetostratigraphy of Paleogene North American land mammal "ages": an update, in: *Geochronology, Time Scales and Global Stratigraphic Correlation* (W. A. Berggren, D. V. Kent, M.-P. Aubry, and J. Hardenbol, eds.), SEPM (Society for Sedimentary Geology) Special Publication No. 54, Tulsa, Oklahoma, pp. 305–315.
Redline, A. D., 1997, Revision of the Wind River faunas, early Eocene of central Wyoming. Part 13. Systematics and phylogenetic pattern of early Eocene *Hyopsodus* (Mammalia: Condylarthra), *Ann. Carnegie Mus.* **66**:1–81.
Rose, K. D., 1980, Clarkforkian land-mammal age: Revised definition, zonation, and tentative intercontinental correlations, *Science* **208**:744–746.
Rose, K. D., 1981a, The Clarkforkian land-mammal age and mammalian faunal composition across the Paleocene-Eocene boundary, *Univ. Michigan Pap. Paleontol.* **26**:1–197.
Rose, K. D., 1981b, Composition and species diversity in Paleocene and Eocene mammal assemblages: An empirical study, *J. Vert. Paleontol.* **1**:367–388.
Rose, K. D., 1999, Fossil mammals from the early Eocene Fisher/Sullivan site, in: Early Eocene Vertebrates and Plants from the Fisher/Sullivan Site (Nanjemoy Formation), Stafford County, Virginia (R. E. Weems and G. J. Grimsley, eds.), Virginia Division of Mineral Resources Publication No. 152, Charlottesville, Virginia, pp. 133–138.
Russell, D. E., and Zhai, R., 1987, The Paleogene of Asia: Mammals and stratigraphy, *Mem. Mus. Nat. Hist. Nat.* (Sér. C) **52**:1–488.
Russell, D. E., Phélizon, A., and Louis, P., 1992, *Avenius* n. gen. (Mammalia, Primates?, Microsyopidae) de l'Éocène inférieur de France, *C. R. Acad. Sci., Paris* (Sér. II) **314**:243–250.
Schiebout, J. A., 1979, An overview of the terrestrial early Tertiary of southern North America–fossil sites and paleopedology, *Tulane Stud. Geol. Paleontol.* **15**:75–93.
Schiebout, J. A., 1981, Effects of sea level changes on the distribution and evolution of early Tertiary mammals, *Trans. Gulf Coast Assoc. Geol. Soc.* **31**:383–387.
Schoch, R. M., 1998, Late Paleocene land-mammals from the Williamsburg Formation (Black Mingo Group) of South Carolina, *Trans. Am. Philos. Soc.* **88**:229–245.
Smith, R., Smith, T., and Sudre, J., 1996, *Diacodexis gigasei* n. sp., le plus ancien Artiodactyle (Mammalia) belge, proche de la limite Paléocène-Éocène, *Bull. Inst. Roy. Sci. Nat. Belgique* (Sci. Terre) **66**:177–186.
Smith, T., 1997, Les insectivores s.s. (Mammalia, Lipotyphla) de la transition Paléocène-Éocène de Dormaal (MP7, Belgique): Implications biochronologiques et paléogéographiques, *Mém. Trav. E.P.H.E., Inst. Montpellier* **21**:687–696.
Smith, T., and Smith, R., 1995, Synthèse des données actuelles sur les vertébrés de la transition Paléocène-Éocène de Dormaal (Belgique), *Bull. Soc. Belge Géol.* **104**:119–131.
Steurbaut, E., 1998, High-resolution holostratigraphy of middle Paleocene to early Eocene strata in Belgium and adjacent areas, *Palaeontographica* (Abt. A) **247**:91–156.
Tauxe, L., Gee, J., Gallet, Y., Pick, T., and Bown, T., 1994, Magnetostratigraphy of the Willwood Formation, Bighorn Basin, Wyoming: New constraints on the location of the Paleocene/Eocene boundary, *Earth Planet. Sci. Lett.* **125**:159–172.
Tong, Y., 1997, Middle Eocene small mammals from Liguanqiao Basin of Henan Province and Yuanqu Basin of Shanxi Province, central China, *Palaeontol. Sinica* (Ser. C) **26**:1–256 (in Chinese, English summary).
Wilf, P., Beard, K. C., Davies-Vollum, K. S., and Norejko, J. W., 1998, Portrait of a late Paleocene (early Clarkforkian) terrestrial ecosystem: Big Multi Quarry and associated strata, Washakie Basin, southwestern Wyoming, *Palaios* **13**:514–532.

Wilson, J. A., 1967, Early Tertiary mammals, in: *Geology of Big Bend National Park, Brewster County, Texas* (R. A. Maxwell, J. T. Lonsdale, R. T. Hazzard, and J. A. Wilson, eds.), University of Texas Publication No. 6711, Austin, Texas, pp. 157–169.

Wilson, R. W., 1949, Early Tertiary rodents of North America, *Carnegie Inst. Washington, Contrib. Paleontol.* **584**:67–164.

Winkler, D. A., 1983, Paleoecology of an early Eocene mammalian fauna from paleosols in the Clarks Fork Basin, northwestern Wyoming (U.S.A.), *Palaeogeogr. Palaeoclimatol. Palaeoecol.* **43**:261–298.

Woodburne, M. O., and Swisher, C. C., III, 1995, Land mammal high-resolution geochronology, intercontinental overland dispersals, sea level, climate, and vicariance, in: *Geochronology, Time Scales and Global Stratigraphic Correlation* (W. A. Berggren, D. V. Kent, M.-P. Aubry, and J. Hardenbol, eds.), SEPM (Society for Sedimentary Geology) Special Publication No. 54, Tulsa, Oklahoma, pp. 335–364.

Chapter 4

Paleocene-Eocene Microvertebrates in Freshwater Limestones of the Willwood Formation, Clarks Fork Basin, Wyoming

JONATHAN I. BLOCH and GABRIEL J. BOWEN

JONATHAN I. BLOCH and GABRIEL J. BOWEN: Earth Sciences Department, University of California, Santa Cruz, Santa Cruz, CA 95064 • Department of Geological Sciences and Museum of Paleontology, University of Michigan, Ann Arbor, MI 48109, carpo@umich.edu.

Eocene Biodiversity: Unusual Occurrences and Rarely Sampled Habitats, edited by Gregg F. Gunnell, Kluwer Academic/Plenum Publishers, New York, 2001.

1. Introduction

Terrestrial sediments from the Bighorn Basin and contiguous Clarks Fork Basin of northwestern Wyoming have yielded the best record of late Paleocene and early Eocene mammalian evolution known anywhere in the world (see Gingerich, 1980). Fossils preserved in sediments of the Fort Union and Willwood Formations, mostly avulsion belt and overbank deposits (Kraus and Aslan, 1993; Kraus, 1997), document a radiation of archaic mammals and the existence of the modern orders Marsupialia, Carnivora, Dermoptera, and Lipotyphla in the Paleocene; the first appearance of rodents in North America in the latest Paleocene; and the earliest records of Artiodactyla, hyaenodontid Creodonta, Perissodactyla, and Primates in the earliest Eocene (Gingerich, 1989).

In 1975, Philip Gingerich discovered a partial dentary of a new species of the enigmatic plesiadapiform *Carpolestes* (Proprimates) in a piece of orange-weathering limestone at University of Michigan locality SC-29 (late Clarkforkian North American land mammal age) in the Willwood Formation of the Clarks Fork Basin. Further collecting of limestone fragments revealed a dentary of a new species of the insectivore *Leptacodon* (Lipotyphla). Since this discovery, University of Michigan field parties have collected over 100 fossiliferous limestones from as many as 30 localities in the Clarks Fork Basin (Figure 1). These limestones have produced many exceptionally preserved late Paleocene and early Eocene vertebrates. The limestones in the University of Michigan collections are only a small sample of the limestones in the Clarks Fork Basin that undoubtedly will continue to produce extraordinary fossils far into the future. Over 50 species of mammals and birds are now known from the limestones. Hundreds of undescribed specimens of mammals, birds, reptiles, and amphibians in the UM collections will almost certainly increase limestone biodiversity. Skulls, jaws, and partial skeletons are found in association for some of these species (Figure 2). Species which are common in the limestone fauna, especially the smaller mammals and birds, are extremely rare in other parts of the Willwood Formation. Many of the species found in the limestones are new, and are not represented in the surrounding mudstones. Previous publications have resulted in descriptions of six new bird species (Gingerich, 1987; Houde, 1986, 1987, 1988; Houde and Olsen, 1989, 1992) and nine new mammal species (Rose and Gingerich, 1987; Gingerich, 1987; Beard and Houde, 1989; Thewissen and Gingerich, 1989; Bloch *et al.*, 1998; Bloch and Gingerich, 1998) from the limestone fauna. The limestones represent at least one, if not several, unique depositional environments that may be sampling from different habitats than the traditionally sampled overbank paleosols.

A complete analysis of Clarks Fork Basin limestone geology, faunal assemblage and paleoenvironment is currently in progress (Bloch, 1995, 1997, 1998,

1999; Bloch and Gingerich, 1994, 1998; Bloch *et al.*, 1998; Bowen and Bloch, 1999). The fossils recovered from the limestones provide a new, as yet largely unexploited, "preservational window" that will contribute to a better understanding of species diversity and community structure through the Paleocene-Eocene transition, a period of major climatic change (Kennett and Stott, 1991; Zachos *et al.*, 1993; Thomas and Shackleton, 1996) and subsequent faunal change in the ocean (Steineck and Thomas, 1996) and on land (Rose, 1981; Gingerich, 1989; Maas *et al.*, 1995; Gunnell *et al.*, 1995; Gunnell, 1998; Morgan *et al.*, 1995; Clyde and Gingerich, 1998). Additionally (although not discussed here) exceptional preservation and association of fossil material will allow for more accurate analyses of phylogenetic relationships as well as a more detailed understanding of functional morphology for many of the late Paleocene and early Eocene small mammal groups, and thus promises to contribute to a more resolved understanding of mammalian evolution.

This chapter includes (1) a review of both the geology of limestone localities and the lithology of the limestones, (2) a summary of the faunal assemblages of the limestones, (3) discussion of depositional environments that could account for the formation of the limestones and accumulation of their unique fauna, and (4) discussion of the limestone as a unique faunal assemblage and the implications for studying mammalian paleoecology and evolution through the Paleocene-Eocene transition.

2. Geology

The Bighorn Basin, surrounded by the Beartooth, Absaroka, Owl Creek, Bighorn and Pryor mountains, is located in northwestern Wyoming. The study area is located in the Clarks Fork Basin which occupies several hundred square miles in the northwestern corner of the Bighorn Basin. Alluvial sediments of the Fort Union and Willwood Formations in the Clarks Fork Basin preserve an exceptionally continuous sequence which yields latest Cretaceous through early Eocene faunas. These formations consist of sandy channel deposits, sandy and silty levee and crevasse splay deposits, and mudstones representing floodplain accumulation (Bown, 1979; Gingerich, 1987). In particular, the latest Paleocene-early Eocene Willwood Formation consists of about 700 m of meander belt sheet sandstones, overbank deposits, and avulsion deposits (Kraus and Aslan, 1993; Kraus, 1997). In this alluvial context, at least two kinds of cumulative floodplain paleosols, red and gray paleosols, formed on the overbank mudstone deposits. The red mudstones are thought to represent well-drained paleosols in which the B horizon has been rubified (Kraus, 1997). Gray mudstones within floodplain paleosol sequences are generally interpreted as either the leached A horizon of a composite soil profile or as incipient soil (enceptisol). Kraus (1997) suggests that gray mudstones with mottles, pedogenic slikensides, and isolated accumulation of calcium carbonate may represent soils that formed under hydromorphic conditions.

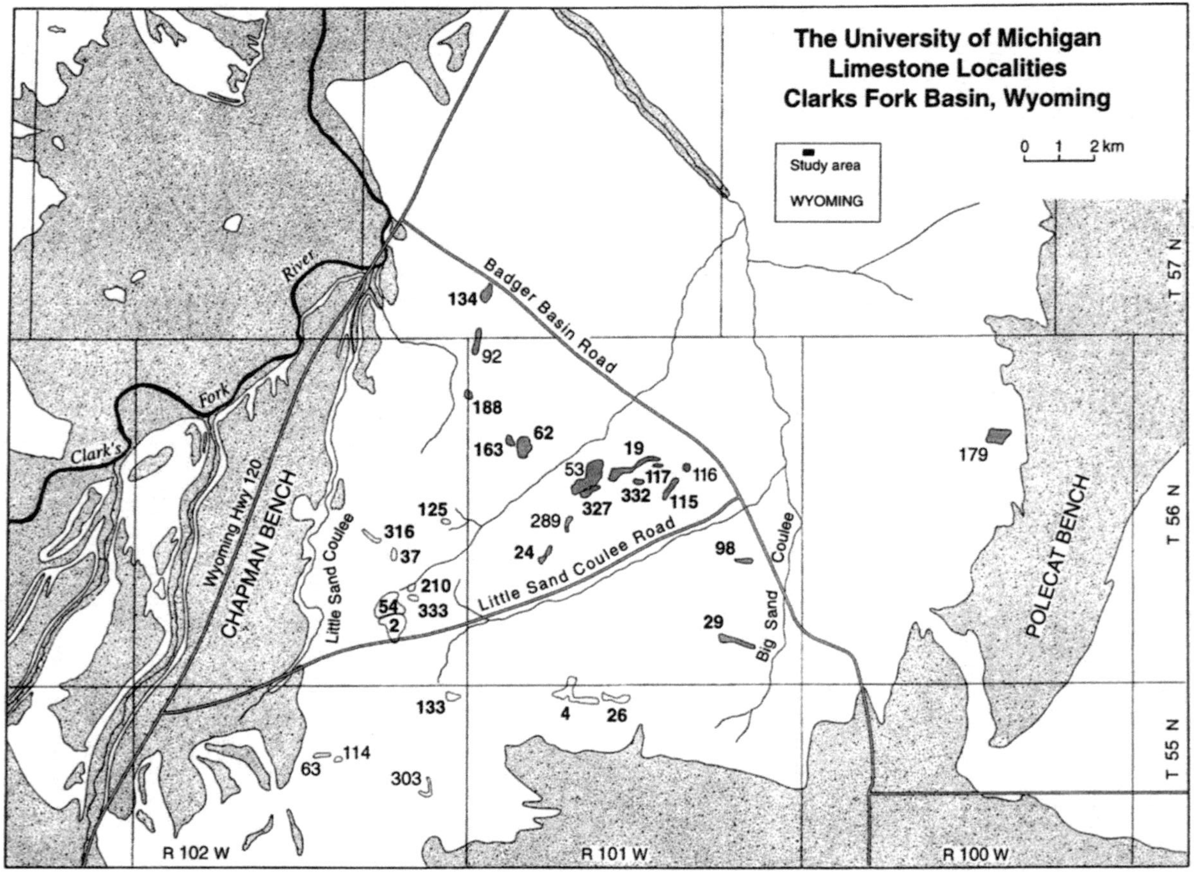

FIGURE 1. Map of limestone-bearing University of Michigan fossil vertebrate localities (30 locals.) in the Clarks Fork Basin. Localities producing limestones sampled in this study are in bold type. Localities shaded dark gray are Clarkforkian age and all localities shaded light gray are Wasatchian age. Modified from Rose (1981: figure 2).

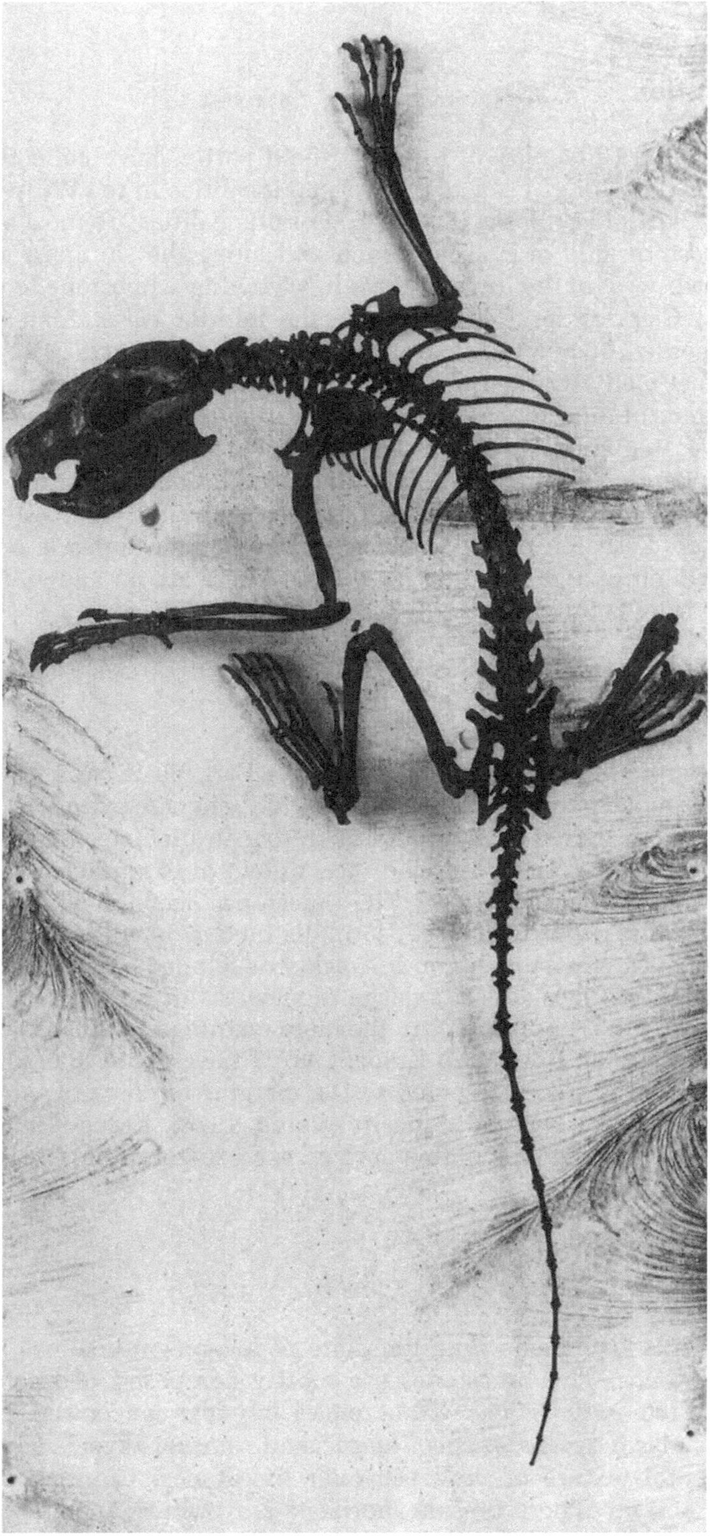

FIGURE 2. Photo of *Plesiadapis cookei* skeleton (× ⅓) from a freshwater limestone from UM locality SC-117 (Cf-2).

2.1. Localities

Since 1975, University of Michigan field parties have collected over 100 fossiliferous limestones from as many as 30 localities in the Willwood Formation of the Clarks Fork Basin (Figure 1). Fossil localities are most abundant on the southeastern side of Polecat Bench and along the southern rim of Sand Coulee, north-west of the town of Powell, Wyoming. Limestone localities span the earliest Clarkforkian (Cf-1) through the middle Wasatchian (Wa-4) land mammal ages (ca. 56.5–53.3 Ma; Butler *et al.*, 1981; Clyde *et al.*, 1994), with representation at many stratigraphic intervals throughout this time period (Figure 3). Abundance of limestones is temporally variable, with many more limestone in early Wasatchian localities, just after the Paleocene-Eocene boundary (ca. 55.5 Ma; Berggren *et al.*, 1995), than either Clarkforkian or middle Wasatchian localities. This study will focus on limestones from Cf-2, Cf-3, Wa-1, and Wa-2 faunal zones. Research on Cf-1, Wa-3, and Wa-4 limestones is on-going, the results of which will be reported elsewhere. There are no known fossiliferous limestones from faunal zone Wa-0 in the Clarks Fork Basin.

2.2. Stratigraphy

Fossiliferous limestones from the Clarks Fork Basin are of two different endmember morphologies (Figure 4): crudely shaped cylinders a few centimeters to more than 1 m high, and 10–50 cm in diameter (Gingerich, 1987; Kraus, 1988); and flat-lying lenses up to 2 m by 2 m in surface area, and about 10–20 cm thick (Gingerich, 1987). The limestones occur in abundance at specific stratigraphic levels where they form distinctive, laterally extensive bands. The limestones are most common in weakly developed paleosols that form on levee or crevasse-splay deposits (stage 1 paleosols of Bown, 1985; Bown and Kraus, 1987), with no limestones in the more mature paleosols (stage 3 through stage 5; Bown, 1985; Bown and Kraus, 1987; Kraus, 1988). Bands of cylindrical nodules can be traced laterally to the margins of channel sandstones but do not occur within channel deposits (Kraus, 1988). Kraus (1988) noted the presence of a thin envelope of fine-grained sand surrounding the lower part of all cylindrical nodules.

2.3. Lithology

The Clarks Fork Basin limestones are pedogenic grainstones, packstones, and wackestones. The limestones are mostly composed of microcrystalline calcite mud (80–95% $CaCO_3$) with a remarkably low concentration of primary terrigenous clastic grains. Inspection of hand samples reveals some variation in the internal texture of the freshwater limestones. Lamination is weakly defined to absent. Where present, horizons are often defined by planes with high concentrations of fossils or organic matter but with no increased

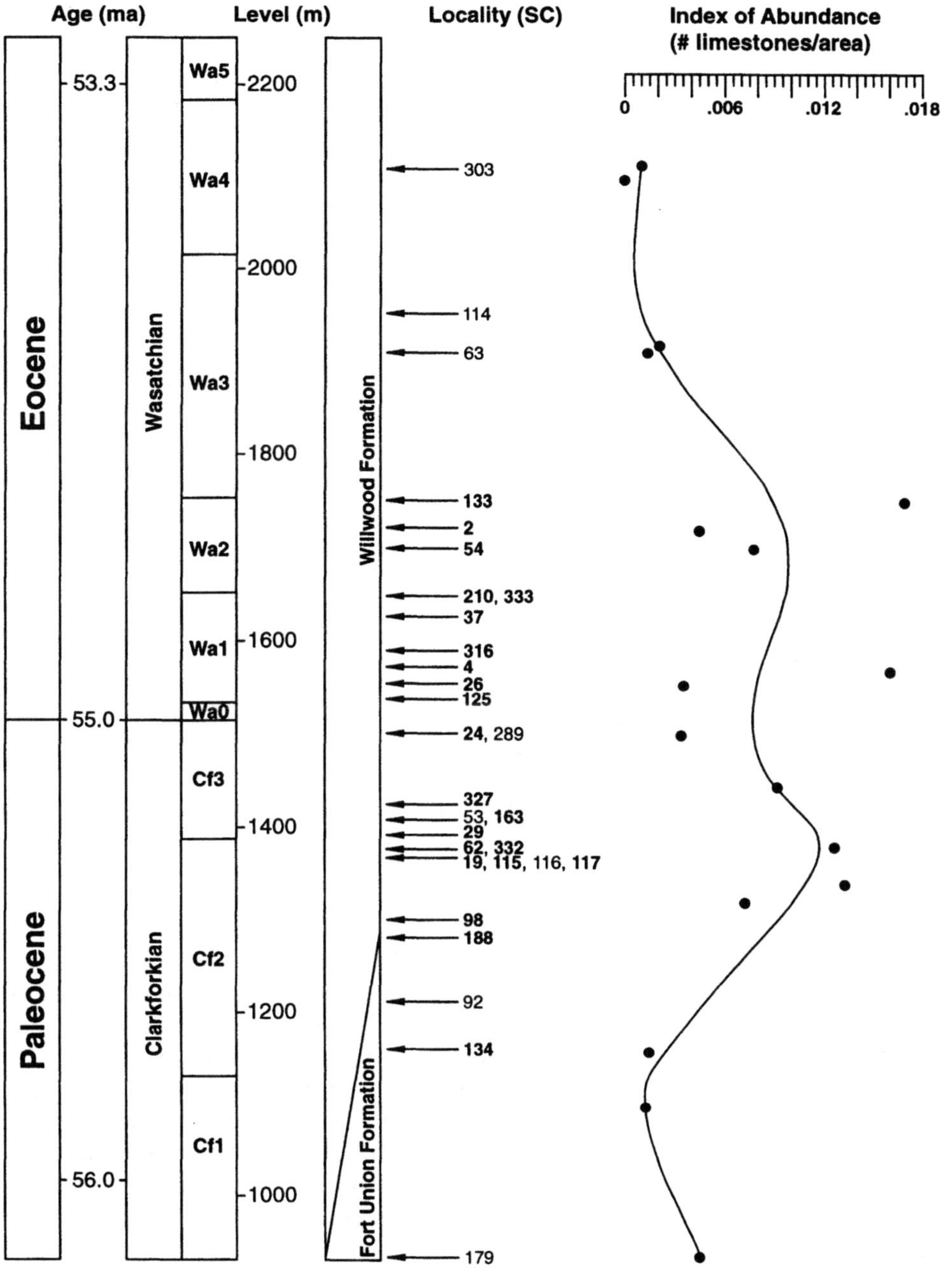

FIGURE 3. Stratigraphic section in Clarks Fork Basin (in part, as the Clarks Fork Section extends down to the K-T boundary). UM limestone-bearing fossil vertebrate localities, from Figure 1, are shown to the right of the section. Localities producing limestones sampled in this paper are in bold type. The Fort Union-Willwood contact is gradational and occurs approximately at the top of the Cf-1 subdivision of the Clarkforkian North American Land Mammal Age. Ages are pinned to meter levels using paleomagnetic calibrations (55.9 mya and 53.4 mya) and the isotopic excursion at the P-E boundary (model 1 of Wing *et al.*, 1999). The index of abundance is the number of limestones per area (m^2) of surveyed transect. Note that the abundance of limestones in a given locality is temporally variable, with more limestones in late Clarkforkian and early Wasatchian localities, just before and after the Paleocene-Eocene boundary (about 1520 m), than either earlier Clarkforkian or middle Wasatchian localities.

FIGURE 4A. Cylindrical limestone from UM locality SC-29 (Cf-3). **4B.** Flat-lying limestone from UM locality SC-19 (Cf-2).

concentration of clastic material. Most limestones lack internal lamination entirely and display abundant bioturbation features including branched and cross cutting networks of spar or mud filled rhizoliths, mud and micrite filled burrows, and rounded micritic grains. While the presence of bone horizons and laminations in some limestones provides evidence that their precipitation was episodic, the pervasive bioturbation of most samples precludes any generalized statement concerning the depositional rhythm characteristic of the accumulations. Fossil gastropods are relatively rare and not associated with bone horizons, but commonly are undamaged and contain little or no micritic infill (Figure 5A). Whole fossil bird eggs, when preserved, are also commonly undamaged with no micritic infill.

In thin section the micrite appears muddy, at least in part due to the presence of iron compounds and manganese which seem most concentrated at the walls of voids and burrows but are distributed throughout the matrix. Although gastropod shells and the interior of broken long bone shafts are primarily filled with spar, weak geopetal development is seen as a thin micritic lining covering the bottom surface of some spar-filled voids.

The majority of the allochems preserved in the limestones are calcite cylinders consistent in size, morphology and crystal structure with calcified root cells (Klappa, 1978, 1979, 1980; Beeze-Deak *et al.*, 1997). These calcite

FIGURE 5A. Thin section photomicrograph of a well-preserved, spar-filled gastropod. Calcified root cells are locally abundant in the surrounding matrix. **5B.** Photomicrograph of calcified root cells. Organization of cells within the aggregate is similar to that seen in the secondary xylem of modern woody and herbaceous dicots. Large diameter cells correspond in size and arrangement to vessel cells. (Long dimension is 2.5 mm in A, 0.66 mm in B).

cylinders are observed in articulation within some root-shaped branching voids and are distributed individually throughout the groundmass. Together with calcified seeds and other biologically derived allochems, the root cells (Figure 5B) comprise up to 95% of the clasts in some of the limestones. The remaining detrital components are allocthonous and are angular to subangular, well to moderately well sorted, grains of plagioclase, microcline, quartz and mica. Grain size ranges from 20 to 400 μm (course silt to medium sand) with most samples containing predominantly fine and very fine sand.

Cements are clear calcite spar, and typically two growth generations are represented. The earlier of these is a thin, phreatic drusy or prismatic spar and the later is blocky spar, which in almost all cases completely fills voids. In some samples a thin layer of hematite separates early and late cement generations. These cement generations seem to be analogous to the two isotopically distinct spar generations reported from pedogenic B-horizon carbonate nodules from the Clarks Fork Basin by Bao *et al.* (1998).

3. Fauna

3.1. Previous Work

Previous publications on freshwater limestones of the Clarks Fork Basin have focused on a limited portion of the limestone fauna (Houde, 1986, 1987, 1988; Gingerich, 1987; Gunnell and Gingerich, 1987; Rose and Gingerich, 1987; Beard, 1989a, 1989b, 1990; Beard and Houde, 1989; Gunnell, 1989; Houde and Olsen, 1989, 1992; MacPhee *et al.*, 1989; Thewissen and Gingerich, 1989; Kay *et al.*, 1990; Kay *et al.*, 1992; Gingerich and Gunnell, 1992; Rose *et al.*, 1993; Bloch and Gingerich, 1994, 1998; Bloch, 1995, 1997, 1998; MacPhee *et al.*, 1995; Bloch *et al.*, 1998). These publications have resulted in descriptions of six new bird species and nine new mammal species over the past eleven years. An exception was published by Gingerich (1987), who dissolved most of a limestone lens from University of Michigan locality SC-29 and described the entire molluscan and vertebrate fauna. The faunal list was composed of eight gastropod species and at least eighteen vertebrate species. Despite the fact that SC-29 is one of the most productive Clarkforkian fossil localities known to date, the fauna recovered from lens J sharply contrasts with that of the rest of the locality. Taxa preserved in that limestone, but not in the surrounding mudstone, include abundant terrestrial gastropods, amphibians, lizards, a new genus of heron (representing the oldest heron yet known), two insectivore genera, two 'bat' genera (if bats, these represent the oldest known in the fossil record—but see Hand *et al.*, 1994; Simmons and Geisler, 1998), a dermopteran, two genera of small plesiadapiform primates, and a small carnivore. Lens J did not preserve the medium or larger mammals (>200 grams) that represent at least 85% of the fauna recovered from the floodplain paleosols which are the dominant fossiliferous lithology at SC-29. Aquatic organisms such as fish, crocodilians, and turtles are also not preserved in lens J. Gingerich interpreted the gastropod fauna (of which all were land snails; most were air-breathing pulmonates) as suggestive of a humid microenvironment including decaying wood or leaves. The amphibians and reptiles, which seemed to indicate proximity to leaf litter, were consistent with the habitat indicated by the gastropods. Gingerich noted that the vertebrate fossils showed signs of mammalian predation or scavenging. Broken eggshell, possibly avian in origin, was preserved throughout lens J, although it appeared to have come from eggs too small to have represented a heron.

Houde (1988) and Houde and Olsen (1989, 1992) described the avifauna from several Clarks Fork Basin limestones, most of which were completely different from the avifauna of stratigraphically contiguous mudstones. Houde noted that the co-occurrence of small vertebrates with the eggshell accumulations might indicate some degree of transportation and concentration (as lithornithids, the birds preserved in the limestone, were not predatory). Houde further reasoned that, lacking signs of abrasion, the eggshells were not transported very far. He concluded that eggshell within a single limestone represented the contents of a single lithornithid bird nest that had been transported a short distance.

3.2. Analysis of Mammalian Fauna

3.2.1. Data

For this study, abundance data of mammalian taxa in limestones and surrounding mudstones from 22 localities of Cf-2, Cf-3, Wa-1 and Wa-2 faunal zones were gathered (Tables 1–4). The mudstone sample was split into surface-collected samples, quarried samples, and washed samples when possible. Only the surface-collected samples are used in this study. The data are compiled from previous publications and from the University of Michigan Museum of Paleontology vertebrate collections (J. I. Bloch; P. D. Gingerich; unpublished data). Abundance of taxa are reported as total number of specimens. The mudstone data are preliminary in that portions of the collection (labeled "Miscellaneous") have not yet been studied and thus are not included in this paper.

3.2.2. Taxonomic Distributions

As documented by Rose (1981), the surface-collected assemblages from Clarkforkian mudstones are dominated by phenacodontid condylarths, with *Ectocion* being the most common taxon (Figure 6). In contrast, condylarths are almost unrepresented in the Clarkforkian limestone faunas, which are instead dominated by either proprimates (Cf-2) or lipotyphlan insectivores (Cf-3). While proprimates are more abundant in Cf-2 than Cf-3 limestones, rodents, multituberculates, and marsupials are common in all of the Clarkforkian limestones and are only rare elements of the surface-collected mudstones.

Surface-collected assemblages from Wasatchian are also dominated by condylarths. However, unlike the Clarkforkian assemblages, immigrant species of Perissodactyla, Artiodactyla, and Primates are also very abundant.

Condylarths are nearly absent in Wasatchian limestones, which are dominated by proprimates (Wa-1), and by lipotyphlan insectivores, rodents and proprimates (Wa-2; Figure 7).

3.2.3. Body Size Distributions

The surface-collected mudstone sample is biased against preservation and collection of the small mammal component of the fauna, with most specimens representing species 1–22 kg in body mass. In contrast, most of the specimens in the limestones represent small taxa with a mode of about 60 grams (Figure 8). Thus, the limestones yield a much better sample of small mammals than the mudstones.

While the number of small mammal specimens is much higher in the limestones than the mudstones, the number of small mammal genera represented

Table 1. Faunal list for Cf-2. Includes: SC-134, SC-188, SC-98, SC-115, SC-117, SC-19, SC-62, and SC-332.

Taxon		Limestone		Mudstone	
Order	Genus	N	% (n=91)	N	% (n=331)
Carnivora	*Didymictis*	-	-	12	3.6
	Uintacyon	1	1.1	-	-
	Viverravus	-	-	4	1.2
	Indeterminate	-	-	1	0.3
Condylarthra	*Apheliscus*	-	-	1	0.3
	Chriacus	-	-	3	0.9
	Dissacus	-	-	3	0.9
	Ectocion	-	-	108	32.6
	Haplomylus	-	-	6	1.8
	Phenacodus	-	-	33	9.7
	Thryptacodon	-	-	3	0.9
	Tricentes	-	-	-	-
Creodonta	*Dipsalidictis*	-	-	9	2.7
	Dipsalodon	-	-	1	0.3
Dermoptera	*Plagiomene*	-	-	1	0.3
	Worlandia	-	-	4	1.2
Dinocerata	*Probathyopsis*	-	-	4	1.2
Lipotyphla	*Ceutholestes*	2	2.2	-	-
	Leptacodon	18	19.8	-	-
	Indeterminate	7	7.7	4	1.2
Marsupialia	*Peradectes*	11	12.1	3	0.9
Multituberculata	*Ectypodus*	10	11.0	1	0.3
-	*Microcosmodon*	-	-	1	0.3
Notoungulata	*Arctostylops*	-	-	2	0.6
Palaeanodonta	*Palaeanodon*	-	-	1	0.3
Pantodonta	*Coryphodon*	-	-	15	4.5
	Cyriacotherium	-	-	2	0.6
	Titanoides	-	-	5	1.5
Proprimates	*Arctodontomys*	-	-	1	0.3
	Carpolestes	17	18.7	6	1.8
	Chiromyoides	-	-	2	0.6
	Ignacius	4	4.4	1	0.3
	Phenacolemur	2	2.2	4	1.2
	Plesiadapis	4	4.4	70	21.2
	Tinimomys	7	7.7	2	0.6
Proteutheria	*Apatemys*	1	1.1	-	-
	Labidolemur	-	-	1	0.3
	Palaeoryctes	1	1.1	-	-
	Prodiacodon	1	1.1	1	0.3
Rodentia	*Acritoparamys*	-	-	1	0.3
	Microparamys	-	-	1	0.3
	Paramys	1	1.1	1	0.3
	Indeterminate	4	4.4	4	1.2
Tillodontia	*Azygonyx*	-	-	9	2.7

Table 2. Faunal list for Cf-3. Includes: SC-29, SC-163, SC-327, and SC-24.

Taxon		Limestone		Mudstone	
Order	Genus	N	% (n=86)	N	% (n=115)
Carnivora	*Didymictis*	-	-	10	8.7
	Viverravus	4	4.7	1	0.8
Creodonta	*Prolimnocyon*	1	1.2	-	-
Chiroptera	Cf. *Icaronycteris*	4	4.7	-	-
Condylarthra	*Apheliscus*	-	-	1	0.8
	Chriacus	-	-	1	0.8
	Dissacus	-	-	2	1.7
	Ectocion	1	1.2	59	51.3
	Haplomylus	-	-	2	1.7
	Phenacodus	-	-	14	12.2
	Thryptacodon	-	-	1	0.8
Dermoptera	*Plagiomene*	-	-	2	1.7
	Worlandia	3	3.5	-	-
Dinocerata	*Probathyopsis*	-	-	3	2.6
Lipotyphla	*Leptacodon*	16	18.6	1	0.8
	Limaconysus	1	1.2	-	-
	Plagioctenodon	2	2.3	-	-
	Wyonycteris	12	14.0	-	-
	Indeteriminate	12	14.0	-	-
Marsupialia	*Peradectes*	5	5.8	-	-
Multituberculata	*Ectypodus*	10	11.6	1	0.8
Pantodonta	*Coryphodon*	-	-	1	0.8
Proprimates	*Carpolestes*	2	2.3	2	1.7
	Ignacius	-	-	1	0.8
	Phenacolemur	-	-	4	3.5
	Tinimomys	2	2.3	-	-
Proteutheria	*Labidolemur*	1	1.2	-	-
	Palaeosinopa	-	-	1	0.8
Rodentia	*Acritoparamys*	-	-	2	1.7
	Microparamys	-	-	1	0.8
	Indeterminate	10	11.6	-	-
Tillodontia	*Azygonyx*	-	-	5	4.4

is not significantly different (Figure 9). They are different, however, in that limestones seem to be sampling a constant number of taxa in each of the faunal zones, while the diversity seems to increase in the mudstone sample. Thus, while the mudstones document an increase in diversity of mammalian taxa, certainly the result of the introduction of immigrant taxa at the beginning of the Wasatchian, the small mammal component of the fauna seems to stay about the same. If true, this difference will be important for understanding the impact of climate change on mammalian communities during the Paleocene-Eocene transition.

4. Depositional Environment

The lithology and faunal elements preserved in the limestones are not consistent with any of the traditionally recognized depositional environments in a fluvial system, including channel lag, abandoned channel, levee, crevasse splay, or floodplain deposits. Previous work on fossiliferous limestones from the Clarks Fork Basin focused on description of new fossil material, with little emphasis on depositional environment. While the mode of formation is poorly understood, several possible depositional environments have been suggested. Depositional environments reviewed here include: deposition as steinkerns of the boles of trees (Kraus, 1985, 1987, 1988; Gingerich, 1987), deposition in shallow lakes or pools (Gingerich, 1987), and deposition in the burrows of large

Table 3. Faunal list for Wa-1. Includes: SC-125, SC-26, SC-4, SC-333, SC-210, SC-37, and SC-316.

Taxon		Limestone		Mudstone	
Order	Genus	N	%(n=105)	N	%(n=855)
Artiodactyla	*Diacodexis*	1	0.9	66	7.7
Carnivora	*Didymictis*	-	-	11	1.3
	Miacis	-	-	3	0.4
	Viverravus	1	0.9	13	1.5
	Uintacyon	-	-	9	1.1
Condylarthra	*Apheliscus*	-	-	10	1.2
	Copecion	-	-	4	0.5
	Dissacus	-	-	4	0.5
	Ectocion	-	-	87	10.2
	Hapalodectes	-	-	1	0.1
	Haplomylus	-	-	62	7.3
	Hyopsodus	-	-	160	18.7
	Pachyaena	-	-	1	0.1
	Phenacodus	-	-	12	1.4
	Thryptacodon	-	-	15	1.8
Creodonta	*Acarictis*	-	-	1	0.1
	Arfia	-	-	7	0.8
	Dipsalidictis	-	-	6	0.7
	Prolimnocyon	-	-	2	0.2
	Prototomus	-	-	7	0.8
Dermoptera	*Plagiomene*	1	0.9	-	-
Lipotyphla	*Leipsanolestes*	-	-	1	0.1
	Leptacodon	1	0.9	-	-
	Palaeictops	-	-	1	0.1
	Prodiacodon	-	-	1	0.1
	Indeterminate	4	3.8	3	0.4
Marsupialia	*Peradectes*	1	0.9	-	-
Mutituberculata	*Ectypodus*	1	0.9	3	0.4
	Neoliotomus	-	-	6	0.7
Palaeanodonta	*Palaeanodon*	-	-	4	0.5
Pantodonta	*Coryphodon*	-	-	3	0.4
Perissodactyla	*Cardiolophus*	-	-	9	1.1
	Hyracotherium	4	3.8	152	17.8
Primates	*Cantius*	-	-	111	13.0
	Teilhardina	-	-	4	0.5
Proprimates	*Arctodontomys*	-	-	1	0.1
	Chalicomomys	6	5.7	-	-
	Ignacius	15	14.3	3	0.4
	Niptomomys	1	0.9	-	-
	Phenacolemur	24	22.9	19	2.2
	Tinimomys	45	42.9	-	-
Proteutheria	*Apatemys*	-	-	3	0.4
	Labidolemur	-	-	1	0.1
	Palaeoryctes	-	-	4	0.5
Rodentia	*Acritoparamys*	-	-	11	1.3
	Paramys	-	-	20	2.3
	Indeterminate	-	-	5	0.6

vertebrates (Gingerich, 1987). Deposition in a spring environment is also considered (although never previously suggested).

4.1. Tree Trunk

Accumulation of unusual sediment and terrestrial fossils within the trunks of trees has been documented several times (Behrensmeyer and Hook, 1992). These discoveries include early reptiles from Paleozoic trees of Nova Scotia (Dawson, 1882; Carroll, 1967; Carroll *et al.*, 1972; Rust *et al.*, 1984), plant remains in the trunks of lycopsids from the Carboniferous of Scotland (Walton, 1935), and mammals and birds in trees from the Miocene of Rusinga Island, Kenya (Walker and Teaford, 1989).

The hypothesis that freshwater limestones from the Clarks Fork Basin represent fossilized trees, or the calcareous steinkern of the boles of trees, is an interpretation that is generally accepted in the literature (for example, see Behrensmeyer and Hook, 1992) as the result of work by Kraus (1985, 1987, 1988) and Gingerich (1987) who found that: cylindrical nodules in the Fort Union Formation (Tiffanian) were proximal, and morphologically similar to, calcified tree trunks suggesting that these nodules, as well as those found in

Table 4. Faunal list for Wa-2. Includes: SC-54, SC-133, and SC-2.

Taxon		Limestone		Mudstone	
Order	Genus	N	%(n=37)	N	%(n=1777)
Artiodactyla	*Diacodexis*	4	10.8	131	7.4
Carnivora	*Didymictis*	-	-	5	0.3
	Miacis	-	-	7	0.4
	Vassacyon	-	-	2	0.1
	Viverravus	-	-	29	1.6
	Uintacyon	2	5.4	8	0.5
Condylarthra	*Apheliscus*	1	2.7	25	1.4
	Copecion	-	-	5	0.3
	Dissacus	-	-	1	0.1
	Ectocion	-	-	74	4.2
	Haplomylus	-	-	264	14.9
	Hyopsodus	-	-	468	26.3
	Pachyaena	-	-	2	0.1
	Phenacodus	-	-	15	0.8
	Thryptacodon	-	-	10	0.6
Creodonta	*Arfia*	-	-	21	1.2
	Dipsalidictis	-	-	2	0.1
	Galecyon	-	-	1	0.1
	Oxyaena	-	-	6	0.3
	Palaeonictis	-	-	1	0.1
	Prolimnocyon	-	-	9	0.5
	Prototomus	2	5.4	14	0.8
	Indeterminate			1	0.1
Lipotyphla	*Leptacodon*	4	10.8	-	-
	Plagioctenodon	-	-	1	0.1
	Palaeictops	-	-	4	0.2
	Prodiacodon	-	-	1	0.1
	Indeterminate	6	16.2	13	0.7
Marsupialia	*Peradectes*	1	2.7	3	0.2
Mutituberculata	*Ectypodus*	-	-	2	0.1
	Neoliotomus	-	-	2	0.1
	Indeterminate	-	-	1	0.1
Palaeanodonta	*Palaeanodon*	-	-	5	0.3
Pantodonta	*Coryphodon*	-	-	12	0.7
Perissodactyla	*Cardiolophus*	-	-	15	0.8
	Homogalax	-	-	2	0.1
	Hyracotherium	1	2.7	271	15.3
Primates	*Cantius*	3	8.1	179	10.1
	Teilhardina	-	-	13	0.7
Proprimates	*Arctodontomys*	2	5.4	23	1.3
	Ignacius	2	5.4	6	0.3
	Niptomomys	-	-	1	0.1
	Phenacolemur	1	2.7	32	1.8
	Tinimomys	-	-	1	0.1
Proteutheria	*Apatemys*	-	-	3	0.2
	Didelphodus			8	0.5
	Eoryctes	-	-	1	0.1
	Labidolemur	-	-	5	0.3
	Palaeoryctes	-	-	3	0.2
	Palaeosinopa	-	-	4	0.2
	Wyolestes	-	-	4	0.2
Rodentia	*Acritoparamys*	-	-	12	0.7
	Cf. *Franimys*	-	-	1	0.1
	Lophioparamys	-	-	1	0.1
	Microparamys	3	8.1	1	0.1
	Paramys	-	-	15	0.8
	Indeterminate	4	10.8	6	0.3
Tillodontia	*Esthonyx*	1	2.7	25	1.4

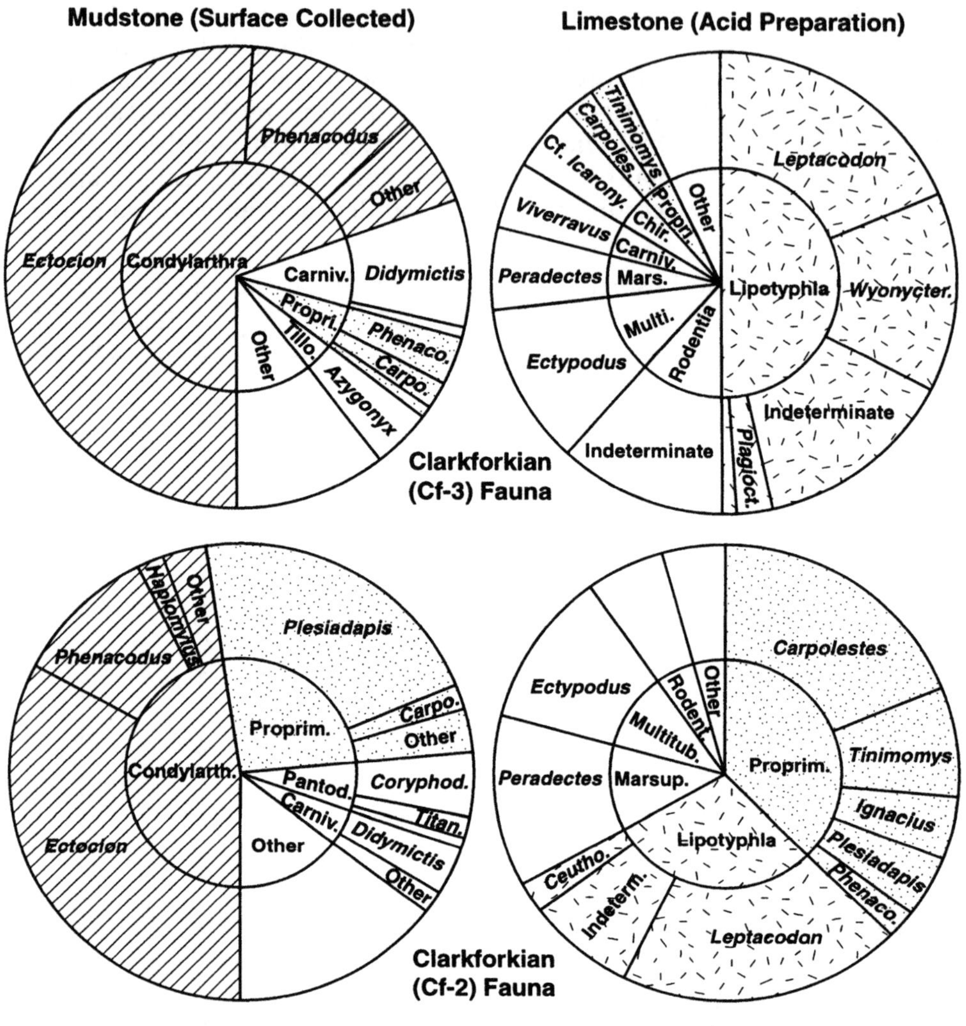

FIGURE 6. Relative abundance of mammalian taxa from freshwater limestones in the *Plesiadapis cookei* (Cf-2) and *Phenacodus-Ectocion* (Cf-3) faunal zones of the Clarkforkian land mammal age. Based on pooled samples (Tables 1 and 2). Width of each sector reflects the percentage of the mammalian fauna represented by each taxon, as determined by number of specimens.

the overlying Willwood Formation (Clarkforkian-Wasatchian) were also tree trunks (Kraus, 1987); in thin section, some of the limestones contain very small quantities of filamentous lignitic material which is similar to the material comprising cell walls in the calcified tree trunks (Kraus, 1987); the morphology of the nodules resemble an upright tree (Kraus, 1987; Gingerich, 1987); orange calcareous rinds surrounding some tree trunks in the upper parts of some Willwood sheet sandstones are compositionally and texturally similar to the limestones (Kraus, 1987); the spacing and distribution of limestones in a single locality is consistent with that of a modern forest (Gingerich, 1987); and the invertebrate, reptilian, avian, and mammalian faunas preserved in the limestone are consistent with accumulation in hollow tree trunks (Gingerich, 1987). However, Gingerich (1987) also noted that the limestones are different from typical tree trunk deposits in not preserving any trace of wood or even woody plant debris. He also noted that none of the limestones show any evidence of any rooted connection to an underclay or underlying soil (although Kraus, 1987 found some of the limestones to be rooted).

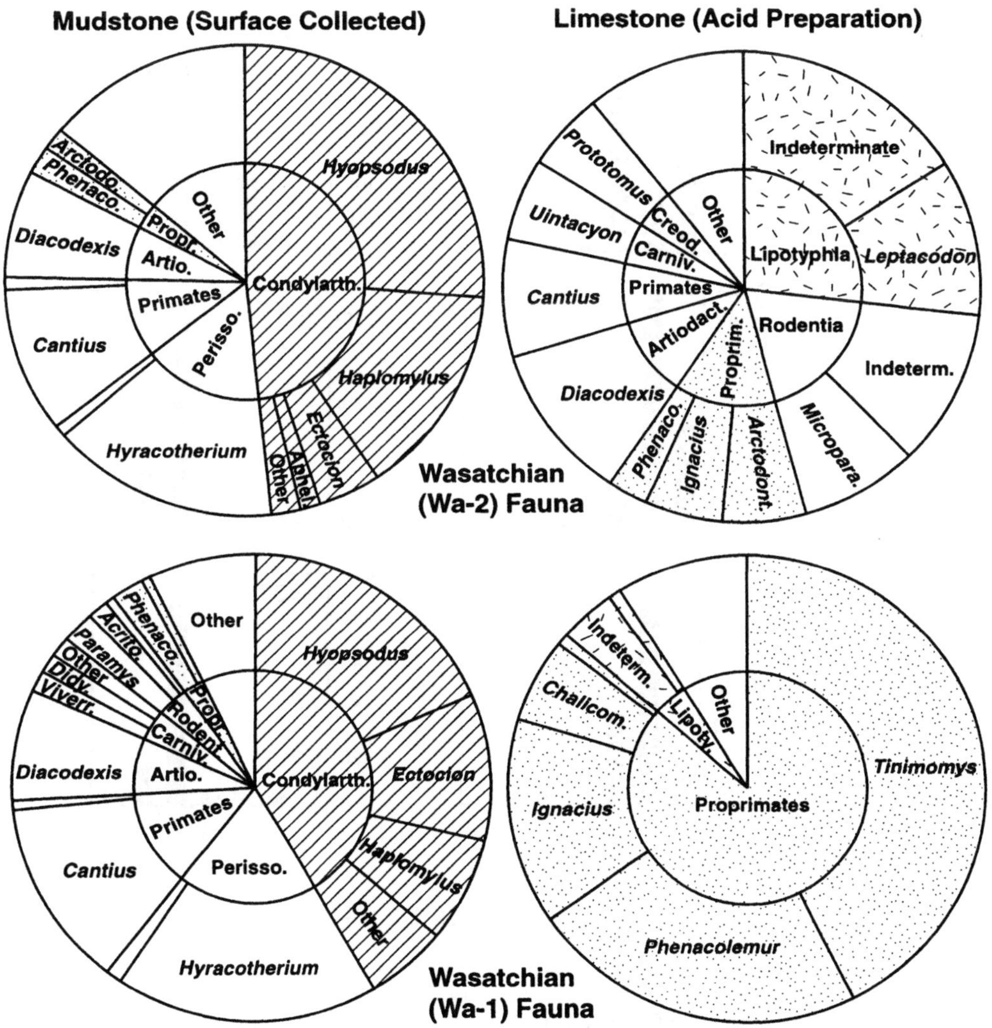

FIGURE 7. Relative abundance of mammalian taxa from freshwater limestones in the *Cardiolophus ralstoni* (Wa-1) and *Arfia shoshoniensis* (Wa-2) faunal zones of the Wasatchian land mammal age. Based on pooled samples (Tables 3 and 4). Width of each sector reflects the percentage of the mammalian fauna represented by each taxon, as determined by number of specimens.

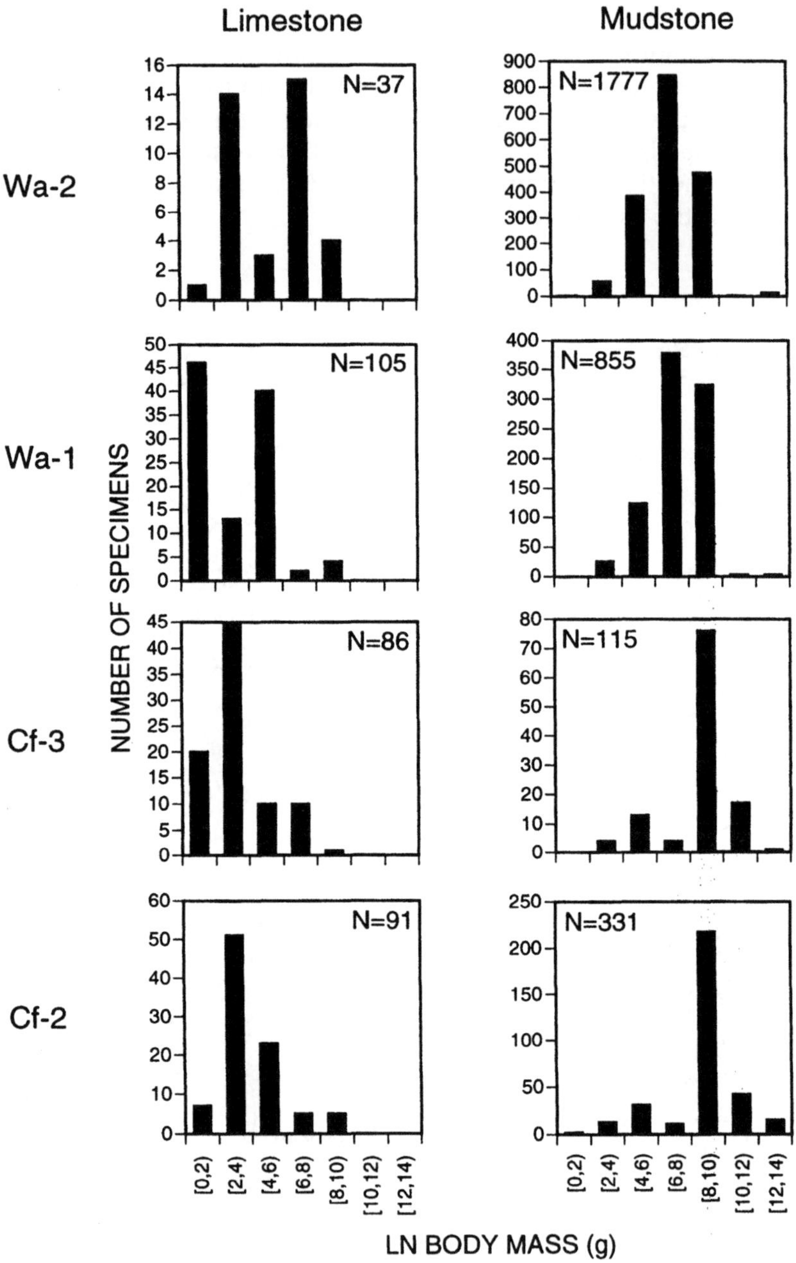

FIGURE 8. Bar graphs representing body size distributions of mammalian specimens through time in both the limestones and mudstones. The x-axis represents body size categories in log units and the y-axis represents number of specimens. The plots are stacked with the youngest faunal zones on top and the oldest on the bottom. Mudstones are plotted in the right column and limestones in the left.

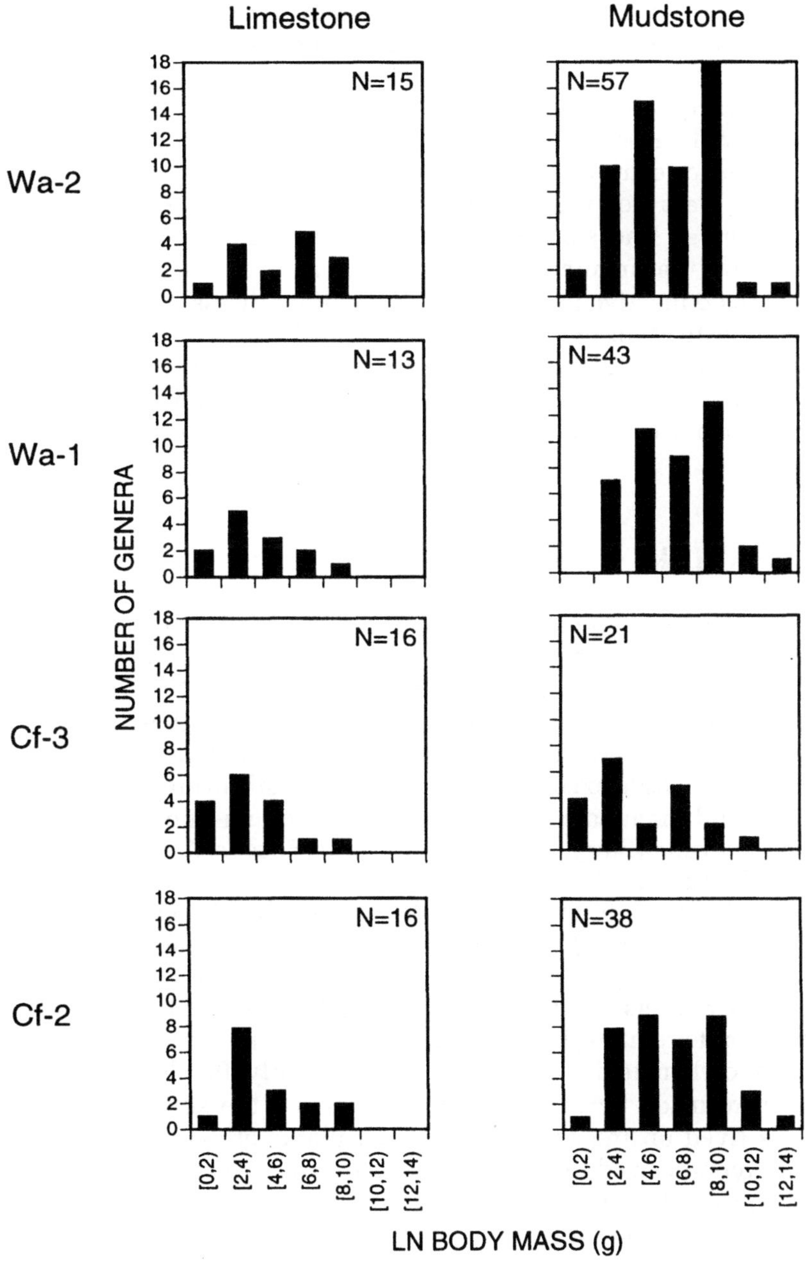

FIGURE 9. Histograms representing body size distributions of mammalian genera through time in both the limestones and mudstones. The x-axis represents body size categories in log units and the y-axis represents number of genera. The plots are stacked with the youngest faunal zones on top and the oldest on the bottom. Mudstones are plotted in the right column and limestones in the left. Note that the limestones seem to be sampling a constant number of taxa in each of the faunal zones, while diversity seems to increase in the mudstone sample.

Kraus (1987) postulated that the trunks of the trees were rapidly buried by levee or crevasse-splay deposits that would have killed the tree. As the tree died it shrank, allowing sand to filter down around the stump. Calcium carbonate precipitated, induced by microbial decay, out of the ground water with sand and silt also incorporated into the developing nodule. Organisms such as terrestrial snails lived among the decaying wood and, upon their deaths, were deposited in the forming nodule. Calcite recrystalization then destroyed all traces of cellular structure in the nodules. Gingerich (1987) noted one of the mammalian specimens showed signs of burial while enveloped in connective tissue (reflection without disassociation of mandibles) suggesting mastication by predators at or shortly after death, a notion further corroborated by the presence of bite marks on some of the specimens. Gingerich speculated that a predator or scavenger's habitual feeding site located above an open tree trunk would explain the concentration of vertebrate remains.

4.2. Shallow Lake or Pond

Freshwater limestones thought to represent shallow lakes or ponds have been documented in some floodplain environments (Jacob, 1973; Beaumont, 1979; Demico *et al.*, 1987; Emry, 1990; Stucky *et al.*, 1990). Aquatic invertebrates and algal skeletal remains are common in most of these freshwater limestones and mammals are only rarely preserved. Discoveries of fossil mammals in this context include exceptionally complete fossils of amphibians, small reptiles, birds, and mammals in limestones from the middle Eocene Sheep Pass Formation, Nevada (Emry, 1990) and the early Eocene Wind River Formation, Wyoming (Stucky *et al.*, 1990).

Gingerich (1987) described a tabular or sheet-like limestone from the Willwood Formation ("Bird Quarry limestone"). Gingerich reasoned that the shape of the lens, preservation of distinct burrows, presence of numerous anastomosing plant rootlets, rarity of gastropods (all terrestrial), and evident diversity of vertebrate remains suggested a pedogenic origin involving precipitation of calcium carbonate in a shallow intermittently flooded depression. The burrows were made by organisms that inhabited the soil, and the terrestrial gastropods lived on the soil surface or on plants growing in the depression. Gingerich hypothesized that vertebrates may have been attracted to the depression as a source of water and, through interaction with mammalian predators, deposited in the developing limestone.

4.3. Vertebrate Burrow

Cylindrical structures in Cenozoic floodplain sediments have often been interpreted as representing vertebrate burrows (Barbour, 1892; Wood and Wood, 1933; Lugn, 1941; Voorhies, 1974, 1975; Martin and Bennett, 1977;

Bown, 1982; Bown and Kraus, 1983; Hunt *et al.*, 1983; Breithaupt and Duvall, 1986; C. Sundell, 1996, 1997, 1998; K. Sundell, 1997, 1998). Some of these structures preserve concentrations of vertebrates (Behrensmeyer and Hook, 1992). These include snakes (Breithaupt and Duvall, 1986; C. Sundell, 1997), mammalian carnivores (Hunt *et al.*, 1983; K. Sundell, 1997), rodents (Voorhies, 1974,1975; C. Sundell, 1996, 1997), insectivores (C. Sundell, 1997), and artiodactyls (K. Sundell, 1997; C. Sundell, 1998).

Gingerich (1987), in his description of a Willwood limestone that contained a skull of *Carpolestes*, hypothesized that "a flood plain hole or depression 0.5 m in diameter and 0.7 m or more in depth probably originated either as a large tree rotted away or possibly as a vertebrate burrow was flooded by a rising water table" (Gingerich, 1987:p.281). Kraus (1988) argued that the Willwood limestones do not represent burrows because they are too large (Bromley *et al.*, 1975), differ morphologically from described burrows of rodents and insectivores that are highly branched and horizontal to subvertical (Voorhies, 1975) or coiled (Voorhies, 1975; Bown, 1982), and differ lithologically from vertebrate burrows, which are usually filled with clastic material and not limestone.

4.4. Spring

Fossil vertebrate concentrations are sometimes found in lithologies representing spring deposits (Hibbard and Riggs, 1949; Colbert, 1968; Hibbard, 1970; Saunders, 1977, 1988; Haynes, 1985; Archer, 1987). Most fossiliferous spring deposits, with the exception of a Miocene occurrence in Australia (Archer, 1987) and a Jurassic occurrence in Wyoming (Colbert, 1968) are found in rocks of Pliocene and Pleistocene age. Non-hydrothermal spring deposits have not yet been documented from the terrestrial Paleogene of North America.

Spring deposits are lenticular bodies composed of two dominant microfacies, a cylindrical tube (corresponding to the spring feeder or vent) and a flat lying lens (corresponding to the pond above the vent; Haynes, 1985). Faunal elements preserved in artesian spring deposits include small mammals, birds, reptiles, and amphibians with aquatic taxa rare or absent (Hibbard and Riggs, 1949; Archer, 1987) and large mammals on the periphery of the deposit that became mired in the wet substrate surrounding the spring (Saunders, 1988).

Limestones from the Clarks Fork Basin are similar to described spring deposits in having both cylindrical tube and flat-lying morphologies and in being fossiliferous. They are different, however, in that the vent feeder of Pliocene and Pleistocene springs is generally filled with sand or gravel from an immediately underlying unit that is acting as an aquifer. The vent is typically filled with microfossils that have been abraded as a result of turbulence as the water rises into the overlying pond. No such vent or abrasion of fossils has been found in the Clarks Fork Basin limestones.

4.5. Discussion

Based on previous work (Gingerich, 1987; Kraus, 1985, 1987, 1988) it is likely that at least some of the limestones represent deposition in the hollowed-out trunks of trees. However, current work on limestone lithology and taphonomy (Bowen and Bloch, 1999) indicates that many of the limestones represent precipitation of low Mg calcite from low-energy, ponded water at the surface of hydromorphic paleosols. Fossil assemblages contained within freshwater limestones likely represent fauna derived from unique, localized microenvironments on the distal floodplain (Figure 10). Accumulation of fossil material occurred episodically, and rapid burial of bones within calcium carbonate mud may have encouraged preservation of delicate specimens. The deposits, however, were subject to pedogenic reworking involving extensive bioturbation, permineralization of plant materials and organic matter decay. Physical and chemical processes occurring within the soil zone produced pedogenic grainstones, packstones and wackestones and coincidentally contributed to the disarticulation of fossil specimens (Figure 11). Deposition in vertebrate burrows or in artesian springs cannot be completely ruled out, but there is little evidence to support either of these depositional models as the origin for the Clarks Fork Basin freshwater limestones (as outlined above).

5. Contribution of Limestone Fauna

5.1. Latest Paleocene Thermal Maximum

The Paleocene-Eocene transition is a period of major climatic change with at least a 4°C increase in temperature over less than ten thousand years at high latitudes and in the deep ocean (Kennett and Stott, 1991; Zachos *et al.*, 1993). Evidence for climatic warming includes a shift in $\delta^{18}O$ (–2‰ to –3‰ excursion) in the tests of benthic and planktonic foraminifera (Kennett and Stott, 1991; Thomas and Shackleton, 1996; Schmitz *et al.*, 1996); an extinction of benthic foraminifera (Thomas and Shackleton, 1996) and deep-sea ostracodes (Steineck and Thomas, 1996); an increase in clay-minerals indicative of subtropical humid climates (kaolinite and smectite) in Antarctica (Robert and Kennett, 1994); an increase in percent of entire-margined plant species (Hickey, 1980; Wing *et al.*, 1991; 1999) and an increase in plant diversity (Bown *et al.*, 1994; Wing *et al.*, 1999). The Paleocene-Eocene transition is also marked by a major excursion in $\delta^{13}C$ in ocean sediments (Kennett and Stott, 1991; Thomas and Shackleton, 1996; Stott *et al.*, 1996; Schmidtz *et al.*, 1996), in the terrestrial record in mammalian tooth enamel and pedogenic carbonates (Koch *et al.*, 1992, 1995), and in terrestrial organic carbon (Stott *et al.*, 1996). Recent work suggests that the carbon-isotope excursion at the Paleocene-Eocene boundary (Late Paleocene Thermal Maximum; LPTM) may have been, in part, the result of a release of a large amount of methane from marine sediments (Dickens *et al.*, 1997).

5.2. Sampling Across the LPTM

The Paleocene-Eocene transition marks one of the most dramatic periods in mammalian evolution, with the first appearances of several modern orders of mammals in North America (Gingerich, 1989) and related changes in mammalian communities (Clyde and Gingerich, 1998; Gunnell, 1998). An understanding of depositional processes and careful sampling of the fossil record are essential for reconstructing changing paleoenvironments in this important interval of mammalian evolution. Traditional methods of surface collecting, typically in mudstones representing ancient soils deposited in floodplain environments, are strongly biased taxonomically and preservationally. This bias is due to the combined effects of individual collector biases, destruction of fragile fossils during erosion, and taphonomic filters inherent in preservation within a single depositional environment represented by a restricted set of lithologies (Bown and Kraus, 1981; Badgley *et al.*, 1995; Badgley and Behrensmeyer, 1995). Screen washing and quarrying of areas with high concentrations of fossils help remedy the collecting bias (Winkler, 1980, 1981, 1983), and, in some quarries of atypical lithology, provide a glimpse into previously unsampled parts of the fauna (Bartels, 1987; Rose and Bown, 1996; Silcox and Rose, this volume).

Past studies of faunal composition and mammalian diversity (Rose, 1981; Gingerich, 1989; Wing *et al.*, 1995; Maas *et al.*, 1995) as well as analyses of faunal turnover, community structure and mechanisms/timing of extinction for the late Paleocene and early Eocene records of the Bighorn Basin (Krause, 1986; Badgley and Gingerich, 1988; Maas *et al.*, 1988; Badgley, 1990; Bown and Kraus, 1993; Maas and Krause, 1994; Badgley and Behrensmeyer, 1995; Gunnell *et al.*, 1995; Morgan *et al.*, 1995; Wing *et al.*, 1995; Clyde and Gingerich, 1998) are based on collections that are biased against the preservation of small and very small mammals. All of these studies, with the exception of Clyde and Gingerich (1998), use combined samples from several taphonomic settings that are not distributed uniformly through the section. Clyde and Gingerich (1998) analyzed fossil material collected from a single taphonomic mode (surface-collected overbank assemblages; isotaphonomic approach of Behrensmeyer *et al.*, 1992) to remove potential biases related to sampling in different facies or different collection procedures. Using this same isotaphonomic approach, careful sampling of the limestone fauna across the Paleocene-Eocene transition and analysis of combined mudstone and limestone faunal assemblages, is helping to remedy this small mammal bias and is allowing for more inclusive paleoecological analyses across the Paleocene-Eocene boundary.

5.3. New Taxa

Discovery of new taxa allows for a more complete picture of community structure and can affect our understanding of the phylogenetic history of life. In particular, discovery of new taxa can strengthen prior phylogenetic patterns,

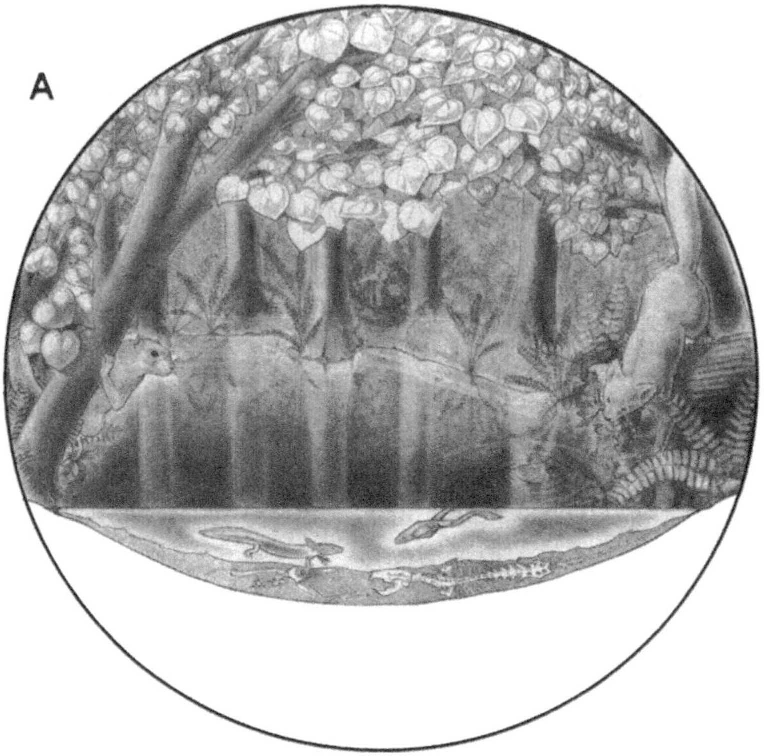

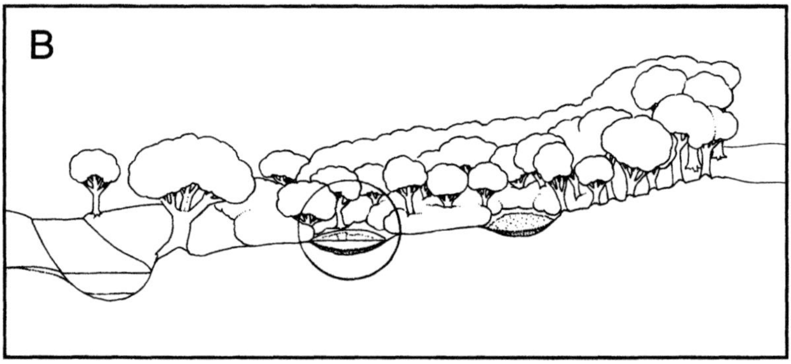

FIGURE 10A. Reconstruction of limestone depositional environment. Many of the limestones probably represent accumulations in a low energy pond. **10B.** Sedimentological model showing position of limestone forming pond (circled area) relative to the channel. Fossil assemblages contained within freshwater limestones likely represent fauna derived from unique, localized microenvironments on the distal floodplain.

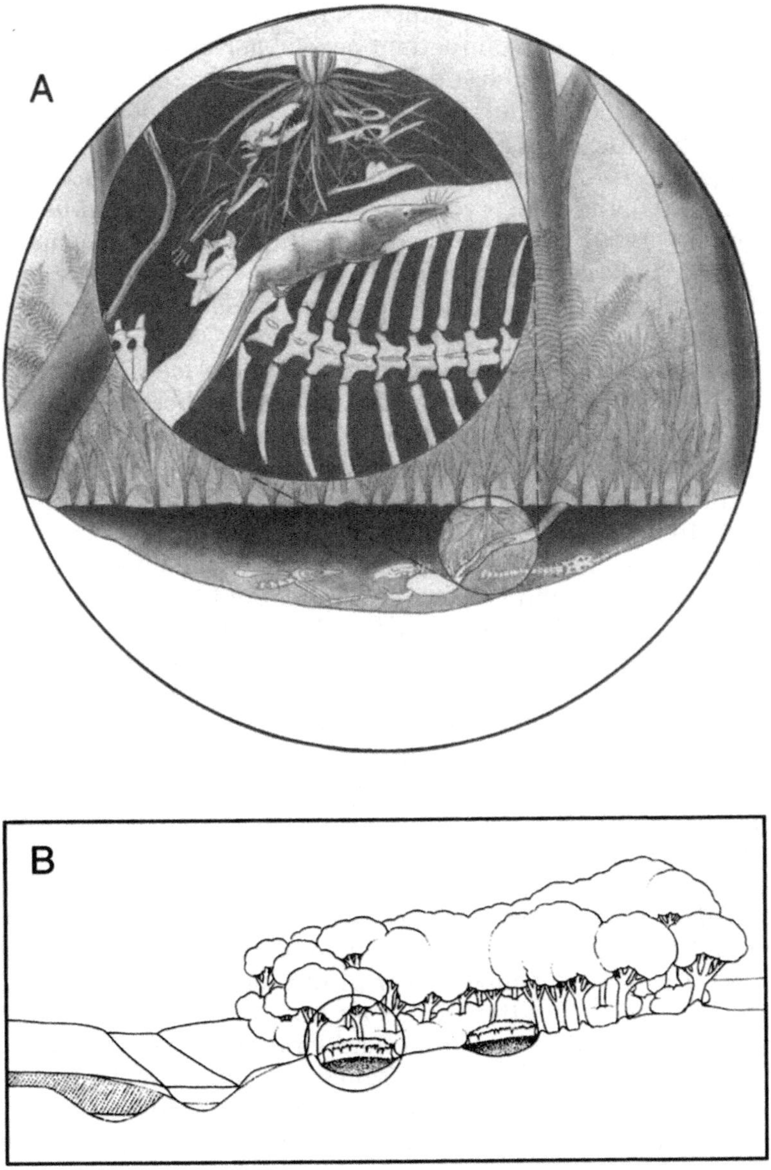

FIGURE 11A. Reconstruction of bioturbation of buried calcium carbonate and entombed vertebrate skeletons by plant roots, invertebrates, and vertebrate burrowing. **11B.** Sedimentological model showing position of limestone deposit (circled area) relative to the, now more proximal (Figure 10B), channel.

overthrow prior tree topologies, widen or close previously recognized stratigraphic gaps implied by phylogeny, and in some cases reveal new gaps (Weishampel, 1996). Increased diversity can also affect our understanding of biogeography and faunal correlation if new forms are found in more than one geographic area.

Despite an intensive collecting history, fossils from Clarks Fork Basin limestones are still yielding many important new taxa. New mammals are all small in size and include the ceutholestid insectivore *Ceutholestes dolosus* (Rose and Gingerich, 1987), the nyctitheriid insectivores *Leptacodon rosei* and *Limaconyssus habrus* (Gingerich, 1987), the adapisoriculid? insectivore *Wyonicteris chalix* (Gingerich, 1987), the geolabidid insectivore *Batodonoides vanhouteni* (Bloch *et al.*, 1998), the palaeoryctid insectivore *Eoryctes melanus* (Thewissen and Gingerich, 1989), the carnivore *Viverravus bowni* (Gingerich, 1987; although Polly, 1997 synonymized this species with *Viverravus laytoni*), the micromomyid plesiadapiform *Chalicomomys antelucanus* (Beard and Houde, 1989; although Rose and Bown, 1996 synonymized *Chalicomomys* with *Micromomys*), and the carpolestid plesiadapiform *Carpolestes simpsoni* (Bloch and Gingerich, 1998). New species of birds include the heron *Calcardea junnei* (Gingerich, 1987), the paleognathous birds *Lithornis promiscuus* and *Lithornis plebius* (Houde, 1986, 1988), and the coly-like birds *Sandcoleus copiosus*, *Anneavis anneae*, and *Chascacocolius oscitans* (Houde, 1987; Houde and Olsen, 1989, 1992).

As noted above, many of the new species from the limestones are of very small mammals. For example, the geolabidid *Batodonoides vanhouteni* has molar teeth indicating a body size smaller than that of any mammal known to date (Bloch *et al.*, 1998). *Batodonoides vanhouteni* is the oldest known species assigned to the genus, which was known previously from the middle Eocene of California (Novacek, 1976; Kelly *et al.*, 1991; Walsh, 1991, 1996) and thus represents a significant range extension. A fossil of *B. vanhouteni* was also found in a quarry deposit in the southern Bighorn Basin (Bloch *et al.*, 1998; Silcox and Rose, this volume) and thus is not restricted to the limestone faunas. The fact that the genus (the teeth of which are less than a millimeter in length) is only known from two localities outside of southern California is almost certainly an artifact of sampling. Discovery of *Batodonoides* in the Wasatchian of Wyoming and the early-to-late Uintan of southern California may be further evidence for a westward faunal dispersal during the early-to-middle Eocene (see Lillegraven, 1979), but this idea is far from proven.

Many new mammalian taxa from Clarkforkian and Wasatchian limestones remain undescribed, including insectivores, marsupials and plesiadapiforms. The final analysis of these taxa will increase the diversity of small mammal taxa, allow for a more complete picture of faunal change and community dynamics over the Paleocene-Eocene boundary, allow for more precise evaluation of timing of immigration events, and increase our ability to correlate interbasin and intercontinental faunas.

5.4. Exceptional Preservation

Despite the diversity of small mammals known from the Willwood Formation, most of the species are known only from isolated teeth and jaw fragments. Upper and lower dentitions, and cranial and postcranial remains are rarely associated (Bown and Beard, 1990; Winkler, 1983). New morphology for previously established mammal groups will allow for a re-evaluation of phylogenetic relationships for many of the small mammal groups of the late Paleocene and early Eocene based on more complete morphological data sets. Functional aspects of the organisms, such as locomotor behavior, will be easier to assess with increased knowledge of postcranial morphology.

Freshwater limestones from the Clarks Fork Basin are unique in having exceptional preservation and association of skeletal material that allows for study of previously undocumented morphology for many forms of Paleocene and Eocene mammal. These include skulls and/or partial skeletons of: proprimates *Tinimomys*, *Phenacolemur*, and *Ignacius* (Beard, 1989a, 1989b, 1990; Kay *et al.*, 1992), and *Plesiadapis* (Gunnell and Gingerich, 1987; Gingerich and Gunnell, 1992); the carnivores *Uintacyon*, *Didymictis*, and *Viverravus* (Bloch, unpublished); the dermopteran? *Plagiomene* (MacPhee *et al.*, 1989); the palaeoryctid insectivore *Eoryctes* (Thewissen and Gingerich, 1989); the marsupial *Peradectes* (Bloch, unpublished); the perissodactyl *Hyracotherium* (Bloch, unpublished), the notoungulate? *Arctostylops* (Bloch, 1999); a multituberculate, several rodents, and a diversity of insectivores including several nyctitheriids (Bloch, unpublished).

5.5. Uniform Sampling of Limestones: Filling in the Small Mammal Gap or Sampling from Different "Communities"?

Preservation of small, delicate, vertebrate fossils in freshwater limestones is explained by the hard nature of the limestone matrix, which makes it resistant to the destructive forces of weathering, chemical buffering, and rapid burial in a low energy environment. Any of the biases introduced to the paleosol sample through weathering and surface prospecting are eliminated by the careful dissolving of limestone matrix and collecting even the most microscopic and delicate fossils. The small mammals, often destroyed or overlooked, are retrieved using this method.

This does not explain, however, the diversity of new taxa and lack of certain key taxa in the limestone faunal samples. If the limestones were merely sampling certain body size ranges of the surrounding fauna, this would not explain why condylarths and primates, common elements of the surrounding mudstone fauna (in body size ranges represented in the limestone) are almost completely absent from the assemblage. Furthermore, it does not explain why there are so many new taxa of body sizes which, although rare, are present in the surrounding mudstone. One possibility is that the limestones are sampling a different, contemporaneous, "community." Concentration of small mammal remains in the limestones can be explained, at least in part, as predation accumulations by carnivorous mammals and predatory birds (Andrews, 1990; Fisher, 1981). As documented by Mellett (1974), carnivores can catch their prey in one region and carry them in their digestive tracks a considerable distance before leaving their feces and regurgitates in another depositional environment. A detailed taphonomic study of the limestone assemblages (Bloch and Boyer, unpublished) should provide the data needed to address these questions.

6. Summary

Fossiliferous freshwater limestones are found in weakly developed paleosols of the late Paleocene and early Eocene Willwood Formation in the Clarks Fork Basin of Northwestern Wyoming. The limestones, or carbonate mudstones, are distinctly different from surrounding mudstone and sandstone lithologies in having lenticular and columnar shapes; being nearly wholly composed of calcium carbonate; and preserving a large number of root cells, invertebrate burrows, pulmonate gastropods, and microvertebrates.

The limestone fauna is different from that preserved in the mudstone in having an abundance of proprimates and lipotyphlan insectivores as opposed to a clear dominance of condylarths in the latter. The limestones yield a much better sample of small mammals than the mudstones and, while the mudstones document an increase in diversity of mammalian taxa over the Paleocene-Eocene transition, the limestones document that the small mammal component of the fauna seems to stay about the same.

The most likely conditions for the deposition of the Clarks Fork Basin limestones is in the hollowed out trunks of trees or in shallow pools. However, deposition in vertebrate burrows or in artesian springs cannot be ruled out.

The Paleocene-Eocene transition is a period of major climatic change and, subsequently, is one of the most dramatic periods in mammalian evolution. Discovery of new taxa and exceptionally complete fossil material will allow for more rigorous analyses of mammalian diversity and community structure through this time interval. It also has the possibility of influencing our understanding of the phylogenetic history of some mammal groups and almost certainly will effect our understanding of biogeography and faunal correlation in the Paleogene.

It is possible that, through mammalian and/or avian predators as well as a more distal position on the floodplain, the microfauna preserved in the limestones represents a sample from one or more distal habitats not normally sampled by the mudstone lithologies. A better understanding of the these distal habitats and their faunas could be very useful in understanding larger faunal dynamics and evolution across the Paleocene-Eocene transition.

ACKNOWLEDGMENTS. We thank D. L. Fox, G. F. Gunnell, P. Wilf, and an anonymous reviewer for improvements to the manuscript. We also thank Drs. P. D. Gingerich, B. H. Wilkinson, C. W. Badgley, D. C. Fisher, K. D. Rose, P. Houde, M. D. Uhen, W. C. Clyde, and W. J. Sanders for helpful discussions. D. Erickson made the illustration in Figure 2 and D. Boyer made illustrations in Figures 10–11. D. Boyer, J. Stopar, P. Rose, and J. Anderson helped dissolve limestone. We especially thank G. F. Gunnell for inviting us to participate in this symposium. Field and laboratory research was supported by grants from the National Science Foundation to Philip D. Gingerich (EAR-8918023) and from the Scott Turner Fund, Department of Geological Sciences, University of Michigan, to J. I. B.

References

Andrews, P., 1990, *Owls, Caves and Fossils*, British Museum (Natural History), London.

Archer, M., 1987, Riversleigh scene 1: crossing the threshold of eternity, in: *The Antipodean Ark* (S. Hand, and M. Archer, eds.), Angus and Robertson, London, pp. 74–76.

Badgley, C., 1990, A statistical assessment of last appearances in the Eocene record of mammals, in: *Dawn of the Age of Mammals in the northern part of the Rocky Mountain Interior* (T. M. Bown, and K. D. Rose, eds.), *Geological Society of America Special Paper* **243**:153–167.

Badgley, C., and Gingerich, P. D., 1988, Sampling and faunal turnover in early Eocene mammals, *Palaeogeography, Palaeoclimatology, Palaeoecology* **63**:141–157.

Badgley, C., Bartels, W. S., Morgan, M. E., Behrensmeyer, A. K., and Raza, S. M., 1995, Taphonomy of vertebrate assemblages from the Paleogene of northwestern Wyoming and the Neogene of northern Pakistan, *Palaeogeography, Palaeoclimatology, Palaeoecology* **115**: 157–180.

Badgley, C., and Behrensmeyer, A. K., 1995, Preservational, paleoecological and evolutionary patterns in the Paleogene of Wyoming-Montana and the Neogene of Pakistan, *Palaeogeography, Palaeoclimatology, Palaeoecology* **115**:319–340.

Barbour, E. H., 1892, Nature, Structure, and phylogeny of *Daimonelix*, *Geological Society of America Bulletin* **8**:305–314.

Bartels, W. S., 1987, Fossil reptile assemblages and depositional environments of selected early Tertiary vertebrate bone concentrations, Bighorn Basin, Wyoming, Ph.D. Dissertation, The University of Michigan (unpublished).

Bao, H., Koch, P. L., and Hepple, R. P., 1998, Hematite and calcite coatings on fossil vertebrates, *Journal of Sedimentary Research* **68(5)**:727–738.

Beard, K. C., 1989a, Postcranial anatomy, locomotor adaptations, and paleoecology of Early Cenozoic Plesiadapidae, Paromomyidae, and Micromomyidae (Eutheria, Dermoptera), Ph.D. Dissertation, The Johns Hopkins University (unpublished).

Beard, C. K., 1989b, Evidence for gliding behavior in the early Eocene dermopteran *Ignacius graybullianus*, *Journal of Vertebrate Paleontology* **9(3)**:13A.

Beard, C. K., 1990, Gliding behavior and palaeoecology of the alleged primate family Paromomyidae (Mammalia, Dermoptera), *Nature* **345**:340–341.

Beard, C. K., and Houde, P., 1989, An unusual assemblage of diminutive plesiadapiformes (Mammalia, ?Primates) from the early Eocene of the Clarks Fork Basin, Wyoming, *Journal of Vertebrate Paleontology* **9(4)**:388–399.

Beaumont, E. A., 1979, Depositional environments of Fort Union sediments (Tertiary, northwest Colorado) and their relation to coal, *Am. Assoc. Petroleum Geologists Bull.* **63**:194–217.

Beeze-Deak, J., Langohr, R., and Verrecchia, E. P., 1997, Small scale secondary $CaCO_3$ accumulations in selected sections of the European loess belt. Morphological forms and potential for paleoenvironmental reconstruction, *Geoderma* **76**:221–252.

Behrensmeyer, A. K., Damuth, J. D., Dimichele, W. A., Potts, R., Sues, H., and Wing, S. L., 1992, *Terrestrial Ecosystems Through Time*, The University of Chicago Press, Chicago.

Behrensmeyer, A. K., and Hook, R. W., 1992, Paleoenvironmental contexts and taphonomic modes, in: *Terrestrial Ecosystems Through Time* (A. K. Behrensmeyer, J. D. Damuth, W. A. DiMichele, R. Potts, H. Sues, and S. L. Wing, eds.), The University of Chicago Press, Chicago, pp. 15–93.

Berggren, W. A., Kent, D. V., Swisher III, C. C., and Aubry, M. P., 1995, A revised Cenozoic geochronology and chronostratigraphy, in: *Geochronology Time Scales and Global Stratigraphic Correlation, SEPM Special Publication*, Volume 54, pp. 129–212.

Bloch, J. I., 1995, A diminutive insectivore (Mammalia, Lipotyphla) from a Wasatchian early Eocene limestone of the Clarks Fork Basin, Wyoming, *Journal of Vertebrate Paleontology* **15(3)**:19A.

Bloch, J. I., 1997, New Eocene palaeoryctid and re-evaluation of Paleocene *Palaeoryctes* (Mammalia: Insectivora) from western North America, *Journal of Vertebrate Paleontology* **17(3)**:33A.

Bloch, J. I., 1998, Paleocene-Eocene microvertebrates in freshwater limestones of the Clarks Fork Basin, Wyoming, *Journal of Vertebrate Paleontology* **18(3)**:28A.

Bloch, J. I., 1999, Partial skeleton of *Arctostylops* from the Paleocene of Wyoming: arctostylopid-notoungulate relationship revisited, *Journal of Vertebrate Paleontology* **19(3)**:32A.

Bloch, J. I., and Gingerich, P. D., 1994, New species of *Carpolestes* (Mammalia, Proprimates) from Clarkforkian late Paleocene limestones of the Clarks Fork Basin, Wyoming: Teeth, Skulls, and Femur, *Journal of Vertebrate Paleontology* **14(3)**:17A–18A.

Bloch, J. I., and Gingerich, P. D., 1998, New species of *Carpolestes* (Mammalia, Proprimates) from the late Paleocene of northwestern Wyoming, *Contributions from the Museum of Paleontology, University of Michigan* **30(4)**:131–162.

Bloch, J. I., Rose, K. D., and Gingerich, P. D., 1998, New species of *Batodonoides* (Lipotyphla, Geolabididae) from the early Eocene of Wyoming: smallest known mammal?, *Journal of Mammalogy* **79(3)**:804–827.

Bowen, G. J., and Bloch, J. I., 1999, Petrology and geochemistry of early Cenozoic fossiliferous floodplain limestones from the Bighorn Basin, Wyoming: Implications for deposition and taphonomy, *GSA Abstracts with Programs* **31(7)**:A-458.

Bown, T. M., 1979, Geology and mammalian paleontology of the Sand Creek Facies, Lower Willwood Formation (lower Eocene), Washakie County, Wyoming, *The Geological Survey of Wyoming* Memoirs **2**:1–151.

Bown, T. M., 1982, Ichnofossils and rhizoliths of the nearshore fluvial Jebel Qatrani Formation (Oligocene), Fayum Province, Egypt, *Palaeogeography, Palaeoclimatology, Palaeoecology* **40**:255–309.

Bown, T. M., 1985, Maturation sequences in the lower Eocene alluvial paleosols, Willwood Formation, in: *Field Guidebook to Modern and ancient Fluvial Systems in the United States: Fort Collins, Third International Fluvial Sedimentology Conference* (R. M. Flores, and M. Harvey, eds.), pp. 20–26.

Bown, T. M., and Beard, K. C., 1990, Systematic lateral variation in the distribution of fossil mammals in alluvial paleosols, lower Eocene Willwood Formation, Wyoming, in: *Dawn of the Age of Mammals in the northern part of the Rocky Mountain Interior* (T. M. Bown, and K. D. Rose, eds.), *Geological Society of America Special Paper* **243**:135–151.

Bown, T. M., and Kraus, M. J., 1981, Vertebrate fossil-bearing paleosol units (Willwood Formation, lower Eocene, northwest Wyoming, U.S.A.): Implications for taphonomy, biostratigraphy, and assemblage analysis, *Palaeogeography, Palaeoclimatology, Palaeoecology* **34**:31–56.

Bown, T. M., and Kraus, M. J., 1983, Ichnofossils of the alluvial Willwood Formation (Lower Eocene), Bighorn Basin, Northwest Wyoming, U. S. A., *Palaeogeography, Palaeoclimatology, Palaeoecology* **43**:95–128.

Bown, T. M., and Kraus, M. J., 1987, Integration of channel and floodplain suites, 1. Developmental sequence and lateral relations of alluvial paleosols, *Journal of Sedimentary Petrology* **57**:587–601.

Bown, T. M., and Kraus, M. J., 1993, Time-stratigraphic reconstruction and integration of paleopedologic, sedimentologic, and biotic events (Willwood Formation, lower Eocene, northwest Wyoming, U. S. A.), *PALAIOS* **8**:68–80.

Bown, T. M., Rose, K. D., Simons, E. L., and Wing, S. L., 1994, Distribution and stratigraphic correlation of upper Paleocene and lower Eocene fossil mammal and plant localities of the Fort Union, Willwood, and Tatman Formations, Southern Bighorn Basin, Wyoming, *U. S. G. S. Special Paper* **1540**:1–103.

Breithhaupt, B. H., and Duvall, D., 1986, The oldest record of serpent aggregation, *Lethaia* **19**:181–185.

Bromley, R. G., Curran, H. A., Frey, R. W., Gutschick, R. C., and Suttner, L. J., 1975, Problems in interpreting unusually large burrows, in: *The Study of Trace Fossils* (R. W. Frey, ed.), Springer-Verlag, New York, pp. 351–376.

Butler, R. F., Gingerich, P. D., and Lindsay, E. H., 1981, Magnetic polarity stratigraphy and biostratigraphy of Paleocene and lower Eocene continental deposits, Clarks Fork Basin, Wyoming, *Journal of Geology* **89**:299–316.

Carroll, R. E., 1967, Labyrinthodonts from the Joggins Formation, *Journal of Paleontology* **41**:111–142.

Carroll, R. E., Belt, E. S., Dineley, D. L., Baird, D., and McGregor, D. C., 1972, Vertebrate Paleontology of eastern Canada, *Proceedings Twenty-Fourth International Geological Congress, Field Guide for Excursion* **A59**:1–113.

Clyde, W. C., and Gingerich, P. D., 1998, Mammalian community response to the latest Paleocene thermal maximum: An isotaphonomic study in the northern Bighorn Basin, Wyoming, *Geology* **26(11)**:1011–1014.

Clyde, W. C., Stamatakos J., and Gingerich, P. D., 1994, Chronology of the Wasatchian Land-Mammal Age (early Eocene): magnetostratigraphic results from the McCullough Peaks Section, Northern Bighorn Basin, Wyoming, *Journal of Geology* **102**:367–377.

Colbert, E. H., 1968, *Men and Dinosaurs: the search in field and laboratory*, E. P. Dutton, New York.

Dawson, J. W., 1882, On the results of recent explorations of erect trees containing animal remains in the coal formation of Nova Scotia, *Philosophical Transactions of the Royal Society, London*, **173**:621–659.

Dickens, G. R., Castillo, M. M., and Walker, J. C. G., 1997, A blast of gas in the latest Paleocene: Simulating first-order effects of massive dissociation of oceanic methane hydrate, *Journal of Geology* **25**:259–262.

Demicco, R. V., Bridge, J. S., and Cloyd, K. C., 1987, A unique freshwater carbonate from the upper Devonian Catskill Magnafacies of New York State, *Journal of Sedimentary Petrology* **57**:327–334.

Emry, R. J., 1990, Mammals of the Bridgerian (middle Eocene) Elderberry Canyon Local Fauna of eastern Nevada, in: *Dawn of the Age of Mammals in the northern part of the Rocky Mountain Interior* (T. M. Bown, and K. D. Rose, eds.), *Geological Society of America Special Paper* **243**:187–210.

Fisher, D. C., 1981, Taphonomic interpretation of enamel-less teeth in the Shotgun Local Fauna (Paleocene, Wyoming), *Contributions from the Museum of Paleontology, University of Michigan* **25**:259–275.

Gingerich, P. D., 1980, History of early Cenozoic vertebrate paleontology in the Bighorn Basin, in: *Early Cenozoic Paleontology and Stratigraphy of the Bighorn Basin* (P. D. Gingerich, ed.), *University of Michigan Papers on Paleontology* **24**:7–24.

Gingerich, P. D., 1987, Early Eocene bats (Mammalia, Chiroptera) and other vertebrates in freshwater limestones of the Willwood Formation, Clark's Fork Basin, Wyoming, *Contributions from the Museum of Paleontology, The University of Michigan* **27(11)**:275–320.

Gingerich, P. D., 1989, New earliest Wasatchian mammalian fauna from the Eocene of northwestern Wyoming: composition and diversity in a rarely sampled high-floodplain assemblage, *University of Michigan Papers on Paleontology* **28**:1–97.

Gingerich, P. D., and Gunnell, G. F., 1992, A new skeleton of *Plesiadapis cookei, A Quarterly Newsletter of the Exhibit Museum, The University of Michigan* **6(1)**:1–3.

Gunnell, G. F., 1989, Evolutionary History of Microsyopoidea (Mammalia, ?Primates) and the relationship between plesiadapiformes and primates, *University of Michigan, Papers on Paleontology* **27**:1–157.

Gunnell, G. F., 1998, Mammalian faunal composition and the Paleocene/Eocene Epoch/Series boundary: Evidence from the Northern Bighorn Basin, Wyoming, in: *Late Paleocene-Early Eocene Climatic and Biotic Events in the Marine and Terrestrial Records*, (M. P. Aubry, S. Lucas, and W. A. Berggren, eds.), Columbia University Press, New York, pp. 409–427.

Gunnell, G. F., and Gingerich, P. D., 1987, Skull and partial skeleton of *Plesiadapis cookei* from the Clark's Fork Basin, Wyoming, *American Journal of Physical Anthropology* **72(2)**:206A.

Gunnell, G. F., Morgan, M. E., Maas, M. C., and Gingerich, P. D., 1995, Comparative paleoecology of Paleogene and Neogene mammalian faunas: trophic structure and composition, *Palaeogeography, Palaeoclimatology, Palaeoecology* **115**:265–286.

Hand, S., Novacek, M., Godthelp, H., and Archer, M., 1994, First Eocene bat from Australia, *Journal of Vertebrate Paleontology* **14(3)**:375–381.

Haynes, V. C. J., 1985, Mastodon-bearing springs and late quaternary geochronology of the lower Pomme de Terre Valley, *The Geological Society of America, Special Paper* **204**:1–35.

Hibbard, C. W., and Riggs, E. S., 1949, Upper Pliocene vertebrates from Keefe Canyon, Meade County, Kansas, *Bulletin of the Geological Society of America* **60**:829–860.

Hibbard, C. W., 1970, Pleistocene mammalian faunas from the Great Plains and Central Lowland provinces of the United States, in: *Pleistocene and Recent Environments of the Central Great Plains*, (W. Dort, and J. K. Jones, eds.), Lawrence University Press, Kansas, pp. 395–433.

Hickey, L. J., 1980, Paleocene stratigraphy and flora of the Clarks Fork Basin, in: *Early Cenozoic Paleontology and Stratigraphy of the Bighorn Basin, Wyoming, 1880–1980* (P. D. Gingerich, ed.), *University of Michigan Papers on Paleontology* **24**:33–49.

Houde, P. W., 1986, Ostrich ancestors found in the northern hemisphere suggest new hypothesis of ratite origins, *Nature* **324**:563–565.

Houde, P. W., 1987, Histological evidence for the systematic position of *Hesperornis* (Odontornithes: Hesperornithiformes), *AUK* **104**:125–129.

Houde, P. W., 1988, Paleognathous birds from the early Tertiary of the northern hemisphere, *Publications of the Nuttall Ornithological Club* **22**:1–148.

Houde, P. W., and Olsen, S. L., 1989, Small arboreal nonpasserine birds from the early Tertiary of western North America, in: *Acta XIX Congressus Internationalis Ornithologici.*, (H. Ouellet, ed.), University of Ottawa Press, Ottawa, pp. 2030–2036.

Houde, P. W., and Olsen, S. L., 1992, A radiation of coly-like birds from the Eocene of North America (Aves: Sandcoleiformes new order), *Natural History Museum of Los Angeles County, Science Series* **36**:137–160.

Hunt, R. M. Jr., Xue, X. X., and Kaufman, J., 1983, Miocene burrows of extinct beardogs: Indication of early denning behavior of large mammalian carnivores, *Science* **221**:364–366.

Jacob, A. F., 1973, Depositional environments of Paleocene Tongue River Formation, western North Dakota, *Am. Assoc. Petroleum Geologists Bull.* **57**:1038–1052.

Kay, R. F., Thorington, R. W. J., and Houde, P., 1990, Eocene plesiadapiform shows affinities with flying lemurs not primates, *Nature* **345**:342–344.

Kay, R. F., Thewissen, J. G. M., and Yoder, A. D., 1992, Cranial anatomy of *Ignacius graybullianus* and the affinities of the Plesiadapiformes, *American Journal of Physical Anthropology* **89**:477–498.

Kelly, T. S., Lander, E. B., Whistler, D. P., Roeder, M. A., and Reynolds, R. E., 1991, Preliminary report on a paleontologic investigation of the lower and middle members, Sespe Formation, Simi Valley Landfill, Ventura County, California, *PaleoBios* **13**:1–13.

Kennett, J. P., and Stott, L. D., 1991, Abrupt deep sea warming, paleoceanographic changes and benthic extinctions at the end of the Paleocene, *Nature* **353**:319–322.

Klappa, C. F., 1978, Biolithogenesis of Microcodium: elucidation, *Sedimentology* **25**:489–522.

Klappa, C. F., 1979, Calcified filaments in quaternary calcretes: organo-mineral interactions in the subaerial vadose environment, *Journal of Sedimentary Petrology* **49(3)**:955–968.

Klappa, C. F., 1980, Rhizoliths in terrestrial carbonates: classification, recognition. genesis and significance, *Sedimentology* **27**:613–629.

Koch, P. L., Zachos, J. C., and Gingerich, P. D., 1992, Correlation between isotope records in marine and continental carbon reservoirs near the Paleocene/Eocene boundary, *Nature* **358**: 319–322.

Koch, P. L., Zachos, J. C., and Dettman, D. L., 1995, Stable isotope stratigraphy and paleoclimatology of the Paleogene Bighorn Basin (Wyoming, USA), *Palaeogeography, Palaeoclimatology, Palaeoecology* **115**:61–89.

Kraus, M. J., 1985, Sedimentology of early Tertiary rocks, northern Bighorn Basin, in: *Field Guide to Modern and Ancient Fluvial Systems in the United States: Third International Fluvial Sedimentology Conference Guidebook*, (R. M. Flores and M. Harvey eds.), pp. 26–33.

Kraus, M. J., 1987, Integration of channel and floodplain suites, II. Vertical relations of alluvial paleosols, *Journal of Sedimentary Petrology* **57(4)**:602–612.

Kraus, M. J., 1988, Nodular remains of early tertiary forests, Bighorn Basin, Wyoming, *Journal of Sedimentary Petrology* **58(5)**:888–893.

Kraus, M. J., 1997, Lower Eocene alluvial paleosols: Pedogenic development, stratigraphic relationships, and paleosol/landscape associations, *Palaeogeography, Palaeoclimatology, Palaeoecology* **129**:387–406.

Kraus, M. J., and Aslan, A., 1993. Eocene hydromorphic paleosols: significance for interpreting ancient floodplain processes, *Journal of Sedimentary Petrology* **63(3)**:453–463.

Krause, D. W., 1986, Competitive exclusion and taxonomic displacement in the fossil record: the case of rodents and multituberculates in North America, *Contributions to Geology, University of Wyoming, Special Paper* **3**:95–117.

Lillegraven, J. A., 1979, A biogeographical problem involving comparisons of later Eocene terrestrial vertebrate faunas of western North America, in: *Historical biogeography, plate tectonics, and the changing environment*, (J. Gray, and A. J. Boucot, eds.), Oregon State University Press, pp. 333–347.

Lugn, A. L., 1941, The origin of *Daimonelix*, *Journal of Geology* **49**:673–696.

Maas, M. C., and Krause, D. W., 1994, Mammalian turnover an community structure in the Paleocene of North America, *Historical Biology* **8**:91–128.

Maas, M. C., Krause, D. W., and Strait, S. G., 1988, The decline and extinction of Plesiadapiformes (Mammalia: ?Primates) in North America: displacement or replacement?, *Paleobiology* **14**:410–431.

Maas, M. C., Anthony, M. R. L., Gingerich, P. D., Gunnell, G. F., and Krause, D. W., 1995, Mammalian generic diversity and faunal turnover of the late Paleocene and early Eocene of the Bighorn and Crazy Mountain Basins, Wyoming and Montana, *Palaeogeography, Palaeoclimatology, Palaeoecology* **115**:181–207.

MacPhee, R. D. E., Beard, K. C., Flemming, C., and Houde, P., 1995, Petrosal Morphology of *Tinimomys graybulliensis* is Plesiadapoid, Not Microsyopoid, *Journal of Vertebrate Paleontology* **15**:67A.

MacPhee, R. D. E., Cartmill, M., and Rose, K. D., 1989, Craniodental morphology and relationships of the supposed Eocene Dermopteran *Plagiomene* (Mammalia), *Journal of Vertebrate Paleontology* **9(3)**:329–349.

Martin, L. D., and Bennett, D. K., 1977, The burrows of the Miocene beaver *Palaeocastor*, western Nebraska, U. S. A., *Palaeogeography, Palaeoclimatology, Palaeoecology* **22**:173–193.

Mellett, J. S., 1974, Scatological origin of microvertebrate fossil accumulations, *Science* **185**:349–350.

Morgan, M. E., Badgley, C., Gunnell, G. F., Gingerich, P. D., Kappelman, J. W., and Maas, M. C., 1995, Comparative paleoecology of Paleogene and Neogene mammalian faunas: body-size structure, *Palaeogeography, Palaeoclimatology, Palaeoecology* **115**:287–317.

Novacek, M. J., 1976, Insectivora and Proteutheria of the later Eocene (Uintan) of San Diego County, California, *Natural History Museum of Los Angeles County Contributions in Science* **283**:1–52.

Polly, P. D., 1997, Ancestry and species definition in paleontology: a stratocladistic analysis of Paleocene-Eocene Viverravidae (Mammalia, Carnivora) from Wyoming, *Contributions from the Museum of Paleontology, The University of Michigan* **30**:1–53.

Robert, C., and Kennett, J. P., 1994, Antarctic subtropical humid episode at the Paleocene-Eocene boundary: clay mineral evidence, *Journal of Geology* **22**:211–214.

Rose, K. D., 1981, The Clarkforkian Land-Mammal Age and mammalian faunal composition across the Paleocene-Eocene boundary, *University of Michigan, Papers on Paleontology* **26**:1–196.

Rose, K. D., Beard, K. C., and Houde, P., 1993, Exceptional new dentitions of the diminutive Plesiadapiformes *Tinimomys* and *Niptomomys* (Mammalia), with comments on the upper incisors of Plesiadapiformes, *Annals of the Carnegie Museum* **62**:351–361.

Rose, K. D., and Bown, T. M., 1996, A new plesiadapiform (Mammalia: Plesiadapiformes) from the early Eocene of the Bighorn Basin, Wyoming, *Annals of the Carnegie Museum* **65**:305–321.

Rose, K. D., and Gingerich, P. D., 1987, A new insectivore from the Clarkforkian (earliest Eocene) of Wyoming, *Journal of Mammalogy* **68(1)**:17–27.

Rust, B. R., Gibling, M. R., and Legun, A. S., 1984, Coal deposition in an anastomosing fluvial system: The Pennsylvanian Cumberland Group south of Joggins, Nova Scotia, Canada, *Special Publication of the International Association of Sedimentologists*, **7**:105–120.

Saunders, J. J., 1977, Late Pleistocene vertebrates of the western Ozark Highland, Missouri, *Illinois State Museum Reports of Investigations* **33**:1–118.

Saunders, J. J., 1988, Fossiliferous spring sites in southwestern Missouri, *Bulletin of the Buffalo Society of Natural Sciences* **33**:127–149.

Schmitz, B., Speijer, R., and Aubry, M. P., 1996, Latest Paleocene benthic extinction event on the southern Tethyan shelf (Egypt): Foraminiferal stable isotope record, *Journal of Geology* **24**:347–350.

Simmons, N. B., and Geisler, J. H., 1998, Phylogenetic relationships of *Icaronycteris, Archaeonycteris, Hassianycteris*, and *Palaeochiropteryx* to extant bat lineages, with comments on the evolution of echolocation and foraging strategies in Microchiroptera, *Bulletin of the American Museum of Natural History* **235**:1–182.

Steineck, P. L., and Thomas, E., 1996, The latest Paleocene crisis in the deep sea: Ostracode succession at maud Rise, Southern Ocean, *Journal of Geology* **24**:583–586.

Stott, L. D., Sinha, A., Thirty, M., Aubry, M. P., and Berggren, W. A., 1996, The transfer of ^{12}C changes from the ocean to the terrestrial biosphere across the Paleocene-Eocene boundary: criteria for terrestrial-marine correlations, in: *Correlations of the early Paleogene in Northwest Europe, Geological Society of London Special Publication* **101**:401–411.

Stucky, R. K., Krishtalka, L., and Redline, A. D., 1990, Geology, vertebrate fauna, and paleoecology of the Buck Spring Quarries (early Eocene, Wind River Formation) Wyoming, *Geological Society of America, Special Paper* **243**:169–186.

Sundell, C. A., 1996, White River Oligocene mammalian burrows and advanced tooth wear in rodents, *Journal of Vertebrate Paleontology* **16(3)**:68A.

Sundell, C. A., 1997, Orellan burrows and associated fauna from Converse Co, WY, *Journal of Vertebrate Paleontology* **17(3)**:80A.

Sundell, C. A., 1998, Orellan vertebrate burrows from Douglas, WY; their structure, inhabitants and paleoecological implications, *Journal of Vertebrate Paleontology* **18(3)**:81A.

Sundell, K. S., 1997, Oreodonts: large burrowing mammals of the Oligocene, *Journal of Vertebrate Paleontology* **17(3)**:80A.

Sundell, K. S., 1998, *Hyaenodon*: nemesis of burrowing oreodonts, *Journal of Vertebrate Paleontology* **17(3)**:80A.

Thewissen, J. G. M., and Gingerich, P. D., 1989, Skull and endocranial cast of *Eoryctes melanus*, a new palaeoryctid (Mammalia: Insectivora) from the early Eocene of western North America, *Journal of Vertebrate Paleontology* **9(4)**:459–470.

Thomas, E., and Shackleton, N. J., 1996, The Paleocene-Eocene benthic foraminiferal extinction and stable isotope anomalies, in: *Correlations of the Early Paleogene in Northwest Europe, Geological Society of London Special Publication* **101**:401–411.

Voorhies, M. R., 1974, Fossil pocket mouse burrows in Nebraska, *American Midland Naturalist* **91**:492–498.

Voorhies, M. R., 1975, Vertebrate burrows, in: *The Study of Trace Fossils*, (R. W. Frey ed.), Springer-Verlag, New York, pp. 325–350.

Walker, A. W., and Teaford, M., 1989, The hunt for *Proconsul, Scientific American* **260**:76–82.

Walsh, S. L., 1991, Eocene mammal faunas of San Diego County, in: *Eocene geologic history: San Diego Region, Pacific Section* (P. L. Abbott and J. A. May, eds.), *Society of Economic Paleontologists and Mineralogists* **68**:161–178.

Walsh, S. L., 1996, Middle Eocene mammal faunas of San Diego County, California, in: *The Terrestrial Eocene-Oligocene Transition in North America*, (D. R. Prothero and R. J. Emry, eds.), Cambridge University Press, Cambridge, pp. 76–109.

Walton, J., 1935, Scottish Lower Carboniferous plants: The fossil hollow trees of Arron and their branches (*Lepidophloios wunshinnus* Caruthers), *Transactions of the Royal Society of Edinburgh* **58**:313–337.

Weishampel, D. B., 1996, Fossils, phylogeny, and discovery: a cladistic study of the history of tree topologies and ghost lineage durations, *Journal of Vertebrate Paleontology* **16(2)**:191–197.

Wing, S. L., Bown, T. M., and Obradovich, J. D., 1991, Early Eocene biotic and climatic change in interior western North America, *Journal of Geology* **19**:1189–1192.

Wing, S. L., Alroy, J., and Hickey, L. J., 1995, Plant and mammal diversity in the Paleocene to early Eocene of the Bighorn Basin, *Palaeogeography, Palaeoclimatology, Palaeoecology* **115**:117–155.

Wing, S. L., Bao, H., and Koch, P. L., 1999, An early Eocene cool period? Evidence for continental cooling during the warmest part of the Cenozoic, in: *Warm Climates in Earth History* (B.

T. Huber, K. MacCleod, and S. L. Wing, eds.), Cambridge University Press, Cambridge, pp. 97–237.

Winkler, D. A., 1980, Taphonomy and faunal sampling in the early Wasatchian of the Clarks Fork Basin, in: *Early Cenozoic Paleontology and Stratigraphy of the Bighorn Basin, Wyoming* (P. D. Gingerich, ed.), *University of Michigan Papers on Paleontology* **24**:81–85.

Winkler, D. A., 1981, Paleoecology and taphonomy of an early Eocene mammalian fauna in the Clarks Fork Basin, Masters Thesis, University of Michigan (unpublished).

Winkler, D. A., 1983, Paleoecology of an early Eocene mammalian fauna from the paleosols in the Clarks Fork Basin, northwestern Wyoming, *Palaeogeography, Palaeoclimatology, Palaeoecology* **43**:261–298.

Wood, H. E., and Wood, A. E., 1933, *Daimonelix* in the Pleistocene of Texas, *Journal of Geology* **41**:824–833.

Zachos, J. C., Lohmann, K. C., Walker, J. C. G., and Wise, S. W., 1993, Abrupt climate change and transient climates during the Paleogene: a marine perspective, *Journal of Geology* **101**:191–213.

Chapter 5

Unusual Vertebrate Microfaunas From the Willwood Formation, Early Eocene of the Bighorn Basin, Wyoming

MARY T. SILCOX and KENNETH D. ROSE

1. Introduction

The Bighorn Basin of northwestern Wyoming is well known for its extensive and richly fossiliferous Paleocene and Eocene alluvial deposits, which have preserved events in the early Cenozoic history of mammals in unusually fine

MARY T. SILCOX • Department of Anthropology, Penn State University, University Park PA 16802 e-mail: mts1ø@psu.edu KENNETH D. ROSE • Department of Cell Biology and Anatomy, Johns Hopkins University School of Medicine, Baltimore MD 21205, kdrose©welchlink.welch.jhu.edu.

Eocene Biodiversity: Unusual Occurrences and Rarely Sampled Habitats, edited by Gregg F. Gunnell, Kluwer Academic/Plenum Publishers, New York, 2001.

detail (e.g., Bown and Rose, 1987; Rose, 1990; Bown *et al.*, 1994a, b; Silcox and Rose, 1997; Gingerich, 1989; O'Leary, 1996). Fieldwork in this region dates back to 1880, when Jacob Wortman uncovered the first fossilized mammalian remains (Gingerich, 1980). Since that time, field crews headed by Sinclair, Granger, Jepsen, Gilmore, Simons, Gingerich, and Bown and Rose have built extensive collections of early Tertiary vertebrate fossils. Tens of thousands of jaws and hundreds of skeletons provide a detailed record of the morphology of early mammalian taxa, and of the faunal composition of the region as a whole (e.g., Bown *et al.*, 1994b).

To date, over 1500 fossil mammal localities have been established in the early Eocene (Wasatchian) Willwood Formation of the southern Bighorn Basin (Bown *et al.*, 1994b, and unpublished data). The vast majority of these localities have been collected solely by surface prospecting because fossils are usually not sufficiently concentrated to warrant quarrying or screen-washing. Moreover, "most attempts to process these lag deposits by screen washing techniques have proved less fruitful because of the enormous numbers of calcrete glaebules that appear in wash concentrate and because of past failure to identify the beds where fossils occur in place." (Bown and Kraus, 1981: 37). Less than 1% of the >1500 sites are concentrated enough to warrant quarrying and/or screen-washing. This paper deals with four of these concentrations (listed in Table 1). Accumulations of small vertebrate remains are in each case very limited in geographic extent, diminishing significantly within a few tens of meters laterally. These sites are developed in paleosols that differ little from the beds surrounding them. They are unusual, therefore, only in representing dense collections of fossil material, and are not united by unique geological features that would give some clue to how they became more concentrated than other fossil-bearing sites.

Studies looking primarily at surface-prospected sites from the Bighorn Basin have played a very important role in our understanding of the mammalian composition of the early Eocene in North America (e.g., Bown, 1979; Rose, 1981a; Bown *et al.*, 1994b; Maas *et al.*, 1995). These studies suggest that the mammalian fauna in the Wasatchian was dominated by four orders: Condylarthra, Perissodactyla, Primates, and Artiodactyla. These orders are largely represented in the Bighorn Basin by four genera: *Hyopsodus*, *Hyracotherium*, *Cantius*, and *Diacodexis* (see Figure 1). Although interesting variations in the species composition have been noted for different stratigraphic levels (e.g., Bown *et al.*, 1994a) and in different paleosol stages (Bown and Beard, 1990), this fundamental pattern generally holds true at the generic and ordinal levels in Wasatchian strata throughout the Bighorn Basin, with the exception that *Haplomylus* is more common than *Hyopsodus* at some sites at lower stratigraphic levels (Bown, 1979; Rose, 1981a).

Winkler (1983) challenged this overall picture in his study of screen-washed and surface-collected samples from the Willwood Formation in the northern Bighorn Basin. He suggested that previous studies had grossly underestimated the small mammal part of the fauna, and that multituberculates and insectivores were more common in the early Eocene than previously perceived.

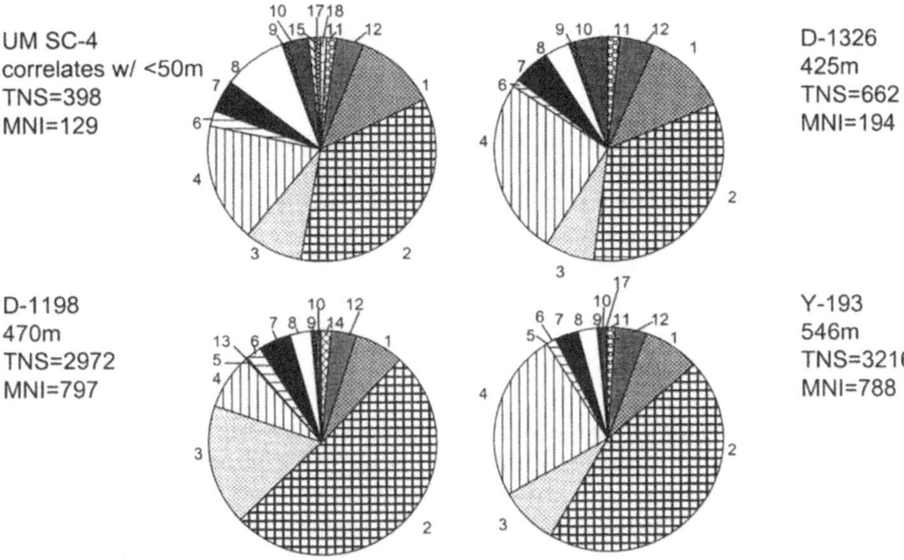

FIGURE 1. Comparison of the ordinal mammalian composition by MNI of four middle Wasatchian surface-prospected localities in the Bighorn Basin. 1 = Primates, 2 = Condylarthra, 3 = Artiodactyla, 4 = Perissodactyla, 5 = Palaeanodonta, 6 = Creodonta, 7 = Carnivora, 8 = Rodentia, 9 = Pantodonta, 10 = Tillodontia, 11 = Proteutheria, 12 = Plesiadapiformes, 13 = Taeniodonta, 14 = Marsupialia, 15 = Multituberculata, 16 = Insectivora, 17 = Mesonychia, 18 = ?Dermoptera (Plagiomenidae). Data from Bown *et al.*, 1994, and Rose, 1981a. Meter levels are measured from the base of the Willwood Formation.

Weaknesses of Winkler's study include its small sample sizes, and the fact that it focussed on the assemblage from only a single fossil site. This paper reassesses Winkler's hypothesis by looking at four sites with much higher sample sizes. Here we provide an overview of these unusual localities and a preliminary consideration of their importance. Previous publications have described particular finds from these sites (Bown and Rose, 1991; Rose and Bown, 1991; Rose, 1995a; Rose and Bown, 1996; Bloch *et al.*, 1998) and some summary compositional details have been given by Bown *et al.* (1994b) and Silcox and Rose (1996). Future treatments will contain more detailed descriptions and analyses.

2. Materials and Methods

All four of the quarry sites considered in this study were carefully surface-prospected. Three of the four sites were quarried by removing and breaking up consolidated mudstone with hand tools. In the case of McNeil Quarry (see Table 1 and below) the weathered paleosol was so friable that excavation tools were largely unnecessary and the sediment was easily sifted manually to locate concretions, which in most cases held bone. All four sites were screen-washed.

Matrix from Rose Quarry and McKinney Quarry was soaked in a mild detergent solution and washed through 1 mm screen. The resulting concentrate was sorted under a magnifying lamp or binocular microscope.

D.G. Kron applied a more rigorous washing protocol to additional matrix from McKinney Quarry, and to samples from the McNeil and Dorsey Creek quarries. After soaking in a mild detergent solution this matrix was given a preliminary wash and sorted in coarse screens, and was subsequently washed through three graded screens, the finest of which had pore diameters of 0.5 mm. Sorting was performed under a binocular dissecting microscope.

2.1. Compositional Analysis

For three of the quarries under consideration (McNeil, Rose, and McKinney quarries) tallies were completed for both TNS (total number of specimens) and MNI (minimum number of individuals; see Tables 2–4). Data for the surface locality surrounding Rose Quarry (D-1460 surface; Bown *et al.*, 1994b) is included in Table 2 for comparison with the quarry sample from the same site.

For Dorsey Creek Quarry, preparatory work and sorting are still at an early stage. A preliminary faunal list and TNS count based on currently available jaw fragments and some isolated teeth are given in Table 5. The list does not include unprepared jaws, most isolated teeth, or any postcrania.

There is some debate about which metric, MNI or TNS, better represents the original mammalian fauna (e.g., Grayson, 1978). For this reason both counts are given in Tables 2–4. Badgley (1986) indicated that MNI is the more appropriate measure when it is suspected that the sample has not been transported or disturbed. Previous taphonomic work in the Willwood Formation (Bown and Kraus, 1981) has provided support for the view that material is usually not transported very far from its initial site of burial. Support for little transport at these quarry sites is provided by several instances in which matching right and left sides have been found, and the excellent quality of the preservation of very small teeth and bones. Moreover, there is no significant evidence of the abrasion that would be expected if the elements had undergone considerable water transport. For these reasons MNI is used as the metric of choice in pie charts (see Figures 2, 5, and 7).

For most taxonomic groups, specimen counts and MNI calculations do not include non-dental remains. The one exception is the Palaeanodonta, which might otherwise be overlooked because of their extreme dental reduction. Postcrania have been identified for only a small proportion of Willwood taxa, and many remain unallocated, particularly for very small species. Therefore, inclusion of postcrania in the TNS counts or MNI calculations would inevitably bias the value toward taxa with known skeletons.

Rank abundance curves were calculated for Rose Quarry, the surrounding surface site (D-1460 surface) and McKinney Quarry. Rank abundance graphs consist of a plot of the relative frequency of each species against the logarithm of its rank in terms of the number of individuals observed. Rose (1981b) applied

Table 1. Wasatchian Quarry Localities

USGS Locality#	Quarry name	Meter level[1]	Name of discoverer, year discovered	Paleosol stage[2]
D-1460Q	Rose Quarry	411m	K.D. Rose, 1982	1
D-1762Q	McKinney Quarry	414m	K.C. McKinney, 1988	0
D-2035Q	Dorsey Creek Quarry	397m	T.M. Bown and K.D. Rose, 1994 using locality data provided by P. Robinson.	1
D-2037Q	McNeil Quarry	347m	P. McNeil, 1994	1

[1]Measured from the base of the Willwood Formation. See Bown et al., 1994b.
[2]See Bown and Kraus, 1987.

the concept of the broken-stick model (MacArthur, 1957, 1960) to the study of the taxonomic evenness of early Tertiary mammalian assemblages. This model allows the prediction of the relative frequency of the taxa in an assemblage with an even distribution, in which no particular taxon or taxa dominate the fauna. In this case the rank abundance graph describes a gently curved line. Deviations from this prediction tend to entail some taxa being represented more commonly than expected according to the model, a pattern called "type four" (Hutchinson, 1961). Assemblages which show evenness have been interpreted to be more stable ecologically, while those which follow the "type four" pattern, and are thus less even, are thought to be from more unstable environments. An additional possibility, that differences in the shape of the curve result from differences in the types of biases influencing the collection, is also considered below.

3. Synopsis of Quarry Sites

3.1. Rose Quarry (D-1460Q) (Table 2, Figure 2)

This locality was initially found by inspection of a low gray hill that appeared slightly different in color from the surrounding locality (D-1460 surface). Close examination of the surface quickly produced more than 30 small mammal jaws and many bones, representing a dozen taxa, from an area of less than 20 m^2. The unweathered producing layer was soon located just beneath the surface, and the site was quarried extensively during the 1982–1984 field seasons. Subsequent screen-washing of roughly one ton of matrix greatly increased the sample of fossils, but because of microfractures in much of the bone, most remains recovered in this fashion were isolated bones and teeth.

Rose Quarry has produced a large sample of well-preserved micromammal remains, including over a thousand dental and gnathic specimens representing a minimum of 150 individuals in 42 species. Of the 1005 dental/gnathic

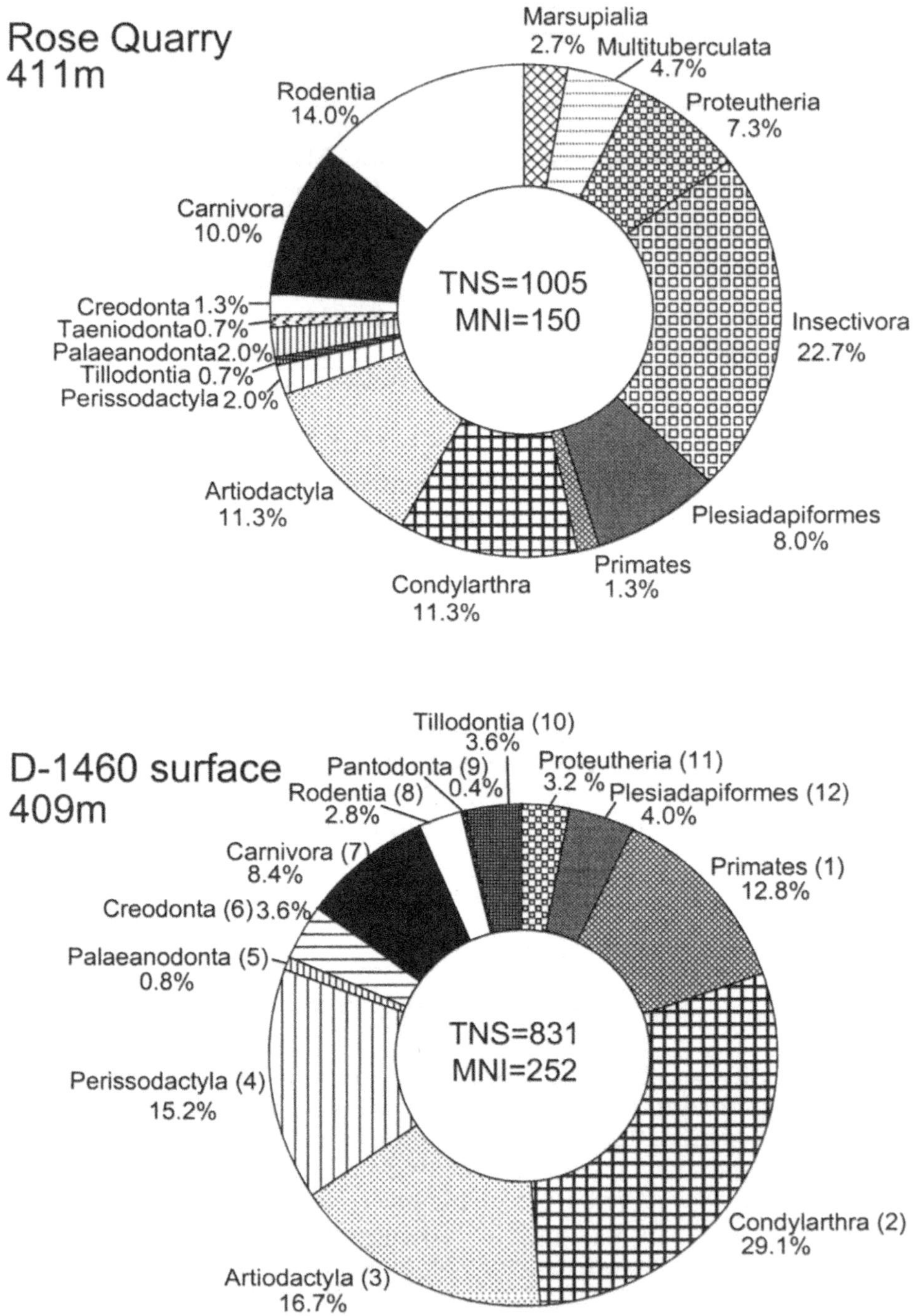

FIGURE 2. A-Pie chart of relative frequencies by MNI of mammalian orders at Rose Quarry (D-1460Q). B-Pie chart of relative frequencies by MNI of mammalian orders at D-1460 surface. Numbers in brackets are given to allow comparison with Figure 1.

Table 2. Rose Quarry and D1460 surface[1] Mammalian Assemblages

	Rose Quarry			D1460 surface		
	TNS	MNI	% MNI	TNS	MNI	% MNI
Multituberculata						
Ectypodus tardus	8	5	3.3			
Parectypodus simpsoni	3	2	1.3			
indet.	2					
Marsupialia						
Peratherium comstocki	3	2	1.3			
Peradectes sp.	5	2	1.3			
Proteutheria						
Apatemys chardini	2	1	0.7			
Apatemys sp.				6	4	1.6
Didelphodus absarokae	9	2	1.3	2	2	0.8
Palaeoryctidae, unident.	4	1	0.7			
cf. *Prodiacodon tauricinerei*	48	7	4.7	2	2	0.8
Palaeanodonta						
Palaeanodon ignavus	2	1	0.7	3	2	0.8
Palaeanodon sp. nov.	1	1	0.7			
Tubulodon sp.	2	1	0.7			
Insectivora						
Macrocranion nitens	90	19	12.7			
cf. *Plagioctenodon savagei*	5	2	1.3			
Nyctitheriidae?, unident.	1	1	0.7			
Talpavoides dartoni	14	7	4.7			
cf. *Talpavus nitidus*	5	4	2.7			
unidentified sp. (USGS 6684)	1	1	0.7			
indet.	27					
Creodonta						
Prototomus sp.	8	2	1.3	2	2	0.8
Tritemnodon sp.				1	1	0.4
Prolimnocyon sp.				7	4	1.6
Oxyaena sp.				5	2	0.8
Carnivora						
Miacis petilus	15	3	2.0	3	1	0.4
Viverravus acutus	16	6	4.0	6	3	1.2
Viverravus? sp. 2	1	1	0.7			
Viverravus? sp. 3, sp. nov.	2	1	0.7			
Didymictis protenus	18	3	2.0	19	6	2.4
Vulpavus australis				15	8	3.2
Vulpavus canavus	3	1	0.7	1	1	0.4
Vassacyon promicrodon				1	1	0.4
Uintacyon rudis				2	1	0.4
indet.	25					

(continued)

Table 2. (Continued)

	Rose Quarry			D1460 surface		
	TNS	MNI	% MNI	TNS	MNI	% MNI
Plesiadapiformes						
Microsyops angustidens	30	6	4.0	20	7	2.8
Microsyops cardiorestes				1	1	0.4
Niptomomys doreenae	3	1	0.7			
Niptomomys cf. *thelmae*	5	2	1.3			
*Phenacolemur*cf. *fortior*	24	3	2.0			
Phenacolemur sp.				4	2	0.8
Primates						
Cantius abditus	5	2	1.3	82	31	12.3
cf. *Copelemur* sp.				2	1	0.4
Taeniodonta						
Ectoganus sp.	1	1	0.7			
Tillodontia						
Esthonyx bisulcatus	4	1	0.7	33	9	3.6
Pantodonta						
Coryphodon sp.				4	1	0.4
Condylarthra						
Hyopsodus sp.	47	7	4.7	214	67	26.6
Phenacodus trilobatus				10	4	1.6
Phenacodus sp.	1	1	0.7			
Thryptacodon antiquus	5	1	0.7	1	1	0.4
Apheliscus insidiosus	55	8	5.3			
Chriacus sp.				1	1	0.4
Artiodactyla						
Diacodexis metsiacus	142	16	10.7	130	40	15.9
Diacodexis robustus				2	1	0.4
Bunophorus sp.	1	1	0.7	2	1	0.4
Perissodactyla						
Hyracotherium cf. *pernix*	28	3	2.0	173	28	11.1
Hyracotherium large sp.				39	8	3.2
Homogalax cf. *protapirinus*				6	2	0.8
Rodentia						
Knightomys sp.	9	2	1.3			
cf. *Paramys copei*	35	4	2.7	10	1	0.4
Paramys cf. *taurus*	116	12	8.0	16	3	1.2
cf. *Acritoparamys atwateri*	10	3	2.0			
indet. (incisors, edent. jaws)	164					
rodent small sp.[2]				3	1	0.4
rodent minute sp.[2]				3	2	0.8
Totals:	**1005[3]**	**150**	**100**	**831**	**252**	**100**

[1]Data for D1460 surface modified from Bown et al., 1994b; includes the adjacent locality D1454, located at the same meter level.

[2]Could be referable to taxa identified at the quarry.

[3]TNS not including indet. rodents, carnivores, insectivores, and multis=782

specimens, 240 are jaw fragments. The productive layer is developed in a stage 1 paleosol (see Bown and Kraus, 1987) that cuts into the surrounding stage 3 paleosol of D-1460, a surface-collected site located at 409 m above the base of the Willwood Formation (Bown *et al.*, 1994b) and about 2 m below the quarry. The surface site has produced 831 dental/gnathic specimens, representing 252 individuals in only 36 species (Table 2, Figure 2B). In spite of the fact that the quarry and surface-prospected localities are essentially contemporaneous, the 36 species from the surface site include several that are not represented at the quarry and, conversely, the quarry sample includes taxa (mostly very small species) not at the surface site. Specifically, 19 species that have been identified from the quarry sample are missing from the surface collection, and 15 species found at the surface site are missing from the quarry.

A χ^2 test comparing the quarry and surface collections at the ordinal level found that they differ significantly ($p < 0.0005$).* The same result was found using either TNS or MNI data. In addition there are marked contrasts in the relative abundance of taxa present in both assemblages. *Apheliscus insidiosus* and *Macrocranion nitens*, two of the most common elements of the quarry collection (accounting for 5.3% and 12.7% respectively) are completely unknown from the surface assemblage. Conversely, common elements of the surface collection, like *Hyracotherium* large sp. and *Hyopsodus* sp., are unknown or much more rare in the quarry collection. The contrast between the primates present in each assemblage is particularly interesting. There are three species of small plesiadapiform primates in the quarry assemblage that are missing from the surface collection. On the other hand, the larger adapid euprimate *Cantius* is one of the most common surface-collected taxa but ranks 18th in terms of MNI at the quarry, accounting for only 1.3% of the total assemblage. The possible implications of the various contrasts between these two collections are discussed below.

In contrast to McKinney Quarry (discussed below) and most surface-prospected localities, fish, turtle, and crocodilian remains are virtually unknown from Rose Quarry; lizard remains, however, are common and predominantly represent glyptosaurine anguids. Besides copious mammalian dental material, this quarry has yielded an abundance of small mammal postcrania, including more than 1000 identifiable elements. If these elements were included in the calculation of MNI at Rose Quarry there would be a few marked changes. For example, use of the most common postcranial element, astragali of *Diacodexis* (n = 57), would almost double the relative abundance of that taxon (MNI = 29) in the quarry sample, compared to its abundance based on dental evidence (MNI = 16).

Highlights of the collection from Rose Quarry include well-preserved jaws of diverse lipotyphylan insectivorans including previously rare forms such as

*χ^2 tests comparing different assemblages were calculated at the ordinal level. The structure of the χ^2 test does not allow for any cell counts of zero. For this reason, orders that were unrepresented in one of the two collections were not included in the calculation. Both the application of the test at the ordinal (rather than the specific or generic) level, and the removal of unrepresented orders will make the test more conservative (that is, less likely to find a significant difference).

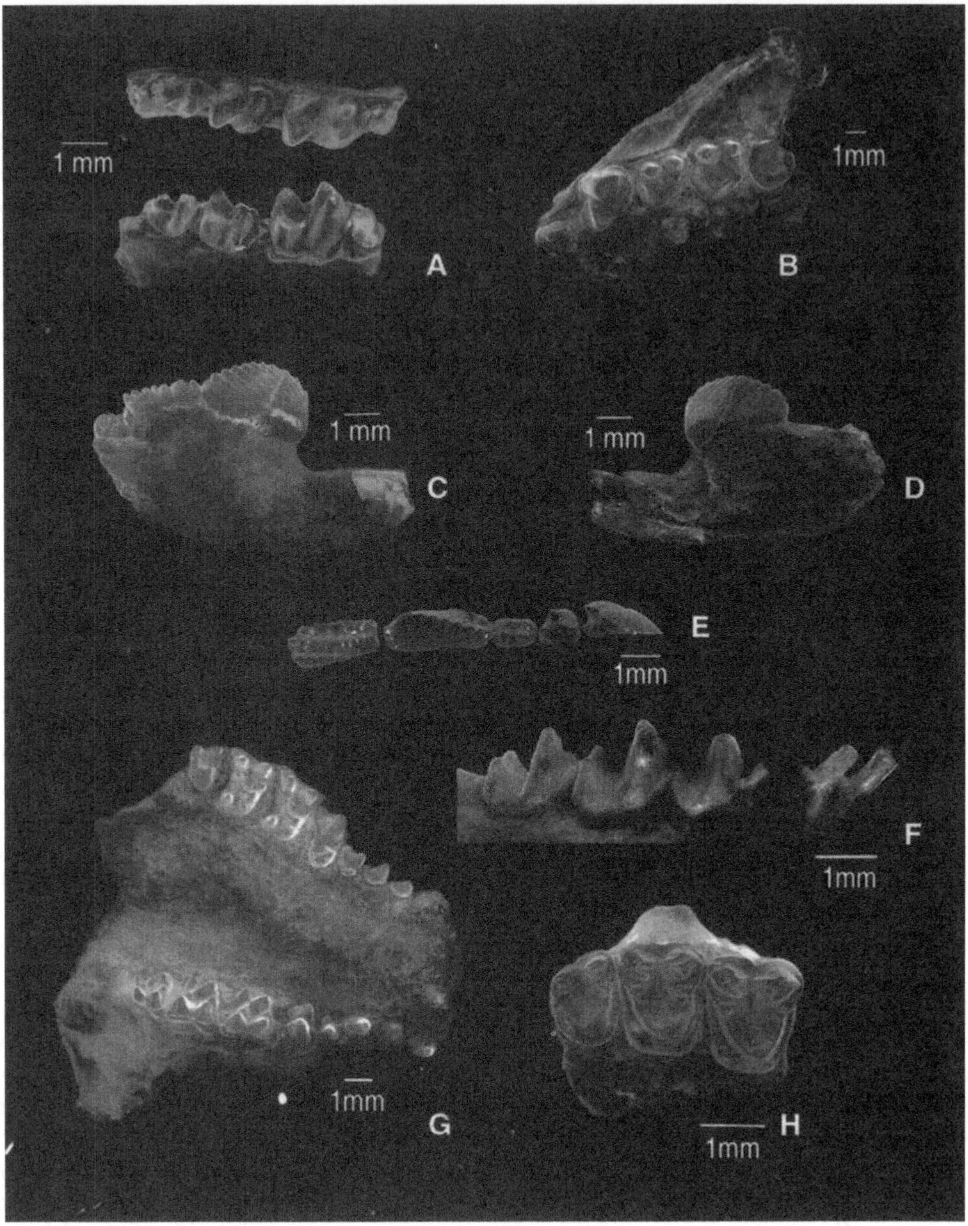

FIGURE 3. Small mammal fossils from Willwood quarries.
A—*Eolestes simpsoni*, USNM 495039, occlusal and buccal views of right dentary with P_4(talonid), M_1-M_3, from McKinney Quarry.
B—*Apheliscus* sp., USNM 494896, left maxilla with P^3-M^3 from McNeil Quarry.
C—*Ectypodus tardus* USNM 494903, lingual view of left dentary with incisor base and P_4-M_2 (M_2 erupting) from McNeil Quarry.
D—*Ectypodus tardus*, USNM 494904, lingual view of right dentary with dI_1, P_4, and erupting I_1, from McNeil Quarry.
E—*Parectypodus simpsoni*, USNM 494912, occlusal view of right maxilla with P^1-P^4, M^1, from McNeil Quarry.
F—*Talpavoides dartoni*, USNM 495001, buccal view of right dentary with C, P_1, P_3-M_2, from Dorsey Quarry. Teeth are within dentary, which is still embedded in matrix.
G—*Macrocranion nitens*, USNM 495248, palate with RP^1-M^3, LC-M^3, from McNeil Quarry.
H—*Niptomomys* cf. *thelmae*, USNM 495249, right maxilla with M^{1-3} from Rose Quarry.

Talpavoides dartoni (Figure 3F), a number of diminutive carnivorans (Figure 4A,B), and the presence of two species of *Niptomomys* (*N. doreenae* and *N.* cf. *thelmae*; Figure 3H). This is the only locality at which two species of *Niptomomys* are known to be present. There are no omomyid euprimates known from this site. Rose Quarry is at 411 m above the base of the Willwood Formation, within a period of faunal turnover (Biohorizon B interval—see Schankler, 1980; Bown and Kraus, 1993; Bown *et al.*, 1994a,b) in which there are substantial changes in the omomyid fauna (Bown and Rose, 1987). This could partially explain the absence of omomyids (see discussion below).

3.2. McKinney Quarry (D-1762Q) (Table 3, Figure 5)

An unusual concentration of black bone eroding from a light-colored outcrop of sandy-clay led to the discovery of this locality. The productive layer was quarried and screen-washed for several days over the course of three field seasons (1988–90), and has been worked intermittently since then. The quarry site is situated on a small hill surrounded by a moderately productive locality (D-1762 surface). Fossils are concentrated in a thin zone within an incipiently developed paleosol (stage 0; Bown *et al.*, 1994b), whereas the surrounding locality is mostly in stage 1 paleosols. Both sites are located at 414 m above the base of the Willwood Formation.

McKinney Quarry is notable for producing not only well-preserved micromammal remains (Bown and Rose, 1991; Rose and Bown, 1991), but also abundant aquatic vertebrate fossils including scales, vertebrae and other elements of fishes (*Amia* and *Lepisosteus*; Figure 6E), shell fragments of turtles including trionychids, and crocodilian teeth and bones (cf. *Allognathosuchus*). This abundance reflects the close proximity of the site to a channel in Eocene times, further indicated by the immaturity of the paleosol. The current collection includes 642 identifiable mammalian dental/gnathic specimens, representing 106 individuals (MNI) in 39 species. Most of the remains are isolated teeth, with only 40 fragmentary jaws (not 73, as reported by Bown *et al.*, 1994b: 22). Postcranial material is not as plentiful as at Rose Quarry, although notable specimens include an atlas from the pantodont *Coryphodon*, and two sizes of pantolestid calcanei (two pantolestid species, *Palaeosinopa veterrima* and *Palaeosinopa incerta* have been identified based on dental remains).

Like Rose Quarry, this locality is within the Biohorizon B interval. Although the two quarries are essentially contemporaneous (<5 m apart stratigraphically, which is probably <10,000 years; Bown and Kraus, 1993), their assemblages are significantly different, (χ^2 test $p < 0.0005$). The most striking contrast in faunal composition between McKinney Quarry and Rose Quarry is the difference in proportions of euprimates in each assemblage. At both sites the adapid *Cantius* is relatively rare, comprising less than 2% of the assemblage by MNI. Omomyid euprimates, on the other hand, form a significant part of the known collection at McKinney Quarry (over 10% by MNI), while they are completely unknown from Rose Quarry, even though the sample from the

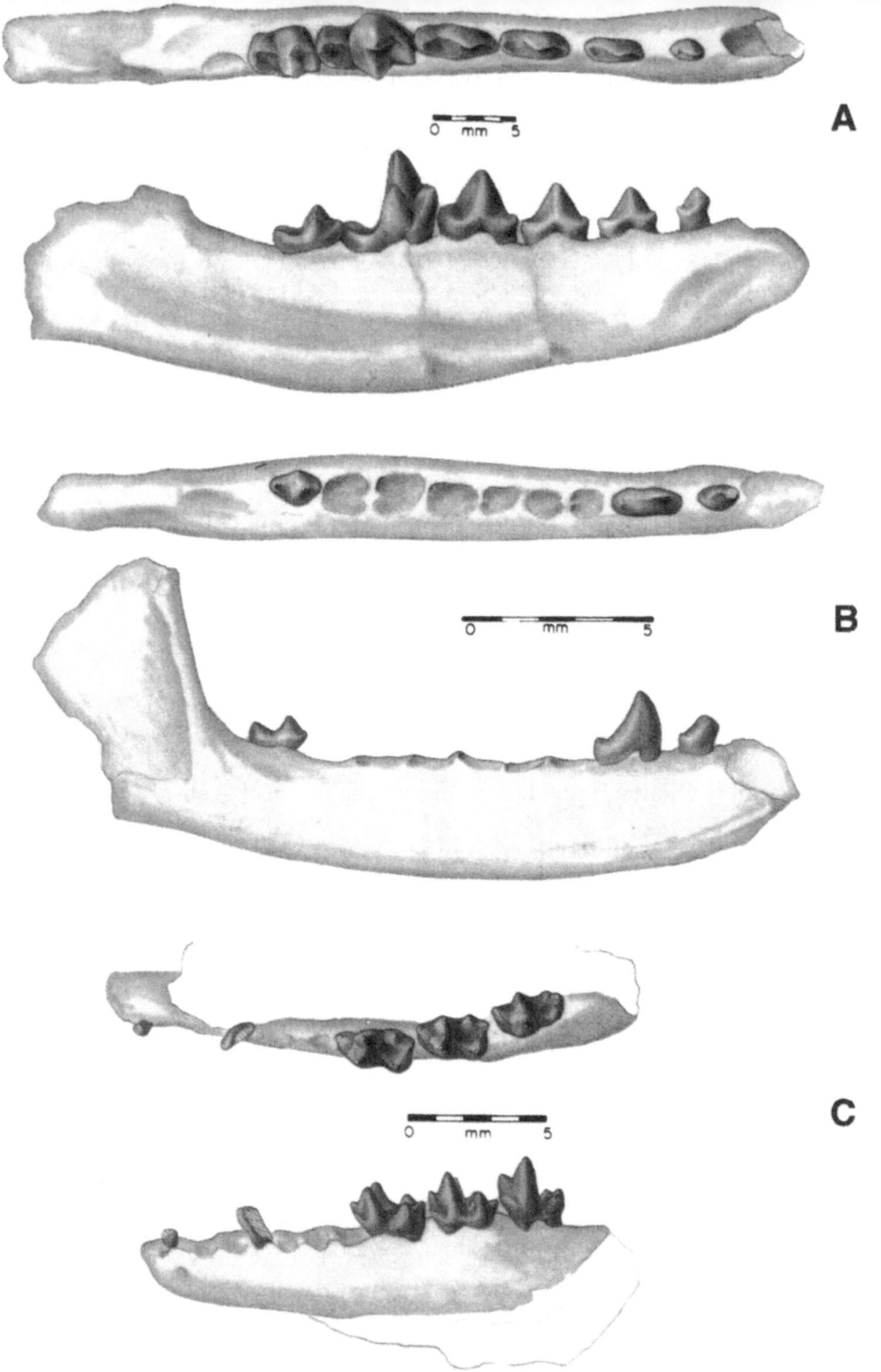

FIGURE 4. Small mammal jaws from Rose Quarry.
A—*Miacis* cf. *petilus*, USNM 495250 (formerly USGS 6649), left dentary w/ P_1-M_2 in occlusal and lingual views.
B—*Viverravus* sp. nov.?, USNM 495251 (formerly USGS 6719), right dentary with P_{1-2}, M_2 in occlusal and buccal views.
C—*Peratherium comstocki*, USNM 495252 (formerly USGS 6686), left dentary w/ M_{1-3} and parts of two anterior teeth in occlusal and buccal views.

Table 3. McKinney Quarry Mammalian Assemblage

	TNS	MNI	% MNI
Marsupialia			
marsupial indet.	1	1	0.9
Proteutheria			
cf. *Apatemys rodens*	2	1	0.9
Apatemys cf. *bellulus*	3	2	1.9
Palaeosinopa veterrima	1	1	0.9
Palaeosinopa incerta	2	1	0.9
Pantolestidae?, unident.	2		
cf. *Palaeictops bicuspis*	11	2	1.9
cf. *Prodiacodon tauricinerei*	15	2	1.9
cf. *Palaeoryctes* sp. 1	1	1	0.9
cf. *Palaeoryctes* sp. 2	1	1	0.9
Didelphodus absarokae	14	3	2.8
Insectivora			
cf. *Plagioctenodon savagei*	1	1	0.9
Talpavoides dartoni	1	1	0.9
Macrocranion nitens	27	5	4.7
Eolestes simpsoni	2	1	0.9
Nyctitheriidae?, unident.	2	1	0.9
unident. (or indet.) insectivores	3		
Creodonta			
Oxyaena sp.	1	1	0.9
Prototomus robustus	3	1	0.9
Carnivora			
Miacis cf. *petilus*	9	2	1.9
Viverravus cf. *acutus*	13	3	2.8
Vulpavus cf. *australis*	6	3	2.8
Carnivora indet.	6		
Plesiadapiformes			
Microsyops angustidens	7	2	1.9
Niptomomys cf. *thelmae*	2	1	0.9
Phenacolemur cf. *jepseni*	23	4	3.8
Primates			
Steinius vespertinus	28	7	6.6
Arapahovius advena	10	3	2.8
cf. *Absarokius metoecus*	1	1	0.9
Cantius abditus	9	2	1.9
?omomyid incisor	1		
Tillodontia			
Esthonyx bisulcatus	3	1	0.9
Pantodonta			
Coryphodon cf. *lobatus*	8	2	1.9

(continued)

Table 3. (Continued)

	TNS	MNI	% MNI
Condylarthra			
Hyopsodus ?simplex	68	9	8.5
Hyopsodus cf. *miticulus*	38	9	8.5
Chriacus sp.	1	1	0.9
Phenacodus cf. *vortmani*	1	1	0.9
Artiodactyla			
Diacodexis metsiacus	44	5	4.7
Diacodexis robustus	6	2	1.9
Perissodactyla			
Hyracotherium cf. *pernix*	105	8	7.5
Isectolophidae, unident.	2	1	0.9
Rodentia			
Microparamys sp. (and/or *Knightomys*)	8	2	1.9
cf. *Paramys copei*	14	3	2.8
cf. *Paramys taurus*	56	8	7.5
unident. incisors	80		
Totals:	**642**	**106**	**100**

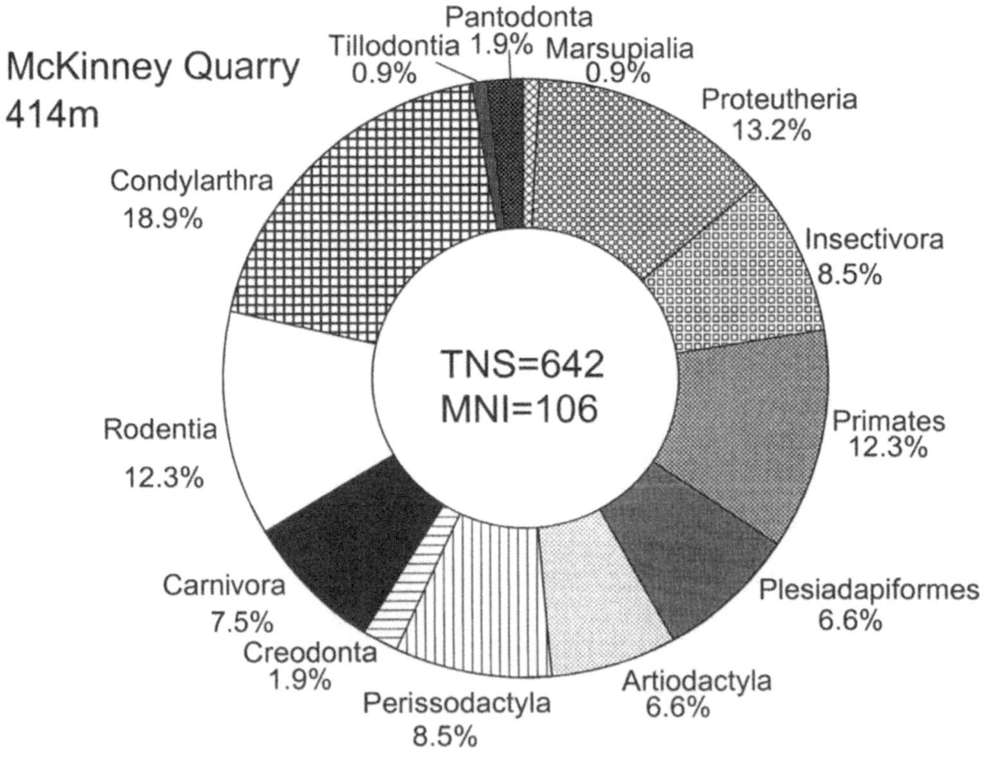

FIGURE 5. Pie chart of relative frequencies by MNI of mammalian orders at McKinney Quarry.

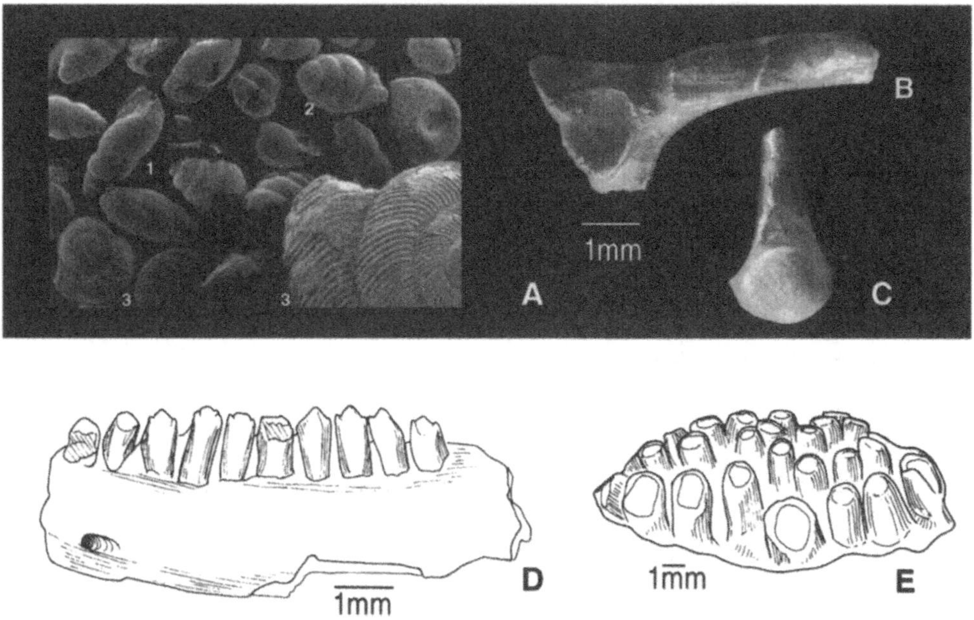

FIGURE 6. Non-mammalian specimens from Willwood quarries.
A—Pulmonate gastropods from WW-12, 1-ellobiids, 2-pupillid, 3-discids.
B—Anuran ilium, USNM 495253, from WW-12.
C—Anuran distal humerus, USNM 495254, from WW-12.
D—?*Parasauromalus* jaw, USNM 495044, from McKinney Quarry.
E—Holostean fish (*Amia*), USNM 495297, palatal teeth from McKinney Quarry.

able to two species, *Steinius vespertinus* and *Arapahovius advena*, which are very rare or absent at other localities in the Bighorn Basin, and unknown from elsewhere. This unusual situation may be partly explained by the position of both quarries within Biohorizon B. In this stratigraphic interval in the Willwood Formation the omomyid species known from the lower part of the formation (*Teilhardina* and the *Tetonius-Pseudotetonius* lineage) had disappeared, and the dominant taxa from the upper part of the section (*Absarokius* species) had not yet immigrated into the region (Bown and Rose, 1987). Therefore, the absence at both sites of the common omomyids from the rest of the Willwood section is probably related to their position within Biohorizon B. The presence of the generally rare omomyids *S. vespertinus* and *A. advena* at McKinney Quarry requires an additional explanation, perhaps related to microhabitat differences between this locality and most other Biohorizon B sites.

3.3. McNeil Quarry (D-2037Q) (Table 4, Figure 7)

This quarry was discovered as a result of a localized concentration of bones of *Ectocion* and *Phenacodus*. Following collection of all elements from the

Table 4. McNeil Quarry Mammalian Assemblage

	TNS	MNI	%MNI	Juvenile specimens[1]
Multituberculata				
Ectypodus tardus	19	4	8.0	2
Parectypodus simpsoni	37	9	18.0	
Proteutheria				
cf. Leptictidae, unident	1			1
cf. *Prodiacodon tauricinerei*	3	1	2.0	
Insectivora				
Macrocranion nitens	32	6	12.0	2
cf. *Plagioctenodon* sp.	5	1	2.0	
Nyctitheriidae?, unident.	2	1	2.0	
Carnivora				
Viverravus acutus	3	1	2.0	
Viverravus cf. *politus*	1	1	2.0	
Condylarthra				
Hyopsodus sp.	6	3	6.0	3
Phenacodus trilobatus	1	1	2.0	
Ectocion osbornianus	17	6	12.0	7
Apheliscus insidiosus	5	2	4.0	1
Artiodactyla				
Diacodexis metsiacus	1	1	2.0	
Dichobunidae, unident.	2	1	2.0	
Perissodactyla				
Hyracotherium cf. *pernix*	3	2	4.0	1
Rodentia				
Knightomys sp.	28	8	16.0	1
medium-sized species	5	2	4.0	2
Totals:	**171**	**50**	**100**	**20 (8.6%)**

[1]specimens with deciduous and/or erupting permanent teeth.

surface, examination of the deposit quickly produced several jaws of smaller mammals, including multituberculates.

McNeil Quarry is the most unusual of the four quarries, both geologically and paleontologically. The fossil-producing bed is in a localized pocket of poorly consolidated, weathered paleosol. Lithologically the site is unusual among vertebrate localities in the Bighorn Basin because the sediment is rich in gypsum and jarosite. This probably formed as a result of the oxidation of pyrite, which in turn reflects the presence of a concentration of organic matter in waterlogged conditions during soil formation. (Kraus, 1998, and pers. comm., 1998) Overlaying the quarry bed is a thin sandstone layer, indicating relatively rapid burial of the vertebrate concentration. This undoubtedly contributed to the preservation of the organic remains (M.J. Kraus, pers. comm., 1998). The quarry covers an area of only around 10 m^2. It was worked for a total of five days over four field seasons, after which the productive layer was worked out. Much of the sediment was subsequently screen-washed. The quality of preservation is generally high, with a significant proportion of dentaries and snouts that contain anterior teeth (Figure 3G).

The composition of this quarry is highly unusual for a Wasatchian deposit. The mammalian assemblage is relatively depauperate, with only 17 species. About one-third of the 171 known specimens are attributable to neoplagiaulacoid multituberculates (*Ectypodus tardus* and *Parectypodus simpsoni*; Figure 3C,D,E), and they account for 13 of the 50 individuals by MNI (Table 4). The only Wasatchian localities that have produced more multituberculate specimens are some of the Four Mile quarries (McKenna, 1960; Krause, 1982). However, those sites have produced nearly an order of magnitude more identifiable specimens, so that at McNeil Quarry the relative frequency of multituberculates is much higher (26% by MNI compared to less than 10%—see Figure 8). The assemblage is also unusual in showing a relatively large number of juvenile specimens (8.6% by TNS).

In addition to gnathic specimens, there are also many postcrania. In contrast to Rose Quarry, however, relatively few pertain to small mammals. Instead, most of the postcrania are from a small number of individuals of *Ectocion* and *Phenacodus*. Many of the bones belong to a single skeleton of *Phenacodus trilobatus*, whose nearly complete lower dentition and badly preserved skull were also found. Unfortunately, the *Phenacodus* elements show signs of modern root damage and are partly preserved in gypsum, which makes the bone very fragile.

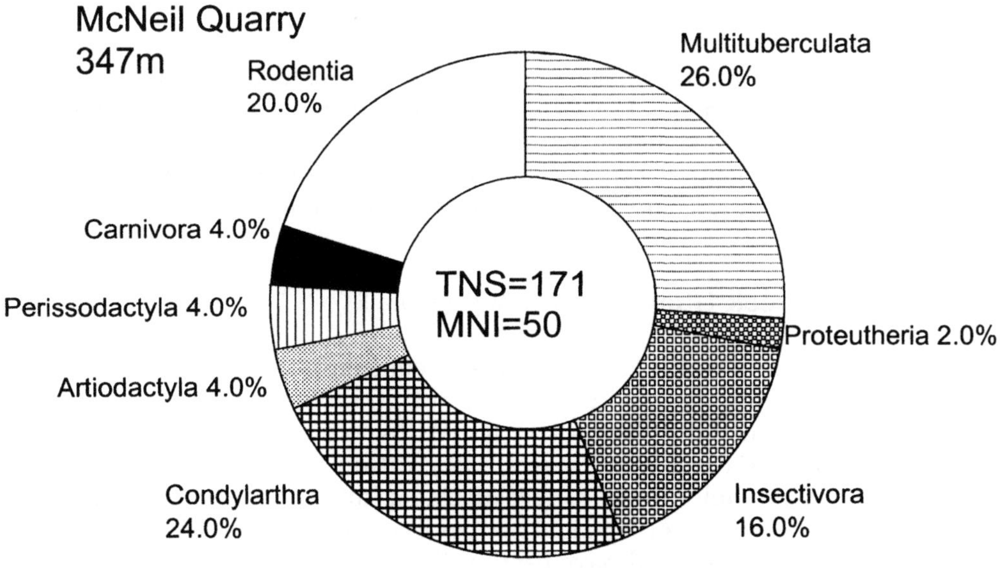

FIGURE 7. Pie chart of relative frequencies by MNI of mammalian orders at McNeil Quarry.

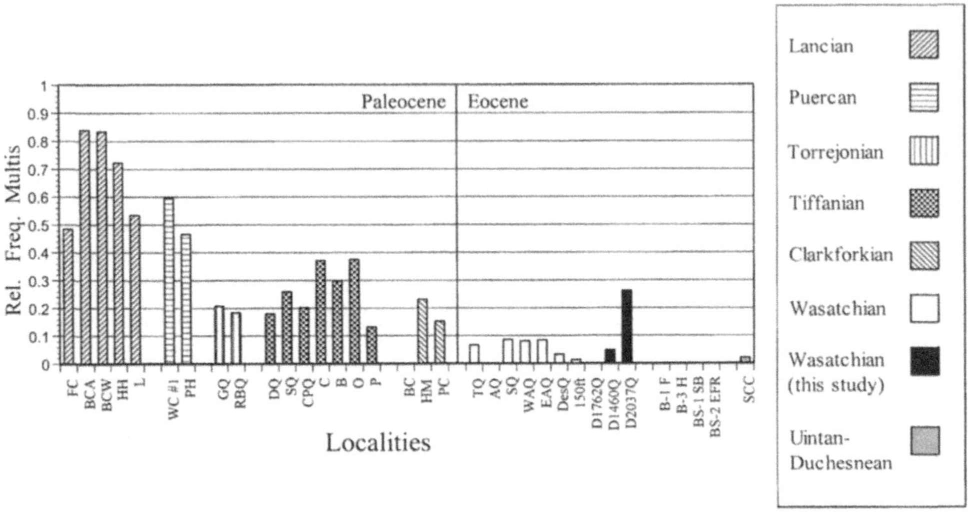

FIGURE 8. Changing relative frequencies of multituberculates through time. Locality Names: FC—Flat Creek, BCA—Bug Creek Anthills, BCW—Bug Creek West, HH—Harbicht Hill, L—Lance, WC#1—Worm Coulee #1, PH—Purgatory Hill, GQ—Gidley Quarry, RBQ—Rock Bench Quarry, DQ—Douglass Quarry, SQ—Scarritt Quarry, CPQ—Cedar Point Quarry, C—Circle, B—Brisbane, O—Olive, P—Princeton Quarry, BC—Bear Creek, HM—Holly's microsite, PC—Paint Creek, AQ—Anthill Quarry, WAQ—West Alheit Quarry, EAQ—East Alheit Quarry, SQ—Sand Quarry, TQ—Timberlake Quarry, DesQ—Despair Quarry, 150 ft—Bighorn Basin 150 ft, B-1F—B-1 Fault, B-3H—B-3 Hawk, BS-1SB—BS-1 Steele Butte, BS-2EFR—BS-2 East Fork River, SCC—Swift Current Creek. Note that the Bridgerian is not included in the legend since there are no multituberculates from the sites representing this period (localities with "B" prefixes). Data from Krause, 1986, and this study.

Located at 347 m, the site is situated stratigraphically between two periods of faunal turnover in the Bighorn Basin (Biohorizons A and B; Schankler, 1980; Bown and Kraus, 1993; Bown *et al.*, 1994a), implying that basin-wide patterns of change cannot explain its unusual compositional features. Possible alternative explanations are discussed in the taphonomic section.

3.4. Dorsey Creek Quarry (D-2035Q) (Table 5)

This locality was discovered as a surface concentration of small vertebrate remains on the crest of a hill in the valley of USGS locality D-1514. The quarry is located at 397 m above the base of the Willwood Formation. Excavation of unweathered matrix produced several jaws within a few hours of its discovery, including the holotype of a new plesiadapiform, *Picromomys petersonorum* Rose and Bown, 1996. The site has been actively quarried by three or four people at a time for several days each of the last four field seasons (1995–8). About a ton of matrix has also been screen-washed. Geologically this site is a relatively weakly developed paleosol (stage 1) situated in a crevasse-splay, implying very rapid deposition (M.J. Kraus, 1997 and pers. comm., 1998). Although more needs to be understood about the concentrating mechanisms for this unusually rich quarry (see below), the apparent swiftness with which the deposit was formed suggests that it may present a picture of the local mammalian fauna that has been little affected by time-averaging, which does influence surface-lag accumulations and possibly other quarry samples.

More than 200 jaws have been recovered from this locality, which implies that it is likely to be one of the richest Wasatchian quarries known from the southern Bighorn Basin. A detailed compositional study has not yet been undertaken for Dorsey Creek Quarry, since many specimens remain unprepared, and considerable concentrate has yet to be sorted. A preliminary survey of the prepared jaws and teeth, detailed in Table 5, indicates high proportions of rodents (14.8%) and the enigmatic condylarth *Apheliscus insidiosus* (9.0%). Lipotyphlan insectivorans are very common (42.8%) and diverse, accounting for almost a third of the species present. The high quality of preservation has led to the discovery of a number of new taxa (see below), as well as some of the first known anuran bones from the Willwood Formation. Unfortunately, although some well-preserved postcranial material has been recovered, the delicate nature of the bone makes it unlikely to survive the screen-washing process.

4. Taphonomy

While a detailed consideration of the taphonomy of these quarries is outside the scope of this article, some preliminary remarks are essential to consider the possible importance of these sites to more general questions.

There are two basic classes of taphonomic processes that need to be considered in looking at any fossil concentration: abiotic and biotic. The former class can be easier to study, in that geologic information can be brought to bear to support or falsify particular hypotheses. Given the complexity of inferring the results of behavior in fossil organisms, it can be difficult to produce falsifiable hypotheses about biotic factors.

4.1. Abiotic Processes

In the case of small mammal concentrations, the role of water must be considered (Dodson, 1973; Korth, 1979). Particularly, small mammal bones can be carried by fast-moving water because they have a relatively low specific gravity. A concentration may be formed at a place in the channel where the water velocity slows. This possible mode of origin can be tested in a number of different ways. First, such a mechanism would predict that the deposit be found in, or near, a channel. Second, the concentration should consist of bones small enough to be moved, and should be enriched in more easily transported

Table 5. Dorsey Creek Quarry Mammalian Assemblage[1]

	TNS	%TNS
Multituberculata		
Parectypodus simpsoni	1	0.5
Ectypodus tardus	10	5.3
Marsupialia		
Peradectes cf. *protinnominatus*	1	0.5
Peratherium comstocki	1	0.5
Proteutheria		
Apatemys bellulus	3	1.6
cf. *Prodiacodon tauricinerei*	4	2.1
Palaeanodonta		
Palaeanodon ignavus	1	0.5
Insectivora		
Talpavoides dartoni	18	9.5
cf. *Talpavus nitidus*	3	1.6
Wyonycteris sp.	4	2.1
Nyctitheriidae? sp. A	4	2.1
Nyctitheriidae? sp. B	2	1.1
cf. *Centetodon patratus*	1	0.5
Centetodon neashami	3	1.6
Macrocranion nitens	27	14.3
Batodonoides vanhouteni	4	2.1
unident. insectivores	15	7.9
Carnivora		
Viverravus acutus	6	3.2
Didymictis sp.	2	1.1
Plesiadapiformes		
Microsyops angustidens	5	2.6
Niptomomys thelmae	1	0.5
Phenacolemur cf. *fortior*	2	1.1
Picromomys petersonorum	3	1.6
Primates		
Cantius abditus	1	0.5
Condylarthra		
Apheliscus insidiosus	17	9.0
Hyopsodus sp.	9	4.8
Phenacodus vortmani	1	0.5
Artiodactyla		
Diacodexis metsiacus	6	3.2
Perissodactyla		
Homogalax sp.	1	0.5
Hyracotherium sp.	5	2.6
Rodentia		
small species (>1 species?)	5	2.6
medium species (>1 species?)	14	7.4
large species	9	4.8
Totals:	**189**	**100**

[1]Preliminary data.

elements. According to Dodson (1973) the parts of a mouse skeleton which are the most susceptible to being moved are the thoracic vertebrae and maxillae, and the least transportable are the lower jaws and incisors. Third, the process of transport can be damaging to specimens, so abrasion should be observed in at least some cases (e.g., see Korth, 1979). Finally, if the deposit were formed in this way the matrix should be enriched in larger-sized grains, which would also fall out of slow-moving water.

None of the four quarries under consideration is located in a channel deposit, and in no case are the recovered specimens particularly abraded. All four did form rather close to channels. Three of the four (Rose, McNeil, and Dorsey Creek quarries) are located in stage 1 paleosols (see Table 1; Bown and Kraus, 1987). Although this type of paleosol is immature, it forms in an area that is distinctly separate from the actual channel. McKinney Quarry is in an even more weakly developed stage 0 paleosol, immediately adjacent to the channel. However, even this site fails to meet the other two criteria for this mode of origin. The deposit is relatively rich in difficult-to-transport elements, such as lower jaws and rodent incisors, and contains some relatively large post-cranial elements (e.g., *Coryphodon* metapodials) which are unlikely to have been carried along by a current (Voorhies, 1969; Dodson, 1973). Although a detailed study of grain size has not been performed, observation of samples under a binocular dissecting microscope suggests that the majority of the deposit is a very fine-grained mudstone, with only occasional inclusions of larger-grained sand, though McKinney Quarry contains more sand than the other three quarries.

The proximity of McKinney Quarry to the Wasatchian channel does suggest another abiotic possibility to test, however. Immature paleosols were halted in their development as a result of a rapid rate of sedimentation in Wasatchian time. Previous studies (Bown and Beard, 1990) have suggested that the more frequent episodes of deposition associated with immature paleosols improved the probability of preservation of remains. It is possible, therefore, that a concentration could be formed simply as a result of very rapid deposition favoring preservation and fossilization of untransported material. Bown and Kraus (1981) provided support for a more general version of this model to explain the high quality of preservation seen in other parts of the section. In a stage 0 paleosol like the bed which contains McKinney quarry, formed as a result of channel avulsion (Kraus, 1997), there would have been an even more rapid rate of sedimentation than normally found in Willwood deposits, leading to an even higher probability of preservation of passively accumulated remains. The result could be the formation of a concentration simply through the higher probability of preservation of elements being incorporated as the deposit formed. There are two obvious tests of this idea. First, if the localization of the material to the original site was passive, the collection should not be particularly size-biased. Second, parts of the same bed, or deposits formed in the same way, should be comparably concentrated.

In the case of McKinney Quarry the first criterion is met. The known assemblage includes a size range of taxa and elements that stretches from some of the largest mammals known at the time (e.g., *Coryphodon*) to very small taxa (e.g., diminutive rodents and insectivorans). No other stage 0 deposits have been quarried, however, which means that the second test is impossible to assess at this time. It is worth noting, however, that this site seems to be less concentrated than the other quarries under consideration. This would suggest that McKinney Quarry requires the least exceptional or extraordinary explanation of the sites being analyzed here. This hypothesis does not explain, however, the absence of *Apheliscus* (common at Rose, Dorsey Creek, and McNeil quarries), or the unusual omomyid fauna. These features could be related to local variations in faunal distributions, or could require a more complex taphonomic scenario.

A variation on this idea needs to be considered for Dorsey Creek Quarry. This site formed as a splay deposit, resulting from channel avulsion, which also implies very rapid deposition (M.J. Kraus, pers. comm., 1998). This locality does not pass the two tests suggested above, however. Preliminary analysis of the assemblage does indicate a size bias in favor of smaller forms. The bed that contains the quarry has been followed and examined for small mammal remains some distance from the quarry. No other concentrations have been found along strike. Therefore, some other factor needs to be considered to explain the concentration of small mammals at the quarry site. In spite of this, the rapid deposition inevitable in a splay deposit probably did aid in the preservation of mammalian material, even if it cannot be implicated as the original concentrating mechanism.

Another class of abiotic processes that may be considered is catastrophes. Flash floods, sandstorms, or volcanic ash falls, for example, could eradicate a family grouping or herd of animals, leading to an unusual concentration. Bone beds that are heavily enriched in a particular taxon are most plausibly explained in this way (e.g., Williamson, 1998). There are two types of tests for this hypothesis. First, most scenarios that involve a catastrophe require a concentration of animals of the same species, so that the deposit should represent an undiverse assemblage. Second, the age profile for a catastrophic accumulation typically differs from that produced in an attritional setting. Attritional assemblages are generally enriched in very young and very old animals, which are more susceptible to predation and other types of mortality. Catastrophic assemblages follow the age profile of the original population. This implies that they differ from attritional profiles in being more enriched in mature, but not old, individuals (Voorhies, 1969).

Only one of the four quarries could be considered species-poor. McNeil Quarry, with only 17 species, is impoverished relative to the other quarry sites under consideration. However, the presence of 17 species, most of which are represented by more than one individual, is still more speciose than might be expected in a catastrophic assemblage. Although it can be difficult to provide precise age estimates for individuals of extinct taxa, the material from McNeil Quarry is unusually enriched in specimens with deciduous or erupting teeth. Although suggestive of some kind of unusual origin (see below), this does not meet the expectation for a catastrophic assemblage.

4.2. Biotic Processes

Although the abiotic mode of origin for the concentration at McKinney Quarry is still hypothetical, there are no aspects of that assemblage that suggest a particular biotic process. This is not the case, however, for Rose, McNeil, and Dorsey Creek quarries. Generally, biotic processes that are involved in the formation of microvertebrate concentrations can be sub-divided into two basic classes: predatory accumulations and non-predatory accumulations.

Andrews (1990:42) noted that "predators are important accumulators of small mammal bone." He divided predators into three groups: owls, diurnal raptorial birds, and mammalian carnivores. These different types of animals may accumulate bones in concentrations of their scat (mammals) or regurgitated pellets (birds), or by bringing carcasses back to a den or nest, or by storing food (Andrews and Evans, 1983). Bones accumulated by predators tend to get broken or corroded. The extent of breakage and corrosion varies among extant species of owls, raptors, and carnivores examined by Andrews (1990). Thus in modern samples where the potential predators and their breakage patterns are known, it may be possible to identify the predator that made the accumulation. However, known predators are so variable in the accumulations they leave that identifying a fossil accumulation as predator-related, or suggesting which kind of predator was responsible, is much more problematic when the predators involved may belong to extinct species, genera, or orders. For example, both Andrews (1990) and Dodson and Wexlar (1979) found that even different species of owl produced different types of preservation in the bones removed from their pellets, complicating the identification of a particular concentration as the remains of owl pellets. Nevertheless, some general observations are informative (based largely on Mayhew, 1977; Andrews and Evans, 1983; Hoffman, 1988; Andrews, 1990):

A. Owls tend to ingest whole animals, or at least more of the carcass than diurnal raptors do, so bones are less usually damaged.
B. Diurnal raptors break up the bones ingested, but leave more of the carcass behind. Concentrations of their excreta are intermediate in the level of preservation between owls and mammalian carnivores.
C. Mammalian carnivores tend to ingest much of the carcass, including the skeleton. Bones are usually broken and corroded because of the action of teeth (obviously missing in birds) and because the degree of digestion is greater since bones pass through the entire digestive tract. In contrast to carnivore scat, bird pellets containing bones and other indigestible parts of the prey animal are expelled before they pass through the whole digestion process (Hoffman, 1988). The effect of digestion in extant carnivores ranges from mild, with complete dentitions and minimally broken jaws in scat, to moderate rounding with minimal chemical corrosion, to significant abrasion and corrosion. In most cases, jaws and especially skulls are severely broken up. It is worth noting, however, that mandibles and maxillae are among the

> most common elements to be found in carnivore scat (Andrews and Evans, 1983). Only some modern carnivores, such as canids, leave significant tooth puncture marks on bones.

It seems quite probable that predators have been instrumental in microvertebrate fossil accumulations (e.g. Mellett, 1974). Some of the fossils from Rose, McNeil, and Dorsey quarries are consistent with the types of breakage patterns observed in modern predator accumulations, but this is not definitive evidence that predators were responsible for these concentrations. Still, the prevalence of well-preserved jaws and snouts of small mammals in the McNeil Quarry sample suggests that an avian predator, possibly owls, played a part in this small-mammal accumulation. Other features of this sample are consistent with this hypothesis. Particularly, the presence of a high proportion of juveniles is very suggestive, since young mammals make easier prey targets. Of the 171 specimens uncovered from McNeil quarry, 8.6% represent juvenile individuals (Table 4). This is an unexpectedly high proportion for a crevasse-splay deposit (Badgley, 1986).

The size range of taxa preserved at McNeil Quarry complicates this interpretation, however. The assemblage at McNeil Quarry includes *Phenacodus* and *Ectocion*, which are both large taxa relative to the other species represented at the site. Although the *Phenacodus* material apparently all belongs to a partial skeleton representing a single individual, and thus could be explained as an animal that died in the same location as a predatory bird accumulation, *Ectocion* is much more common (12% by MNI). The relative abundance of *Ectocion* is difficult to explain, particularly in light of the lack of toothmarks on these fossils, making a role for a non-avian predator less likely. The taphonomic history of this site seems to be somewhat complex, therefore, and requires further investigation.

Many of the remains from Rose and Dorsey Creek quarries are more suggestive of carnivore accumulation, and tooth marks have been observed on several specimens from Rose Quarry. Although neither of these sites is particularly rich in juvenile material, the size bias towards smaller forms does support this notion.

Definitively accepting this explanation is problematic, however, when considered in the context of the other type of biotic concentrating mechanism: non-predatory assemblages. Non-predatory behaviors can be suggested which would concentrate small bones, but falsifying most of these ideas is difficult.

The only recourse in these cases, then, is to attempt to find a modern pattern of behavior that would produce a concentration that closely matches a fossil situation. This is a less than ideal solution to the problem since there is no reason to believe that modern forms exhaust all the possible patterns of behavior. One possibility is that one or more members of the fauna concentrated bones while constructing a den or nest, in a manner comparable to modern packrats. If such an accumulation were formed in a channel levy, then a sheet flood of the type that produced the splay deposit that contains Dorsey Creek Quarry could redeposit the concentration of bones to more lamellar arrangement, as seen at that site. This possibility merits further consideration, when that quarry is more comprehensively studied.

In summary, these considerations suggest that a predatory origin is unlikely for McKinney Quarry, given its wide size range and lower degree of concentration relative to the other sites. If the scenario given above, involving rapid burial of essentially untransported material, is correct, then this site could be relatively representative of the original fauna at that place and time. It is plausible, if not probable, that avian or mammalian predators were involved in the concentration of the other three accumulations. This in turn suggests that their assemblage composition may not be a completely accurate representation of the fauna.

5. Compositional Comparisons

There are two different types of data that can be used to examine the hypothesis that surface-collected sites are impoverished in the small mammal component of the original fauna. First, the discovery of new taxa, or specimens preserving new information, can provide unequivocal evidence that at least certain types of data are missing from surface sites. Second, the composition of new assemblages can be considered and compared to what has previously been known.

The high quality of preservation found in these quarry deposits has led to the discovery of a number of new taxa. Already published new forms include the omomyid primate *Arapahovius advena* Bown and Rose, 1991, from McKinney Quarry, the insectivore *Batodonoides vanhouteni* Bloch, Rose and Gingerich, 1998, from Dorsey Creek Quarry, and the plesiadapiform *Picromomys petersonorum* Rose and Bown, 1996, also from Dorsey Creek Quarry. All three of these taxa are notable for their small size. *P. petersonorum* is the smallest known plesiadapiform, and smaller than any known euprimate (Rose and Bown, 1996; with the possible exception of a euprimate calcaneus reported from southeastern China; Gebo *et al.*, 1998). *B. vanhouteni* may be the smallest mammal known, either living or extinct (Bloch *et al.*, 1998). Two specimens from Rose Quarry also document a probable new species of *Viverravus*, which may be even smaller than *Viverravus rosei* Polly, 1997 (see Figure 4B). In addition, preliminary analysis of the fauna at Dorsey Creek Quarry has indicated that it includes other new species of insectivorans. The minuteness of these forms probably explains why they have never been uncovered by prospecting in surface-lag deposits, since they would be both difficult to find and likely to be destroyed by weathering soon after being exposed.

In addition to adding new species, the excellent preservation at these sites is resulting in new morphological discoveries in some previously described small taxa. An example of this is the discovery of specimens of the omomyid primate *Steinius vespertinus* that include anterior alveoli (Rose and Bown, 1991). This taxon was first described by Matthew (1915) as ?*Omomys* based on a mandible with M_1-M_3. Discovery of better preserved remains from McKinney

Quarry preserving alveoli indicating a lower dental formula of 2-1-4-3 provided evidence that the taxon belongs near the base of the omomyid radiation (Rose and Bown, 1991; Rose, 1995b).

Other new details of morphology which have been uncovered include the first anterior lower dentition of the condylarth *Apheliscus* from Dorsey Creek Quarry, containing an inclined, spatulate incisor which is unlike the more vertical and pointed incisors of *Hyopsodus* (Gazin, 1968). Also new from these quarries are the first known upper dentitions of the plesiadapiform *Niptomomys thelmae* (from Rose Quarry, Figure 3H), and of the multituberculate *Parectypodus simpsoni* (from McNeil Quarry, Figure 3E). Excellent specimens of previously poorly known insectivoran species have been uncovered from all of the quarries including the best reported specimens of *Eolestes simpsoni* (Figure 3A), *Centetodon neashami*, *Centetodon patratus*, *Talpavoides dartoni* and *Wyonycteris*. Of additional interest are the abundant specimens of the erinaceomorph insectivore *Macrocranion nitens* from all four quarries. Well-preserved jaws, some with anterior teeth (Figure 3G), allow a new appreciation of the level of morphological variation in this otherwise rare taxon. Other notable finds include well-preserved marsupial dentitions, including some with anterior teeth, from Rose Quarry and Dorsey Creek Quarry (Figure 4C), a number of nearly complete miacoid dentitions from Rose Quarry (Figure 4A,B), and many snouts of multituberculates, rodents, and insectivorans from McNeil Quarry (Figure 3G).

Well-preserved specimens of probably new non-mammalian taxa have the potential to add substantially to our understanding of the herpetofauna from this time period. Particularly noteworthy are exceptional dentaries of small iguanids (Figure 6D), such as *Parasauromalus*, from Dorsey Creek and McKinney quarries, and the first remains of anurans from the Willwood formation from Dorsey Creek Quarry and the recently discovered locality WW-12 (see below; Figure 6B,C).

Although these finds of new taxa and rare specimens give some indication that surface localities may be missing a component of the overall mammalian composition, isolated instances do not indicate the relative importance of this component to the total make-up of the Wasatchian fauna. To consider this question, it is necessary to look at the composition of localities as a whole.

As discussed above, the mammalian composition of Willwood Wasatchian localities collected by surface-prospecting is dominated by four taxa: *Hyopsodus*, *Hyracotherium*, *Cantius*, and *Diacodexis* (Bown, 1979; Rose, 1981a; Bown *et al.*, 1994b), representing the orders Condylarthra, Perissodactyla, Primates, and Artiodactyla, respectively. In most studied Wasatchian localities (e.g., Figure 1) these four orders, represented principally by these four genera, make up 50–70% of the minimum individual count. This observation is true in both relatively poorly sampled and poorly fossiliferous localities, and in very rich sites with thousands of identifiable specimens. The surface locality D-1460 surrounding Rose Quarry, follows this typical pattern (Table 2, Figure 2B).

Taxa belonging to these four orders account for only 32% and 46%, respectively, of the identified specimens (Figures 2 and 5) in the assemblages from

Rose and McKinney quarries. Furthermore, at Rose Quarry the majority of the condylarth specimens do not belong to the genus *Hyopsodus*, but instead represent the genus *Apheliscus*, which is rare in surface localities (and absent at D-1460 surface—see Table 2). At McKinney Quarry, most of the condylarths are *Hyopsodus* (and *Apheliscus* is unrepresented; see Table 3), but the primate fauna at this locality differs from that of most Wasatchian sites in being dominated by omomyids rather than adapids. In this case, Rose Quarry is more like typical surface localities, in that all the known euprimates from this quarry are adapids, although they are less frequent at the quarry than in surface-prospected sites. These observations make it clear that even in areas of apparent similarity between surface and quarry localities, there are some substantial differences.

More obvious contrasts with surface assemblages are evident in the relative proportions of proteutherians, lipotyphylan insectivorans, rodents, marsupials and plesiadapiforms (the last group is included separately from primates or dermopterans in all tables and figures since their ordinal position is in debate). In surface localities these groups typically form a minor part of the overall composition (see Table 2, Figures 1 and 2B, and Bown *et al.*, 1994b), usually less than 10% of identifiable specimens. At McKinney Quarry, however, the combined total percentage for these taxa is 42%, while at Rose Quarry the sum is 55% (see Tables 2 and 3, and Figures 2 and 5). Since these four groups include most of the Wasatchian species of small size, their abundance is strongly suggestive of a size bias in the composition of surface-prospected localities. χ^2 tests at the ordinal level comparing D-1460 surface to McKinney and Rose quarries were both highly significant ($p < 0.0005$).

As noted earlier, there are also some marked differences between Rose Quarry and McKinney Quarry (χ^2 test at the ordinal level, $p < 0.0005$), in spite of the fact that these quarries are nearly contemporaneous and less than 15 km apart. The differences in the condylarth and primate compositions are described above. In addition, finer-scale differences are apparent in a comparison between the composition lists for the two sites. For example, Rose Quarry has 13 specimens, out of 1005 dental/gnathic mammalian remains, belonging to the Multituberculata, whereas this group is unknown from McKinney Quarry ($n = 642$). McKinney Quarry includes remains of a number of taxa of larger body size, such as the pantodont *Coryphodon* and the creodont *Oxyaena*, which are missing from the fauna at Rose Quarry. In general McKinney Quarry includes more large and medium-sized forms than occur at Rose Quarry which, as already noted, appears to be biased towards smaller forms.

Given the apparent size bias at Rose Quarry, and the unexplained, unusual features of the composition of McKinney Quarry, it is clear that neither can be argued to be representative of the general composition of the fauna during this time period. In spite of these cautions, there are a number of points of similarity between these two sites that may be meaningful to understanding generalities about the composition of the mammalian fauna in the Wasatchian. As mentioned above, small-bodied forms are both common and diverse in the quarry localities, unlike surface localities, suggesting that the latter may be biased against small forms. The high relative frequency of certain of these taxa at both McKinney Quarry and Rose Quarry suggests that they were much more common than is indicated by surface collections. The two most obvious examples are the erinaceomorph insectivoran *Macrocranion nitens* (12.7% at Rose Quarry and 4.7% at McKinney Quarry) and rodents (14.0% at Rose Quarry and 12.3% at McKinney Quarry). At D-1460 surface *M. nitens* is completely unknown, and the rodents account for only 2.8% of the fauna. In other well-sampled surface localities in this part of the section *M. nitens* never accounts for as much as 1%, and rodents are usually well below 5% of the fauna (e.g. see Bown *et al.*, 1994b; note that rodents are more common at some Wasatchian assemblages discussed in Rose, 1981a). This observation strengthens the assertion that surface collections are generally size-biased against small taxa. A similar situation in terms of the rodent fauna is apparent in Clarkforkian deposits, in that surface deposits from this time period generally produce fewer rodents than quarry sites (Rose, 1981a; Krause, 1986).

Preliminary study of the sample from Dorsey Creek Quarry indicates that *M. nitens* (14.3%) and rodents (14.8%) are also very common at that locality. These two taxa also make up a significant part of the McNeil Quarry assemblage, despite its other peculiarities (Table 4, Figure 7). As mentioned above the highest relative frequency for a single group in the McNeil Quarry assemblage is for multituberculates. Such an abundance of multituberculates is unprecedented in Wasatchian deposits, and makes this quarry look much more like a Tiffanian or even Torrejonian assemblage than one from the early Eocene (Figure 8).

The condylarth *Apheliscus* is similarly rare at surface sites and moderately common in three of these quarry assemblages (Dorsey Creek, McNeil, and Rose quarries), which suggests that it may also have been more common than previously suspected. The absence of this taxon at McKinney Quarry is somewhat puzzling in this light, and may support a microhabitat difference between this site and the other quarries, as suggested by other faunal contrasts as well as its apparent closer proximity to a channel.

Another area of similarity between Rose Quarry and McKinney Quarry is the shape of the rank abundance curves calculated for the two localities (Figure 9). The rank abundance curves for both quarry assemblages closely follow the prediction given by the broken stick model (χ^2 tests for each quarry, $p > 0.25$), indicating a very evenly distributed fauna, which could imply a relatively stable environment. However, the rank abundance curve for the surface-prospected locality D-1460 shows a typical "type four" pattern, with the most common taxa being represented at higher frequency than predicted by the model (not surprisingly *Hyopsodus*, *Diacodexis*, *Cantius*, and *Hyracotherium* in that order). A χ^2 test comparing the distribution of individuals across taxa indicates a highly significant difference ($p < 0.0005$) between the observed and expected counts for the broken-stick model. These indications could suggest a less stable ecological setting. However, the beds that make up D-1460 surface are adjacent to and essentially contemporaneous with the quarry. Although the surface-prospected locality is from a somewhat more mature paleosol (stage 3) than the quarry (stage 1), surface assemblages of comparative maturity (e.g. D-1198, stage 1—see Figure 1) are more similar to the assemblage from D-1460 surface than to the Rose Quarry collection. Thus it seems unlikely that the quarry and surface assemblages come from significantly different environments.

This suggests an important caveat to the use of rank abundance curves. This technique should be used cautiously for comparing samples that are biased in favor of certain parts of the fauna, either in terms of size or in terms of an over-representation of particular taxa for taphonomic reasons. At the very least, samples should only be compared to one another if they are thought to have formed under similar conditions and were collected using the same techniques. In these cases the assemblages should engender similar types of biases. For example, the differences between Rock Bench Quarry and Cedar Point Quarry noted by Rose (1981b) are likely to be more meaningful than the contrasts noted above because the sites are very similar in geologic context and were collected in the same way.

Another type of analysis which may be inappropriate for size-biased collections is the evaluation of cenograms (Legendre, 1989). Although this methodology can be valuable for providing environmental data for sites that lack other indicators, it was not applied in this case because of concerns about biases in these quarry samples.

6. Future Work

While excavation at Rose, McKinney, and McNeil quarries has concluded because the productive layer is played out, Dorsey Creek Quarry is still actively producing (over 75 jaws were found in the 1998 field season). Considerable preparatory work and sorting remain before the site can be completely studied.

In view of the interesting and unique nature of the information that these types of concentrations can provide, finding further quarry sites has become a particular focus of continuing prospecting efforts in the southern Bighorn Basin. In 1996 another prospective quarry site (locality WW-12) was located by a surface concentration of small vertebrate remains. Screen-washing of matrix from this site has revealed a particularly interesting molluscan component of the fauna, unknown from any other site—a diversity of microgastropods recovered from fine concentrate left in 0.5 mm mesh screens. Included are terrestrial pupillids and discids and aquatic ellobiids and physids (Figure 6A), many of which probably belong to new taxa (J.H. Hartman, pers. comm., 1998). It is notable that similarly processed fine concentrate from McNeil and Dorsey Creek quarries contains no minute gastropods, strongly suggesting microhabitat differences between this locality and the other quarry sites. The site has also revealed some of the first anuran material identified from the Willwood formation (Figure 6B, C; Dorsey Creek Quarry has also produced anuran material, see above). Future work will include quarrying and additional screen washing at WW-12.

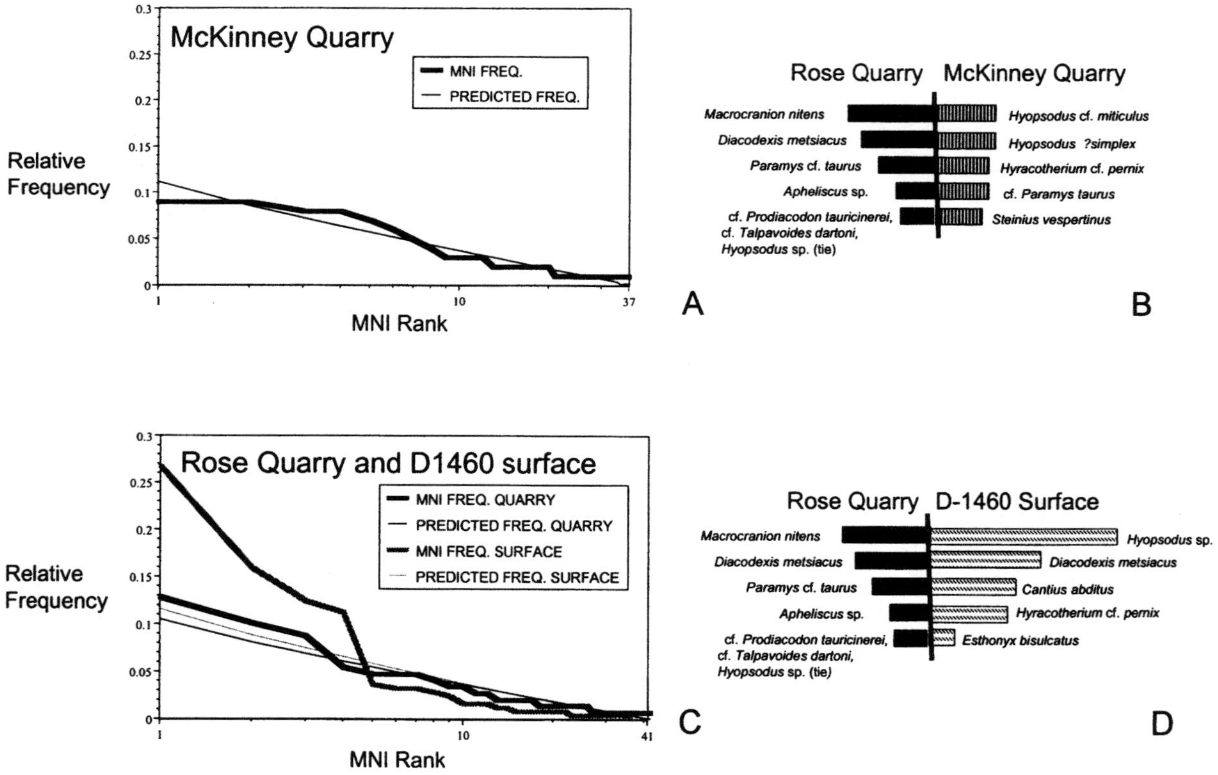

FIGURE 9. A—Rank abundance curve for McKinney Quarry.
B—Comparison of top 5 mammalian taxa from Rose Quarry and McKinney Quarry, based on %MNI.
C—Rank abundance curve for Rose Quarry and D-1460 surface.
D—Comparison of top 5 mammalian taxa from Rose Quarry and D-1460 surface, based on %MNI.

7. Conclusions

Study of Wasatchian microvertebrate quarry assemblages calls into question the reliability of surface-prospected sites for a complete understanding of the mammalian composition during this time period. It is true that the current poor understanding of concentrating and depositional mechanisms for these quarry localities adds necessary caveats to their interpretation. However, the new species that are being discovered, the differing relative frequencies between quarry and surface collections, and the similarities in make-up between otherwise very different quarry assemblages support the assertion that certain parts of the fauna are missing from surface collections. *Macrocranion nitens*, rodents, and *Apheliscus* appear, in particular, to have been more common than previously supposed.

This provides partial support for Winkler's (1983) hypothesis that surface-prospected sites are missing part of the original fauna (see also Wolff, 1975). The two groups that Winkler considered to be particularly under-represented are the Insectivora and the Multituberculata. Support for under-representation of the former is quite strong, given that this grouping forms a large part of all of the quarry assemblages under consideration here, and that a number of new taxa of insectivorans have been uncovered from these sites. In the case of multituberculates the situation is less clear. This group obviously forms an important component of the fauna at McNeil Quarry, but this site is probably heavily biased. Multituberculates are also fairly common at Rose and Dorsey Creek quarries, but are missing completely at McKinney Quarry. This raises the likelihood that multituberculates were locally somewhat more common than had previously been supposed, but did not actually form a major portion of the fauna in the early Eocene.

The conclusion that surface prospected assemblages may be misleading in some ways could seem rather discouraging, considering how important that type of site is to the study of early Tertiary forms. However, this underscores the importance of continued field efforts even in areas as intensively studied as the Bighorn Basin of Wyoming.

ACKNOWLEDGMENTS. We take this opportunity to thank all of the people who have worked at the quarry localities discussed in this paper, including K.C. Beard, T.M. Bown, M. Brown, F. Bungenstock, A. Chew, N. Egi, D.L. Gebo, R.E. Heinrich, L. Hlusko, D.G. Kron, G. Krovitz, K.C. McKinney, P. and L. McNeil, V. Morgan, J.C. Mussell, V.E. Noble, D.T. Rasmussen, I. Raufuss, J.J. Rose, S. Senturia, M. Shekelle, and M. Shimamura, without whose efforts these collections would not exist. Special thanks are owed to D.G. Kron, whose screen-washing protocol has literally added a new dimension to the remains that have been recovered. We are grateful to G. Argyros and W.S. Bartels for identifying some of the fossil herpetological material and to J.H. Hartman and B. Roth for identifications of the gastropods from WW-12. Our appreciation goes to T.M. Bown and M.J. Kraus for geological insights into these localities. Elaine Kasmer expertly prepared the drawings in Figures 4 and 6. T.M. Bown, J.C. Mussell, and two anonymous reviewers read an earlier version of this paper, and suggested many improvements. We would like to express our gratitude to G.F. Gunnell for inviting us to participate in this symposium. This research was conducted under BLM permits, and with support from the USGS and NSF grants BSR-8215099, BSR-8500732, BSR-8918755 and IBN-9419776 to KDR.

References

Andrews, P., 1990, *Owls, caves and fossils*, University of Chicago Press, Chicago.

Andrews, P., and Nesbit Evans, E. M., 1983, Small mammal bone accumulations produced by mammalian carnivores, *Paleobiology* **9(3)**:289–307.

Badgley, C., 1986, Counting individuals in mammalian fossil assemblages from fluvial environments, *Palaios* **1**:328–338.

Bloch, J. I., Rose, K. D., and Gingerich, P. D., 1998, New species of *Batodonoides* (Lipotyphla, Geolabididae) from the Early Eocene of Wyoming: smallest known mammal?, *Journal of Mammalogy* **79**:804–827.

Bown, T. M., 1979, Geology and mammalian paleontology of the Sand Creek Facies, Lower Willwood Formation (Lower Eocene), Washakie County, Wyoming, *Geological Survey of Wyoming Memoirs* **2**:1–151.

Bown, T. M., and Beard, K. C., 1990, Systematic lateral variation in the distribution of fossil mammals in alluvial paleosols, lower Eocene Willwood Formation, Wyoming, in: *Dawn of the Age of Mammals in the northern part of the Rocky Mountain Interior, North America* (T. M. Bown, and K. D. Rose eds.) Geological Society of America Special Paper No. 243, Boulder, Colorado, pp. 135–151.

Bown, T. M., and Kraus, M. J., 1981, Vertebrate fossil-bearing units (Willwood Formation, Lower Eocene, Northwest Wyoming, U.S.A.): implications for taphonomy, biostratigraphy, and assemblage analysis, *Palaeogeography, Palaeoclimatology, Palaeoecology* **34**:31–56.

Bown, T. M., and Kraus, M. J., 1987, Integration of channel and floodplain suites, I. Developmental sequence and lateral relations of alluvial paleosols, *Journal of Sedimentary Petrology* **57**:587–601.

Bown, T. M., and Kraus, M. J., 1993, Time-stratigraphic reconstruction and integration of paleopedologic, sedimentologic, and biotic events (Willwood Formation, lower Eocene, northwest Wyoming, USA), *Palaios* **8**:68–80.

Bown, T. M., and Rose, K. D., 1987, Patterns of dental evolution in early Eocene anaptomorphine primates (Omomyidae) from the Bighorn Basin, Wyoming, *Paleontological Society Memoir* **23**:1–162.

Bown, T. M., and Rose, K. D., 1991, Evolutionary relationships of a new genus and three new species of omomyid primates (Willwood Formation, lower Eocene, Bighorn Basin, Wyoming), *Journal of Human Evolution* **20**:465–480.

Bown, T. M., Holroyd, P. A., and Rose, K. D., 1994a, Mammal extinctions, body size, and paleotemperature, *Proceedings of the National Academy of Sciences, USA* **91**:10403–10406.

Bown, T. M., Rose, K. D., Simons, E. L., and Wing, S. L., 1994b, Distribution and stratigraphic correlation of upper Paleocene and lower Eocene fossil mammal and plant localities of the Fort Union, Willwood, and Tatman Formations, Southern Bighorn Basin, Wyoming, *United States Geological Survey Professional Paper* **No. 1540**:1–103.

Dodson, P., 1973, The significance of small bones in paleoecological interpretation, *University of Wyoming Contributions to Geology* **12(1)**:15–19.

Dodson, P., and Wexlar, D., 1979, Taphonomic investigations of owl pellets, *Paleobiology* **5(3)**:275–284.

Gazin, C. L., 1968, A study of the Eocene condylarth mammal *Hyopsodus*, *Smithsonian Miscellaneous collections* **153(4)**:1–89.

Gebo, D. L., Dagosto, M., Beard, K. C., and Qi, T., 1998, The smallest primate?, *American Journal of Physical Anthropology* **Supplement 26**:86.

Gingerich, P. D., 1980, History of Early Cenozoic vertebrate paleontology in the Bighorn Basin, in: *Early Cenozoic paleontology and stratigraphy of the Bighorn Basin, Wyoming* (P. D. Gingerich, ed.), University of Michigan Museum, Papers on Paleontology, Ann Arbor, No. 24, pp. 1–6.

Gingerich, P. D., 1989, *New earliest Wasatchian mammalian fauna from the Eocene of Northwestern Wyoming: composition and diversity in a rarely sampled high-floodplain assemblage*, University of Michigan Museum, Papers on Paleontology, Ann Arbor, No. 28.

Grayson, D. K., 1978, Minimum numbers and sample size in vertebrate faunal analysis, *American Antiquity* **43(1)**:53–65.

Hoffman, R., 1988, The contribution of raptorial birds to patterning in small mammal assemblages, *Paleobiology* **14(1)**:81–90.

Hutchinson, G. E., 1961, The paradox of the plankton, *American Naturalist* **95**:137–145.

Korth, W. W., 1979, Taphonomy of microvertebrate fossil assemblages, *Annals of Carnegie Museum* **48(15)**:235–285.

Kraus, M. J., 1997, Lower Eocene alluvial paleosols: Pedogenic development, stratigraphic relationships, and paleosol/landscape associations, *Palaeogeography, Palaeoclimatology, Palaeoecology* **129**:387–406.

Kraus, M. J., 1998, Development of potential acid sulfate paleosols in Paleocene floodplains, Bighorn Basin, Wyoming, USA, *Palaeogeography, Palaeoclimatology, Palaeoecology* **144**: 203–224.

Krause, D. W., 1982, Multituberculates from the Wasatchian land-mammal age, early Eocene, of western North America, *Journal of Paleontology* **56(2)**:271–294.

Krause, D. W., 1986, Competitive exclusion and taxonomic displacement in the fossil record: the case of rodents and multituberculates in North America, in: *Vertebrates, phylogeny, and philosophy* (K. M. Flanagan, and J. A. Lillegraven eds.), The University of Wyoming, Laramie, Contributions to Geology Special Paper No. 3, pp. 95–117.

Legendre, S., 1989, Les communauté de mammifères du Paléogène (Eocène supérieur et Oligocène) d'Europe occidentale: structures, milieux et évolution, *Münchner Geowissenschaftliche Abhandlungen, Reihe A Geologie und Paläontologie* **16**:1–110.

Maas, M. C., Anthony, M. R. L., Gingerich, P. D., Gunnell, G. F., Krause, D. W., 1995, Mammalian generic diversity and turnover in the Late Paleocene and Early Eocene of the Bighorn and Crazy Mountain Basins, Wyoming and Montana, USA, *Palaeogeography, Palaeoclimatology, Palaeoecology* **115**:181–207.

MacArthur, R. H., 1957, On the relative abundance of bird species, *Proceedings of the National Academy of Sciences, USA* **43**:293–295.

MacArthur, R. H., 1960, On the relative abundance of species, *American Naturalist* **44**:25–36.

Matthew, W. D., 1915, A revision of the lower Eocene Wasatch and Wind River faunas. Part IV: Entelonychia, Primates, Insectivora (part), *Bulletin of the American Museum of Natural History* **34**:429–483.

Mayhew, D. F., 1977, Avian predators as accumulators of fossil mammal material, *Boreas* **6**:25–31

McKenna, M. C., 1960, Fossil Mammalia from the early Wasatchian Four Mile fauna, Eocene of northwest Colorado, *University of California, Publications in Geological Sciences* **37**:1–130.

Mellett, J. S., 1974, Scatological origin of microvertebrate fossil accumulations, *Science* **185**: 349–350.

O'Leary, M. A., 1996, *Dental evolution in the early Eocene Notharctinae (Primates, Adapiformes) from the Bighorn Basin, Wyoming: documentation of gradual evolution in the oldest true primates*, PhD dissertation, Johns Hopkins University School of Medicine, Baltimore MD.

Polly, P. D., 1997, Ancestry and species definition in paleontology: a stratocladistic analysis of Paleocene-Eocene Viverravidae (Mammalia, Carnivora) from Wyoming. *University of Michigan, Contributions from the Museum of Paleontology* **30(1)**:1–53.

Rose, K. D., 1981a, *The Clarkforkian Land-Mammal Age and mammalian faunal composition across the Paleocene-Eocene boundary*, University of Michigan Museum, Ann Arbor, Papers on Paleontology No. 26.

Rose, K. D., 1981b, Composition and species diversity in Paleocene and Eocene mammals assemblages—an empirical study, *Journal of Vertebrate Paleontology* **1**:367–388.

Rose, K. D., 1990, Postcranial skeletal remains and adaptations in early Eocene mammals from the Willwood Formation, Bighorn Basin, Wyoming, in: *Dawn of the Age of Mammals in the north-*

ern part of the Rocky Mountain Interior, North America (T. M. Bown, and K. D. Rose eds.) Geological Society of America Special Paper No. 243, Boulder, Colorado, pp. 107–133.

Rose, K. D., 1995a, Anterior dentition and relationships of the early Eocene omomyids *Arapahovius advena* and *Teilhardina demissa*, sp. nov., *Journal of Human Evolution* **28**:231–244.

Rose, K. D., 1995b, The earliest Primates, *Evolutionary Anthropology* **3(5)**:159–173.

Rose, K. D., and Bown, T. M., 1991, New fossil evidence on differentiation of the earliest Euprimates, *Proceedings of the National Academy of Sciences, USA* **88**:98–101.

Rose, K. D., and Bown, T. M., 1996, A new plesiadapiform (Mammalia: Plesiadapiformes) from the early Eocene of the Bighorn Basin, Wyoming, *Annals of Carnegie Museum* **65(3)**:305–321

Schankler, D., 1980, Faunal zonation of the Willwood Formation in the central Bighorn Basin,Wyoming. in: *Early Cenozoic paleontology and stratigraphy of the Bighorn Basin, Wyoming* (P. D. Gingerich, ed.), University of Michigan Museum, Papers on Paleontology, Ann Arbor, No. 24, pp. 1–6.

Silcox, M. T., and Rose, K. D., 1996, Remarkable new microfaunal assemblages from the Early Eocene of the Bighorn Basin, Wyoming, *Journal of Vertebrate Paleontology* **16(supplement to 3)**:66A.

Silcox, M. T., and Rose, K. D., 1997, Stasis and gradual change in *Diacodexis* from the Early Eocene of the Bighorn Basin, Wyoming, *Journal of Vertebrate Paleontology* **17(supplement to 3)**:76A.

Voorhies, M. R., 1969, Taphonomy and population dynamics of an Early Pliocene Vertebrate fauna, Knox County, Nebraska, *University of Wyoming Contributions to Geology* **special paper No. 1**:1–69.

Williamson, T. E., 1998, *Meniscotherium* mass-death assemblages, *Journal of Vertebrate Paleontology*, **18(supplement to 3)**:86A.

Winkler, D. A., 1983, Paleoecology of an early Eocene mammalian fauna from paleosols in the Clarks Fork Basin, Northwestern Wyoming (U.S.A.), *Palaeogeography, Palaeoclimatology, Palaeoecology* **43**:261–298.

Wolff, R. G., 1975, Sampling and sample size in ecological analysis of fossil Mammals, *Paleobiology* **1(2)**:195–204.

Chapter 6

Stratigraphy and Taphonomy of Grizzly Buttes, Bridger Formation, and the Middle Eocene of Wyoming

JOHN P. ALEXANDER and BENJAMIN J. BURGER

1. Introduction

For over a century the Bridger Formation of southwestern Wyoming has yielded fossils of many important middle Eocene vertebrate taxa. This formation is of particular interest to the study of ancient primates because their remains are plentiful. In fact, the first fossil primates described from the Western Hemisphere, *Omomys* carteri and *Notharctus tenebrosus* were found here (Leidy, 1869, 1870). Arguably, the premier fossil collecting locale within the Green River Basin is an area known as Grizzly Buttes (Figure 1). Although

JOHN P. ALEXANDER and BENJAMIN J. BURGER • Division of Paleontology, American Museum of Natural History, New York, New York 10024-5192.

Eocene Biodiversity: Unusual Occurrences and Rarely Sampled Habitats, edited by Gregg F. Gunnell, Kluwer Academic/Plenum Publishers, New York, 2001.

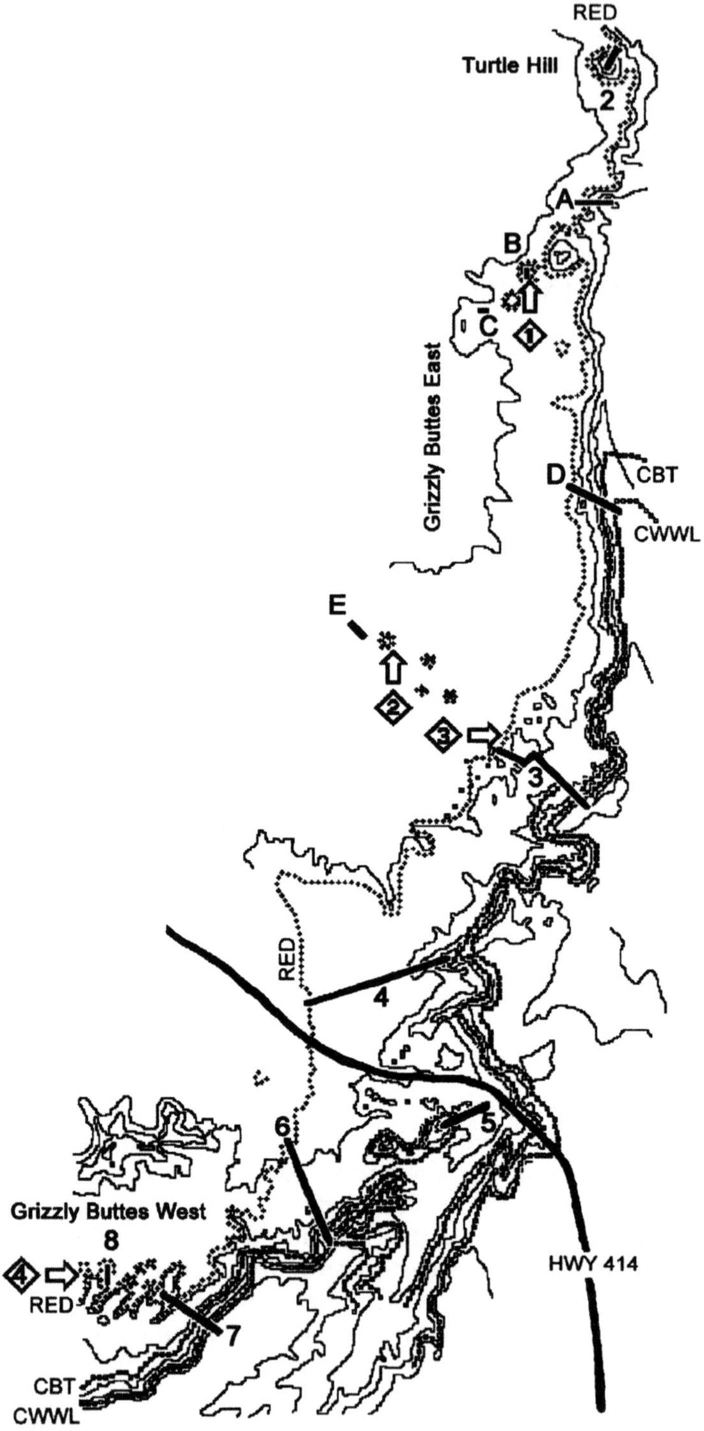

FIGURE 1. Map of Grizzly Buttes, southern Green River Basin, Uinta Co. Wyoming. Diamond symbols indicate principal fossil bearing areas which have produced articulated skeletons. ⟨1⟩ "Cosmodrome", ⟨2⟩ "Windy City", ⟨3⟩: "Forbidden City", ⟨4⟩ "Donner Pantry". Letters A-D and numbers 1–8 indicate position of sections in Figure 2 and Figure 3. RED: Grizzly Buttes Red Bed, CBT: Church Butte Tuff , CWWL: Cottonwood White Layer. HWY 414: Wyoming State Highway 414 = old "Burnt Fork Road". Section 4 = approximately 1 km.

the place name has disappeared from official maps and the local lexicon (West, 1990), it is well established in the paleontological literature. The area has a rich history. Both O. C. Marsh and E. D. Cope collected from these outcrops in the 1870's extending their professional rivalry to the description of Bridgerian mammals (Wheeler, 1961). The area is figured by H. F. Osborn (1910) as the "famous" locality of Grizzly Buttes.

Standing within the Blacks Fork Member of the Bridger Formation, the Grizzly Buttes are a set of low hills eroding out of dissected cliff exposures extending in an arc to the southeast of the towns of Lyman and Mountain View, Wyoming. Field parties from many institutions probably collected here between 1870 and 1900, but it was Walter Granger of the American Museum of Natural History (AMNH) who first documented Grizzly Buttes as a distinct area in his unpublished field report on the expedition of 1903. He identified Grizzly Buttes East as the exposures running 7 km along the eastern side of Smith's Fork of the Green River and Grizzly Buttes West as the outlying hills and walls now known as Crooked Canyon which run 6 km westward to the mouth of Little Dry Creek. The dividing line between the two is Wyoming State Highway 414, known to Granger as the "Henry's Fork road". Granger stated "It is, in fact, probably the richest spot (in fossil remains) in the entire Bridger bad lands".

The sheer volume of vertebrate fossils collected by the Granger expeditions lead to a body of scholarship in which many type specimens were named. These works include several major monographs (Hay, 1908; Matthew, 1909; Gregory, 1920) and numerous papers describing specimens decades after their original discovery (Simpson, 1931). As these earlier collections refer to stratigraphic or aerial distinctions of no greater resolution than "upper" or "lower" beds or Grizzly Buttes "East" or "West", emphasis will be placed on more recent collections. Since 1983, field parties from the AMNH have returned to Grizzly Buttes and recovered many vertebrate skeletons (Alexander, 1992; Wallace, 1994). Among these specimens are well preserved skulls, jaws and skeletons of notharctid and omomyid primates, creodonts, carnivores, rodents, palaeanodonts, perissodactyls and artiodactyls.

Remarkably, the stratigraphic relationships within Grizzly Buttes has never been closely examined. Walter Granger, in his field notes of 1903, noted that attempts to identify local marker beds and to map the stratigraphy of this area were abandoned after a few days due to the complexity of the problem. This deficiency has presently been addressed. Utilizing standard survey methods the first detailed geologic cross sections of Grizzly Buttes (Figures 2–3) were produced for this study in 1996 and 1997. The stratigraphic position of AMNH fossil producing localities are indicated by number on these columns (Table 1). Important beds are described and informally named. The depositional history of each fossiliferous layer is inferred based on field observations of mineralogy, weathering, repose, texture and clast size.

Dr. Emmett Evanoff of the University of Colorado at Boulder has conducted geologic fieldwork in the basin since 1991. He has refined Matthew's stratigraphic system of 1909 by identifying important new stratigraphic marker units

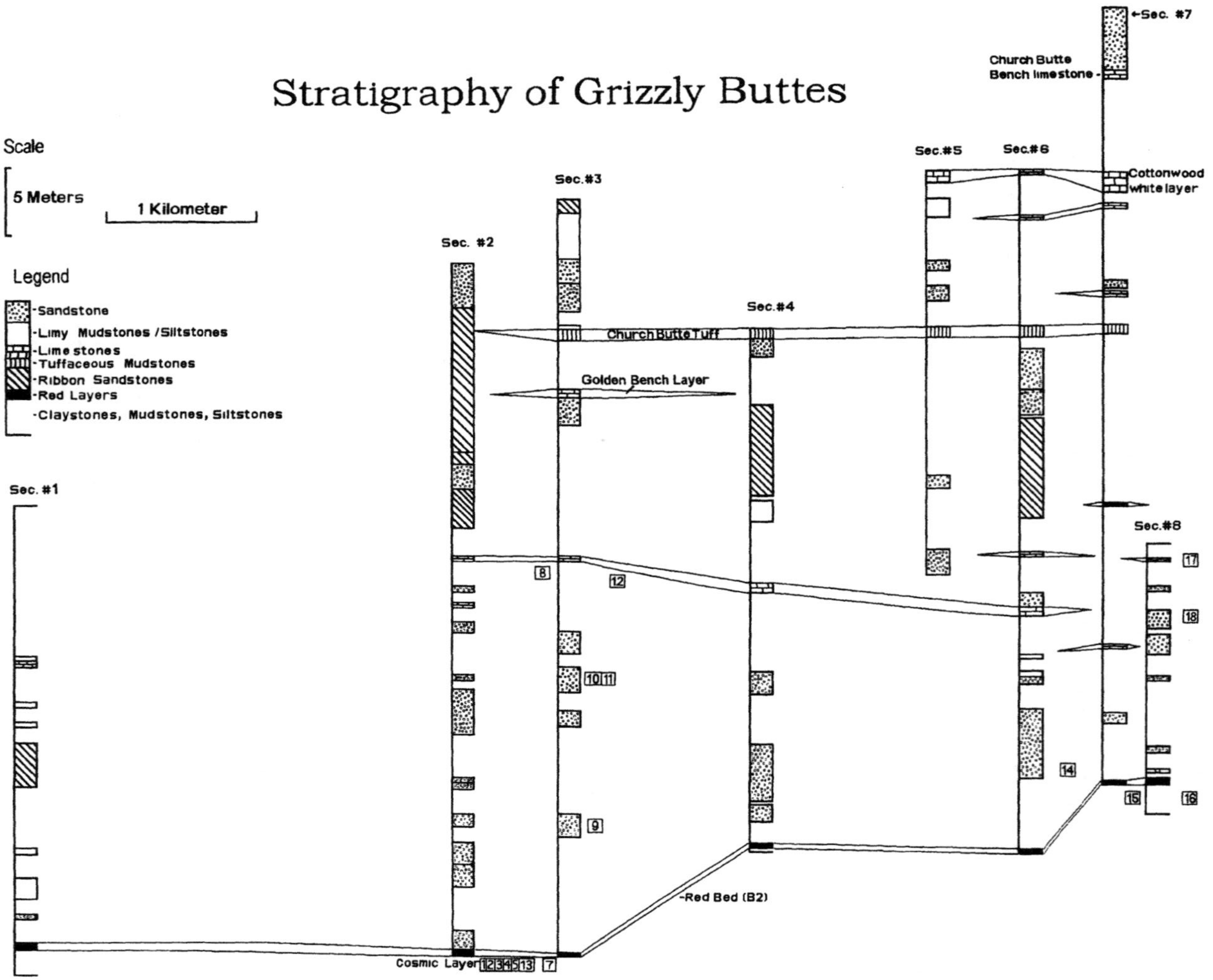

FIGURE 2. Stratigraphic sections, Grizzly Buttes (see text for further discussion).

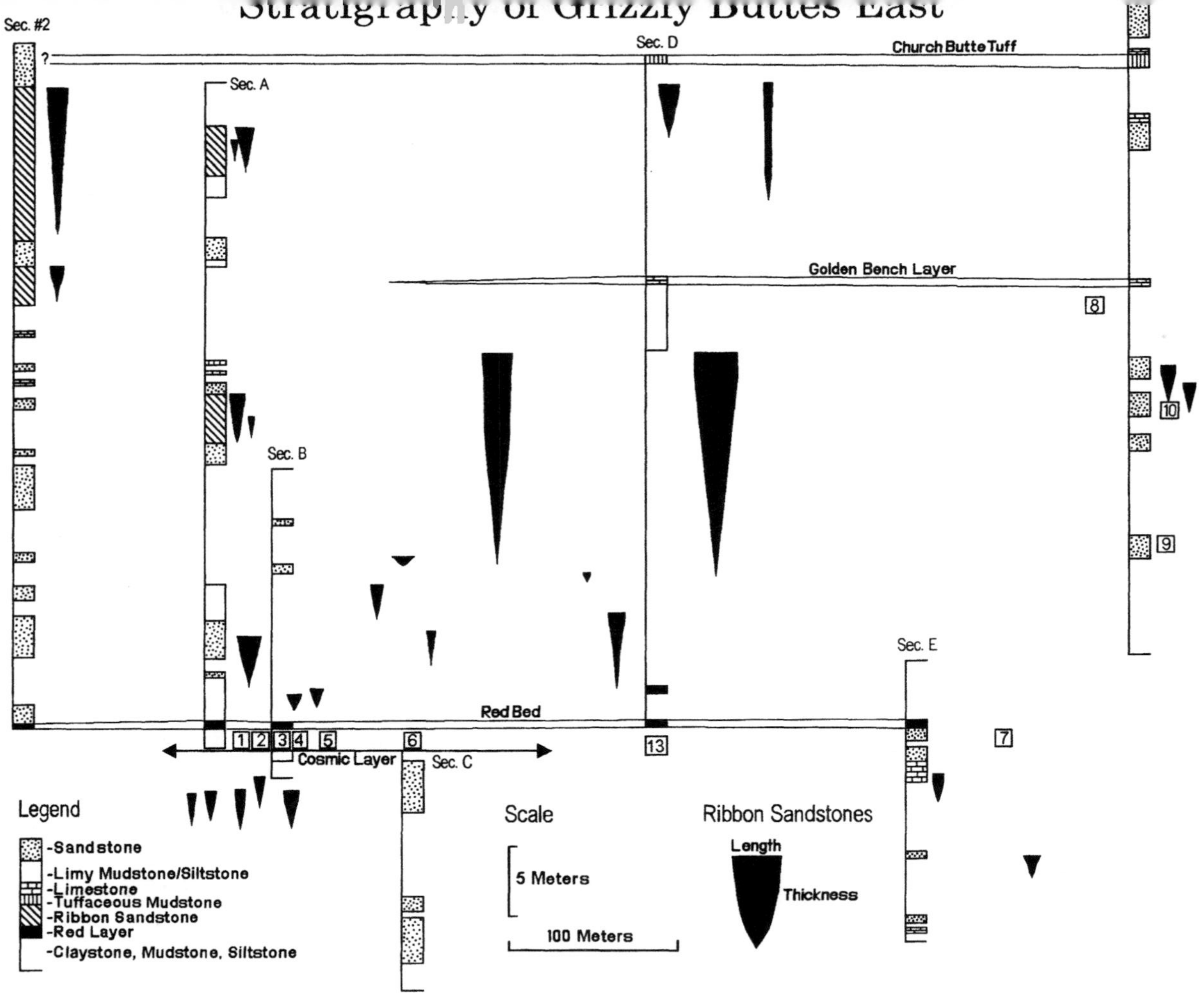

FIGURE 3. Stratigraphic sections, Grizzly Buttes East (see text for further discussion).

Table 1. Grizzly Buttes localities [#] appearing in Figures 2 and 3

Horizon 1: "Cosmic Layer" or "Red Bed" Localities, 47.3 m below Church Butte Tuff

[1] Amphitheatre, Gallery 1-3, Hanging Gallery, Painted Turtle, Pink Floyd, Powerline Hill, Trionyx Hill
[2] Conquistador Hill
[3] Butch Hill
[4] Cosmic Hill
[5] Wrong Hill
[6] Thule Butte
[7] Windy City: Barbs Butte, Dawn of Man, Franks Ridge, Leahanne's Lintel, Stock Yard
[13] Hernia Hill
[15] Corner Hill, Donner Pantry, Outer Wall, Donner Wall
[16] Spanish Saddle
[17] Golden Crown, Gator Bowl

Horizon 2: Crooked Canyon Localities, 33.6 m below Church Butte Tuff

[14] Crooked Crotch, Crooked Crotch North, Wasp Waste, Horseshoe Corner

Horizon 3: Forbidden City/Woodburne Localities, 26.4 m below Church Butte Tuff

[8] East Berlin, Forbidden City North, Serpent Mound, Tiergarten
[9] Great Wall, Northgate
[10] Hermes Hill
[11] Inner Sanctum
[12] Aslan Gate, Simba Gate

such as the "Church Butte tuff" and the "Golden" limestone (Evanoff *et al.*, 1998). We follow his system.

2. Geology

The Bridger Formation consists of 650 m of mostly horizontal fluvial and lacustrine beds deposited in the Green River Basin, an asymmetrical Laramide downwarp, with roughly a north-south axis from the Wind River Range to the Uinta Mountains (Evanoff and Ivy, 1992). Its sedimentary sources are airborne and fluvially transported pyroclastics from the Absaroka volcanic field 300 km to the north (Bradley, 1964; Smedes and Prostka, 1972). Smectite, a clay mineral derived from altered volcanic glass, is common in the mudstones. It causes the rock to swell when wet, weathering outcrops into crumbly "popcorn" surfaces.

Most museum Bridger collections are correlated utilizing W. D. Matthew's (1909) stratigraphic divisions A through E. Bridger A and B are grouped into the Black's Fork Member (Wood, 1934). With a maximum thickness of 137 m (Evanoff and Ivy, 1992), the Bridger B interval is subdivided into units B1 through B5 (Matthew, 1909). It is delineated from the Bridger A below by a thick ostracodal limestone and from the Bridger C above by the Sage Creek white layer, a thick bed of resistant limestone. Grizzly Buttes includes the uppermost B1 beds and much of the B2 interval and falls within the Br-2 biostratigraphic zone (Gunnell and Bartels, 1994). Bridger Formation localities and biostratigraphy are well documented (McGrew and Sullivan, 1970; Gazin, 1976; West, 1976).

Mauger (1977) provided potassium-argon radiometric age determinations on biotite from the Big Island Tuff within the Wilkins Peak Member of the Green River Formation which is laterally correlative with the lower Bridger Formation. Recalibrated dates of 50.4 +/ 1.1 and 50.1 +/ 1.2 MA appear in Krishtalka *et al.* (1987). At about 50 MA, the Bridger B interval falls within the first million years of the Bridgerian Land Mammal Age (Krishtalka *et al.*, 1987).

2.1. Palaeoclimate

Toward the east, the Lower Bridger Formation interdigitates with the Laney Shale Member of the Green River Formation which was laid down in ancient Lake Gosiute. Flora from the Laney Member indicate a warmer climate during the middle Eocene than at present, with an average temperature of 19° C (MacGinitie, 1969). The global thermal maximum for the entire Cenozoic reached its zenith during the earliest middle Eocene (Savin, 1977; Savin and Douglas, 1984). The Grizzly Buttes local fauna flourished during a period of global warming possibly due a carbon dioxide greenhouse effect (Rea *et al.*, 1989).

The arboreal nature of primates makes their variety and abundance good indicators of the degree of forest density and diversity that existed at Grizzly Buttes. The diversity of North American primate genera is never again as great as it is during the time period which brackets the deposition of the Blacks Fork Member (Stucky, 1992). Gunnell and Bartels (1994) produced a cenogram which ranked body weights of the mammalian species found at Grizzly Buttes. The body mass distribution of the Grizzly Buttes fauna compares favorably with that observed in modern closed forest communities in tropical Africa which experience warm, moist conditions. An annual rainfall of 71 cm has been estimated for the early Bridgerian. This rainfall would have peaked in the late spring and early summer (MacGinitie, 1969). The Blacksforkian environment at Grizzly Buttes may have been somewhat wetter than the preceding Gardnerbuttean during which the Laney Shale was deposited (Gunnell and Bartels, 1994).

The paleolatitude of the Green River Basin during the Bridgerian was 5° to 8° south of its modern latitude of 41° north (Roehler, 1993). At this time "paratropical" floras reached as far as 60° north latitude. A paratropical flora is defined as subtropical vegetation augmented by a diversity of deciduous broad-leafed trees and shrubs (Wing, 1987). This includes the juxtaposition of palms and ferns with oak, elm, maple and beech (Brown, 1929, 1934). That such a forest grew at Grizzly Buttes is based on circumstantial evidence from correlated Green River Formation deposits. Pollen from the Green River Formation includes that of conifers such as pine, spruce and fir indicating environmental zonation on the mountains surrounding the basin of at least 2000 m elevation (McGrew and Casilliano, 1970). The fossil wood found at Grizzly Buttes is of poor quality and pollen has yet to be recovered.

2.2. Lithology

2.2.1. Sandstones

The base of Grizzly Buttes lies 60 m below the Church Butte tuff. Above this base the sedimentary sequence of fluvial and lacustrine beds are incised by channel sandstones. Twenty seven ribbon sandstone channels were identified in Grizzly Buttes East (Figure 3). Grain sizes range from very fine sand (4.0 phi) to volcanoclastic pebbles (4.0 phi) which give many of the channels a blue to reddish color.

The Moxa Arch is a nearby monocline to the northwest, quiescent since the early Eocene (Royse, 1993). This arch caused slight thickness and facies changes, and discontinuities in limestone distributions (Murphey, 1995). Crossbedding and troughs in ribbon sandstones indicate a strong southwest paleocurrent flow at Grizzly Buttes with an azimuth of 230° to 240°. This indicates that drainage at Grizzly Buttes was affected by the Moxa Arch to the northwest.

The channels show no abnormal geometry and suggest relatively stable meandering rivers. Some of the smaller channels indicate larger ranges of flow direction from northwest to southeast. One unusual discovery within this sequence are the only hemispheroid stromatolites (cyanobacteria) known within a Bridger Formation channel.

Vertebrate elements were occasionally entrained as bed load in stream channels. *Baena* and *Echmatemys* turtle shells are often found as lag within these paleochannels. In two separate instances, the skulls of the brontothere *Palaeosyops* cf. *paludosus* (AMNH 129396) and the crocodile *Crocodylus affinus* (AMNH 25557), were found lying directly on dense beds of *Plesielliptio*, a freshwater mussel that requires clear, flowing water (Hanley, 1974).

2.2.2. Cosmic Layer

The most important fossiliferous bed at Grizzly Buttes is the "cosmic" layer (Alexander, 1992b), which was previously identified as a volcanic ash in error (Alexander, 1994b). The base of this locally traceable layer lies 47.3 m below the Church Butte tuff. The cosmic layer is interpreted to be the result of high sedimentation due to a series of flooding events along the wooded shoreline by a persistent lake to the south. This is evidenced by a lack of euhedral biotite crystals, a sharp lower contact, uniform thickness, and the presence of fossilized wood.

Although sand channels of later age truncate the cosmic layer on Turtle Hill and Wrong Hill, no allochthonous channel sandstones are formed within the cosmic layer. Rapid deposition of the cosmic layer is further indicated by the excellent preservation of many vertebrate skeletons. The inundation of the forest floor by at least temporary standing water is indicated by the presence of the snail *Biomphalaria* which lives today in shallow freshwater environments in tropical South America (Thompson, 1984).

The sediment was sufficiently silty and structureless to support the burrowing habits of amphisbaenian (limbless) lizards when the forest floor was dry. A skeleton of the rhineurid *Spathorhynchus fossorum* (AMNH 25556) was found within cosmic layer Unit 3 on Cosmic Hill which includes elements not previously described by Berman (1973).

The cosmic layer has four subunits. Unit 1 is a hard silver siltstone 0.7 m thick with calcareous pillows. It forms a prominent bench which caps Thule Butte (Figure 2, Section 4). Fossils include vertebrates, wood, and *Biomphalaria*. Unit 2 is only found in the area around Butch Hill referred to as the "Cosmodrome". It consists of a 0.9 m thick hard gray siltstone which breaks down readily in water. Cosmic Quarry at the base of Butch Hill in Unit 2 has yielded the skeleton of one female (AMNH 129382) and articulated remains of two additional *Notharctus tenebrosus* individuals (Alexander, 1992, 1994) and a lizard, cf. *Saniwa* (AMNH 29992). A partial skull, mandible and associated postcranial elements of a juvenile *N. tenebrosus* (AMNH 129384) were found approximately 20 m west of Cosmic Quarry.

Unit 3 varies lithologically from north to south. On Turtle Hill (Figure 2, Section 3) the layer is a brown to red mudstone 0.45 m thick with geothite crystals. In the most fossiliferous area it consists of an olive green to silver pillow forming limy siltstone. In 1988 the most complete skeleton of *N. tenebrosus* yet described (AMNH 127167) was found in Unit 3.

In 1996, one of only four known skeletons of the rare palaeanodont *Metacheiromys marshi* (AMNH 131777) was found in Unit 3 on Butch Hill. The first of these four skeletons, is the type of the *M. osborni* (AMNH 11549) a junior synonym of *M. marshi* (Schoch, 1984). AMNH 11549 was found in "lower beds" at Grizzly Buttes East in 1903 (Osborn, 1904), probably in the immediate vicinity of the 1996 specimen. AMNH 131777 was found by chance when the overburden was removed from an isolated skull and mandible of an unusually robust example of *Thinocyon* cf. *T. velox* (AMNH 131831). In 1993, an articulated skeleton of *Thinocyon velox* was also found in this unit, about 60 m east of the *Metacheiromys* site. In 1997, an apparent nest of the eggs of an as yet unidentified turtle was found a meter from this skeleton.

Skeletons of *Notharctus*, proviverrine creodonts, and the rare serrated-toothed ziphodont crocodile *Pristichampsus vorax* (AMNH 29993) have been found at "Windy City", a site 5 km to the southwest (Figure 1). Here Unit 3 is olive green but less limy. It correlates to a relatively thick limestone measured on Round Top Hill (Figure 2, Section 6). There is a facies change toward the southwest, indicative of low energy lake deposition.

Unit 4 of the cosmic layer is an olive-gray, very fine grained sandstone (3.5–4.0 phi). It forms calcareous pillows and a prominent bench. At Butch Hill it is 1.4 m thick. The cosmic layer is continuous to the southwest, passing through “Horrible Hill” (MPM Locality 2416) (West, 1979) and other hills which flank Crooked Canyon. It reappears along the opposite wall of Crooked Canyon where erosion has exposed the top of the bedding plane at the “Donner Pantry” and “Spanish Saddle” localities. This horizontal table covers a sixteenth of a section or about 100,000 square meters. This exposure is equivalent to a two meter ribbon of vertical cross section of the cosmic layer following the contour of the landscape for 500 kilometers. “Donner Pantry” has more exposure of the cosmic layer than the rest of Grizzly Buttes combined. Surface collection at this locality has produced a diverse assemblage of fossil mammals including skeletons of *Notharctus*, *Smilodectes*, *Thinocyon*, *Orohippus*, and *Antiacodon*, and a complete skull and mandible of *Hyrachyus*.

The cosmic layer has undergone compressive plastic flow, causing elements to drift out of anatomical position during the process of fossilization. Fragments with sharp, but ancient contacts are often found displaced by many centimeters. This has led to the discovery of articulated skeletons when only isolated elements are in evidence on the surface. Recent expeditions have been successful despite intensive collection in this area for more than a century. Many well preserved specimens attributed to Grizzly Buttes now residing in museum collections may have been obtained from these lacustrine strata. Indeed, matrix adhering to a specimen of *Mesonyx obtusidens* (AMNH 19203) collected by George Olsen from “lower beds” at Grizzly Buttes East in 1922 appears to have come from the cosmic layer.

The new notharctid sample forms a reference set of individuals from a narrow stratigraphic horizon against which the collections of Hayden, Carter, Marsh, Cope and Granger (which have only crude stratigraphic and aerial resolution) can be compared. It also permits the study of morphological variation among penecontemporaneous individuals. The cosmic layer has demonstrated the local lithosympatry of *N. tenebrosus* and *N. pugnax* (Alexander, 1994b).

2.2.3. Red Bed

Grizzly Buttes has more bench forming siltstones and abundant channel sandstones than is typical for the rest of the Bridger Formation. There are relatively fewer smectitic layers at Grizzly Buttes, making those present more noticeable, and therefore good stratigraphic markers. Lying conformably upon the cosmic layer is a highly smectitic dark brown mudstone that is dramatically exposed throughout the area. At various locations in the southern Green River Basin large amounts of goethite give this bed a red to port wine stain color. This “Red Bed” marks the transition between Bridger the B1 and B2 levels of Matthew (1909).

The Red Bed is interpreted to be transitional to a more terrestrial environment that is drier or better drained as it represents a period of incipient soil development. It thickens from 0.6 m elsewhere to 1.7 m near Butch Hill. It contains localized claystone clasts and ranges from clay to very fine grained sand (10.0 3.5 phi). The Red Bed is abundantly fossiliferous. Complete shells of the turtles *Echmatemys, Baena, Baptemys*, and *Arctotrionyx* are commonly found within the Red Bed, lying directly on the contact with the cosmic layer. Dentitions of *Notharctus* and *Hyopsodus* also are common.

2.2.4. Above Red Bed

Higher in the section is an upper smectitic layer. Fossils such as a skull of the alligator *Allognathosuchus* (AMNH 25566), abundant crocodile scutes, and remains of soft shelled turtles (Trionychidae), document a wetter, marshier environment. It is similar in exposed color to the Red Bed layer. It ranges in lithology from a clay clastic conglomerate to a clay free mudstone. On Turtle Hill (Figure 2, Section 1), it contains well rounded, multicolored claystone clasts, of pebble size (2.0 to 4.0 phi). Geothite crystals are common on the surface. To the south of Butch Hill, the unit is represented by a breccia of highly angular, brown claystone clasts, with mudstone matrix, indicating occasional desiccation. Above the upper smectitic layer the presence of channel sandstones indicates a drier or better drained environment.

At the "Forbidden City" locality there occurs a set of siltstone beds lithologically similar to the cosmic layer, but 20 m higher. R. M. West and M. O. Woodburne collected in the western half of this area in 1970 and named it the "Woodburne Locality" (MPM Locality 2415), (West, 1979). Well preserved articulated skeletons are found where ever this lithology is encountered. These include two skeletons each of *Smilodectes gracilis* (AMNH 131762, 131763) and *Notharctus tenebrosus* (AMNH 131764, 131945), and the first known skull of *Omomys carteri* (AMNH 130000). Fish bones and the bivalve *Plesielliptio* are common, indicating a more clear water environment. Fossilized wood also occurs.

2.2.5. Snow White Layer

The "snow white" layer is a thick highly smectitic tan mudstone which stands out against darker units in the area of "Forbidden City". Geothite is present, but does not stain the bed red. Toward the north, the layer is truncated by channels and structureless brown sandstones. The "snow white" layer represents a wetter or less well drained environment. Thickening of this bed toward the south indicates the expansion of the lake from that direction. An as yet, undescribed partial articulated bird skeleton was recently found in this layer.

2.2.6. Golden Layer

The "golden limestone" is the thinnest (0.4 m) of those few limestones that occur in the Bridger B. It weathers to a distinct yellow color. It outcrops below the Cottonwood White Layer just south of "Butch Hill", and can be traced 10 km southwest into Crooked Canyon in Grizzly Buttes West where it caps a hill known as "Golden Crown".

2.2.7. Church Butte Tuff

Channels in the beds between the golden limestone and the Church Butte tuff again represent a drier or better drained environment. The Church Butte tuff is named for Church Buttes 20 km to the north. It is one of only a few lithified tuffs that occur within the Bridger Formation. The layer is extremely flaky, dark brown to dark gray and easily recognized from a distance. It serves as the datum of stratigraphic correlation between measured sections in Grizzly Buttes because it represents a single eruptive event, (as yet undated). It is restricted to the south, and is exposed relatively high on the outcrops. The presence of large euhedral crystals of biotite and hornblende and abundant Ca plagioclase feldspars indicate absence of reworking and fluvial transport. Radiometric dating of this layer will be extremely helpful in reconstructing the stratigraphy and paleontology of the formation.

3. Taphonomy

The productive lacustrine layers at Grizzly Buttes have several unusual taphonomic characteristics. They represent discrete depositional lake shore flooding events so that there is isochroneity of the specimens incorporated within them. They are more abundantly fossiliferous than adjacent units. The faunal list of Grizzly Buttes has been tabulated for both the "old" AMNH collection, made in the early 1900's and the "new" collection made in the 1990's (Table 2). The percentage breakdown of the most significant orders of the fauna appears as a pie chart (Figure 4). Thirteen mammalian orders are represented in these collections. Several mammal orders are presumed to be present at Grizzly Buttes, but are not present in the AMNH sample. Their absence can readily be explained by either their original rarity in the fauna such as tillodonts or their minute size and difficulty of recovery as is the case for marsupials. Several orders have never been reported from Bridger B beds but appear in both younger and older faunas, and are assumed to be present elsewhere in the Intermontane West during this interval. These include the orders Dinocerata (Wheeler, 1961), Dermoptera and Multituberculata (McKenna, 1990; Stucky, 1992).

Hyopsodus accounts for fully one third of the entire fauna. The aggregate percentage of all primate species accounts for just under 20% of the fauna and places second to *Hyopsodus*. Perissodactyls and rodents represent 17% and 12% of the fauna respectively. Each of the other orders represent from 6% to less than 1% of the fauna. *Hyopsodus* falls at the small end of mass distribution for Grizzly Buttes while the perissodactyls fall in the mid to upper range of body mass (Gunnell and Bartels, 1994).

Several factors are at work which affect what ultimately finds its way into the collection. One of these is collection bias. Large animals leave behind large and obvious bones in the fossil record that are more readily noticed in the field than minute elements from small animals. (Behrensmeyer and Hill, 1980).

Table 2. Faunal List for Grizzly Buttes specimens in AMNH collection.

- PISCES
  - Lepisosteiformes
    - Lepisosteidae
      - *Lepisosteus* sp.
  - Amiiformes
    - Amiidae
      - *Amia* sp.
  - Teleostei
    - Teleostei indet.
      - Ictaluridae indet.
- AMPHIBIA
  - Urodela indet.
- REPTILIA
  - Testudines
    - Baenidae
      - *Baena arenosa*
    - Trionychidae
      - *Amyda* sp.
    - Emydidae
      - *Echmatemys* sp
    - Dermatemydidae
      - *Baptemys* sp.
  - Lacertilia
    - Anguidae
      - cf. *Proglyptosaurus*
      - *Saniwa* sp.
  - Amphisbaena
    - Rhineuridae
      - *Spathorhynchus fossorum*
  - Crocodilia
    - Crocodylidae
      - *Crocodylus affinus*
    - Pristichampsidae
      - *Pristichampsus vorax*
    - Alligatoridae
      - *Allognathosuchus* sp.
- AVES
  - Neognatha
    - sp. indet.
- MAMMALIA
  - Mammalia indet.
  - Pantolesta
    - Pantolestidae
      - Pantolestidae indet.
      - *Pantolestes* cf. *natans*
      - *Pantolestes longicaudus*
      - *Pantolestes* sp.
  - Insectivora
    - Insectivora indet.
  - Erinaceoidae
    - Talpidae
      - Talpidae indet.
      - *Scenopagus edenensis*
    - Nyctitheriidae
      - Nyctitheriidae indet.
      - *Marsholestes dasypelix*
      - *Nyctitherium serotinum*
  - Primates
    - Microsyopidae
      - *Microsyops elegans*
      - *Microsyops* sp.
    - Notharctidae
      - Notharctidae indet.
      - *Notharctus* sp.
      - *Notharctus tenebrosus*
      - *Notharctus pugnax*
      - *Smilodectes gracilis*
    - Omomyidae
      - *Omomys carteri*
      - *Omomys* sp.
      - *Anaptomorphus* sp.
      - *Uintanius ameghini*
  - Creodonta
    - Creodonta indet.
    - Hyaenodontidae
      - Hyaenodontidae indet.
      - *Limnocyon verus*
      - *Limnocyon* sp.
      - *Machaeroides aeothen*
      - Proviverrinae indet.
      - *Proviverra grangeri*
      - *Proviverra minor*
      - *Proviverra pungens*
      - *Proviverra rapax*
      - *Proviverra* sp.
      - *Thinocyon mustelinus*
      - *Thinocyon velox*
      - *Thinocyon* sp.
      - *Tritemnodon agilis*
    - Oxyaenidae
      - *Patriofelis ulta*
  - Carnivora
    - Carnivora indet.
    - Miacidae
      - Miacidae indet.
      - Miacinae indet.
      - *Miacis parvivorus*
      - *Miacis* sp.
      - *Oödectes herpestoides*
      - *Oödectes proximus*
      - *Uintacyon antiquus*
      - *Uintacyon vorax*
      - *Uintacyon* sp.
      - *Vulpavus completus*
      - *Vulpavus ovatus*
      - *Vulpavus palustris*

(*continued*)

Table 2 (Continued)

- *Vulpavus profectus*
- *Vulpavus* sp.
- Viverravidae
  - Viverravidae indet.
  - *Viverravus* sp.
  - *Viverravus gracilis*
  - *Viverravus sicarius*
- Pholidota
  - Metacheiromyidae
    - *Metacheiromys marshi*
    - *Metacheiromys dasypus*
  - Epoicotheriidae
    - Epoicotheriidae indet.
- Condylarthra
  - Hyopsodontidae
    - *Hyopsodus minusculus*
    - *Hyopsodus paulus*
    - *Hyopsodus* sp.
- Tillodontia
  - Esthonychidae
    - *Trogosus* sp.
  - Artiodactyla
    - Homacodontidae
      - *Antiacodon pygmaeus*
      - *Helohyus plicodon*
  - Acreodi
    - Mesonychidae
      - *Mesonyx obtusidens*
  - Perissodactyla
    - Perissodactyla indet.
    - Equidae
      - *Orohippus pumilus*
      - *Orohippus* sp.
    - Brontotheriidae
      - Brontotheriidae indet.
      - *Limnohyops matthewi*
      - *Limnohyops monoconus*
      - *Limnohyops priscus*
      - *Palaeosyops major*
      - *Palaeosyops paludosus*
      - *Palaeosyops* sp.
    - Helaletidae
      - *Helaletes intermedius*
      - *Helaletes nanus*
      - *Helaletes* sp.
      - *Hyrachyus eximius*
      - *Hyrachyus modestus*
  - Rodentia
    - Rodentia indet.
    - Paramyidae
      - *Leptotomus grandis*
      - *Leptotomus parvus*
      - *Leptotomus sp.*
      - *Paramys delicatior*
      - *Paramys delicatus*
      - *Pseudotomus robustus*
      - *Paramys* sp.
      - *Reithroparamys delicatissimus*
      - *Taxymys* sp.
      - *Thisbemys plicatus*
    - Sciuravidae
      - Sciuravidae indet.
      - *Sciuravus nitidus*
      - *Sciuravus* sp.

Collection methods other than surface prospecting are needed to recover the less visible microfauna. These include wet and dry screening of sediment (Hibbard, 1949; McKenna, 1962), and picking or sieving tiny fossils which have been deposited by ants on their hills (Delson, 1971; Robinson, 1994). Neither method was used at Grizzly Buttes by Granger in the early 1900's, but both have been employed in recent collecting. Analysis of hundreds of tiny teeth found on the "1K Anthill" site near Cosmic Hill remains to be finished. It has already produced rare milk teeth of *Hyopsodus*, notharctid primates, and cf. *Omomys*. The 1K Anthill sample is not included in Table 2.

Arboreal notharctid primates represent only 13% of all cataloged mammal fossils within the AMNH Grizzly Buttes collection (Figure 4), yet they are the family most often found as articulated skeletons. They are so well preserved that such delicate elements as sesamoids, hyoids and auditory ossicles are frequently found in articulation. In contrast, the far more common taxon *Hyopsodus paulus* which represents 34% of this collection has only been found as fragmentary dentitions within these beds. The short-limbed *Hyopsodus* was certainly more terrestrial than the notharctids, but it has been interpreted as both fossorial (Gazin, 1968) and semiarboreal in habits (Matthew, 1928). It may have been more analogous to modern hyraxes than lagomorphs as a small bodied generalist herbivore.

Matthew (1909) observed that *Thinocyon* is the most abundant of Bridger "carnivora" in the AMNH collection. It represents about 1% of the collection. This pattern has been sustained in the recent collection. Skeletons of this possibly arboreal, mink-like taxon are found as commonly as those of the notharctid primates. Fossorial forms such as the paleanodont *Metacheiromys* and the rhinuerid *Spathorhynchus* have been found as well preserved skeletons. Aquatic vertebrates, particularly reptiles, are abundant throughout the cosmic layer. Turtle shells fragments often litter the landscape. Elements of the baenid turtle *Baena arenosa* are the most common vertebrate remains found at Grizzly Buttes. Complete shells of this taxon are frequently encountered. Many examples have been described (Hay, 1908; Gaffney, 1972). Remains of aquatic mammal genus *Pantolestes* are also found. Delicate items such as reptile eggs are also occasionally preserved.

There appears to be a taphonomic bias in the cosmic layer toward the preservation of arboreal, fossorial, and aquatic forms as complete skeletons, selectively altering the fossil assemblage between the moment of death of individual organisms and the actual remains found.

3.1. Crocodilian Predation

The *Pristichampsus vorax* skeleton (AMNH 29993) found at "Windy City" is remarkable for a number of reasons. It is unusually complete, including the skull, jaws, and front and hind limb material not previously described that will address longstanding questions about its locomotor behavior (Langston, 1975). It is also very rare toward the center of the Green River Basin, yet apparently abundant around the basin margins (Gunnell and Bartels, 1997). Although the type specimen of *Crocodylus vorax* (YPM 249), later synonymized as *Pristichampsus vorax* (Langston, 1975), comes from elsewhere in the Bridger Basin (Troxell, 1925), the lost holotype of "*Crocodylus ziphodon*" (Marsh, 1871), whose description conforms to that of *P. vorax*, was found by a Yale party at Grizzly Buttes in 1870 (Langston, 1975).

Upon preparation of this skeleton, it was discovered that a nearly complete left hind-himb of the perissodactyl *Helaletes* sp. was lodged entirely within its ribcage. The mammalian material includes parts of the pelvis, several lower lumbar vertebrae, the femur, tibia, calcaneum and a number of pedal elements. Several of these elements exhibit parallel grooves and scratches which match the spacing of the serrations on the teeth of the *Pristichampsus*. Similar scratches are also incised on a partial skull of *Hyopsodus paulus* (AMNH 129391) found elsewhere at Grizzly Buttes.

This circumstance is entirely consistent with the predatory and feeding habits of modern crocodilians. They are known to snatch prey at the water's edge with their jaws, drown their victims, and tear them apart by twisting motions of their bodies. This process is described in detail in the apparent bisection of a specimen of the adapid primate *Europolemur koenigswaldi* by an Eocene crocodile at Messel, Germany (Franzen and Frey, 1993). The *Helaletes* remains found within the Grizzly Buttes *Pristichampsus* appear to be this reptile's last meal.

Amazingly, this same situation was encountered by Troxell (1925) in his description of the type specimen of "*Crocodylus*" *vorax* (YPM 249) from the Bridger Formation. Thoracic vertebrae of a mammal, perhaps the perissodactyl *Hyrachyus* sp. were found in such close proximity with the skull that Troxell suspected that "the crocodilian was feeding on a small rhinoceros or titanothere a short time before both were buried, hence the name *vorax*" (Troxell, 1925: figure 22). Such examples as these may offer a possible explanation for the unusual body size distribution of articulated mammal remains encountered at Grizzly Buttes. In the absence of stream action sorting body parts, predator feeding behavior is a possible mechanism for explaining the discovery of well preserved, and articulated portions of mammalian bodies.

Common terrestrial mammals of medium to large body size such as the perissodactyls *Helaletes*, *Hyrachyus* and *Palaeosyops* are not commonly found as complete skeletons, but rather as articulated body "portions". For example, a complete left pes of *Palaeosyops* (AMNH 129392) truncated at mid-tibia and a remarkably complete and articulated skull and mandible of *Hyrachyus* (AMNH KF 97-6) have been found in isolation.

3.2. Mammal Predation

An entirely different taphonomic process may have been at work to explain the frequency of occurrence of the arboreal mammals. There may be a size bias at work selectively under representing smaller arboreal forms such as *Omomys* and *Microsyops* and favoring the notharctids *Notharctus* and *Smilodectes*. Several of the new notharctid individuals show evidence of predation, exhibiting multiple puncture wounds. These remains are found in a condition that is simply too complete to have been significantly impacted by the feeding habits of mammalian predators and scavengers which tend to reduce prey remains to fragments (Behrensmeyer and Hill, 1980; Andrews, 1990). Heavily chewed bones are not often found at Grizzly Buttes. Concomitantly, the remains of large bodied Bridgerian mammalian predators are relatively rare at Grizzly Buttes, accounting for roughly 1% of the AMNH collections. The coyote-sized hyaenodontid *Tritemnodon agilis* and the wolf-sized mesonychid *Mesonyx obtusidens* are represented by only three and seven

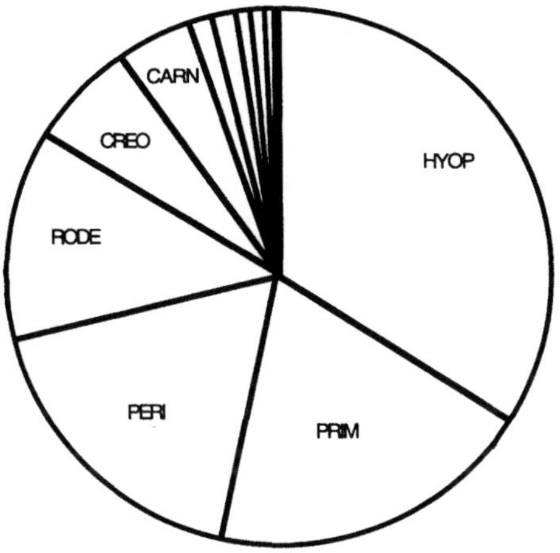

FIGURE 4. Rank order and percentage breakdown of mammalian orders: HYOP—*Hyopsodus sp.*, "Condylarthra" (34.3%); PRIM—Primates (19.5%); PERI—Perissodactyla (17.7%); RODE—Rodentia (12.6%); CREO—Creodonta (6.6%); CARN—Carnivora, Miacoidea (4.3%); Not shown—Pantolesta (1.6%), Artiodactyla (1.4%), "Insectivora" (0.8%), Palaeanodonta (0.7%), Acreodi (0.7%), and Tillodontia (0.2%). The AMNH Grizzly Buttes mammal collection consists of 451 identified specimens collected during the 1991–1996 field seasons and 494 cataloged specimens collected by Walter Granger and others between 1903 and 1906. N = 945. Several hundred additional specimens have been found during the 1997–1999 field seasons.

specimens respectively. Only a single dentition of the panther-sized oxyaenid *Patriofelis ulta* appears in the AMNH Grizzly Buttes collection. The rarity of climax predators in the sample is a logical reflection of their relative rarity in a normal trophic pyramid (Behrensmeyer and Hill, 1980).

3.3. Raptor Predation

It is more likely that the distribution of smaller mammal remains is a result of predation by raptorial birds. The *Omomys* Quarry found 70 m above the base of Bridger C predominantly contains the remains of the primate *Omomys.* It also includes remains of birds including owls (Strigiformes) (Murphey, 1990). This unique accumulation has been interpreted as the fossilized remains of owl pellets, (Murphey *et al.*, 1998). Owl pellets are regurgitated packets of undigested bones and hair of prey consumed by owls (Andrews, 1990). The omomyids *Tetonius* and *Shoshonius* can be inferred to be nocturnal in habit because their enlarged orbits approach the condition of *Tarsius* (Beard and Krishtalka, 1991; Beard and MacPhee, 1994). A nocturnal adaptation would have made *Omomys* particularly susceptible to owl predation (Murphey *et al.*, 1998). Several features of the *Omomys carteri* skull from "Forbidden City" (AMNH 130000) suggest that it too may represent remains from within an owl pellet. These include the preservation of delicate elements such as the premaxillae and nasals, a calcarious veneer adhering to some of its surfaces, and a dentary that is cemented to the palate out of anatomical position (Alexander and MacPhee, 1999a, 1999b).

The most obvious evidence of predation in the primate sample is seen in a male *N. tenebrosus* (AMNH 127167). This specimen bears at least three puncture wounds. A deep puncture occurs on the posterior margin of the left supraorbital process (Figure 5B, arrow). The wound is roughly triangular in cross section with rounded sides 4.0 mm, 3.5 mm and 3.5 mm in length. The edges of the hole are smoothly rounded and bend inward, a condition consistent with the penetration of fresh bone by a tooth or claw. A prominent compressive fracture is found in the identical position on the right supraorbital process. It is roughly circular and 3.0 mm in diameter. It indicates a tooth or claw that pressed about 1.0 mm into fresh bone, but did not puncture the skull. A third "unnatural" hole is found on the right maxilla 6.0 mm inferior to the lacrimal duct. It is oval in cross section and measures 2.0 mm by 1.5 mm. The initial interpretation of these marks on the skull of AMNH 127167 was that this individual's head had been gripped in the maw of a mammalian predator, perhaps the miacoid carnivore *Vulpavus* sp. (Alexander, 1992). The punctures resemble wounds on a partial skull of a juvenile *Australopithecus robustus* found in cave breccia deposits at Swartkrans in South Africa. Two circular holes in the hominid skull were probably the result of the penetration of the lower canines of a leopard, *Panthera pardus*, into each parietal. The upper canines would have penetrated the missing orbits. This damage would have occurred when the individual was dragged up into a tree by a leopard (Brain, 1974).

In addition, there are also apparent wounds on an isolated but complete *N. tenebrosus* mandible AMNH 131765 found only a few hundred meters from AMNH 127167. This specimen is interpreted by canine root diameter to be a female. The left dentary is compressed at m3 indicating a great force was applied to the specimen when the bone was still fresh and unmineralized. The specimen bears a contusion on the buccal boarder of the right dentary which has a complementary puncture just superior to the mandibular foramen on the medial aspect (Figure 5F, right arrow). This puncture trends anteriorly indicating a deep penetration from posterior to the jaw.

The wounds on AMNH 127167 and AMNH 131765 are consistent with the hunting and feeding patterns of raptorial birds (Rettig, 1978; Kerbis, 1997). A skull of a cebus monkey, *Cebus apella* from the Neil Rettig collection of the Field Museum (FMNH) shows damage similar to that found on AMNH 127167. This *Cebus* skull was collected from beneath the nesting site of a breeding pair of harpy eagles (*Harpia harpyjia*) at Moria Creek, Guyana in 1991. Out of the cebid remains found in this study (MNA = 30), only three cebid skulls were recovered (J. Kerbis, personal communication). One of these uncataloged skulls displays deep puncture wounds to the lacrimal region of both orbits, presumably inflicted by the talons of the harpy eagle.

A University of Michigan collection of mammalian remains found beneath nests of the crowned eagle *Stephanoaetus cornatus* from the Kibale Forest in

Uganda was examined in the course of this study (Mitani and Sanders, in prep.). Remains included hyraxes, small bovids, and several species of catarrhine monkey. All are documented as food remains of crowned eagles. One skull of the red colobus monkey, *Colobus badius* (field number 110), displays one talon puncture wound to each orbit and one to the left zygomatic. Of greatest interest is a talon wound low on the left frontal posterior to the postorbital constriction. The margins of this 6.0 mm circular puncture are bent inward indicating an entry wound into fresh bone. It is accompanied by a complimentary exit wound found on the inferior surface of the left sphenoid. This is established on the basis of the displacement of a 3.0 mm oval patch of bone still hinged on the margin of the hole. The entry and exit wounds are not in direct line with each other, but are rather, consistent with the passage of a recurved talon along an arcuate trajectory.

The maxillary wound on AMNH 127167 resembles the exit wound on Kibale 110. This lead to a reexamination of a damaged region within the right lacrimal of AMNH 127167 that is roughly 6.0 mm in diameter. These marks are now interpreted as a recurved talon wound which entered the right orbit and exited the right maxilla. This evidence strongly suggests that AMNH 127167 was the victim of a predatory bird, probably a diurnal eagle, based on body size and presumed diurnal habits of *Notharctus*.

That this specimen is generally complete indicates that the corpse may have been dropped before it could be fed upon. Such instances have been observed with modern eagles (Struhsaker and Leakey, 1990). The observed feeding behavior of eagles such as the Harpy Eagle *Harpia harpyjia* can also explain the relative completeness of primate remains. Adult Harpy Eagles which are efficient feeders, tend to damage and break apart the skulls of monkeys and tree sloths during feeding, while nesting young eagles which are relatively inefficient feeders, will leave complete vertebral columns with the skull and limbs sometimes attached. Sometimes they are held together with skin (Rettig, 1978). Such a carcass might enter the fossil record without clear indication of having been consumed by a predator.

Several multi-year studies of prey remains found in association with nesting pairs of crowned hawk eagles in the Kibale Forest of southwestern Uganda have produced compelling evidence of a taphonomic process at work on arboreal primates and terrestrial mammals such as hyraxes and smaller antelopes. Arboreal primates (monkeys) constitute the predominant prey species at well over 80% of the remains, (Brown, 1982; Skorupa, 1988; Struhsaker and Leakey, 1990). These large eagles (3.4–4.1 kg) selected prey according to species, age and sex. Among the larger red colobus monkeys *Colobus badius*, young juveniles and infants were selected. Among smaller bodied monkey species, adult males were selected. “Eagle prey selectivity by species generally supports the hypothesis that polyspecific associations among monkeys are effective deterrents against predation”, (Struhsaker and Leakey, 1990). Eagle prey selectivity has a major impact on the populations and sex ratios of a number of monkey species (Struhsaker and Leakey, 1990). Although the predator/prey ratio for the Kibale eagles is much higher than that of the resident felids, their annual take of prey is lower (Struhsaker and Leakey, 1990).

Body size is a major consideration in prey selection among eagles, but it isn't absolute. Eagles are capable of killing species much larger than themselves such as the bushbuck (*Tragelaphus scriptus*). At 30 kg, young bushbucks can exceed six times the weight of an adult Crowned Eagle. If an eagle is incapable of bringing a victim back to its nest whole, it will dismember it and carry away selected pieces such as limbs, leaving the axial skeleton behind, (Brown, 1971; Daneel, 1979; Jarvis *et al.*, 1980; Gargett, 1990).

A similar study of crowned eagles was conducted in Cape Province, South Africa (Jarvis *et al.*, 1980). Here, the eagles live in forest patches surrounded by open grassland where the rock hyrax (*Procavia* capensis) is common and becomes the most common prey species (42%), followed by blue duiker (*Cephalophus monticola*, 16%), juvenile bushbucks (*Tragelaphus scriptus*, 12%), and vervet monkeys (*Cercopithecus pygerythrus*, 8%). In Kenya, where rock hyraxes are abundant, they are the preferred prey of crowned eagles (Brown, 1976).

Another study involving black eagles, *Aquila verreauxii*, was conducted in Matobo National Park, Zimbabwe, a savanna area with numerous rock koppies (Barry and Barry, 1996). Here, the rock hyrax, *Procavia capensis*, and the yellow-tailed hyrax, *Heterohyrax brucei*, constitute over 90% of the prey species. *Procavia* (4.53 kg), is as large or larger than *Aquila verreauxii* making them a more difficult to kill (Gargett, 1990). *Heterohyrax* (3.25 kg) was preyed on more heavily than *Procavia*. This may be due to its smaller size or its more diurnal habits which would have made it more susceptible to predation (Barry and Barry, 1996). Because of their greater weight, *Procavia* are often consumed at the kill site or decapitated (Gargett, 1990). Fewer skulls and jaws of *Procavia* were found at eagle nesting sites than those of *Heterohyrax*.

Such a taphonomic process would have a major effect on the fossil record in a fauna where eagles were once at work. This could explain why some species at Grizzly Buttes are commonly found as jaws or skull fragments and others as skeletons. There is a suite of taxa in the Grizzly Buttes fauna with body sizes comparable to these African taxa (Gunnell and Bartels, 1994). The ungulates *Hyopsodus paulus* (0.22 kg), *Antiacodon pygmaeus* (1.67 kg), *Orohippus pumilus* (7.33 kg), *Helaletes nanus* (15.52 kg), and the primates *Notharctus tenebrosus* (3.43 kg) and *Smilodectes gracilis* (1.88 kg) would have been likely targets for any contemporary eagles. The small body size of *Hyopsodus* may have resulted in the more complete dismemberment and destruction of its remains because predators may have been more apt to consume it whole than other prey mammals. This phenomenon might explain the paucity of its postcranial elements in the fossil record.

The fossil remains of large raptors like the owl *Protostrix saurodosis* have been described from the Bridger Formation (Wetmore, 1933). It is reasonable to believe that raptorial birds like hawks or eagles were present in the paratropical forest of Grizzly Buttes. Eagles specializing in primate predation are found in many tropical environments today. These include the Harpy Eagle of South America, the Crowned Eagle of Africa and the Philippine Eagle. Izor (1985) observed that monkeys were the only arboreal mammals that Harpy Eagles selected disproportionately to their availability. Although these are opportunistic predators, arboreal primates may be the optimal prey for their hunting behaviors (Skorupa, 1989). The large bird which may have inflicted the contusions observed on AMNH 127167 remains as yet unknown. In light of the data available for modern primate hunting birds, the apparent ratio of species of *Notharctus* and *Smilodectes* at Grizzly Buttes may more reflect the feeding habits of such birds than their actual numbers when living. These taxa are equally common at other Bridger B samples such as Church Buttes, but are not as frequently encountered as skeletons.

3.4. Intrusion

Intrusive burrowing of *Spathorhynchus* is one of several forms of bioturbation. *Metacheiromys* was certainly a burrowing animal, filling a niche comparable to the pangolin or the armadillo (Simpson, 1931). Although the *M. osborni* skeleton (AMNH 131777) is complete, the elements were found completely disarticulated in a dense concentration less than 50 cm across. This situation is consistent with the animal having died within a subterranean burrow which filled with water, resulting in the maceration of the corpse and the jumbling of the remains. There was however, no obvious intrusive contact surrounding the skeleton observed during excavation.

The fossilized nesting burrows of insects are a common feature at Grizzly Buttes. These resemble modern nest cells of the order Apoidea. These fossilized "bee cells" are roughly the size and shape of olive pits. Fossil cells often preserve the distinctive spiral closure of sealed cells and the caved in lid of chambers from which grown bees emerged. They result from the sedimentary filling or crystallization of egg chambers excavated by non-hiving bees. Modern bee species dig these chambers as shallow as 5 cm and as deep as 3 m beneath the surface (Rozen, 1984a; 1984b; 1993). Bees often seal these chambers with waxy substances and deposit a ball of ingested pollen upon which an egg is laid. The pollen ball is food for the growing bee larva. It may be possible to extract fossil pollen from unhatched bee cells. There is a direct relationship between the size of a bee cell and the insect that pupates within it. Differences in the size and form of these fossilized cells suggests that a number of bee species lived at Grizzly Buttes. This variety is also an indicator of the richness and complexity of the flowering plants and trees that the adult bees would have pollinated. When a young bee emerges from a cell, it cuts off the top with its mandibles, and fills in the empty chamber with sediment as it works its way to the surface.

Such empty cells are uncommon at Grizzly Buttes. Unopened cells in which the larva died without emerging are abundant. These are identifiable because they are complete and contain geode-like crystals rather than sediment. These gregarious bees will often dig egg chambers in large concentrations (Rozen, 1984a; 1984b; 1993). Several such concentrations have been found at Grizzly Buttes. One site, in sediments laterally equivalent to the cosmic layer, contains hundreds of bee cells within a single 10 cm horizon. Such a concentration indicates that the horizon was above the water table at the time of the intrusion. One explanation for the general failure of the cells to hatch is the possibility that the water table rose after the eggs had been laid. This is consistent with the interpretation of the cosmic layer and other beds having been formed by lake shore inundation.

4. Notharctid Primates

Although elements tend to be complete, many Bridger specimens have suffered severe plastic deformation. Gregory (1920) failed to account for this when describing the type skull of *Notharctus osborni* (AMNH 11466) (Granger and Gregory, 1917). This specimen was determined to be a junior synonym of *N. tenebrosus* by Robinson (1957). This resulted in an unnaturally long and narrow reconstruction based on a spurious resemblance to *Lemur catta* (Gregory, 1920). Gazin (1958) recognized that the profile of *Smilodectes gracilis* (USNM 21815) had been dorsoventrally distorted. He corrected for this in a dashed line reconstruction (Gazin, 1958, Pl. 2: figure 3). This correction was ignored in later reconstructions, such as those found in Simons (1964). These errors have been perpetuated for decades. Reconstructions of male and female *N. tenebrosus* and *S. gracilis* developed and illustrated by J. Alexander attempt to rectify this situation (Figures 6–7). The rostrae of *Notharctus* and *Smilodectes* are more reduced than generally believed (Figure 6).

Notharctus tenebrosus and *Notharctus pugnax* are frequently found together in the same layer at Grizzly Buttes (Alexander, 1994b). Of 114 notharctid specimens in the AMNH Grizzly Buttes collection, 77 have been attributed to the genus *Notharctus*. *N. tenebrosus* (n = 59), the smaller of the two species of *Notharctus* is numerically the most common species. *N. pugnax* (n = 10), the large bodied species, is the least common. The ratio between *N. tenebrosus* and *N. pugnax* is approximately 6 : 1. Eight specimens in the sample were attributable only to *Notharctus* sp.

According to arguments made by Shoshani (1992), the numerical prevalence of a ratio of 5 : 1 or greater of one taxon (species A) over another closely related taxon (species B) that shares the same habitat such as a diverse forest suitable for euryphagous species can be used to predict that species A is phylogenetically more primitive, and more of an environmental generalist than species B. Shoshani used the case of five species pairs syntopically and sympatrically sharing the same habitats, in particular the hyraxes *Heterohyrax brucei* and *Procavia capensis*. The case of *Notharctus tenebrosus* and *N. pugnax* is problematic as the later species appears to be gnathically less derived than the former. Perhaps *N. pugnax* may have been more specialized in its diet. The ratio of *Notharctus tenebrosus* to *Smilodectes gracilis* however, fits perfectly into the Shoshani paradigm. The smallest bodied notharctid at Grizzly Buttes *S. gracilis* (n = 12) is less common than *N. tenebrosus*. *S. gracilis* is the most derived of all known notharctid species, demonstrating a number of autapomorphies of the skull and dentition (Alexander, 1993; Rose *et al.*, 1999). Although *N. tenebrosus* with its completely fused dentaries is more derived than other species of *Notharctus* (Beecher, 1983; Alexander, 1993), it is far more similar to all other known notharctids, and may be interpreted as more of an ecological generalist than *Smilodectes*. Larger samples are needed to make these inferences statistically meaningful.

The ratio of one Grizzly Buttes species to another that is actually found in the assemblage is of course subject to taphonomic processes. Since the cosmic layer constitutes a *lagerstätten* deposit it is not subject to the sort of fluvial processes which sort bones in quarry samples according to size (Shotwell, 1958). This is further aided by the fact that the notharctids are of similar body size (Shoshani, 1992).

The new notharctid material permits a brief comparison of the gross skeletal morphology and individual variation of these taxa. *Smilodectes* has a proportionately larger lower body than *Notharctus*, so the identification of lower limb elements is problematic. The overlap in size of *N. tenebrosus* and *S. gracilis* postcranials has probably led to their general confusion. The taxonomic identification of all lower Bridger Formation notharctid postcranials not associated with dentitions may require reevaluation. Several of the new skeletons have well preserved hands and feet. AMNH 127167 has the most complete manus (Figure 5D) of any Eocene primate found in North America (Hamrick and Alexander, 1996). The *Notharctus* skeletons (AMNH 131764, 131772) and *Smilodectes* skeletons (AMNH 131762, 131763) preserve partial extremities.

The postcranials of female *N. tenebrosus* are considerably shorter in length and more gracile than those of males. For example, the fibula of AMNH 131764 (a male) is 111.5 mm long with a mid shaft diameter of 4.5 mm and shows developed muscle scars, while the comparable dimensions of the female AMNH 129382 are 93.5 mm and 3.5 mm respectively. The juvenile *N. tenebrosus* skeleton (AMNH 131772) has most epiphyses unfused and yet has a completely adult dentition indicating a prolonged period of postcranial development for this species. AMNH 131772 is identified as a female on the basis of the root diameter of the adult canine root (Alexander, 1994a). This specimen also has partially articulated feet (note hallux, Figure 5E) aiding the attribution of isolated juvenile postcranial material to this species. Remains of developmentally younger individuals of *N. tenebrosus* have also been found. AMNH 131830 from "Butch Hill" consists of a mandible fragment with dp4 and m1 in the crypt, associated unerupted M1 and M2 and an associated tibia without epiphyses. AMNH 129384 is the most important of these juvenile specimens. It consists of a nearly complete mandible with dp3–4, m1–2 and m3 in the crypt, and a partial skull with a clearly separated metopic suture between the frontals. A maxillary fragment which includes DP1–4 is present, as well as several as yet unprepared postcranial elements. By far the developmentally youngest individual is AMNH 131959, a partial pelvis and lower back of an

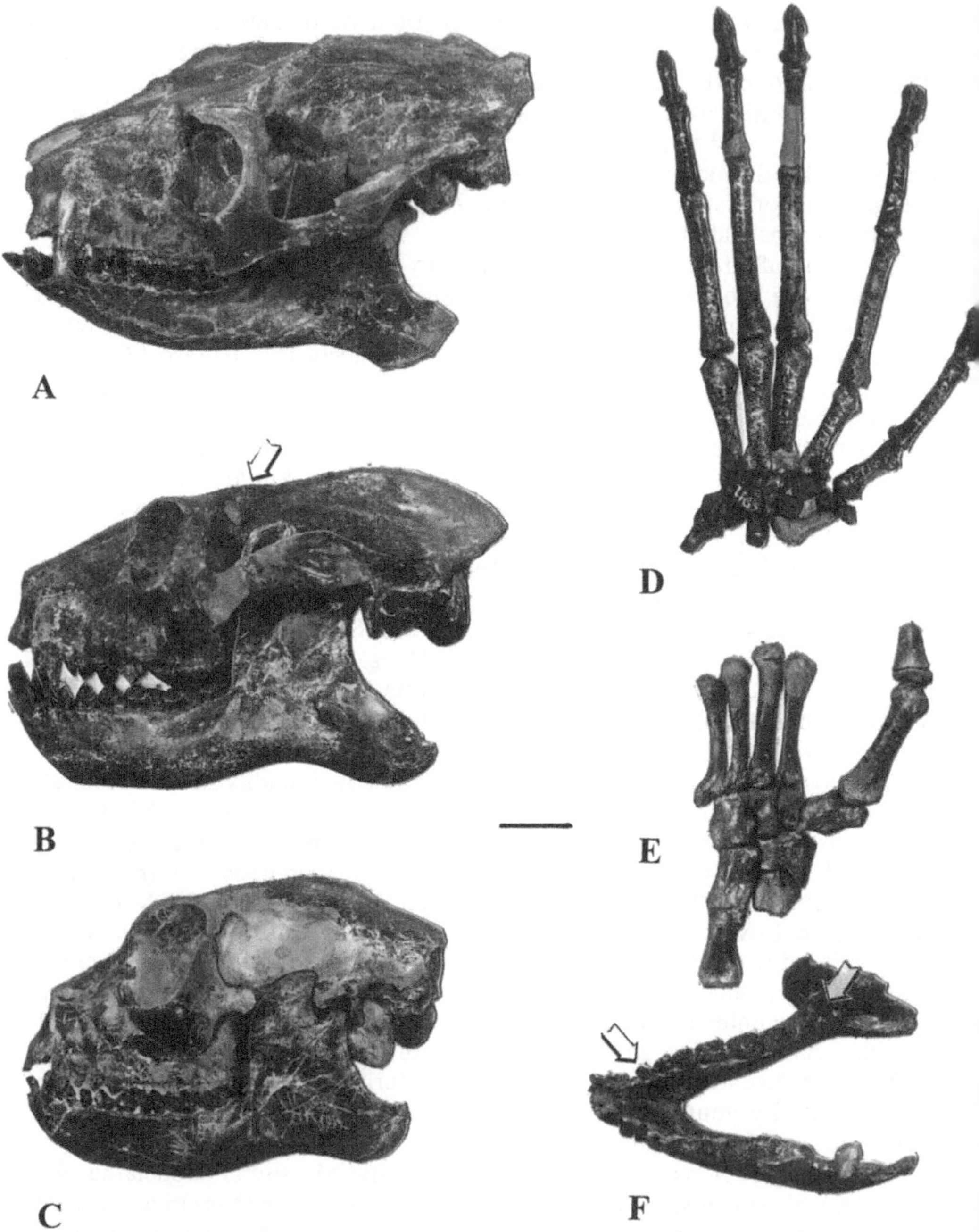

FIGURE 5. Representative notharctid specimens from Grizzly Buttes, Wyoming. A, AMNH 131764, *Notharctus tenebrosus* skull (male, image reversed); B, AMNH 127167, *Notharctus tenebrosus* skull (male); C, AMNH 131762, *Smilodectes gracilis* skull; D, AMNH 127167, *Notharctus tenebrosus* left manus; E, AMNH 131772, *Notharctus tenebrosus* partial right pes (juvenile, image reversed); F, AMNH 131765, *Notharctus tenebrosus* mandible female. Scale bar = 10.0 mm.

unidentified infant of a notharctid. The right ilium and ischium appear to have been separated by a cartilaginous pad suggesting that this individual was a non-ambulatory neonate. Isolated deciduous teeth of notharctids of uncertain attribution have also been found through screen washing.

The sagittal crest of male *N. tenebrosus* (AMNH 131764) terminates posteriorly in a distinctive button-shaped prominence at the apex of the lambdoidal crest anchoring strong temporalis, masseter and trapezius muscles (Figure 5A). This region is missing on AMNH 127167. A muted expression of the sexually dimorphic cranial features seen in *N. tenebrosus* occurs in *S. gracilis* (Alexander, 1994a). USNM 21815 and AMNH 131762 are probably males exhibiting larger canines, greater bizygomatic breadth and better defined temporal lines than USNM 17994 and AMNH 131763 which are probably females. Gazin (1958) observed that the auditory bullae of *S. gracilis* are larger than those of *N. tenebrosus*: that of AMNH 131762 (Figure 5C) (1.0 cm) is twice as voluminous as that of AMNH 131764 (Figure 5B) (0.5 cm). Contrary to Beecher (1983), in the new Grizzly Buttes sample, the mandibular symphysis of both *N. tenebrosus* and *N. pugnax* is always fused, often with the suture completely obliterated. There is a median symphyseal groove on the labial surface, similar to that of *N. pugnax* (Beecher, 1983), that splits halfway down the symphysis into two diverging scars for attachment of the mentalis musculature (Figure 7A). Considerable confusion has existed about the taxonomic status of *N. tenebrosus* and *N. pugnax* (Robinson, 1957). The new synchronous sample has helped to distinguish the two species (Alexander, 1994b). It confirms the type description of *N. pugnax* (Granger and Gregory, 1917), later statistically justified by Gingerich (1979).

A pair of thyrohyals were recovered with the skeleton of a female *N. tenebrosus*, AMNH 129132. Each of these thyrohyals is approximately four times as massive as a thyrohyal found with the hyoid apparatus of a late juvenile *Adapis parisiensis* (NMB QW 1). This element was recently discovered during preparation of the skull of NMB QW 1 (Alexander, 1996). This skull was mentioned by Stehlin (1912), but not figured. A stylohyal and epihyal were also found articulated to the basicranium of *S. gracilis* (AMNH 131762). Hyoids may open a completely new line of evidence for the interpretation of early primate phylogeny (Alexander, 1996), as they differ remarkably among living prosimians and the primitive condition for the apparatus is unknown. The large medially recurved mandibular angles of both *N. tenebrosus* and *S. gracilis* provide for an extensive area of attachment for the superficial masseter and medial pterygoid muscles and may have supported a large hyolaryngeal apparatus, perhaps permitting more resonant vocalization. The mandibular angle accounts for 14% of dentary length in *N. tenebrosus* (AMNH 127167) and 15% in *S. gracilis* (AMNH 131762) (Figure 5C). The mandibular angle of *Pelycodus jarrovii* (AMNH 55514) (Froehlich *et al.*, 1995) extends only 3.0 mm beyond the posterior border of the ascending ramus. Unfortunately the dentary of AMNH 55514 is not complete, preventing a percentage comparison with the other taxa.

Dental abnormalities have been observed in the new AMNH sample indicating a wide range of individual variation within a contemporaneous population. Rosenberger *et al.* (1985) noted that YPM 12956, a male *N. tenebrosus* dentition also from the Bridger B2 level, had lost the crown of its left P1/ and that replacement bone had covered the root in its crypt. A similar condition is seen in AMNH 131764, an adult male. The tip of a second tooth protruding at this locus suggests that the P/2 had failed to replace the dP/2 whose unresorbed root is still in place. In the adult female AMNH 131765 (Figure 5F, left arrow), the right P/1 is completely absent with no evidence of a root or alveolus. The left P/1 in this specimen is hypertrophied and similar in size to the left P/2. The question of the occlusion of upper and lower incisors in notharctids raised by Rosenberger *et al.* (1985) can be addressed in AMNH 131762 (*Smilodectes*) (Figure 5C). The in situ I1/ shows considerable attritional wear on the apical surface, obliterating the mesial prong of the I1/ noted by Greenfield (1995). Prior to this wear stage, it probably would not have been in contact with I/2. As wear progressed on I1/, it first engaged the I/1 and then both the I/1 and the I/2. Rosenberger *et al.* (1985) correctly surmised that the I2/ of *Smilodectes* is so small that it probably did not occlude with I/2 at any stage of wear.

S. gracilis possesses several diagnostic features not previously observed by Gregory (1920) or Gazin (1958) that distinguish it from all other notharctids. The domed frontal bones observed by Gazin (1958) override the nasal root, giving the nasal origin a squared off appearance (Figures 6–7C, D). There is a small foramen 1.0 mm below each orbital rim where the zygomatic ascends to form the postorbital bar (Figure 5C). In *Notharctus* this feature varies from a semicircular notch in AMNH 127167 (Figure 5B) to absent in AMNH 131764 (Figure 5A).

In *Smilodectes* a thin groove runs horizontally on the lateral surface of each dentary (Figure 5C, Figure 6C-D). It varies in length from 3.0 mm in AMNH 131774 to 13.0 mm in AMNH 131763 and can vary between the left and right sides of a single individual; in AMNH 131774 it is occurs only on the left dentary. It is a superficial channel for a blood vessel which runs posteriorly from the mandibular foramen ventral to the first molar. This character is sufficiently diagnostic to permit specific identification of edentulous jaw fragments.

ACKNOWLEDGMENTS. The authors wish to acknowledge the invaluable assistance of the following individuals.

Dr. Emmett Evanoff (UCM), Dr. Peter Robinson (UCM), Paul C. Murphey (UCM), Dr, Eugene S. Gaffney (AMNH) and Dr. Robert M. West for making unpublished stratigraphic and locality information of the Bridger Area available for this study. Dr. Laurie Bryant (BLM) for permitting access to Federal lands. Dr. Julian Kerbis (FMNH) and Dr. William Sanders (UMMP) for access to modern primate collections. Dr. Gregg F. Gunnell for review of this manuscript and body mass estimates of Bridger fauna. Dr. Jerome G. Rozen Jr. (AMNH) for advice on bee cells. Dr. Jeheskel Shoshani (CIS) for discussion of numerical prediction of environmental specialization. Dr. Dawne Adams

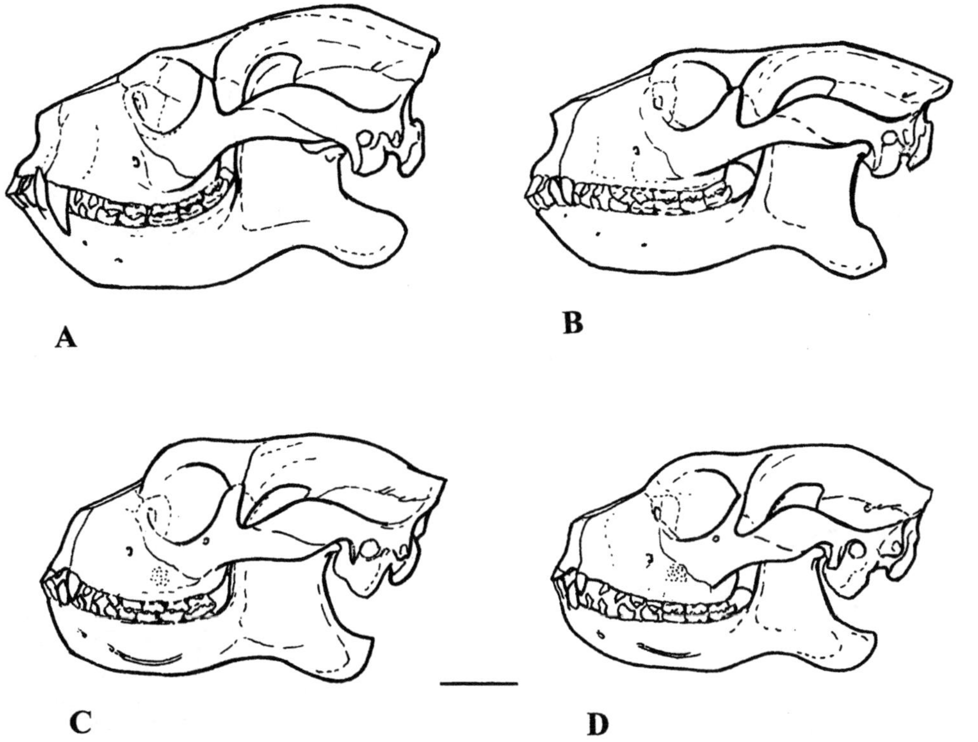

FIGURE 6. Reconstructions of skulls of Bridgerian notharctids in left lateral view. A, male *N. tenebrosus* based on AMNH 127167 and AMNH 131764; B, female *N. tenebrosus* based on AMNH 11466, AMNH 129382, and AMNH 131765; C, male *S. gracilis* based on USNM 17994 and AMNH 131762; D, female *S. gracilis* based on USNM 21815 and AMNH 131763. Scale bar = 10.0 mm.

(Baylor University), Thomas M. Alexander, Peter Ambrose (NPS), Rachel C. Benton (NPS), Dr. Richard Cifelli (OMNH), Dr. Susan Ford (SIU), Dr. Daniel Gebo (NIU), Dr. Marc Godinot (MNHN) Dr. Robert Martin (AIMUZ), Dr. Carl A. Morck, and Dr. Andrew Nelson (UWO) for logistical and financial support of field work. Dr. Robert L. Anemone (SUNY Geneseo), John Bennett (UCM), Patrick Buchmann, Sebastian Fernandez, Kenneth Filchak, Lee Ann Jolly, Linda Levitch, Patrick Light, Leahanne Sarlo, Matthew Turnow and David Wilcotts for assistance in the field. Dr. Eric Delson (CUNY), Dr. Terry Harrison (NYU), Lorraine Meeker (AMNH), Dr. Maureen O'Leary (SUNY Stony Brook) and Chester Tarka (AMNH) for assistance in the preparation of this manuscript. Research reported herein was conducted under BLM permits 209WYPA94, 231WY94 and 306WY94.

Acronyms used in text: AIMUZ—Anthr. Inst. Mus. Urchel, Zurich; AMNH—American Museum of Natural History; BLM—Bureau of Land Management; CUNY—City University of New York; FMNH—Field Museum of Natural History; MPM—Milwaukee Public Museum; MNHN—Museum National d'Historie Naturelle, Paris; NIU—Northern Illinois University; NPS—

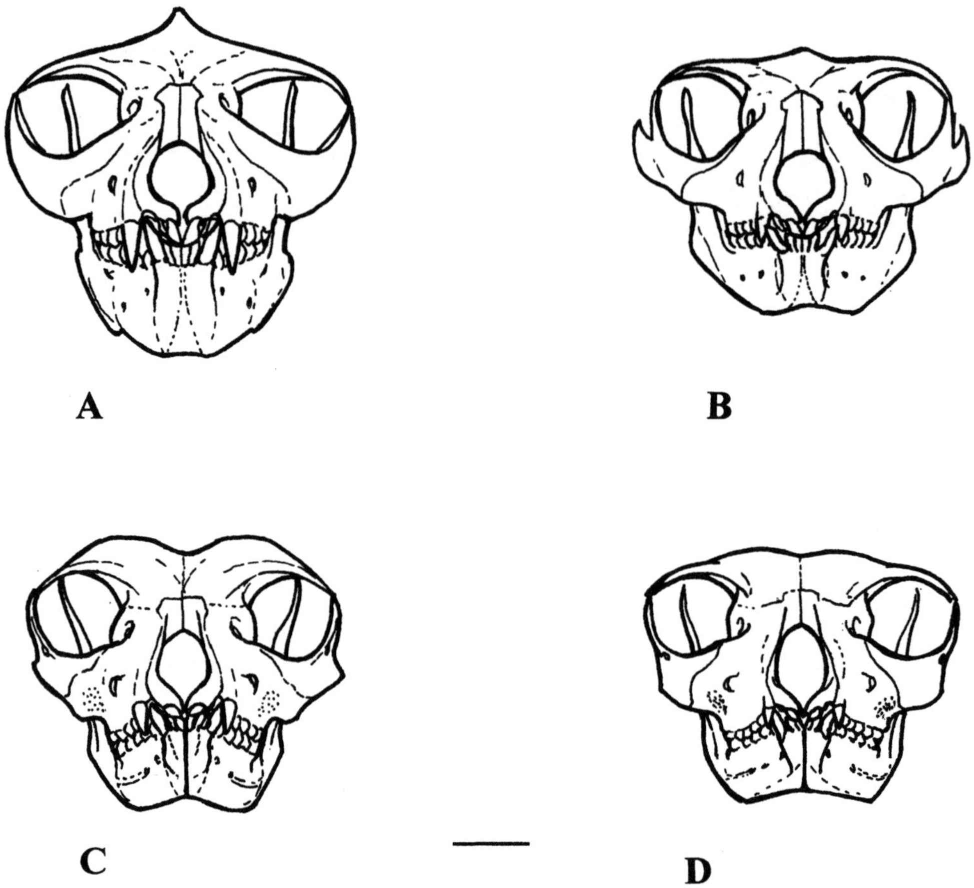

FIGURE 7. Reconstructions of skulls of Bridgerian notharctids in frontal view. Labels as in Figure 1. Scale bar = 10.0 mm.

National Park Service; NYU—New York University; OMNH—Oklahoma Museum of Natural History; SUNY—State University of New York; SIU—Southern Illinois University; UCM—University of Colorado, Boulder; UMMP—University of Michigan Museum of Paleontology; USNM—U.S. National Museum (Smithsonian Institution); UWO—University of Western Ontario; YPM—Yale Peabody Museum.

References

Alexander, J. P., 1992, Alas poor *Notharctus, Natural History* **101(8):**54–59.

Alexander, J. P., 1993, *The splanchnocranial anatomy of the Eocene primate* Notharctus, M. A. Thesis, Hunter College, City University of New York.

Alexander, J. P., 1994a, Evidence for sexual dimorphism in the primate family Notharctidae, *Folia Primatol.* **63:**59–62.

Alexander, J. P., 1994b, Sympatry of two species of *Notharctus* in the Bridger Basin, Wyoming, *Jour. Vert. Paleont.* **14**(supplement to 3):14A.
Alexander, J. P., 1996, Elements of the hyoid apparatus of several Eocene adapiform primates, *Jour. Vert. Paleont.* **16**(supplement to 3):19A.
Alexander, J. P., and MacPhee, R. D. E., 1999a, Skull of *Omomys carteri*, an Eocene omomyid Primate, *American Journal of Physical Anthropology* Supplement 28, p. 83.
Alexander, J. P., and MacPhee, R. D. E., 1999b, Skull morphology of *Omomys carteri*, *Jour. Vert. Paleont.* **19**(supplement to 3):29A.
Andrews, P., 1990, *Owls, Caves and Fossils*, University of Chicago Press, Chicago.
Barry, R. E., and Barry, L. M., 1996, Species composition and age structure of remains of hyraxes (Hyracoidea: Procaviidae) at nests of black eagles, *Jour. Mammal.* **77(3)**:702–707.
Beard, K. C., and Krishtalka, L., 1991, First skull of the early Eocene primate *Shoshonius cooperi* and the anthropoid-tarsier dichotomy, *Nature* **349**:64–67.
Beard, K. C., and MacPhee, R. D. E., 1994, Cranial anatomy of *Shoshonius* and the antiquity of Anthropoidea, in: *Anthropoid Origins* (J. G. Fleagle, and R. F. Kay, eds.), Plenum Press, New York, pp. 55–97.
Beecher, R. M., 1983, Evolution of the mandibular symphysis in Notharctinae (Adapidae, Primates), *Intl. Jour. Primatol.* **4**:99–112.
Behrensmeyer, A. M., and Hill, A., 1980, *Fossils in the Making*, University of Chicago Press, Chicago.
Berman, D. S., 1973, *Spathorhynchus fossorum*, a Middle Eocene Amphisbaenian (Reptilia) from Wyoming, *Copeia* **4**:704–721.
Bradley, W. H., 1964, Geology of the Green River Formation and associated Eocene rocks in southwestern Wyoming and adjacent parts of Colorado and Utah, *U.S. Geol. Surv. Prof. Paper* **496A**:1–86.
Brain, C. K., 1974, A hominid skull's revealing holes, *Natural History* **83(10)**:44–45.
Brown, L. H., 1971, The relationship of the Crowned Eagle *Stephanoaetus cornatus* and some of its prey animals, *Ibis* **113**:240–243.
Brown, L. H., 1976, *Eagles of the World*, Purnell: Cape Town.
Brown, L. H., 1982, The prey of the Crowned Eagle *Stephanoaetus cornatus* in Central Kenya, *Scopus* **6**:91–94.
Brown, R. W., 1929, Additions to the flora of the Green River Formation, *U. S. Geol. Surv. Prof. Pap.* **154**:279–299.
Brown, R. W., 1934, The recognizable species of the Green River Flora, *U. S. Geol. Surv. Prof. Pap.* **185C**:45–77.
Daneel, A. B. C., 1979, Prey size and hunting methods of the Crowned Eagle, *Ostrich* **50**:120–121.
Delson, E., 1971, Fossil mammals of the early Wasatchian Powder River local fauna, Eocene of northeast Wyoming, *Bull. Amer. Mus. Nat. Hist.* **146**:307–364.
Evanoff, E., Brand, L., and Murphey, P., 1998, Bridger Formation (middle Eocene) of southwest Wyoming: widespread marker units and subdivisions of Bridger B through D, *Dakoterra* **5**:115–122.
Evanoff, E., and Ivy, L., 1992, Roadlog Manila, Utah to Fort Bridger, Wyoming, along Utah 43 and Wyoming 414, in: *Field Guide to geologic excursions in Utah and adjacent areas of Nevada, Idaho, and Wyoming* (J. R. Wilson, ed.), Utah Geol. Surv. Misc. Pub. **923**:368–373.
Flynn, J. J., 1986, Correlation and geochronology of middle Eocene strata from the western United States, *Palaeogeogr., Palaeoclimat., Palaeoecol.* **55**:335–406.
Franzen, J. L., and Frey, E., 1993, *Europolemur* completed, *Kaupia* **3**:113–130.
Froehlich, J., Alexander, J. P., and Froehlich, D., 1995, *Pelycodus* premolars and other new specimens from the San Jose Formation of New Mexico, *Jour. Vert. Paleont.* **15**(supplement to 3):30A.
Gaffney, E. S., 1972, The systematics of the North American family Baenidae (Reptilia, Cryptodira), *Bull. Amer. Mus. Nat. Hist.*, **147**:243–319.
Gargett, V., 1990, *The Black Eagle*, Acorn Books, Randburg, South Africa.
Gazin, C. L., 1958, A review of the Middle and Upper Eocene primates of North America, *Smithson. Misc. Coll.* **136**:1–112.

Gazin, C. L., 1968, A study of the Eocene condylarthran mammal *Hyopsodus*, *Smithson. Misc. Coll.* **153**:1–90.

Gazin, C. L., 1976, Mammalian faunal zones of the Bridger Middle Eocene, *Smithson. Contribs. Paleobiol.* **26**:1–25.

Gingerich, P. D., 1979, Phylogeny of Middle Eocene Adapidae (Mammalia, Primates) in North America: *Smilodectes* and *Notharctus*, *Jour. Paleont.* **53**:153–163.

Granger, W., and Gregory, W. K., 1917, A revision of the Eocene primates of the genus *Notharctus*, *Bull. Amer. Mus. Nat. Hist.*, **37**:841–859.

Greenfield, L. O., 1995, The anterior dentition of adapids and anthropoid origins, *Folia Primatol.* **65**:59–84.

Gregory, W. K., 1920, On the structure and relations of *Notharctus*, an American Eocene primate, *Mem. Amer. Mus. Nat. Hist.* **3**:1–243.

Gunnell, G. F., and Bartels, W. S., 1994, Early Bridgerian (middle Eocene) vertebrate paleontology and paleoecology of the southern Green River Basin, Wyoming, Contrib. *Geol., Univ. Wyo.* **30**:57–70.

Gunnell, G. F., and Bartels, W. S., 1997, Basin margin mammalian assemblages from the Wasatch Formation (Bridgerian) of the northeastern Green River Basin, Wyoming: anachronistic taxa and the origin of new genera, *Jour. Vert. Paleont.* **17**(supplement to 3):51A.

Hamrick, M. W., and Alexander, J. P., 1996, The hand skeleton of *Notharctus tenebrosus* (Primates, Notharctidae) and its significance for the origin of the primate hand, *Amer. Mus. Novitates* **3182**:1–20.

Hanley, J. H., 1974, *Systematics, paleoecology and biostratigraphy of nonmarine Mollusca from the Green River and Wasatch Formations (Eocene), southwestern Wyoming and northwestern Colorado*, Ph.D. Thesis, University of Wyoming, Laramie.

Hay, O. P., 1908, The fossil turtles of North America, *Carnegie Inst. Wash. Publ.* **75**:1–568.

Hibbard, C. W., 1949, Techniques of collecting microvertebrate fossils, *Contrib. Mus. Paleont. Univ. Mich.* **8**:7–19.

Izor, R. J., 1985, Sloths and other mammalian prey of the Harpy Eagle, in: *The Evolution and Ecology of Armadillos, Sloths, and Vermilinguas* (G. G. Montgomery, ed.) Smithsonian Institution Press, Washington, D.C., pp. 343–346.

Jarvis, M. J. F., Currie, M. H., and Palmer, N. G., 1980, Food of crowned eagles in the Cape Province, South Africa, *Ostrich* **51**:215–218.

Kerbis, J. C., 1997, Taphonomy of tropical forest environments: monkey eating eagles as agents of bone accumulation, *Jour. Vert. Paleont.* **17**(supplement to 3):56A–57A.

Krishtalka, L., West, R. M., Black, C. C., Dawson, M. R., Flynn, J. J., Turnbull, W. D., Stucky, R. K., McKenna, M. C., Bown, T. M., Golz, D. J., and Lillegraven, J. A., 1987, Eocene (Wasatchian through Duchesnean) Biochronology of North America, in: *Cenozoic Mammals of North America Geochronology and Biostratigraphy* (M. O. Woodburne, ed.), University of California Press, Berkeley, pp. 77–117.

Langston, W., Jr., 1975, Ziphodont Crocodiles: *Pritichampsus vorax* (Troxell), new combination, from the Eocene of North America, *Fieldiana Geology* **33**:291–314.

Leidy, J., 1869, Notice of some extinct vertebrates from Wyoming and Dakota, *Proc. Acad. Nat. Sci. Phila.* **21**:63–67.

Leidy, J., 1870, Descriptions of *Palaeosyops paludosus, Microsus cuspidatus* and *Notharctus tenebrosus, Proc. Acad. Nat. Sci. Phila.* **22**:113–114.

MacGinitie, H. D., 1969, The Eocene Green River Flora of northwestern Colorado and northeastern Utah, *Univ. Calif. Pub. Geol. Sci.* **83**:1–140.

Marsh, O. C., 1871, Notice of some fossil mammals from the Tertiary formation, *Amer. Jour. Sci.* **2**:35–44, 120–127.

Matthew, W. D., 1909, The Carnivora and Insectivora of the Bridger Basin, middle Eocene, *Mem. Amer. Mus. Nat. Hist.* **9**:289–567.

Matthew, W. D., 1928, The evolution of mammals in the Eocene, *Proc. Zool. Soc. London* **1927**:947–985.

Mauger, R. L., 1977, K-Ar ages of biotites from tuffs in Eocene rocks of the Green River, Washakie, and Uinta basins, Utah, Wyoming, and Colorado, *Univ. Wyo. Contrib. Geol.* **15**:17–41.

McGrew, P. O., and Casilliano, M., 1970, Geological history of Fossil Butte National Monument and Fossil Basin, *National Park Service Occasional Papers* 3.

McGrew, P. O., and Sullivan, R., 1970, The stratigraphy and paleontology of Bridger A, *Contrib. Geol. Univ. Wyo.* **9**:66–85.

McKenna, M. C., 1962, Collecting small fossils by washing and screen, *Curator* **3**:221–235.

McKenna, M. C., 1990, Plagiomenids (Mammalia: ?Dermoptera) from the Oligocene of Oregon, Montana, and South Dakota, and middle Eocene of northwestern Wyoming, in: *Dawn of the Age of Mammals in the northern part of the Rocky Mountain Interior, North America* (T. M. Bown, and K. D. Rose, eds.), Geol. Soc. Amer. Spec. Pap. **243**:211–234.

Murphey, P. C., 1990, Depositional setting and fauna of the *Omomys* Quarry, a possible owl site in the Bridger Formation (Middle Eocene) of Southwestern Wyoming, *Jour. Vert. Paleont.* **16**(supplement to 3):55A.

Murphey, P. C., 1995, *Paleontology, distribution and history of a middle Eocene lacustrine deposit in the Bridger Formation, southwestern Wyoming*, Master's Thesis, University of Colorado, Boulder.

Murphey, P. C., Torick, L., Bray, E., Chandler, R., and Evanoff, E., 1998, Taphonomy and paleoecology of the Omomys Quarry, an unusual fossil accumulation from the Bridger Formation, Wyoming, *Jour. Vert. Paleont.* **18**(supplement to 3):65A.

Osborn, H. F., 1904, An armadillo from the Middle Eocene (Bridger) of North America, *Bull. Amer. Mus. Nat. Hist.* **20**:163–165.

Osborn, H. F., 1910, *The Age of Mammals in Europe, Asia, and North America*, Macmillan, New York.

Rea, D. K., Zachos, J. C., Owen, R. M., and Gingerich, P. D., 1989, Global change at the Paleocene-Eocene boundary: climatic and evolutionary consequences of tectonic events, *Palaeogeogr., Palaeoclimat., Palaeoecol.* **79**:117–128.

Rettig, N. L., 1978, Breeding behavior of the harpy eagle *Harpia harpyjia*, *The Auk* **95**: 629–643.

Robinson, P., 1957, The species of *Notharctus* from the Middle Eocene, *Postilla* **28**:1–27.

Robinson, P., 1994, *Myrmekomomys*, a new genus of Micromomyine (Mammalia, ?Microsyopidae) from the lower Eocene rocks of the Powder River Basin, Wyoming, *Contribs. Geol. Univ. Wyo.* **30**:85–90.

Roehler, H. W., 1993, Eocene climates, depositional environments, and geography, Greater Green River Basin, Wyoming, Utah and Colorado, *U. S. Geol. Surv. Prof. Paper* **1506F**:1–74.

Rose, K. D., MacPhee, R. D. E., and Alexander, J. P., 1999, Skull of early Eocene *Cantius abditus* (Primates: Adapiformes) and its phylogenetic implications, with a reevaluation of "*Hesperolemur*" *actius*, *American Journal of Physical Anthropology* **109**:523–539.

Rosenberger, A. L., Strasser, E., and Delson, E., 1985, Anterior dentition of *Notharctus* and the adapid-anthropoid hypothesis, *Folia Primatol.* **44**:15–39.

Royse, F. R., Jr., 1993, An overview of the geologic structure of the thrust belt in Wyoming, northern Utah, and eastern Idaho, in: *Geology of Wyoming*, Memoir no. 5, The Geological Survey of Wyoming, pp. 272–311.

Rozen, J. G., Jr., 1984a, Nesting biology of Diphaglossine bees (Hymenoptera, Colletidae), *Amer. Mus. Novitates* **2786**:1–33.

Rozen, J. G., Jr., 1984b, Comparative nesting biology of the bee tribe Exomalopsini (Apoidea, Anthophoridae), *Amer. Mus. Novitates* **2798**:1–37.

Rozen, J. G., Jr., 1993, Nesting biologies and immature stages of the Rophitine bees (Halictidae) with notes on the cleptoparasite *Biastes* (Anthophoridae) (Hymenoptera: Apoidea), *Amer. Mus. Novitates* **3066**:1–28.

Savin, S. M., 1977, The history of the Earth's surface temperature during the past 100 million years, *Ann. Rev. Earth Planet. Sci.* **5**:319–355.

Savin, S. M., and Douglas, R. G., 1984, Sea level, climate, and the Central American land bridge, in: *The Great American Biotic Interchange* (F. G. Steli, and S. D. Webb, eds.), Plenum, New York, pp. 303–324.

Schoch, R. M., 1984, Revision of *Metacheiromys* Wortman, 1903 and a review of the Palaeanodonta, *Postilla* **192**:1–17.

Shoshani, J., 1992, Ecology and taxonomy: a new approach, using *Heterohyrax* vs *Procavia* as a case study, *Israel Jour. Zool.* **38**:399–416.

Shotwell, J. A., 1958, Intercommunity relationships in Hemphillian (mid-Pliocene) mammals, *Ecology* **39**:271–282.

Simons, E. L., 1964, The early relatives of man, *Sci. Amer.* **7**:1–14.

Simpson, G. G., 1931, *Metacheiromys* and the Edentata, *Bull. Amer. Mus. Nat. Hist.* **55**:295–381.

Skorupa, J. P., 1989, Crowned Eagles *Strephanoaetus coronatus* in rainforest: observations on breeding chronology and diet at a nest in Uganda, *Ibis* **131**:294–298.

Smedes, H. W., and Prostka, H. J., 1972, Stratigraphic framework of the Absaroka Volcanic Super group in the Yellowstone National Park region, *U.S. Geol. Surv. Prof. Paper* **729C**:1–33.

Stehlin, H. G., 1912, Die Saugetiere des Schweizerischen Eocaens. Critischer Catalog der Materialen. Part 7, *Abh. der Schweiz. Pal. Ges.* **38**:1165–1298.

Stucky, R. K., 1992, Mammalian faunas in North America of Bridgerian to Early Arikareean "Ages" (Eocene and Oligocene), in: *Eocene-Oligocene Climatic and Biotic Evolution* (D. R. Prothero, and W. A. Berggren, eds.), Princeton University Press, Princeton, pp. 464–493.

Struhsaker, T. T., and Leakey, M., 1990, Prey selectivity by crowned hawkeagles on monkeys in the Kibale Forest, Uganda, *Behav. Ecol. Sociobiol.* **26**:435–443.

Thompson, F., 1984, *Freshwater snails of Florida, a manual for identification*, University of Florida Press, Gainesville.

Troxell, E. L., 1925, The Bridger Crocodiles, *Amer. Jour. Sci.* **9**:29–72.

Wallace, J., 1994, *The American Museum of Natural History's Book of Dinosaurs and other Ancient Creatures*, Simon and Shuster, New York.

West, R. M., 1976, Paleontology and geology of the Bridger Formation, southern Green River Basin, southwestern Wyoming. Part 1. History of field work and geological setting, *Contribs. Biol. Geol., Milwaukee Publ. Mus.* **7**:1–12.

West, R. M., 1979, Paleontology and geology of the Bridger Formation, southern Green River Basin, southwestern Wyoming. Part 3. Notes on *Hyopsodus, Contribs. Biol. Geol., Milwaukee Publ. Mus.* **25**:1–52.

West, R. M., 1990, Vertebrate Paleontology of the Green River Basin, Wyoming, 1840–1910, *Earth Sciences History* **9**:45–56.

Wetmore, A., 1933, The status of *Minerva antiqua*, *Aquila ferox*, and *Aquila lydekkeri* as fossil birds, *Amer. Mus. Novitates* **680**:1–4.

Wheeler, W. H., 1961, Revision of the Uintatheres, *Bull. Peabody Mus. Nat. Hist.* **14**:1–93.

Wing, S. L., 1987, Eocene and Oligocene floras and vegetation of the Rocky Mountains, *Ann. Missouri Bot. Garden* **74**:176–212.

Wood, H. E., 1934, Revision of the Hyrachyidae, *Bull. Amer. Mus. Nat. Hist.* **67**:181–295.

Chapter 7

Taphonomic Analysis of the Messel Formation (Germany)

JENS LORENZ FRANZEN

1. Introduction

Less than 2% of lagerstaetten with fossil vertebrate remains have yielded anything more than fragments of jaw material and other bones. One such exceptional deposit is the former oil shale mine of "Grube Messel", located about 35 kilometers south east of Frankfurt am Main, Germany (Figure 1). The threat posed to this site by a projected refuse dump is one reason why an intensive excavation program was begun by a variety of scientific institutes in 1975. This work has led to recovery of a rich vertebrate sample with associated preservational data. Thanks to international protest "Grube Messel" was declared a Natural Heritage of all Mankind by UNESCO in 1995. In this way it became

JENS LORENZ FRANZEN • Forschungsinstitut Senckenberg, D-60325 Frankfurt am Main 1, Germany.

Eocene Biodiversity: Unusual Occurrences and Rarely Sampled Habitats, edited by Gregg F. Gunnell, Kluwer Academic/Plenum Publishers, New York, 2001.

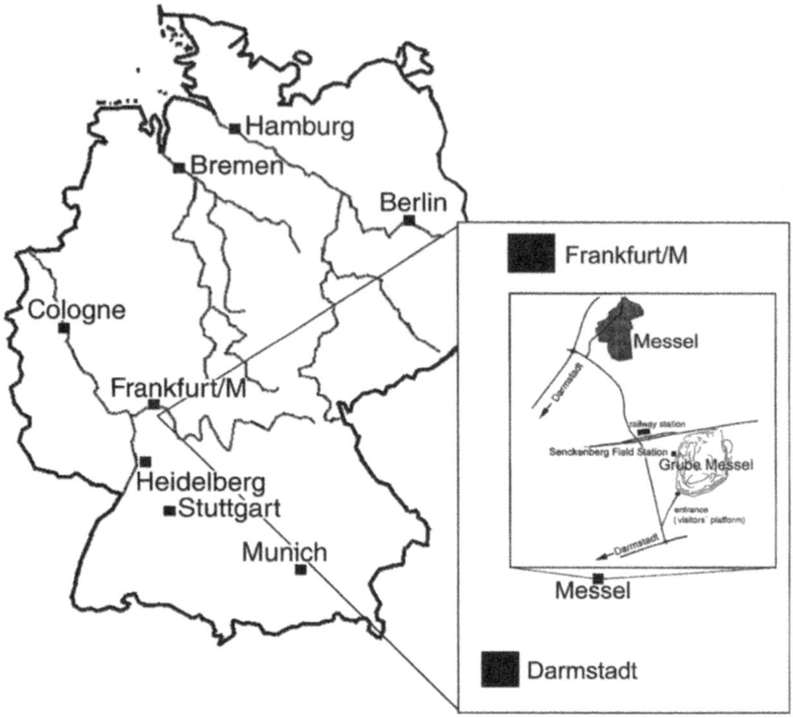

FIGURE 1. Location of "Grube Messel". Drawing Senckenberg-Museum, C. Hemm.

preserved for scientific research forever. Excavations are still being conducted by the Forschungsinstitut Senckenberg and the Hessisches Landesmuseum Darmstadt to this day.

2. Messel Vertebrates

2.1. Fauna

All classes of vertebrates have been discovered from the oil shale of the Messel Formation, which developed from the sediments of a former freshwater lake at the beginning of the middle Eocene, about 49 million years ago (Franzen and Haubold, 1986; Figure 2). Most abundant are the former inhabitants of the lake itself, especially fish. Turtles and crocodiles also lived in and around the lake and are preserved in the Messel oil shales. Animals that thrived in the adjacent surroundings, for instance anurans and snakes, are relatively frequently found. Exceptional discoveries include skeletons of mammals that lived in the forests that stretched away from the lake. Only gregarious mammals, such as early horses, are common but many other mammals are preserved at Messel too. Really puzzling is the fact that birds and bats, normally very rare in the fossil record because of the fragile character of their skeletons, are among the most frequent vertebrates from the Messel Formation. During 25 years of intensive excavation several hundreds of more or less complete skeletons of each of these taxa have been recovered.

2.2. Preservation

As far as the quality of preservation is concerned at Messel, most of the fossil vertebrates are recovered as complete articulated skeletons, although isolated bones and scattered skeletons occur occasionally as well. The most remarkable fact is that in some cases the fossil vertebrates are not only preserved as skeletons but still display the outlines of their entire bodies as black shadows in the oil shale. Among some anuran specimens it is possible even to distinguish the eyes, the liver, and the veins on the medial side of their legs (Figure 3). Birds show the detailed structures of their feathers (Figure 4) while mammals display their fur to the tips of the hairs and even their external ears (Figures 5, 6). It is not only the external morphology that is preserved. In some cases it is even possible to distinguish the outline of the intestines (Figure 7) and in many cases remains of food within the intestinal tract have survived to the present day in an extraordinary state of preservation (Richter, 1992; Figures 8, 9). Thus the paleontologist who wishes to reconstruct what these animals looked like or how they might have lived can rely upon direct evidence instead of vague assumptions and bold speculations.

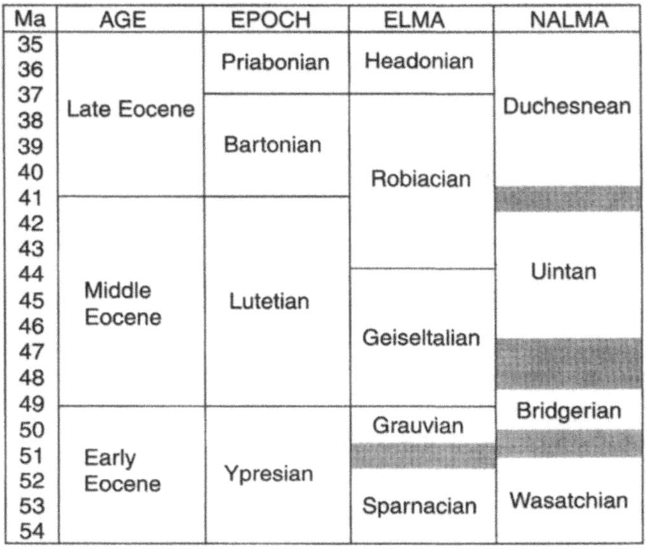

FIGURE 2. Time table displaying a tentative correlation of NALMAS and ELMAS. Drawing Senckenberg-Museum, C. Hemm (after Mertz *et al.*, 2000; Walsh, 1996). NALMA = North American Land Mammal Age, ELMA = European Land Mammal Age.

FIGURE 3. Frog (*Eopelobates*) displaying eyes, liver, and veins on medial side of thighs. Photograph from Senckenberg-Museum, C. Schumacher.

3. Geology and Paleoclimate

3.1. Geological Investigation at Messel

Based on the results of the excavations at Messel and of a drilling program carried out at the mine in 1980 (Weber and Hofmann, 1982), a wide ranging series of research projects has been developed over the years with support from the Deutsche Forschungsgemeinschaft. Recently, other oilshale outcrops in the area surrounding Messel have also been investigated through an additional drilling program carried out with funds from a private sponsor (Harms, 1998). The spectrum of investigations ranges from geochemistry, geology, and sedimentology to specialized processing of the flora and fauna preserved in the sediments. Particularly interesting for this paper are the factors that controlled the extraordinary preservation of the fossils and the origins of these fossiliferous beds (Franzen, 1985; Franzen and Michaelis, 1988).

The lake was no more than 1 square mile in area as indicated by near shore facies at the margins of the present outcrops (Franzen *et al.*, 1982). Most of the area of the lake was evidently more than 10 meters deep as is demonstrated

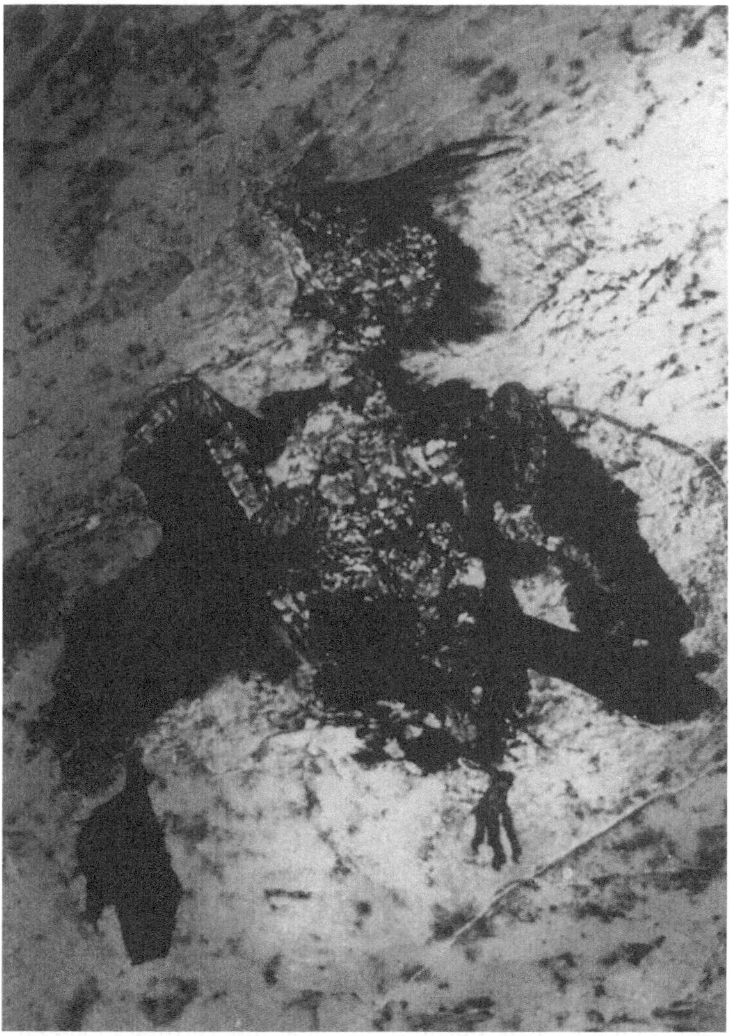

FIGURE 4. Bird showing detailed structure of feathers. Photograph from Senckenberg-Museum, C. Schumacher.

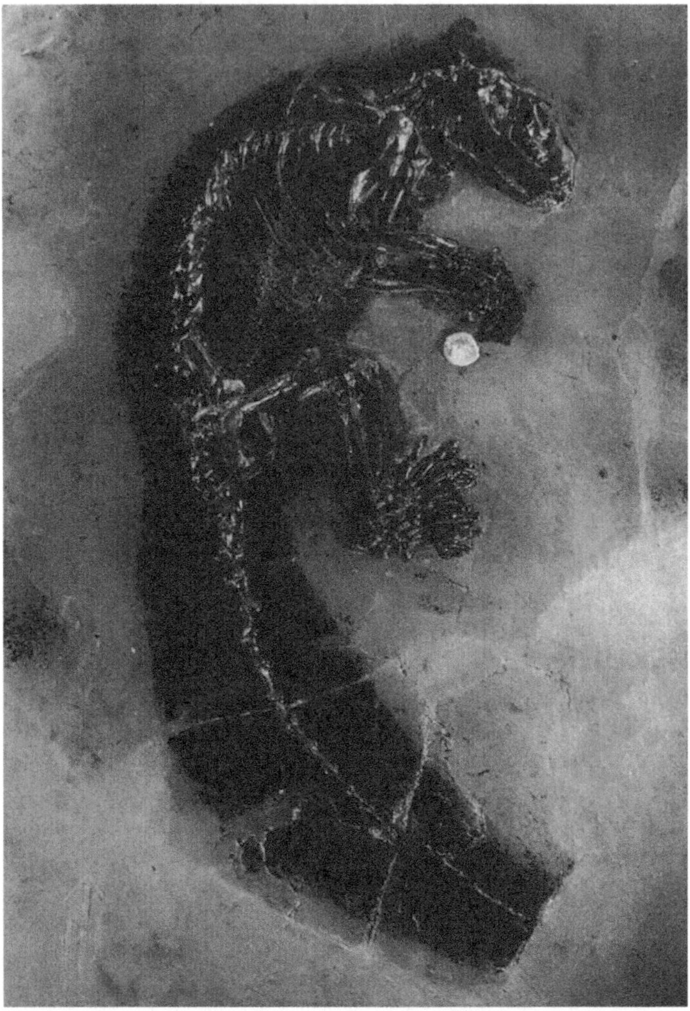

FIGURE 5. Condylarth mammal (*Kopidodon macrognathus*) displaying the complete outline of its soft body including the external ears and a bushy tail. Photograph from Senckenberg-Museum, S. Tränkner.

by the distribution of completely articulated skeletons (Figures 10, 11). Their occurrence depends on the fact that carcasses of drowned mammals do not rise again to the surface as long as the pressure of the water column is high enough. Experiments undertaken by Elder (1985) have shown that the water column has to be at least 10 meters deep to prevent inflation of the bodies by the generation of decompositional gasses. In this way, by examination of the distribution of the skeletons it is possible to trace the 10 meters isobath of Eocene Lake Messel.

The origins of Lake Messel are not entirely clear. The lake may have come into existence as the result of a small tectonic breakdown in context with the

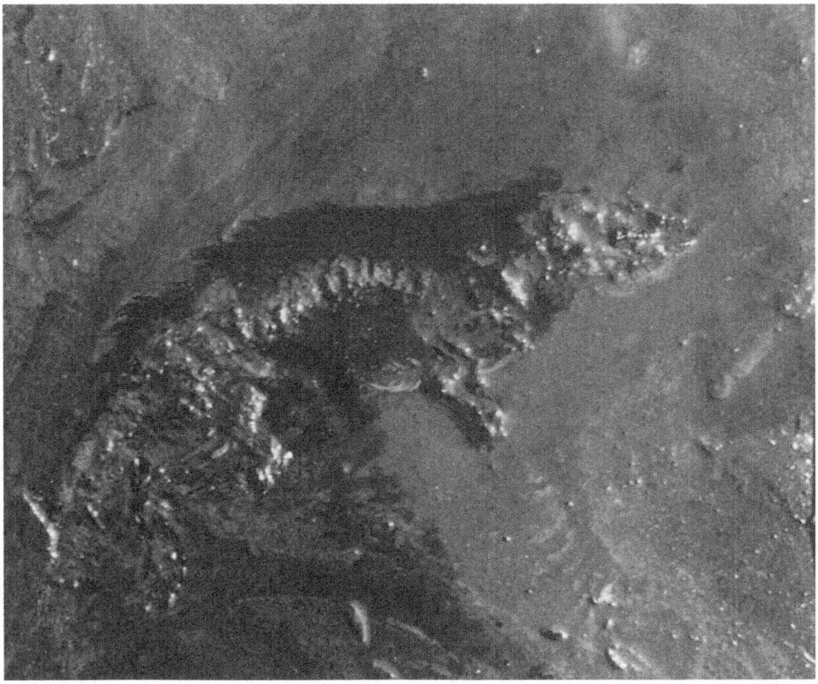

FIGURE 6. Erinaceomorph mammal (*Pholidocercus hassiacus*) immediately after discovery, displaying fur to tips of the hairs. Photograph from Senckenberg-Museum, J. L. Franzen.

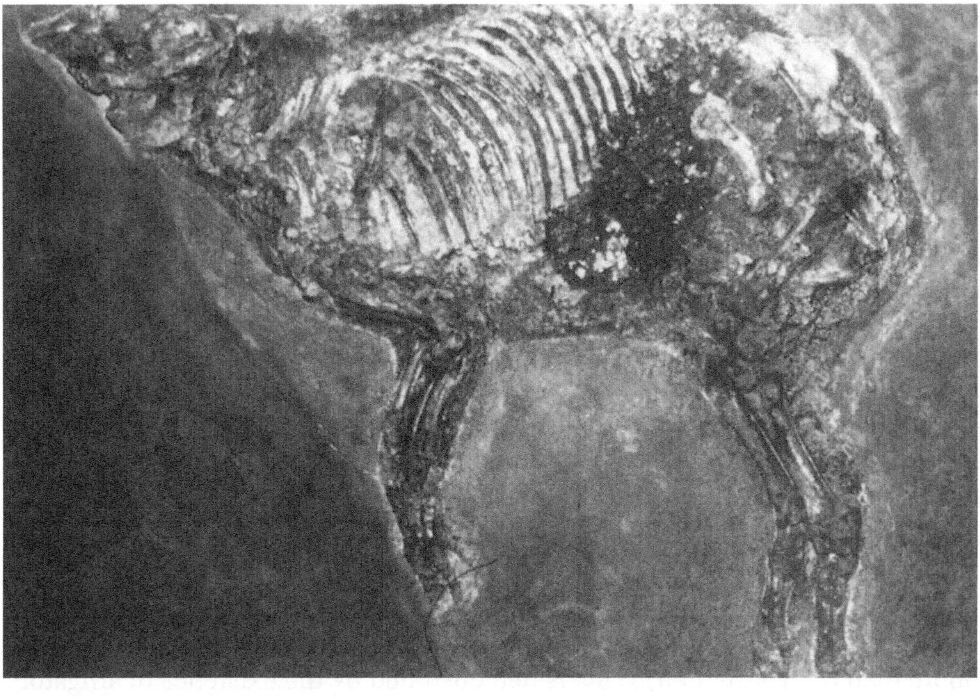

FIGURE 7. Dawn horse (*Hallensia matthesi*) showing outline of the caecum. Photograph from Senckenberg-Museum, J. L. Franzen.

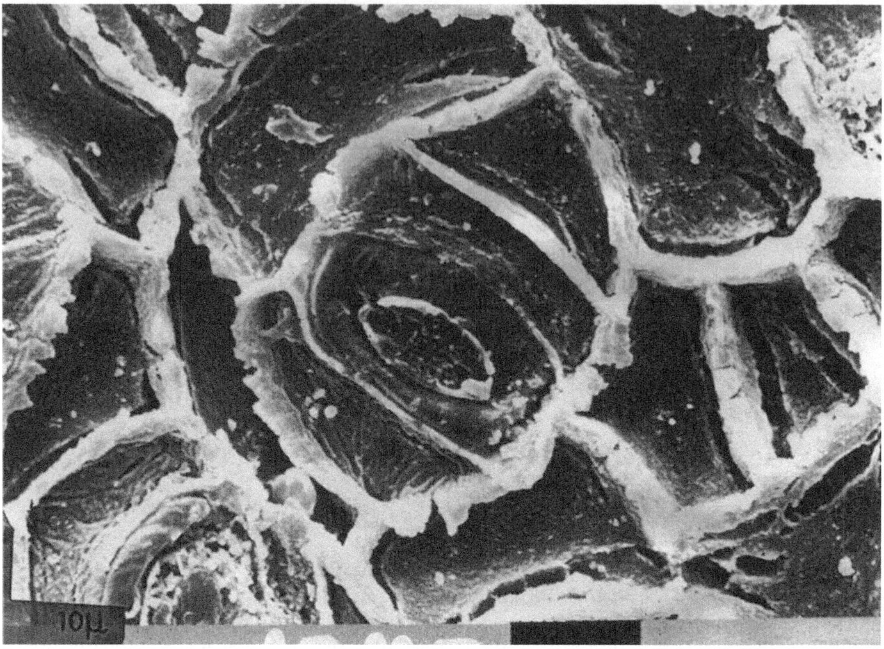

FIGURE 8. Fragment of a leaf out of the digestive tract of a dawn horse (*Propalaeotherium parvulum*); view of underside showing the cell walls of a Lauraceae and a stomatum. Scanning electron micrograph, Senckenberg-Museum, G. Richter.

rift valley of the Oberrheingraben developing at that time (Haupt, 1922; Matthess, 1966; Franzen, 1985), or it may have resulted from a maar explosion as a crater lake (Hummel, 1925; Rietschel, 1994; Pirrung, 1998; Figure 12). The latter explanation seems to be favored by the volcanic origin of most of the sediments (Weber and Zimmerle, 1985) and by the occurrence of contemporary basaltic outcrops in the surrounding area (Lippolt *et al.*, 1975). Also drillings undertaken recently in the Messel area have proved a maar origin at least for two other outcrops (Harms, 1998). On the other hand, the present outcrops of oilshale at Grube Messel being limited by faulted boundaries (Matthess, 1966), point to a tectonic origin of the Eocene lake.

3.2. Paleoclimatic Reconstruction

The climate at the time of the deposition of the Messel oil shales was tropical to subtropical with mean annual temperatures of more than 20°C at least (more than 67° Fahrenheit). This is demonstrated by the existence of alligators and palm-trees in the Messel deposits as well as by oxygen isotope paleotemperatures estimated from tertiary mollusks of the North Sea basin (Berg, 1964; Buchardt, 1978; Schaarschmidt, 1992; Figure 13). With mean annual temperatures of at least 20°C and water depths of at least 10 meters, Lake Messel was thermally stratified (Figure 14) and represented a warm-meromictic subtropical lake type (Goth, 1990).

FIGURE 9. Gut contents of a dawn horse (*Propalaeotherium parvulum*)—grape seeds. Photograph from Senckenberg-Museum, J. L. Franzen.

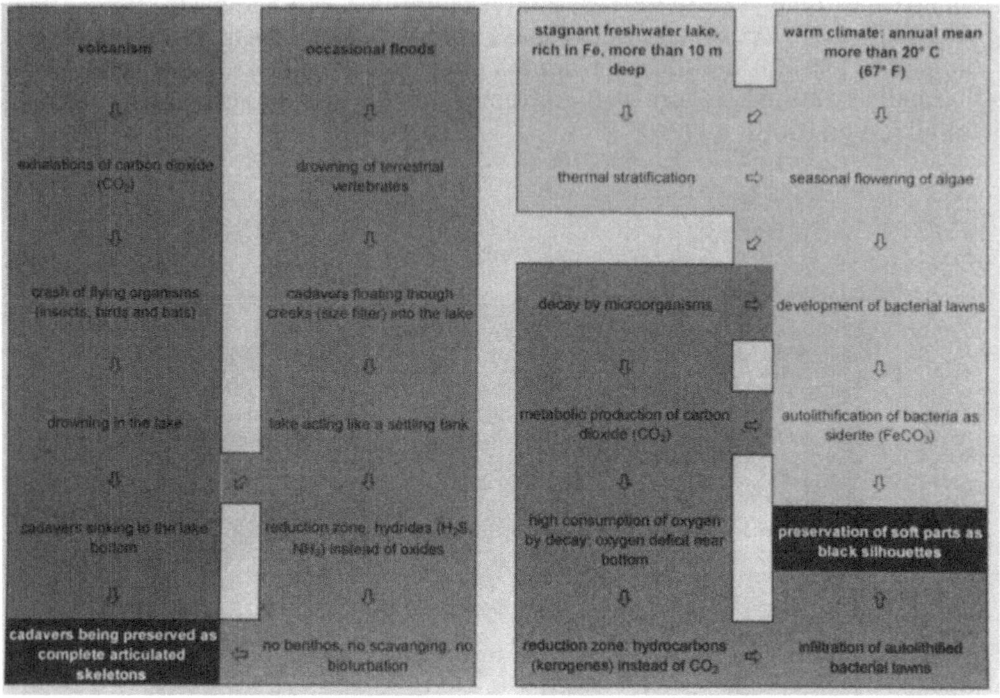

FIGURE 10. Factors involved in the exceptional preservation of articulated skeletons and so called soft parts of Eocene vertebrates from "Grube Messel". Drawing Senckenberg-Museum, D. Hochberger (after J. L. Franzen).

4. Taphonomy

4.1. Skeletal Preservation

The fossil lagerstaette at Messel is a limnic stagnation deposit (Seilacher *et al.*, 1985). Eocene Lake Messel was, from time to time, connected with a river system such that the lake acted as a settling tank for animal carcasses (Franzen, 1985). All of the preservable parts of organisms drifting downstream or once living in the lake were ultimately embedded in the argillaceous sediments of the lake bed. There, anaerobic conditions prevailed because of the low energy environment and a high consumption of oxygen resulting from the decomposition of masses of micro-organisms (mainly algae) that flourished periodically under the tropical to subtropical climate (Goth, 1990). Thus reducing conditions prevailed and the oxides produced by decomposing bacteria (Wuttke, 1983) were at least partly reduced to hydrides such as ammonium (NH4) or other poisonous substances like hydrogen sulfide (H2S), preventing development of any benthic macro-organisms. Therefore there was no bioturbation. Vertebrate carcasses were completely buried at the bottom of the lake and were neither destroyed by scavengers nor disturbed by currents. Nor did they rise to the surface again within the 10 meters isobath, so that within the area of the isobath the carcasses were normally preserved as complete and articulated skeletons.

The Messel deposits consist of finely laminated claystone, the lamination being due to annual climatic fluctuations. They consist of algae-rich layers caused by seasonal blooms which were superimposed on a steady background sedimentation of smectite and other clay minerals (Goth, 1990). The sedimentation rate was low (about 0.1 mm per year, Irion, 1977). It was occasionally interrupted by slumps coming down from the slopes around the lake. As the lens of oilshales was originally about 200 meters thick, it can be calculated that the Messel lake existed for hundreds of thousands of years in a lowland area.

4.2. Unique Preservational Attributes

Clearly Lake Messel acted as a trap or settling tank for animals but there was a filtering system process in existence. No terrestrial mammals of more

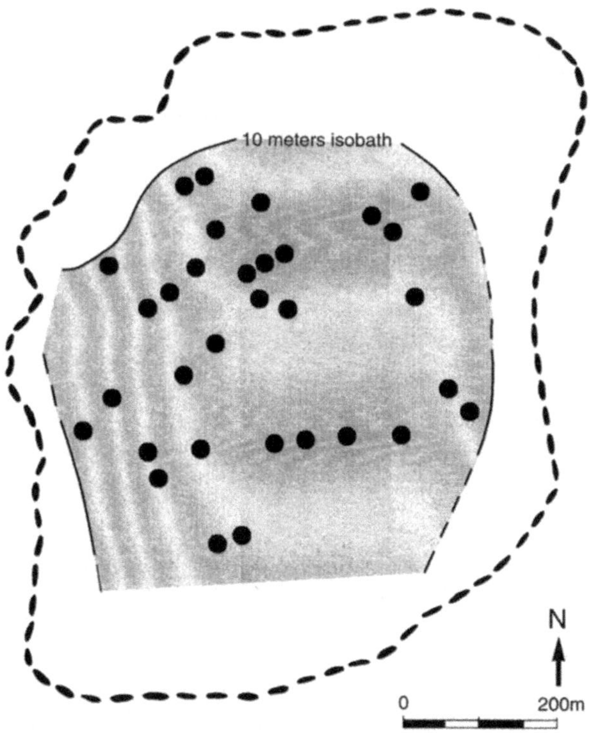

FIGURE 11. Distribution of discoveries of horse skeletons within "Grube Messel" and reconstruction of the 10 meters isobath (based also on orientation of bench marks). Drawing Senckenberg-Museum, C. Hemm (after Franzen *et al.*, 1982 and E. P. J. Heizmann, unpublished).

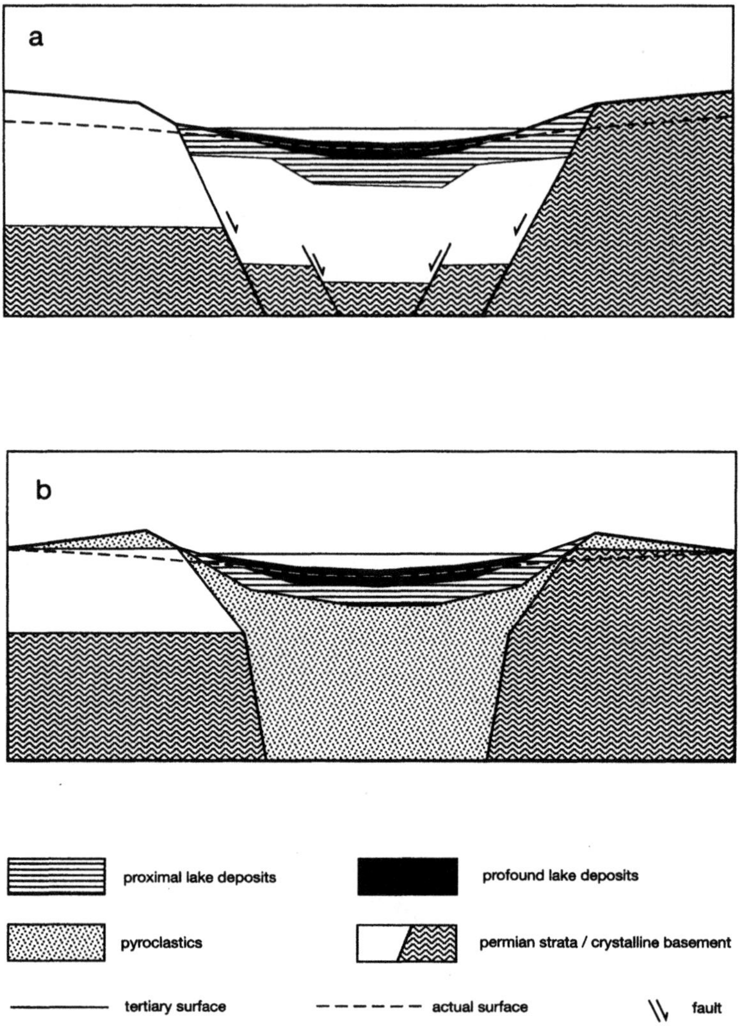

FIGURE 12. Different models of the origin of Eocene Lake Messel. a) as a small tectonic breakdown, b) as a result of a maar explosion. Drawing Senckenberg-Museum, C. Hemm (after Pirrung, 1998).

than 1 meter in length are found at Messel such that larger carcasses must have been filtered out before reaching the lake. Perhaps the rivers were only small creeks. Most vertebrates were drowned by occasional floods as is demonstrated by their relaxed position typical for such a kind of death. But why do flying animals such as birds and bats occur in such great abundance in the Messel shales (see Storch, this volume)? There must have been a specialized trap for these flying animals. There is evidence to suggest that the air directly above the water surface was occasionally polluted by carbon dioxide ($CO2$; Franzen, 1978; Richter and Storch, 1980). As has been documented by many accidents

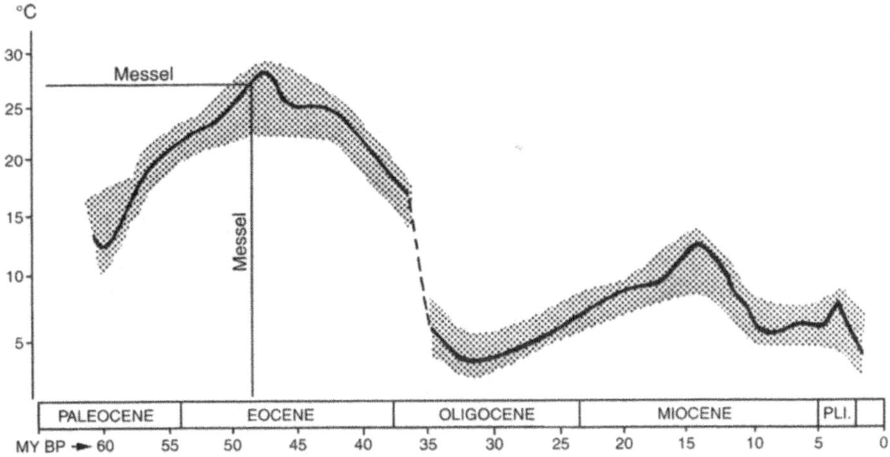

FIGURE 13. Paleotemperatures of the Tertiary North Sea basin (after Buchardt, 1978). Drawing Senckenberg-Museum, C. Hemm (after Buchardt, 1978).

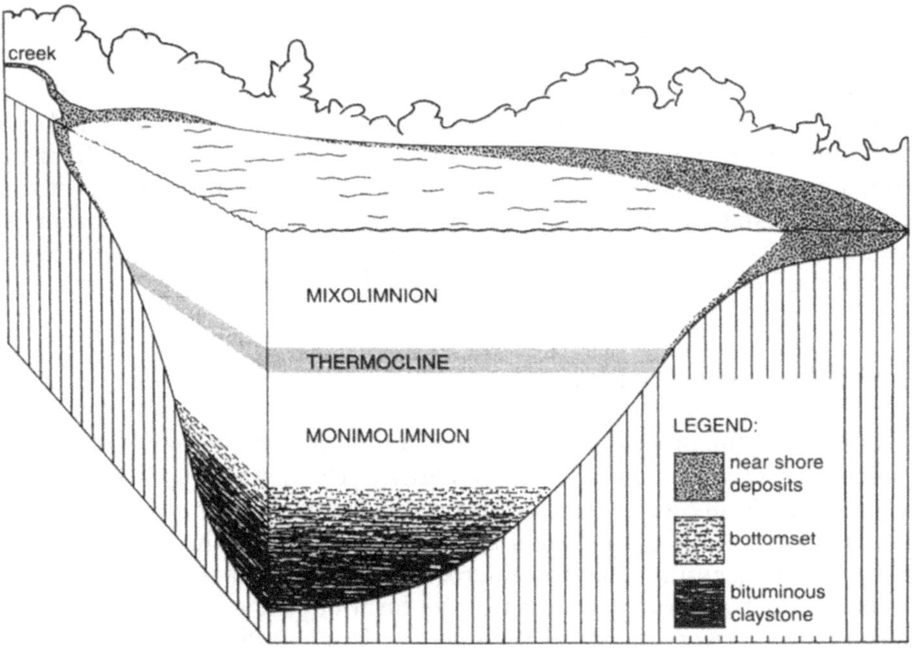

FIGURE 14. Model of the thermally stratified meromictic Eocene Lake Messel. Drawing Senckenberg-Museum, C. Hemm (after Behnke *et al.*, 1986 and Goth, 1990).

today, CO2 cannot be detected by the animals in advance and it leads to loss of consciousness. Hence, flying animals that enter into such a situation crash and drown. The carbon dioxide may have originated from the lake itself, but there is no evidence that supports degassing in the oil shales so volcanic exhalations are a more likely explanation (Franzen *et al.*, 1982; Franzen and Köster, 1994).

Another case of special preservation concerns that of the Messel primates. Prosimians, occurring frequently in other fossil lagerstaetten of that time, are extremely rare at Messel. This is even more strange because a typical biotope for primates, rainforest under subtropical to tropical climatic conditions, was evidently abundant around Eocene Lake Messel. Only eight specimens of primates have been discovered up to now at Messel, and all are fragmentary. This is a special problem because the currents in Eocene Lake Messel were evidently so weak that only complete carcasses, because of their gas content, could float into the lake. So in the case of the fragmentary primate skeletons there must have been a special medium of transport. Primates, living high in the trees, were apparently secure from the normal causes of death at Messel such as floods or volcanic exhalations of carbon dioxide. Instead evidence indicates that the primate specimens preserved in the oilshales at Messel are the result of accidents with crocodiles or other carnivorous animals and that their predators were responsible for the fragmentation of their skeletons and their transport as well (Franzen, 1987, 1988, 1993, 1997).

There are four pieces of evidence that corroborate this hypothesis. First, two Messel primate specimens clearly display bitemarks (Franzen, 1988; Franzen and Frey, 1993). Second, one of these specimens contains the broken tip of a crocodile tooth still embedded in one of the bitemarks (Franzen and Frey, 1993; Figure 15). Third, there is a coprolite of a semiamphibious mammal, presumably *Buxolestes*, containing the fragmentary mandible of a *Europolemur koenigswaldi* specimen (Franzen, 1997; Figure 16). Finally, all primate fragments at Messel are still articulated, which means that there was still soft body tissue, like tendons, ligaments, and skin attached to the bones when they were buried. This makes a simple decomposition argument less plausible as it does not explain the fragmentation of the skeletons nor their transport.

4.3 Soft Body Preservation

What were the processes that resulted in the preservation of body outlines of some Messel vertebrates? Michael Wuttke (1983) working at that time on fossil anurans from Messel for his doctoral dissertation, attempted to document the fine structure of the epidermis in the body outlines. This led to a surprising discovery. Scanning electron microscope studies revealed no histological details, but instead, minute bodies in the form of rods or grains consisting of siderite (FeCO3) were observed (Figure 17). This could only be explained as originating from a dense covering of bacteria that must have formed on the underside of the body when putrefaction had begun soon after the carcass had

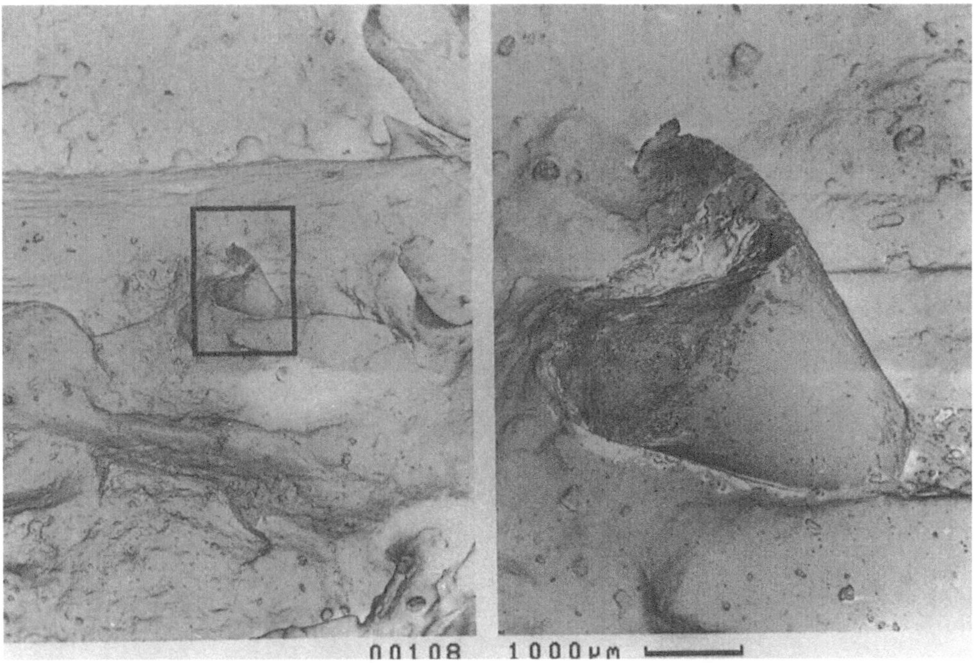

FIGURE 15. Broken tip of a crocodile tooth preserved in a bitemark of *Europolemur koenigswaldi* (Karlsruhe specimen) at two magnifications. Scanning electron micrograph, Senckenberg-Museum, D. Fiege and J. L. Franzen.

FIGURE 16. Coprolite of *Buxolestes* (?) containing a fragment of a mandible of *Europolemur koenigswaldi*. Photograph, Senckenberg-Museum, S. Tränkner.

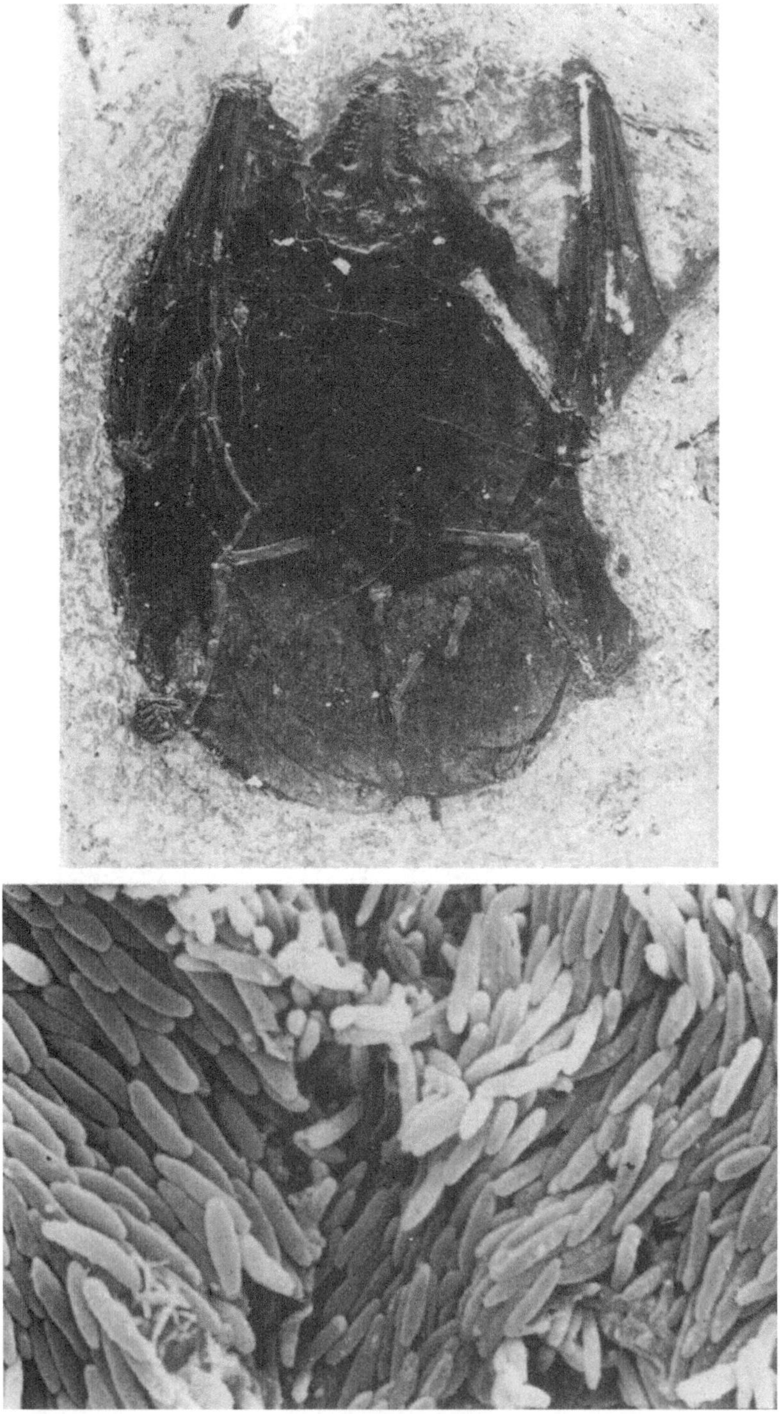

FIGURE 17. Softbody imaging by bacteria ("bacteriography") in a bat (*Palaeochiropteryx tupaiodon*). a) Outline of the soft body including the patagium (flying membrane). b) Bacteria preserving the body outline. Scanning electron micrograph, Senckenberg-Museum, G. Richter.

been deposited on the bottom of Lake Messel. Apparently the bacteria were petrified in the sediments through the production of carbon dioxide (CO_2) and precipitation of iron (Fe), the latter present in the lake water as a result of weathering of igneous rocks and Permian red-beds in the surrounding region. Only later was this thin veneer of lithified bacteria changed into a black silhouette by infiltration and cementation by kerogene from the surrounding sediment. In this way, the soft-body contours of these Eocene vertebrates have been preserved, not directly, but as a picture resulting from a process that could be termed bacteriography.

ACKNOWLEDGMENTS. The help of several colleagues is acknowledged. Drawings have been made by Christine Hemm and Dirk Hochberger, photos were taken except from the author by Christine Schumacher and Sven Tränkner while Dieter Fiege helped with SEM scanning. Two scanning electron micrographs (Figures 8 and 17b) are due to Gotthard Richter.

References

Behnke, C., Eikamp, H., and Zollweg, M., 1986, Die Grube Messel. Paläontologische Schatzkammer und unersetzliches Archiv für die Geschichte des Lebens, Geologie, Bergbaugeschichte, Fossilien, pp. 1–168, Goldschneck-Verlag, Korb.

Berg, D., 1964, Krokodile als Klimazeugen, *Geol. Rundschau* **544:**328–333.

Buchardt, B., 1978, Oxygen isotope paleotemperatures from the Tertiary period in the North Sea area, *Nature* **275:**121–123.

Elder, R., 1985, *Principles of Aquatic Taphonomy with Examples from the Fossil Record*, Unpublished doctoral dissertation, University of Michigan, Ann Arbor.

Franzen, J. L., 1978, Senckenberg-Grabungen in der Grube Messel bei Darmstadt. 1. Probleme, Methoden, Ergebnisse 1976–1977, *Courier Forschungsinstitut Senckenberg* **27:**1–135.

Franzen, J. L., 1985, Exceptional preservation of Eocene vertebrates in the lake deposit of Grube Messel (West Germany), *Phil. Trans. R. Soc. London, B* **311:**181–186.

Franzen, J. L., 1987, Ein neuer Primate aus dem Mitteleozän der Grube Messel (Deutschland, S-Hessen), *Courier Forschungsinstitut Senckenberg* **91:**151–187.

Franzen, J. L., 1988, Ein weiterer Primatenfund aus der Grube Messel bei Darmstadt, in: *Der eozäne Messelsee—Eocene Lake Messel* (J. L. Franzen, and W. Michaelis, eds.), Courier Forschungsinstitut Senckenberg **107:**275–298.

Franzen, J. L., 1997, Ein Koprolith als Leckerbissen. Der siebte Primatenfund aus Messel, *Natur u. Museum* **127(2):**46–53.

Franzen, J. L., and Frey, E., 1993, *Europolemur* completed, *Kaupia* **3:**113–130.

Franzen, J. L., and Köster, A., 1994, Die eozänen Tiere von Messel—ertrunken, erstickt oder vergiftet? *Natur u. Museum* **124(3):**91–97.

Franzen, J. L., and Michaelis, W. (eds.), 1988, *Der eozäne Messelse—Eocene Lake Messel*, Courier Forschungsinstitut Senckenberg **107:**1–452.

Franzen, J. L., Weber, J., and Wuttke, M., 1982, Senckenberg-Grabungen in der Grube Messel bei Darmstadt. 3. Ergebnisse 1979–1981, *Courier Forschungsinstitut Senckenberg* **54:**1–118.

Goth, K., 1990, Der Messeler Ölschiefer—ein Algenlaminit, *Courier Forschungsinstitut Senckenberg* **131:**1–143.

Harms, F.-J., 1998, Zwei Forschungsbohrungen bei Messel niedergebracht, *Natur u. Museum* **128(4):**126–127.

Haupt, O., 1922, Die eocänen Süßwasserablagerungen (Messeler Braunkohlenformation) in der Umgegend von Darmstadt und ihr paläontologischer Inhalt, *Z. deutsch geol. Ges* **73:**175–178.

Hummel, K., 1925, Vulkanisch bedingte Braunkohlenbildung, *Braunkohle* **23**:293–298.
Irion, G., 1977, Der eozäne See von Messel, *Natur u. Museum* **107(7)**:213–218.
Lippolt, H. J., Baranyi, I., and Todt, W., 1975, Die Kalium-Argon-Alter der postpermischen Vulkanite des nord-östlichen Oberrheingrabens, *Aufschluß, Sonderband* **27**:205–212.
Matthess, G., 1966, Zur Geologie des Ölschiefervorkommens von Messel bei Darmstadt, *Abhandlungen des Hessischen Landesamtes für Bodenforschung* **51**:1–87.
Mertz, D. F., Swisher, C. C., Franzen, J. L., Neuffer, F. O., and Lutz, H., 2000, Numerical dating of the Eckfeld maar fossil site, Eifel, Germany: a calibration mark for the Eocene time scale, *Naturwissen schaften* **87**:270–274.
Pirrung, B. M., 1998, Zur Entstehung isolierter alttertiärer Seesedimente in zentraleuropäischen Vulkanfeldern, *Mainzer Naturwissenschaftliches Archiv.*, Beiheft **20**:1–117.
Richter, G., 1992, Fossilized gut contents: analysis and interpretation, in: *Messel. An insight into the history of life and of the Earth* (S. Schaal, and W. Ziegler, eds.), Clarendon Press, Oxford, pp. 285–289.
Richter, G., and Storch, G., 1980, Beiträge zur Ernährungsbiologie eozäner Fledermäuse aus der "Grube Messel," *Natur u. Museum* **110(2)**:353–367.
Rietschel, S., 1994, Messel—ein Maar-See?, *Mainzer Naturwiss. Archiv., Bh.* **16**:213–218.
Schaarschmidt, F., 1992, The vegetation: fossil plants as witnesses of a warm climate, in: *Messel. An insight into the history of life and of the Earth* (S. Schaal, and W. Ziegler, eds.), Clarendon Press, Oxford, pp. 27–52.
Seilacher, A., Reif, W.-E., and Westphal, F., 1985, Sedimentological, ecological and temporal patterns of fossil Lagerstätten, *Phil. Trans. R. Soc. London, B* **311**:5–23.
Walsh, S. L., 1996, Theoretical Biochronology, the Bridgerian/Uintan Boundary, and the "Shoshonian Subage" of the Uintan, in: *The Terrestrial Eocene-Oligocene Transition in North America* (D. R. Prothero, and R. J. Emry, eds.), Cambridge University Press, Cambridge, pp. 52–74.
Weber, J., and Hofmann, U., 1982, Kernbohrungen in der eozänen Fossillagerstätte Grube Messel bei Darmstadt, *Geologische Abhandlungen Hessen* **83**:1–58.
Weber, J., and Zimmerle, W., 1985, Pyroclastic detritus in the lacustrine sediments of the Messel Formation, *Senckenbergiana lethaea* **66**:171–176.
Wuttke, M., 1983, "Weichteil-Erhaltung" durch lithifizierte Mikroorganismen bei mittel-eozänen Vertebratendus den Ölschiefern der 'Grube Messel' bei Darmstadt, *Senckenbergiana lethaea* **64**:509–527.

Chapter 8

Paleobiological Implications of the Messel Mammalian Assemblage

GERHARD STORCH

1. Introduction

Grube Messel near Frankfurt, Germany, is reknowned for the diversity and completeness of its early Middle Eocene mammalian fosssils. The extraordinary preservation allows for an opportunity for unprecedented accuracy in reconstruction of anatomy and behaviour. Preserved elements include completely articulated skeletons, soft body outlines, gut contents, unborn fetuses, and such delicate structures as external ears and flying membranes of bats. Thus, adaptations to particular strategies for locomotion, feeding, defence, orientation, and reproduction can be assessed by a suite of anatomical and inferred behavioral evidence that is unavailable under normal conditions of fossil preservation. Up to now, Messel already has yielded specimens from around 50 mammalian species spanning the systematic units marsupials, pseudorhyncocyonids, pantolestids, apatemyids, paroxyclaenids, insectivores, microchiropterans, primates, pangolins, anteaters, rodents, carnivores, creodontans, perissodactyls, and artiodactyls. The range continues to grow as new

GERHARD STORCH • Forschungsinstitut Senckenberg, D-60325 Frankfurt am Main 1, Germany.

Eocene Biodiversity: Unusual Occurrences and Rarely Sampled Habitats, edited by Gregg F. Gunnell, Kluwer Academic/Plenum Publishers, New York, 2001.

excavations are carried out and new materials are analyzed (for synopsis see Schaal and Ziegler, 1992).

At the beginning of the Eocene, Europe experienced dramatic faunal innovations by the invasion of modern groups such as bats, perissodactyls, and artiodactyls, and the archaic fauna from the preceding Paleocene vanished in the course of the Eocene. A variety of life strategies for both the newcomers and oldtimers among Messel mammals is indicated and many strategies appear very specialized, lacking any known living analogues. This suggests that interpretations based on specimens that are less complete as the result of more normally occurring modes of preservation, greatly underestimate the variety of adaptive strategies employed in ancient and allegedly primitive and generalized mammals. This chapter is centered on the group of the author's major interest, the micromammals.

2. Small Omnivorous-Insectivorous Forest Floor Dwellers

This omnivorous-insectivorous trophic niche was occupied by three small-sized species of the hedgehog-related, erinaceomorph family Amphilemuridae (Storch, 1993, 1996). *Pholidocercus hassiacus*, the Hessian scaly-tail, was a robust well-defended animal with an opportunistic feeding strategy (Figures 1 and 2). The tail is surrounded by a tube of dermal ossifications which overlap like tiles and form a protective cover. Bony dermal scutes are known among living mammals from armadillos but are unknown from insectivores. The outline of the dorsal pelage shows long and rather stiff bristles. The frontal and nasal bones bear a sharply defined field of deep vessel indentations which suggest an original cover by a horny plate or leathery callus. This structure was either part of the defensive strategy against predators or a protective device for digging with the nose and forehead for food. Large and deeply cleft terminal phalanges of hands and feet indicate an able scratch-digger. Preserved gut contents include various kinds of insect cuticles and tissues of pulpy fruits and leaves.

Macrocranion tupaiodon represents a different adaptive type than *Pholidocercus*: It is a slender animal with cursorial adaptations for rapid and maneuverable flight from predators (Figures 3 and 4). The hind limbs, particularly tibia and metatarsals, are conspicuously elongated and bear strong muscle attachments. The first and fifth digits of hands and feet are much shortened and all terminal phalanges are short and blunt. The orbits are very small. A mobile proboscis, inferred from attachments of facial muscles on the rostrum, large external ears, and very long tactile hairs on the muzzle suggest acute olfactory, acoustic, and tactile senses to assess its orientation. The pelage is short and wooly. Gut contents reveal fish as a favorite dish. The lack of any obvious swimming adaptations suggests that the animals foraged along the edge of the Messel lake and may have simply scavenged fish carcasses. Besides, the guts contain fragments of seed shells, lumps of tissue from fruits, insect cuticles, and scarce remains of leaves, stalks, and fungus.

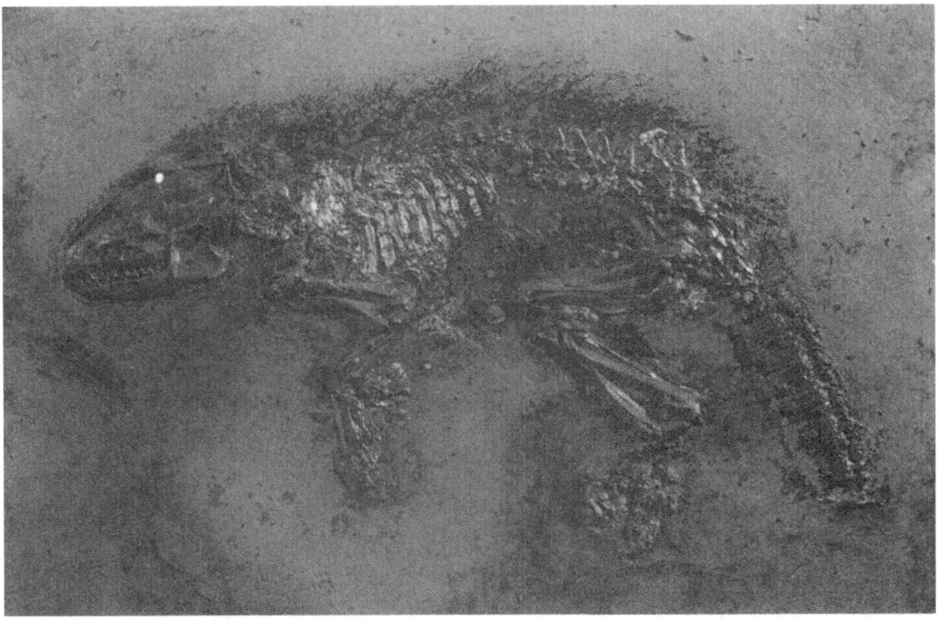

FIGURE 1. Hedgehog-related *Pholidocercus hassiacus*, proximal part of tail broken off (Photograph courtesy of Forschungsinstitut Senckenberg).

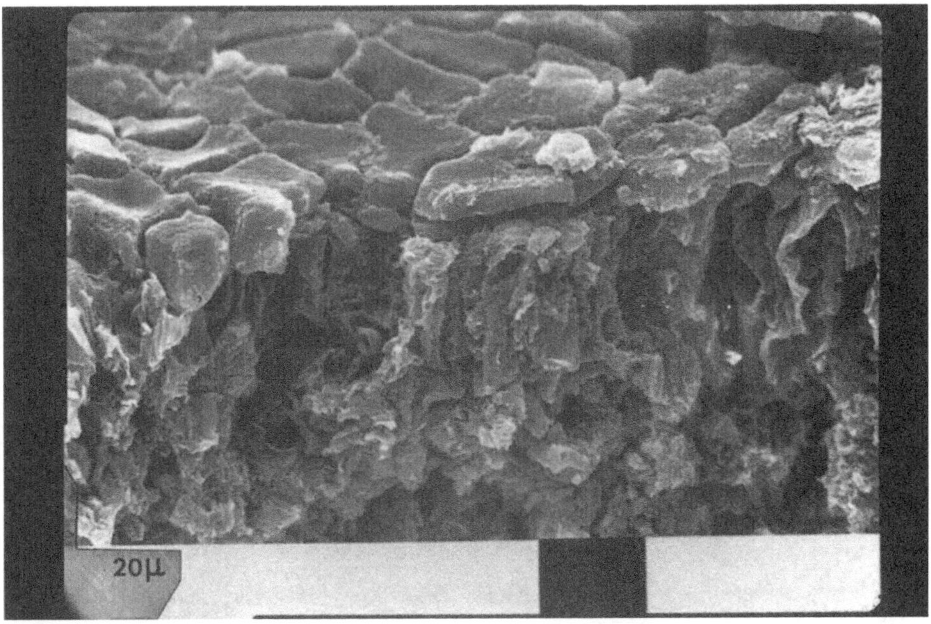

FIGURE 2. *Pholidocercus hassiacus*, leaf tissues from gut contents (Photograph courtesy of Forschungsinstitut Senckenberg).

FIGURE 3. Hedgehog-related *Macrocranion tupaiodon* (Photograph courtesy of Forschungsinstitut Senckenberg).

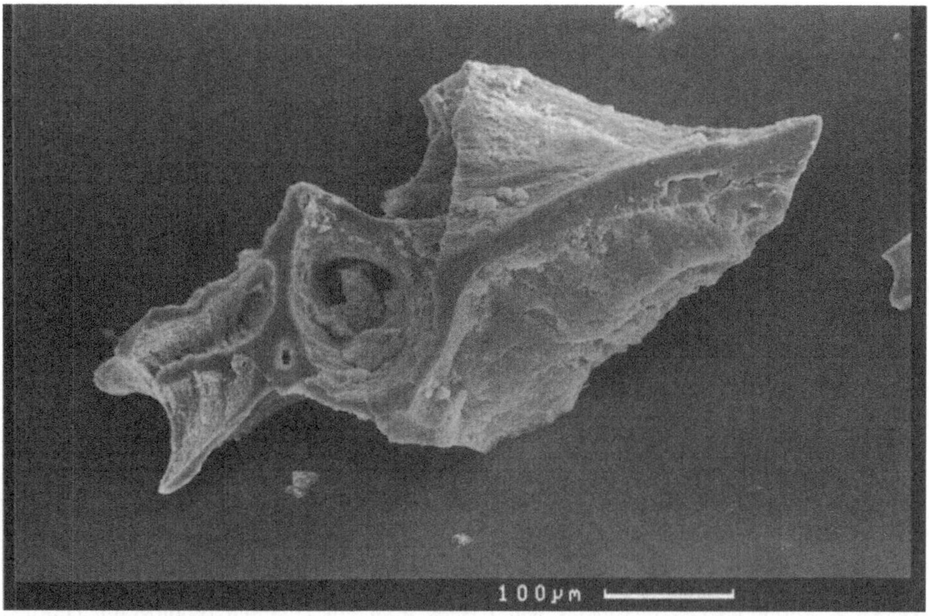

FIGURE 4. *Macrocranion tupaiodon*, fragment of fish vertebra from gut contents (Photograph courtesy of Forschungsinstitut Senckenberg).

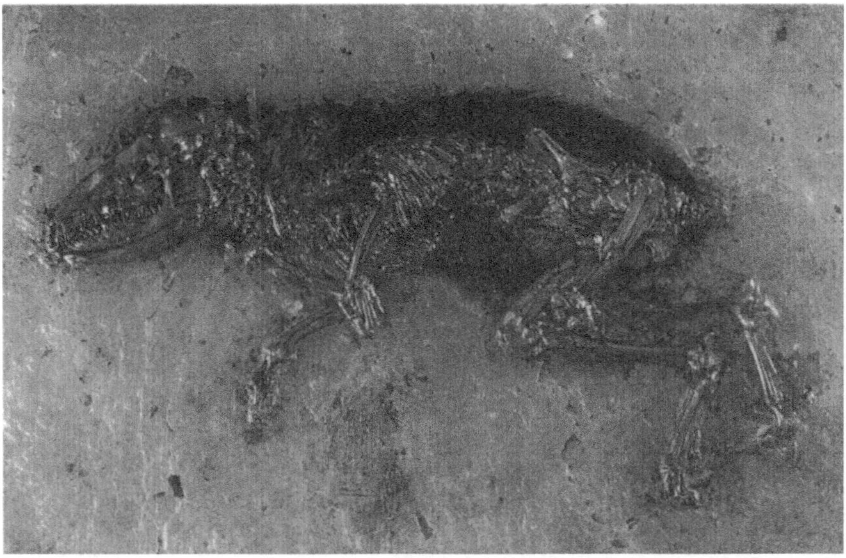

FIGURE 5. Hedgehog-related *Macrocranion tenerum*, the supposedly long tail is broken off (Photograph courtesy of Forschungsinstitut Senckenberg).

Macrocranion tenerum is a gracile animal with striking specializations which combine the survival strategies of the aforementioned species—rapid flight and an efficient protective device (Figures 5–7). A saltatorial and occasionally even ricochetal mode of locomotion is inferred from the extremely elongated hindlimbs, particularly the 2nd through 4th metatarsals, the very short and blunt terminal phalanges, the strong third trochanter on the femur, the extensively fused tibia and fibula, and the long tuber calcanei. The outline of the dorsal integument, on the other hand, reveals that the body is covered with spines (or very strong bristles) similar to extant hedgehogs. Such a combination of strategies is unknown in any other insectivore and is extraordinary for any mammal. Rich gut contents are known only from a single specimen. They consist of thin insect cuticles which all have the same scaly surface design and most probably come from small social insects such as ants. The animals obviously gathered food from the forest floor rather than chased their prey, and a rapid bounding gait likely served as a means of escaping from predators.

Thus, three closely related, small omnivorous-insectivorous forest floor-dwellers could exploit the Messel ecosystem. In addition, they had to share their niches to some extent with the myrmecophagous pangolins and the strange bipedal, carnivorous-omnivorous pseudorhyncocyonids.

3. Aerial Insectivory

Bats are by far the most frequent Messel mammals and currently several hundreds of virtually complete specimens are known. They are assigned to seven

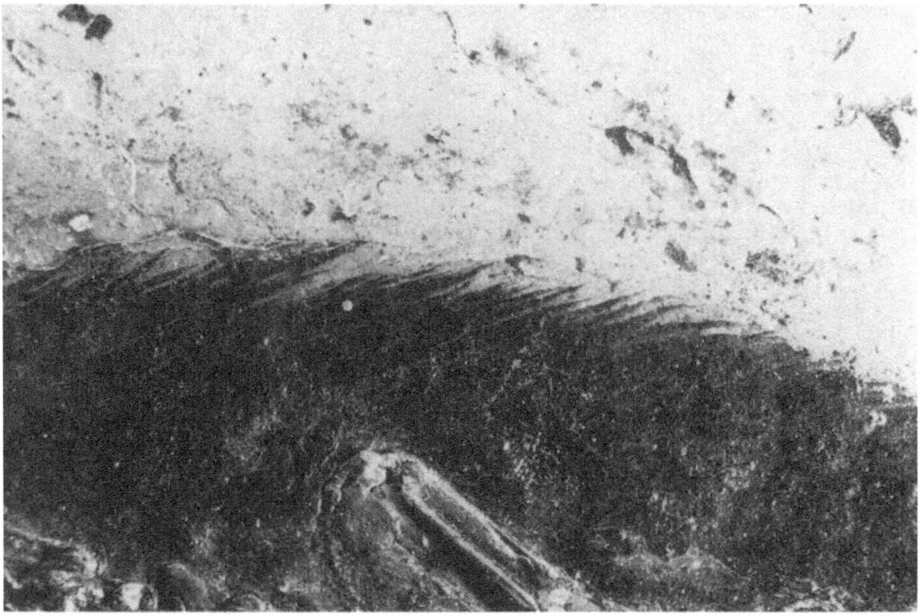

FIGURE 6. *Macrocranion tenerum*, close-up showing spines and strong bristles on back (Photograph courtesy of Forschungsinstitut Senckenberg).

FIGURE 7. *Macrocranion tenerum*, cuticle fragment, presumably of ants, from gut contents (Photograph courtesy of Forschungsinstitut Senckenberg).

species and three genera of the extinct microchiropteran families Archaeonycteridid (two species of *Archaeonycteris*), Palaeochiropterygidae (two species of *Palaeochiropteryx*), and Hassianycteridid (three species of *Hassianycteris*). Adaptive radiations can be assessed independently by analyses of the flight apparatus, gut contents, and inner ear structures (Habersetzer and Storch, 1987, 1990, 1992, 1994).

Soft body outlines of the patagia and articulated skeletons permit the reconstruction of the flight apparatus and of functional parameters that are important in describing the flight characteristics (Figures 8 and 9). Actually, it is rather a process of graphic unfolding than of reconstructing. Particular flight styles of bats are indicated in the relationship of the area of the flight membrane to the body weight. Extant species which have relatively large wing areas and hence low wing-loading, hunt close to the ground and in the foliage. The shape of the wing is another important parameter. Narrow wings are typical for many extant species that fly fast and high. Among Messel bats, *Archaeonycteris* species are characterized by a remarkably long body, broad wings, and moderate wing-loading. Such a combination compares with that in many living vespertilionids, e.g. the mouse-eared bat, *Myotis myotis*, and thus a rather unspecialized flight in medium height in the open spaces between the trees can be concluded. *Palaeochiropteryx* species, on the other hand, have a delicate body, very broad wings, and highly reduced wing-loading. They are extreme specialists, similar to highly specialized horseshoe bats (Rhinolophidae) and Old World leaf-nosed bats (Hipposideridae). A slow and maneuverable flight close to the ground and close to or among the foliage and even hovering capabilities can be assumed. The *Hassianycteris* species have narrow wings, high body weight, and high wing-loading. They compare with extreme specialists among living fast and high flying bats such as mastiff bats (Molossidae) and sheath-tailed bats (Emballonuridae), and can be considered as well adapted to flight in the open air above the forest canopy. It is noteworthy that the variability of functional flight parameters of the seven microbat species from Messel match the span in local tropical bat faunas which are much more rich in species.

Analyses of gut contents prove insectivory for all Messel bats. Chitin is preserved superbly and without surface corrosion. Even the faintest structures of moths' scales and the surface sculptures of insect cuticles are discernible (Figure 10). However, cutting, grinding, and decorticating actions during mastication result in a highly fragmentary state of the chitin remains and there are problems of identifications of the insect taxa involved. In all probability, the *Archaeonycteris* species were specialized in feeding on beetles. Thick cuticle fragments, often with iridescent structural colors, predominate in their guts, while lepidopteran scales are scarce. An enormous amount of *Palaeochiropteryx* guts were examined and they consist almost exclusively of the remains of moths such as scales and cuticles. Particular arrangements of chitin hairs and pores on the cuticles suggest specific moth families. *Hassianycteris* species obviously preferred a mixed diet of rather large prey items, above all beetles and lepidopterans. Their food remains include very thick and heavily

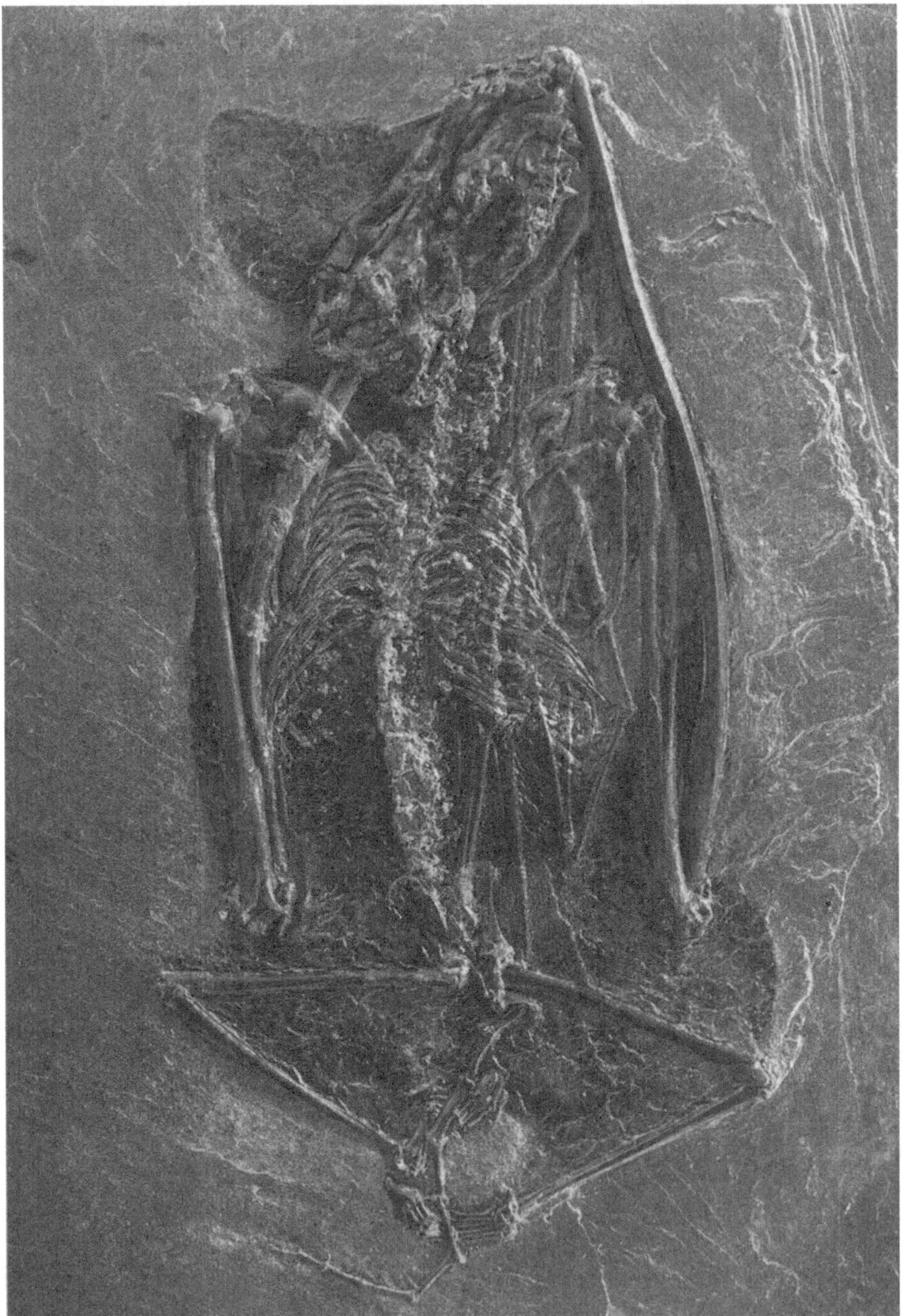

FIGURE 8. *Undescribed species*, showing external ear and uropatagium (Photograph courtesy of Forschungsinstitut Senckenberg).

FIGURE 9. Reconstruction of flight apparatus and hunting spaces of Messel bats (Photograph courtesy of Forschungsinstitut Senckenberg).

sculptured cuticle fragments. Admixtures of gut contents which may occur in all species, consist of lumps of pollen, which were probably ingested while sticking to prey animals, and bat hairs which were likely swallowed while grooming the fur.

A relation between inner ear (cochlea) size and echolocation performance has been shown for various extant species. The evaluation of the acoustic specialization of Paleogene bats can be achieved by comparisons with extant bats of known echolocation and foraging strategies (Figure 11). A plot of the diameter of the cochlea against the basicranial width of extant bats results in a fan-like distribution, ranging from non-echolocating megabats along the lower margin to highly sophisticated echolocaters, the horsehoe bats, on the upper margin; vespertilionids take a central position. The Messel bats occupy in relative cochlea size the corridor between megachiropterans and vespertilionids: In *Hassianycteris* and *Palaeochiropteryx*, there is overlap with the lower range of vespertilionids, and in *Archaeonycteris*, there is overlap with the upper range of megachiropterans. Combined with structural features of the inner ear, this would suggest rather short multiharmonic pulses for *Archaeonycteris*, echolocation sounds within a frequency band of 30–90 kHz for *Palaeochiropteryx*, and intense sound frequencies below 30 kHz for *Hassianycteris*. *Hassianycteris* meets the relatively simple requirements for the acoustic location of insects in the open air where no background echoes produce a disturbing

FIGURE 10. Moth scale from gut contents of *Palaeochiropteryx tupaiodon* (Photograph courtesy of Forschungsinstitut Senckenberg).

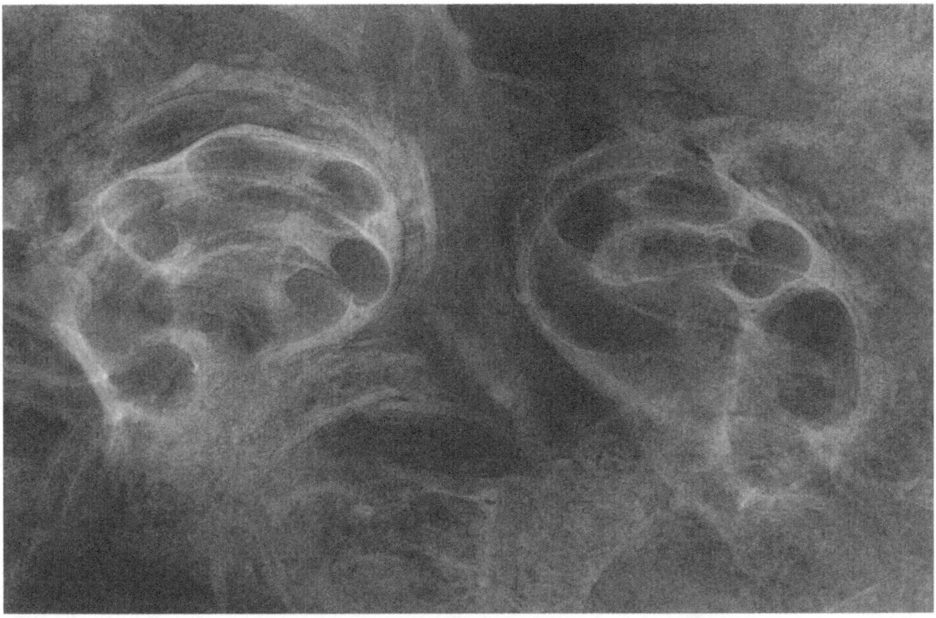

FIGURE 11. *Hassianycteris messelensis*, x-ray photo of cochleae (Figures 11–15 from Koenigswald and Storch, 1998) (Photograph courtesy of Forschungsinstitut Senckenberg).

"noise". *Palaeochiropteryx* is extremely specialized for a hunting space in an area close to the ground and among vegetation. This area is rich in obstacles which reflect multiple disturbing echoes. Rhinolophids and hipposiderids have developed sophisticated acoustic strategies—high-frequency and constant frequency sounds up to 156 kHz—to discriminate small flying prey items from the intense background "noise". However, the cochlea of *Palaeochiropteryx* is acoustically unspecialized and thus would suggest a maneuverable insect hunt in the free space just above the water surface of former lake Messel. On the assumption that clouds of poisonous gas were occasionally emitted from the lake (see Franzen, this volume), this would help to explain the overwhelming abundance of its fossils.

The Messel bat species have managed to occupy the major hunting and flight niches of the Messel ecosystem and that way represent a biologically balanced fauna.

4. Arboreal Life

Climbing abilities of small mammals can be inferred from skeletal characters such as sharp, curved, and laterally compresssed terminal phalanges, and are thus accessible from many fossil sites where postcranial bones are preserved.

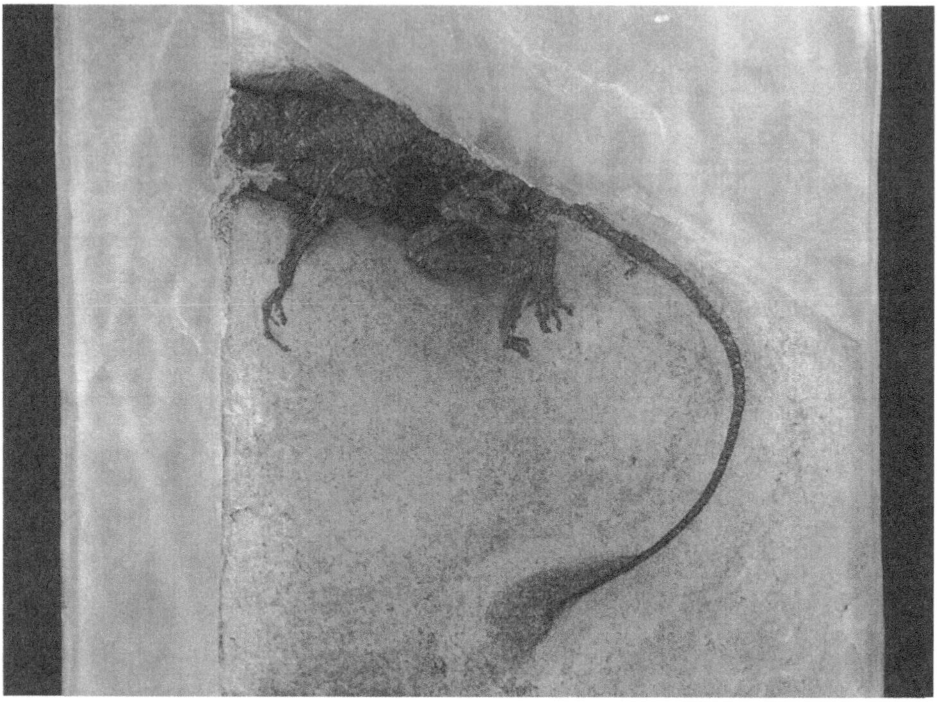

FIGURE 12. The apatemyid *Heterohyus nanus*, skull not preserved (Photograph courtesy of Forschungsinstitut Senckenberg).

Arborealism—living in trees and bushes—can be perfectly performed by such different adaptive types as ambling tree sloths or leaping primates. A mammal may rush through the canopy with powerful leaps, supported by a long bushy tail as rudder, or it may be a safe but leisurely climber, equipped with a prehensile tail. To assess different scansorial habits requires extraordinary preservation. Tails, for instance, when preserved articulated and showing the outlines of fur and skin, can help to reconstruct modes of climbing behaviour. Among Messel mammals, arboreal species are known within marsupials, apatemyids, paroxyclaenids, primates, rodents, and carnivores.

The apatemyid *Heterohyus nanus* (Figure 12) compares in its feeding strategies—woodboring insects as food—with the extant marsupial *Dactylopsila* and the lemuroid primate *Daubentonia*. It obviously occupied the niche of woodpeckers (Picidae) in the Messel ecosystem. Pairs of upper and lower incisors are considerably enlarged to tear open the bark of trees and branches, and the second and third fingers are enormously elongated to extract insect larvae from their bore holes (Koenigswald, 1990). The terminal phalanges have the typical morphology as in other small climbers for anchoring the foot in bark. The tail is strikingly long. A recently discovered specimen exhibits a tail that is naked or sparcely haired over most of its length and bears a terminal tuft of long hairs. The tail was obviously used as a balancing aid and not as a prehensile organ, and *Heterohyus nanus* in all probability was an agile climber.

Kopidodon macrognathus (Figure 13) belongs to the family Paroxyclaenidae within a group of archaic mammals, currently classified as Proteutheria or Pantolesta. Its dentition suggests a diet consisting of vegetable matters, preferably fruits. Skeletal features indicate an arboreal mammal. Hands and feet have extremely narrow but very high claws of uniform morphology and subequal size. The structure of joints suggests great flexibility of the distal segments of fore- and hindlimbs, in particular strong capabilities to rotate hands and feet. The phalanges have small lateral flanges to which strong flexor muscles of the individual fingers were attached. Moreover, the fingers and toes are usually embedded in a remarkable position, being strongly bent toward the palm and sole and thus corroborating the assumption of good grasping abilities (Koenigswald, 1992). The tail is conspicuously bushy like in many scansorial living mammals whose home is the forest canopy.

Ailuravus macrurus is the largest of the Messel rodent species (Figure 14), by far larger than the extant Red squirrel (*Sciurus vulgaris*). The limbs show similar climbing adaptations as *Kopidodon*, including flexible digits of hands and feet to allow a secure grip. From previous reconstructions a leaping, squirrel-like locomotion was concluded, mainly based on the possession of an allegedly long bushy tail. However, well-preserved specimens clearly show a shaggy dorsal pelage but an apparently hairless (or scarcely haired) tail. The proximal vertebrae of the tail are strong, and the distal part of the tail is rolled up in burial position, resembling the condition of the prehensile tail in the Messel marsupial "*Peradectes*". The stout, short, and isometric hindlimbs are also in conflict with the assumption of a leaping mode of locomotion. All gut contents of *Ailuravus* consist exclusively of remains of leaves and most remarkably, one tree species distinctly predominates in each of the guts, the dominating species being different in the individual gut contents, however (Richter, 1987). *Ailuravus macrurus* would compare in its arboreal habits much better with living prehensile-tailed porcupines (*Coendou*) than with tree squirrels.

5. Reproduction and Ontogeny

Strategies of parental/maternal care of extinct mammals can be inferred to some extent from the number of their fetuses. Fortunately, fetuses are preserved *in situ* in various Messel species. The r-strategy among living mammals is characterized by having one or two young which are raised under intense care and protection. The K-strategy, on the other hand, is characterized by more numerous offspring and a higher tribute to predators and other risks and it is most probably the more primitive strategy among eutherians. It is an interesting issue whether early members of extant mammalian "r-groups" already showed this reproductive strategy.

From both species of primitive horses from Messel fetuses are known—from the smaller *Propalaeotherium parvulum* even several pregnant mares—

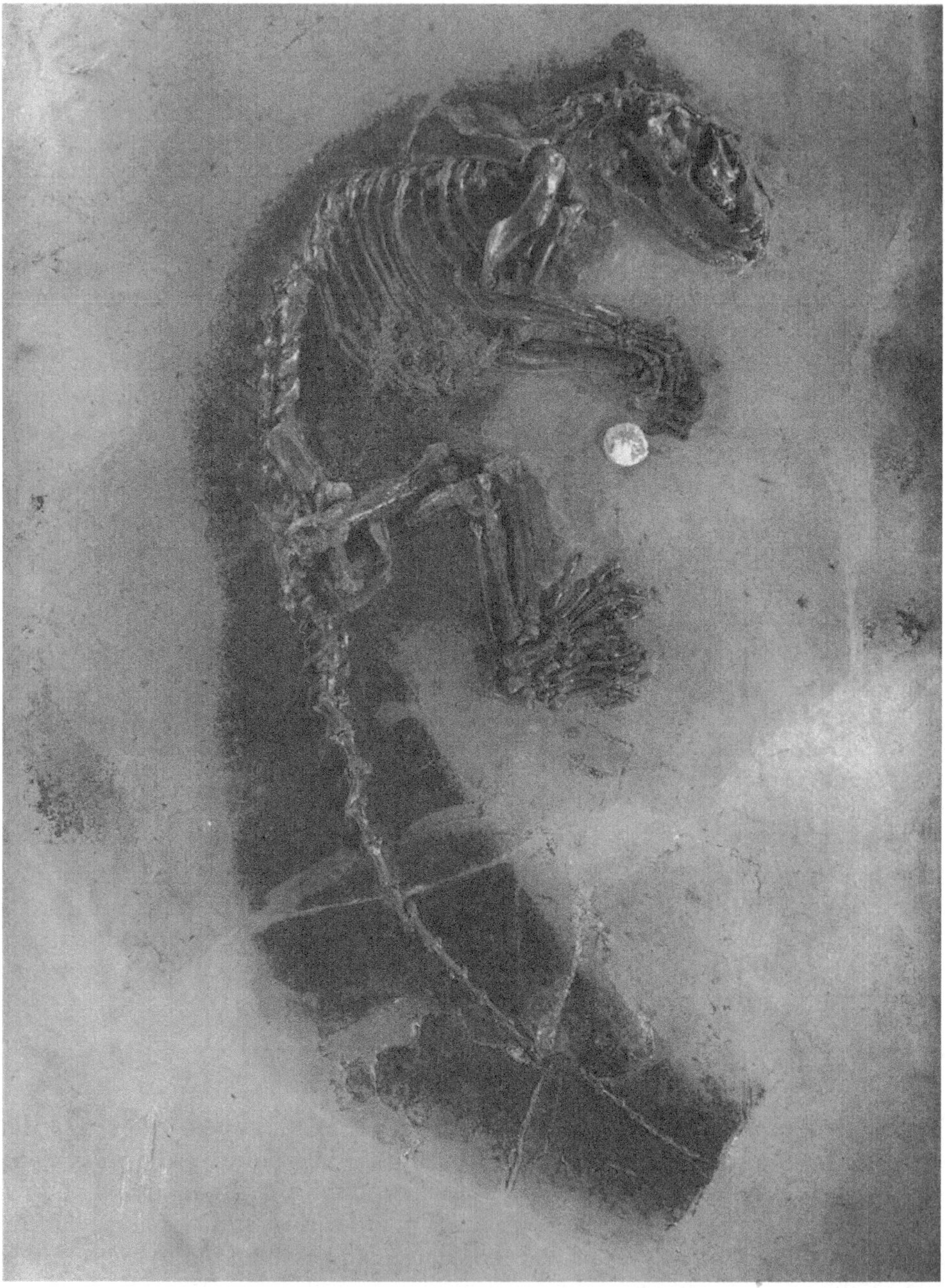

FIGURE 13. The paroxyclaenid *Kopidodon macrognathus* (Photograph courtesy of Forschungsinstitut Senckenberg).

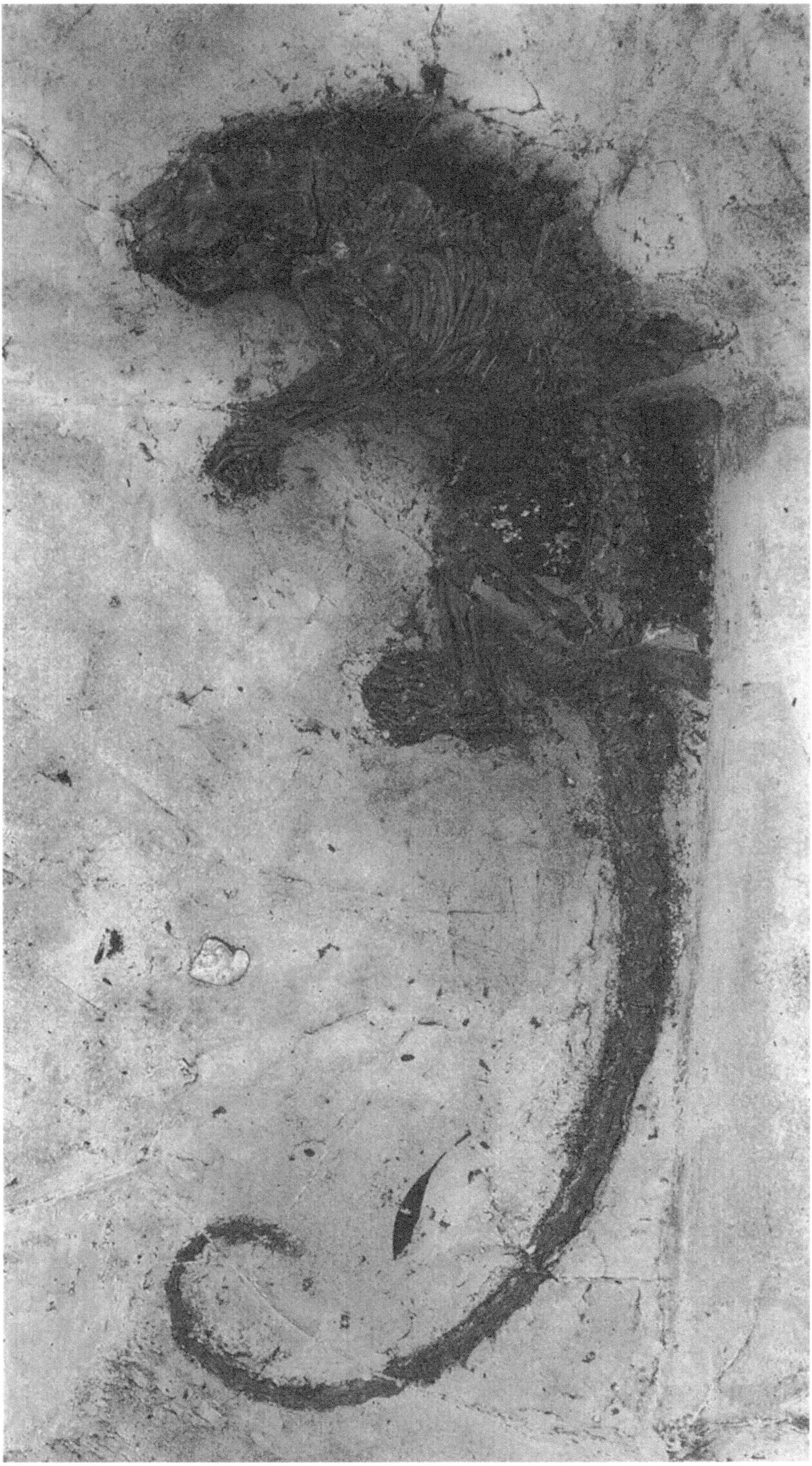

FIGURE 14. The largest Messel rodent, *Ailuravus macrurus* (Photograph courtesy of Forschungsinstitut Senckenberg).

and like living horses they all have only one fetus. Unlike extant horses, however, they did not roam open country where a single young certainly is a favorable condition. Thus far, only one pregnant female of the pangolin *Eomanis waldi* has been found (Figure 15). As is normal in living pangolins, the Messel specimen, too, had a single fetus only. Since a dorsal protective cover of horny scales has been documented in *Eomanis waldi* (Koenigswald *et al.*, 1981), we can assume similar maternal care to that of living pangolins. During early stages of postnatal life, the young would huddle up at sleep and rest under its mother's armour and ride on its mother's back to food sources such as ant and termite nests.

All known pregnant females of the bat *Palaeochiropteryx tupaiodon* have two fetuses, visible at the right and left side of the abdomen and indicating that both ovaries were functional (Figure 16). Extant bats generally produce only one young. Twins occur occasionally or regularly in some mega- and microchiropterans. Multiple births with higher litter size are inconsistent with the requirements of active flight and the more babies a female must carry and feed reduces chances of survival. The largest litters of four to five young are found in the North American Red bat (*Lasiurus borealis*) and many die each spring as a result of being overburdened by the weight of the young (Hill and Smith, 1984).

Early ontogenetic stages are known from several Messel mammals, and the bat *Palaeochiropteryx tupaiodon*, the most abundant species, is even represented by an ontogenetic series including females with fetuses *in situ* and juveniles of different individual age (Sigé *et al.*, 1998). The morphology and the eruption and replacement patterns of the deciduous dentition in this early bat are of particular paleobiological interest. All living bats, mega- as well as microchiropterans, are characterized by unusual and highly modified deciduous teeth of partly uncertain homologies. These styliform, sharp, and more or less hooked teeth allow the young bat to firmly and safely hook onto its mother's fur (in addition to the sharp claws of the feet and the thumb and index of the hand) during the early period of postnatal life. The derived clinging function of this peculiar morphology thus has no longer any relation with feeding but is of vital necessity for the young during maternal care in sustained reversed position at rest or during flight. The specialized condition in extant bats suggests that its evolution would have begun as soon as flight became an effective and generalized mode of locomotion, styliform deciduous teeth are, however, unknown from the fossil record. This does not mean such teeth were actually absent, but more probably that they can be preserved *in situ* only under exceptional conditions, as in the Messel oil shales. Otherwise, isolated hooklets are too small and fragile to be preserved or found with current collecting procedures.

The ontogenetic Messel assemblage of *Palaeochiropteryx tupaiodon* shows early stages in the evolutionary transition of the deciduous dentition of bats, from its primitive role in feeding to its highly advanced condition for clinging (Figures 17 and 18). The posterior deciduous premolars have well-differentiated cusps and ridges and thus a decidedly more molariform (dP4)

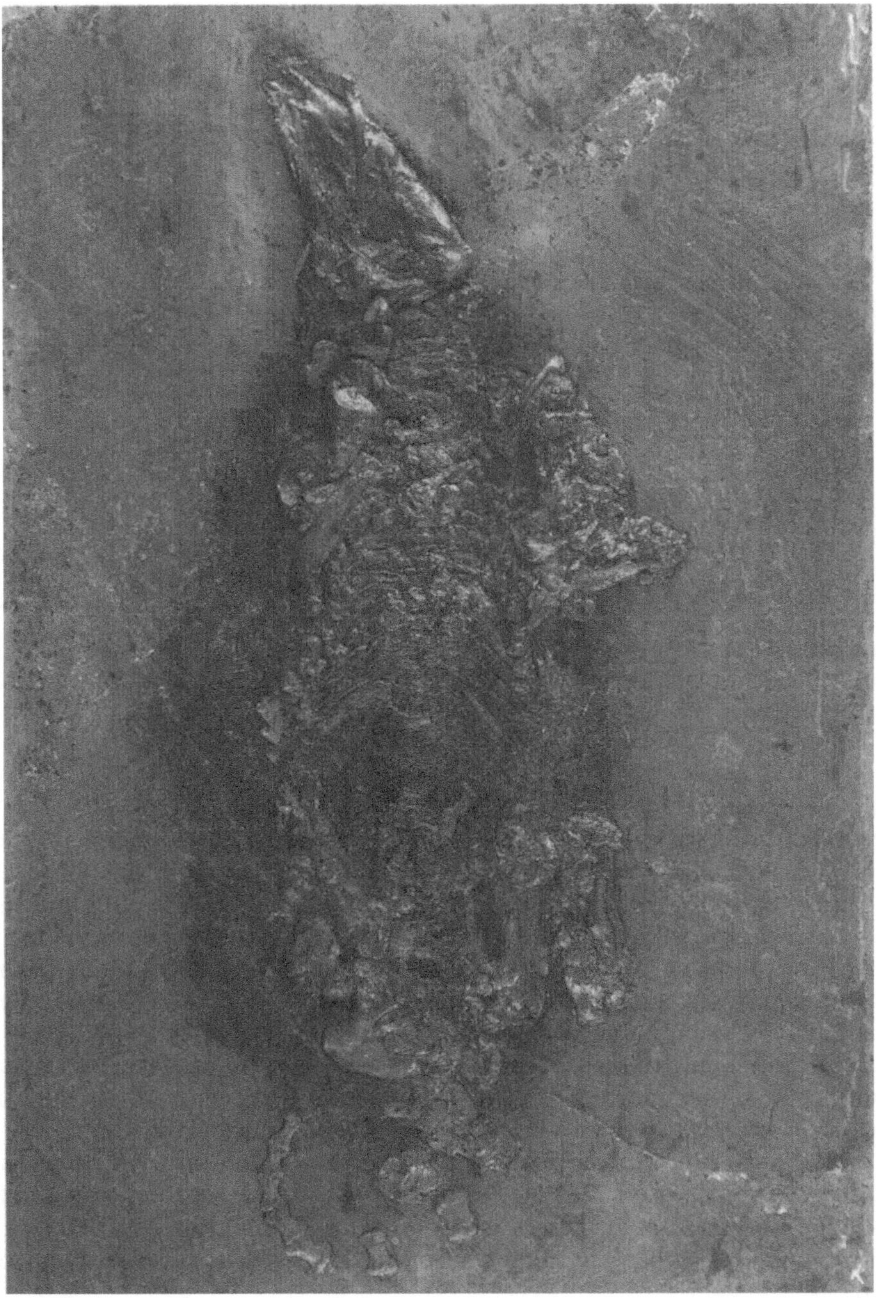

FIGURE 15. The pangolin *Eomanis waldi*, the ovate dark outline and the dorsoventral position in the rock are caused by an original covering of horny scales (Photograph courtesy of Forschungsinstitut Senckenberg).

FIGURE 16. Abdominal region of pregnant bat (*Palaeochiropteryx tupaiodon*), showing two fetuses with calcification of deciduous dentitions (Photograph courtesy of Forschungsinstitut Senckenberg).

and submolariform (dP3) morphology than their permanent successors. These teeth retained the primitive eutherian condition with all its masticatory functions of decorticating, puncturing, cutting, and grinding, and thus contrast significantly with the highly specialized nature of dP3–4 in living bats (in *Palaeochiropteryx*, there is no tooth replacement at the position of the submolariform anterior premolar, presumably the second permanent premolar position among all bats). Parallel to this primitive morphology, the slender and sharp anterior deciduous teeth clearly show early stages to the clinging condition of milk dentitions in living bats. The canines are fusiform and recurved apically, the upper incisors are styliform and slightly hooked, and the pectinate lower incisors have tiny hooked spicules. This specialized anterior dentition, already well established in Messel *Palaeochiropteryx*, was surely initiated before the early Middle Eocene, most probably as soon as active flight evolved. The evolutionary modification to clinging function of the entire milk dentition extending back to the fourth premolar, however, was achieved much later.

The styliform-hooked morphology of deciduous teeth is one essential component of the adaptive complex of clinging to the mother's fur—their precocious eruption during the perinatal period would be another. The early calcification of the complete deciduous dentition is a remarkable feature of living bats. Two of the intra-uterine fetal stages of *Palaeochiropteryx* show early calcification, too, and thus demonstrate that both adaptive components were already realized in this early Middle Eocene species.

ACKNOWLEDGMENTS. Figures 17 and 18 are reprinted from "The deciduous dentition and dental replacement in the Eocene bat *Palaeochiropteryx tupaiodon* from Messel: The primitive condition and beginning of specialization of milk teeth among Chiroptera" by B. Sigé, J. Habersetzer, and G. Storch from *Lethaia*, 1998, 31, pages 349–358.

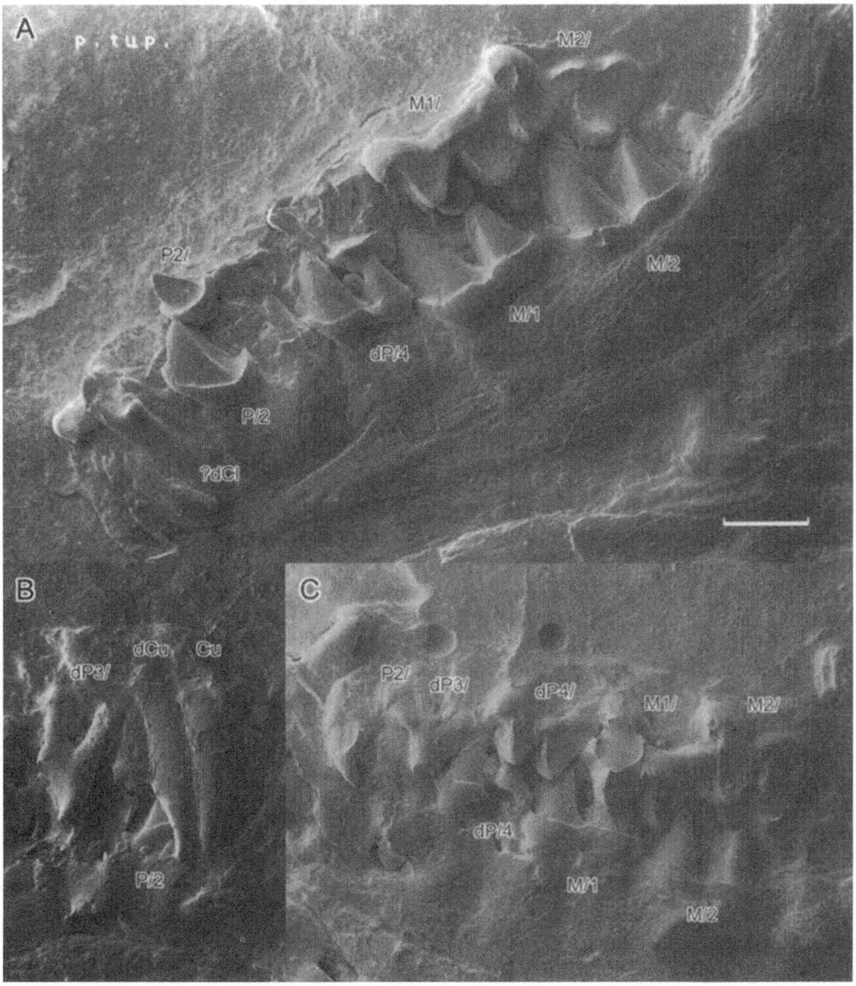

FIGURE 17. Dentition of *Palaeochiropteryx tupaiodon*, showing molariform and submolariform deciduous premolars (Photograph courtesy of Forschungsinstitut Senckenberg).

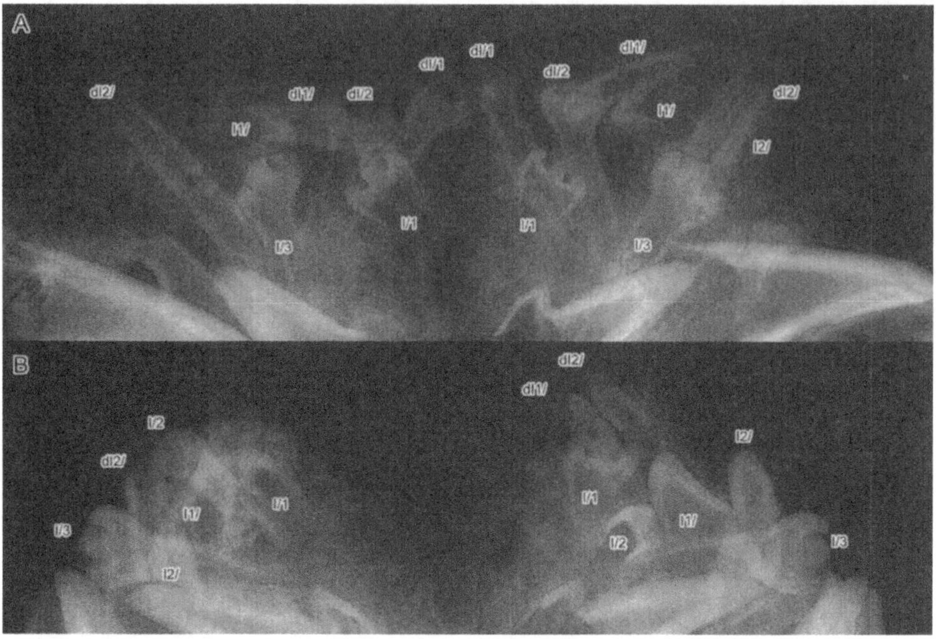

FIGURE 18. Tooth replacement in *Palaeochiropteryx tupaiodon*, showing sharp styliform anterior milk teeth, suited for a clinging function of the young (Figures 17–18 from Sigé *et al.*, 1998) (Photograph courtesy of Forschungsinstitut Senckenberg).

References

Habersetzer, J., Richter, G., and Storch, G., 1994, Paleoecology of early middle Eocene bats from Messel, FRG. Aspects of flight, feeding and echolocation, *Historical Biology* **8**:235–260.

Habersetzer, J., and Storch, G., 1987, Klassifikation und funktionelle Flügelmorphologie paläogener Fledermäuse (Mammalia, Chiroptera), *Courier Forschungsinstitut Senckenberg* **91**:117–150.

Habersetzer, J., and Storch, G., 1990, Ecology and echolocation of the Eocene Messel bats, in: *European bat research 1987* (V. Hanák, and J. Gaisler, eds), Charles University Press, Praha, pp. 213–233.

Habersetzer, J., and Storch, G., 1992, Cochlea size in extant Chiroptera and and middle Eocene microchiropterans from Messel, *Naturwissenschaften* **79**:462–466.

Hill, J. E., and Smith, J. D., 1984, *Bats. A natural history*, British Museum (Natural History), London.

Koenigswald, W. v., 1990, Die Paläobiologie der Apatemyiden (Insectivora s.l.) und die Ausdeutung der Skelettfunde von *Heterohyus nanus* aus dem Mitteleozän von Messel bei Darmstadt, *Palaeontographica* **A210**:41–77.

Koenigswald, W. v., 1992, The arboreal *Kopidodon*, a relative of primitive hoofed animals, in: *Messel. An insight into the history of life and of the Earth*, (S. Schaal, and W. Ziegler, eds). Clarendon Press, Oxford, pp. 233–237.

Koenigswald, W. v., Richter, G., and Storch, G., 1981, Nachweis von Hornschuppen bei *Eomanis waldi* aus der "Grube Messel" bei Darmstadt (Mammalia, Pholidota), *Senckenbergiana lethaea* **61**:291–298.

Koenigswald, W. v., and Storch, G. (eds.), 1998, *Messel. Ein Pompeji der Paläontologie*, Jan Thorbecke Verlag, Sigmaringen.

Richter, G., 1987, Untersuchungen zur Ernährung eozäner Säuger aus der Fossilfundstätte Messel bei Darmstadt. *Courier Forschungsinstitut Senckenberg* **91**:1–33.

Schaal, S., and Ziegler, W. (eds.), 1992, *Messel. An insight into the history of life and of the Earth*, Clarendon Press, Oxford.

Sigé, B., Habersetzer, J., and Storch, G., 1998, The deciduous dentition and dental replacement in the Eocene bat *Palaeochiropteryx tupaiodon* from Messel: The primitive condition and beginning of specialization of milk teeth among Chiroptera, *Lethaia* **31**:349–358.

Storch, G., 1993, Morphologie und Paläobiologie von *Macrocranion tenerum*, einem Erinaceomorphen aus dem Mittel-Eozän von Messel bei Darmstadt (Mammalia, Lipotyphla), *Senckenbergiana lethaea* **73**:61–81.

Storch, G., 1996, Paleobiology of Messel erinaceomorphs, *Palaeovertebrata* **25**:215–224.

Chapter 9

The Eocene Mammalian Fauna of Chambi (Tunisia) in Its Geological Context

JEAN-LOUIS HARTENBERGER, JEAN-YVES CROCHET, CLAUDE MARTINEZ, BERNARD MARANDAT, and BERNARD SIGÉ

1. Introduction

In the Kasserine area of Tunisia there is a Cenozoic basin containing Paleogene continental deposits. Palustrine limestones located at the base of the Paleogene series have yielded the Chambi mammalian fauna. Representatives of the

JEAN-LOUIS HARTENBERGER • Laboratoire de Paléontologie, Institut des Sciences de l'Evolution, CC64, Université Montpellier II, 34095 Montpellier Cédex 5, France. JEAN-YVES CROCHET • Laboratoire de Paléontologie, Institut des Sciences de l'Evolution, CC64, Université Montpellier II, 34095 Montpellier Cédex 5, France. CLAUDE MARTINEZ • ORSTOM Géofluides, Bassins, Eaux, CC58, Université Montpellier II, 34095 Montpellier Cédex 5, France. BERNARD MARANDAT • Laboratoire de Paléontologie, Institut des Sciences de l'Evolution, CC64, Université Montpellier II, 34095 Montpellier Cédex 5, France. BERNARD SIGÉ • Centre de Paléontologie stratigraphique et Paléoécologie, Université Lyon I, 69622, Villeurbanne, France.

Eocene Biodiversity: Unusual Occurrences and Rarely Sampled Habitats, edited by Gregg F. Gunnell, Kluwer Academic/Plenum Publishers, New York, 2001.

following mammalian orders and families have been found: Marsupialia, Peradectidae; Placentalia, Protheutheria indet.; ?Lipotyphla, ?Erinaceomorpha; Chiroptera, Rhinolophoidea, Nycterididae and Vespertilionoidea, Philisidae; Primates, Adapidae; Rodentia, Zegdoumyidae; Macroscelidea, Herodotinae; Hyracoidea, Pliohyracidae. The analysis of this local fauna provides insights into its position with regards to other North African mammalian faunas and its possible relationships with Eocene Eurasian faunas.

2. Paleogeography and Biostratigraphy

The first paleogeographic maps of Tunisia were published By Flandrin (1948). This author described the "Kasserine Island," an area he considered as an emergent land mass during the Eocene in the eastern Tethys Seaway. This paleogeographic concept was clarified by Burollet (1956) in his review of the stratigraphy of Tunisia. He believed that the "Kasserine Island" was present from the Senonian (= late Cretaceous, Coniacian through Maastrichtian) to the Miocene. As shown in Figure 1, the area of the "Kasserine Island" is situated in central Tunisia, north of the Jeffara Mole. One of the highest mountains of this region is the Chambi Massif that reaches an elevation of 1544 meters. The structure of this massif is made up of folded and faulted Cretaceous marine limestones attributed to the Abiod Formation (Burollet, 1956). Paleogene red continental sediments lie unconformably on the folded Cretaceous limestones and are mostly situated north of the Chambi Massif summit. Above the continental Paleogene sequence (which is 100 to 150 meters thick) there are unconformable Miocene deposits. These Miocene rocks have been tentatively correlated with the Beglia Formation (Biely *et al.*, 1972). A general geological setting of the Chambi Massif is shown in Figure 2.

Lacustrine and terrestrial gastropods (mostly internal molds) previously were the only fossils available for dating the Paleogene formation at Chambi (Sassi *et al.*, 1984). Subsequent studies of charophytes in North Africa (Mebrouk, 1993; Mebrouk *et al.*, 1997) have shown that the base of the Paleogene formation is early Eocene in age and the top of the sequence is late Eocene or early Oligocene in age. The first fossil mammals were found in 1985 (Hartenberger *et al.*, 1985).

3. Continental Paleogene Series

As illustrated in Figures 3 through 5, the Paleogene sediments, which are about 100 to 150 meters thick, represent a complex geological history in this area. At the base, fluvial conglomerates predominate with polygenic elements of decimeter size enclosed in a calcitic, sandy cement. These conglomerates are distributed in meter thick lenses and are overlain by thin beds (10 cm) of lacustrine or palustrine limestones. The mammal bones and teeth are found some 30 m above the base of the Paleogene series in a massive lacustrine limestone

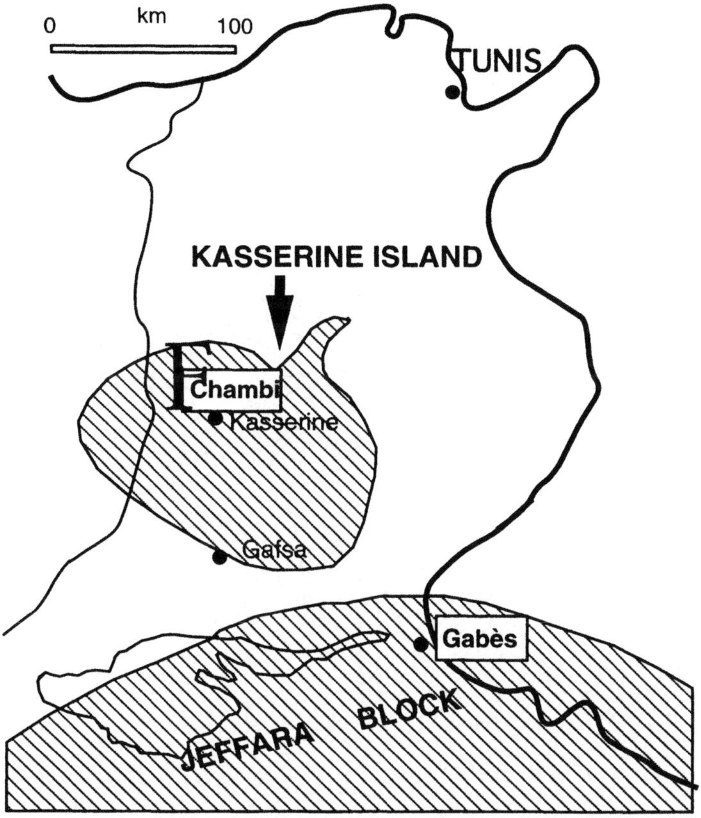

FIGURE 1. Palaeogeographic map of Tunisia during the Eocene with the Chambi mammalian locality indicated (modified from Sassi *et al.*, 1984).

layer with beds ranging from 30 cm to 1 m in thickness. The charophytes mentioned above occur in a marl horizon at the bottom of the limestones. At the top of the fossiliferous limestones, conglomerates again occur, but with smaller clasts than at the base. The upper part of the Paleogene series is made up of sands with rubefied traces indicating an interruption deposition and the onset of weathering. Above these sands thick lacustrine or palustrine limestones containing gypsum crystals are found. These limestones are generally distributed in thinner beds than the sequence containing the vertebrate fossils. Gastropod molds are frequently found in this upper limestone unit and have been identified as *Buliminus* and *Palaeocyclotus* (Sassi *et al.*, 1984). Charophytes also are relatively common in these limestones. The Paleogene series ends with sandy clays containing fossil plants and charophytes. Miocene deposits unconformably overlie the Paleogene sequence.

Continental formations similar to those at Chambi have been described from the Paleogene of southern France and in some cases these "lacustrine limestones" have yielded isolated vertebrate fossils. At a few French localities a

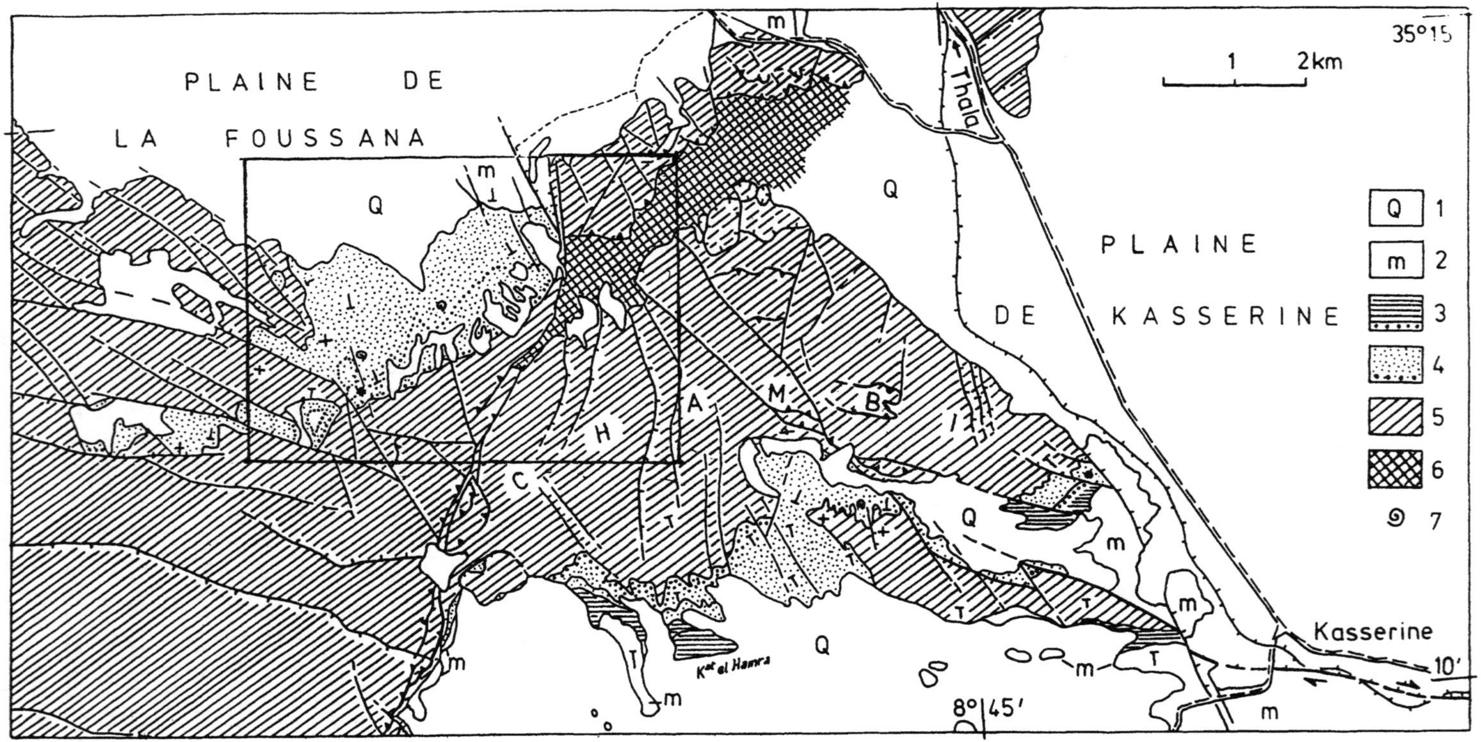

FIGURE 2. Simplified geological map of the Chambi Massif, near Kasserine. Legend: 1. Quaternary (Q); 2. Middle Miocene argillaceous sandstones (m); 3. Early Miocene (red clays with gypsum and layers of conglomerates); 4. Continental formation with layers of conglomerates, clays, and lacustrine limestones with gastropods and vertebrates; 5. Cretaceous; 6. Trias; 7. Main gastropod localities. Star (*) indicates mammalian locality. Square is enlarged in Figure 3.

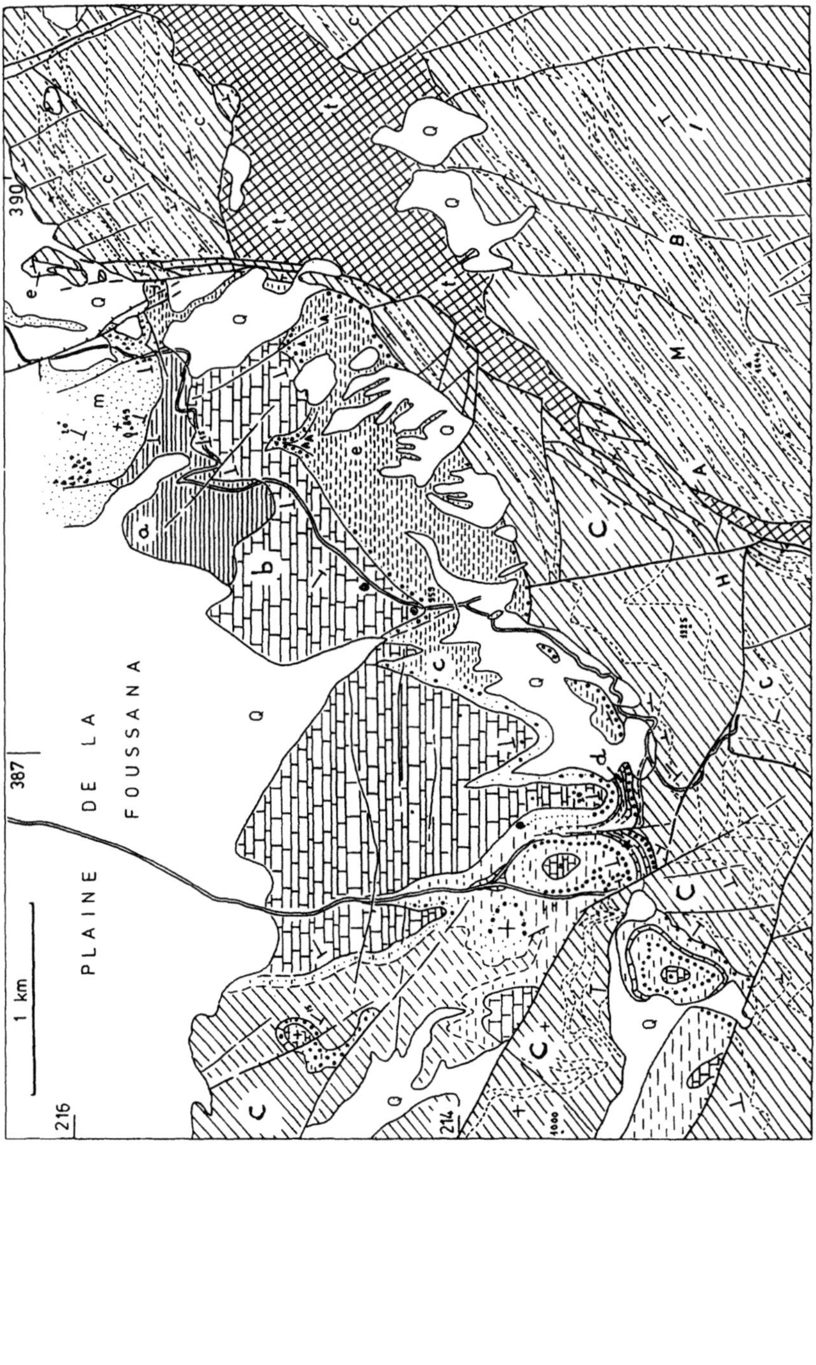

FIGURE 3. Outcrop map (C. Martinez) showing members constituting the Cenozoic basin north of Jebel Chambi (Chambi Massif). Legend: 1. Quaternary (Q); 2. Middle Miocene argillaceous sandstones (m); 3. Early Paleogene continental formation (e)—clays and marlstones with plant fossils (a)—calcareous limestones with gastropods (b)—clays and variegated marls (c)—gypsum, conglomerates, clays, and lacustrine limestones with gastropods and vertebrates (d); 4. Cretaceous (C); 5. Trias (t). Normal, reverse, and thrust faults are indicated. The dashed lines in the Cretaceous roughly follow the banks. Numbers in the margin correspond to Lambert coordinates (Tunisia). Gastropod and mammal localities are indicated by stars (*).

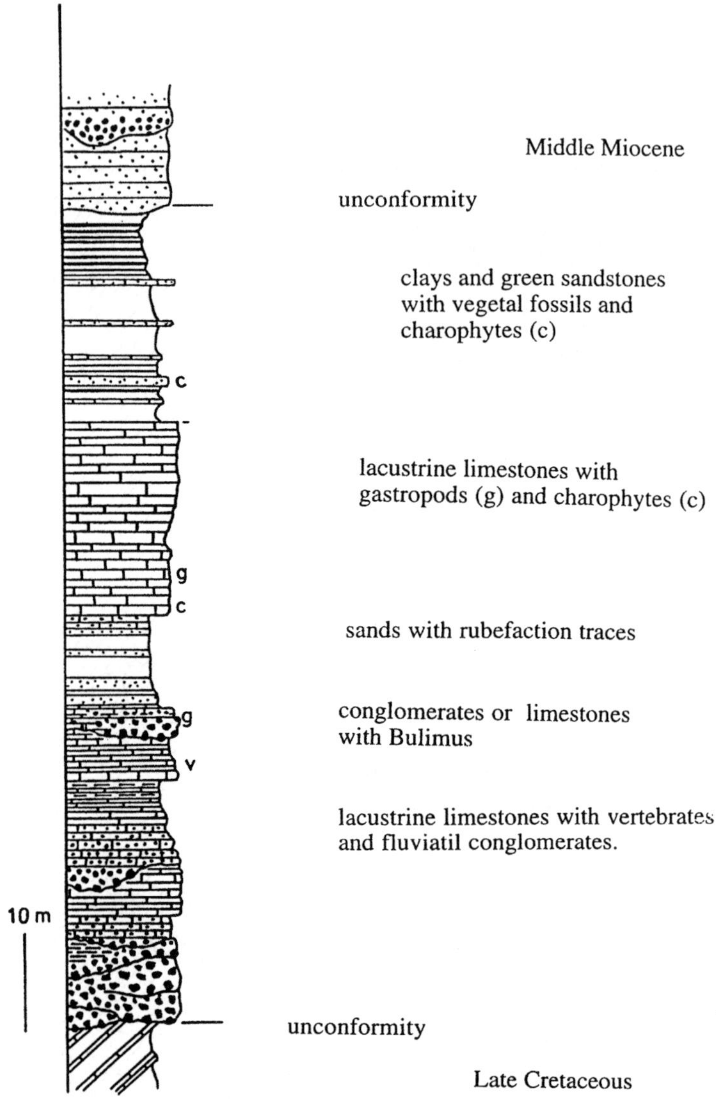

FIGURE 4. Composite geological section showing lithological succession as observed in the northern Chambi area and position of gastropod and vertebrate localities.

diverse mammalian fauna has been recovered from deposits similar to Chambi, especially in Aumelas, near Montpellier (Hartenberger, 1963). These limestones are distributed in various small basins as illustrated in 1/50,000 geological maps of the Montpellier and Saint Martin de Londres areas. In fact all of these lacustrine basins can be considered as part of a larger system, the "Lutetian lake," that covered most of the Languedoc area during the Middle Eocene. A synthetic overview of continental Paleogene formations of Southern France is given in

Cavelier (1984) in which most mammalian localities have been taken into consideration. Considering the only "lacustrine" limestones, Freytet (1973) noted that the carbonate facies of these formations can be of several types including: limestones deposited in fluvial channels; true lacustrine limestones with subaqueous diagenetic evolution; or palustrine limestones resulting from emersion of lacustrine muds frequently affected by soil processes.

At Chambi it seems that only one limestone horizon is fossiliferous and probably corresponds to a channel deposit. Specimens are scattered across a

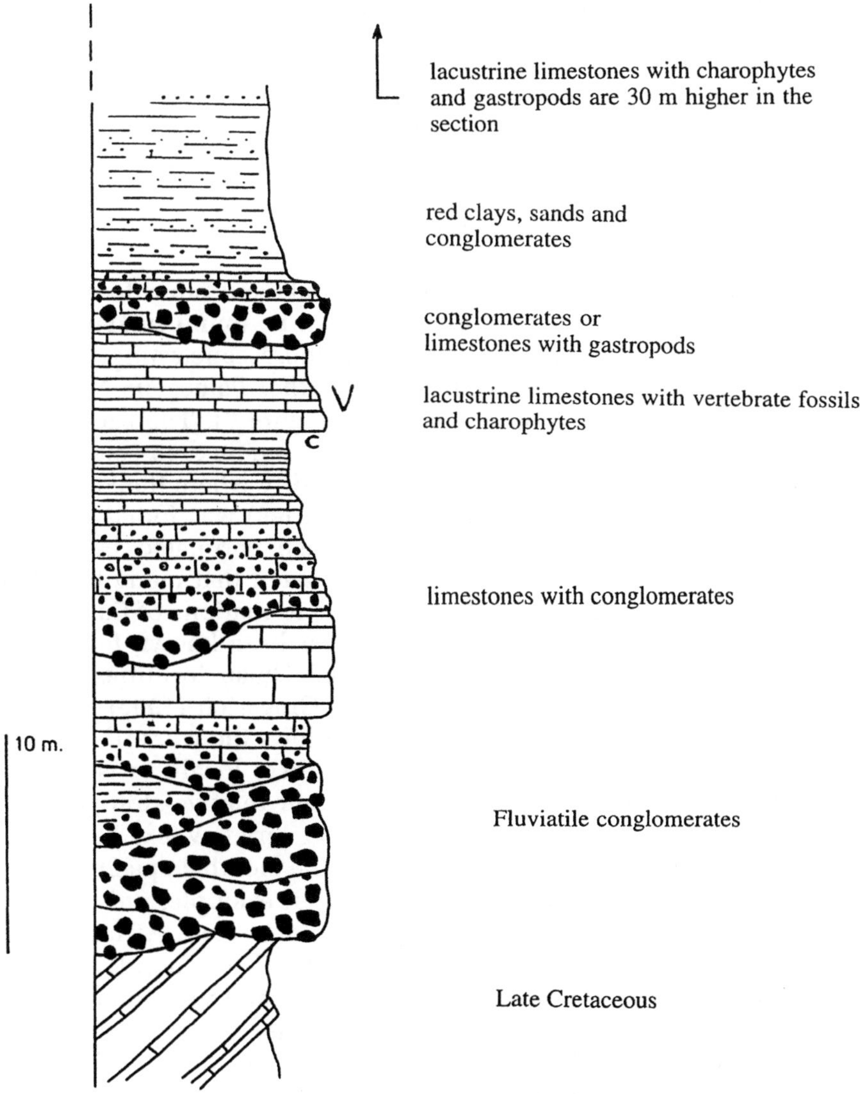

FIGURE 5. Lithological succession as observed near the vertebrate locality.

200 × 50 meter area and up to now it has been impossible to locate concentrations of bone. Fossils are obtained by dissolving limestone blocks in dilute solutions of acetic or formic acid following a method summarized by Orth (1987). At present, 500 kg of limestones have been dissolved. Blocks are selected in the field when small bones occur on the surface or after breaking the limestones open and finding evidence of bone. Finds of teeth in the field are rare but the presence of small bones is relatively frequent. Nevertheless, it appears now that the occurrence of "good" fossils in these limestones is very irregular and many blocks lack any bone at all. As the fossils are very fragile, the acid preparation process takes a great deal of time and requires frequent water washing of the blocks and many applications of glue to the fossils. It takes approximately one month to dissolve a 3 kg block with an 8% acid solution. Only a few relatively complete dentitions have been found (Figure 7) and isolated teeth are frequently broken or worn. Table 1 presents a list of the biotic elements so far discovered from Chambi.

4. Summary of Chambi Vertebrates

An extensive and critical review of the mammals and other fossils found in the Chambi area and their taxonomic status can be found in Hartenberger *et al.* (1997). The main conclusions of these authors concerning the biostratigraphic position of Chambi in comparison to other Paleogene North African and Arabian localities are summarized here (Figure 6). The age of the Paleocene and early Eocene localities of Morocco is based on their selachian faunas (see Gheerbrant *et al.*, 1993), but the precise locality where the Paleocene proboscidean *Phosphatherium* (Gheerbrant *et al.*, 1998) was found is unknown. The Algerian and Tunisian localities have been correlated by charophyte studies (Mebrouk *et al.*, 1998). These authors noted that many taxa found at these localities also are known in the Paleogene formations of Europe where a biochronological distribution of charophytes has been known for a long time (Rivelines, 1986). The positions of Fayum and Oman localities in Figure 6 follow the conclusions drawn respectively by Gingerich (1993) and Pickford *et al.* (1994). The position of Zallah in Libya follows the work of Fejfar (1987).

In fact, the stratigraphic position of all of these localities is in a state of flux and there is no doubt that in the future their relative stratigraphic positions will be revised. One of the problems is that very few taxa have large geographic distributions. This is compounded by the fact that specimens in hand are often fragmentary or poorly preserved. As an example, only two species found at Chambi have been recognized from other North African Paleogene localities: the charophyte *Raskyella sahariana* has been found at Gour Lazib and Glib Zegdou in Algeria (Mebrouk *et al.*, 1997); and the rodent *Zegdoumys sbetlai* is present at Glib Zegdou. At the genus level, the Chambi hyracoid *Titanohyrax* also is present at Glib Zegdou (*T. mongereaui*) and in the Fayum (3 species). The Chambi adapid primate *Djebelemur* also is present in Algeria (*contra* Godinot, 1994). At the family level, affinities have been recognized between Chambi forms and some early Eocene European families of marsupials, primates, chiropterans, and condylarths. It appears that near the Paleocene-Eocene transition, interchanges between Africa and southern European mammalian terrestrial faunas were possible, but the direction of these exchanges cannot be determined at this time. North African, as well as European, mammalian faunas were isolated during most of the Eocene. It is well

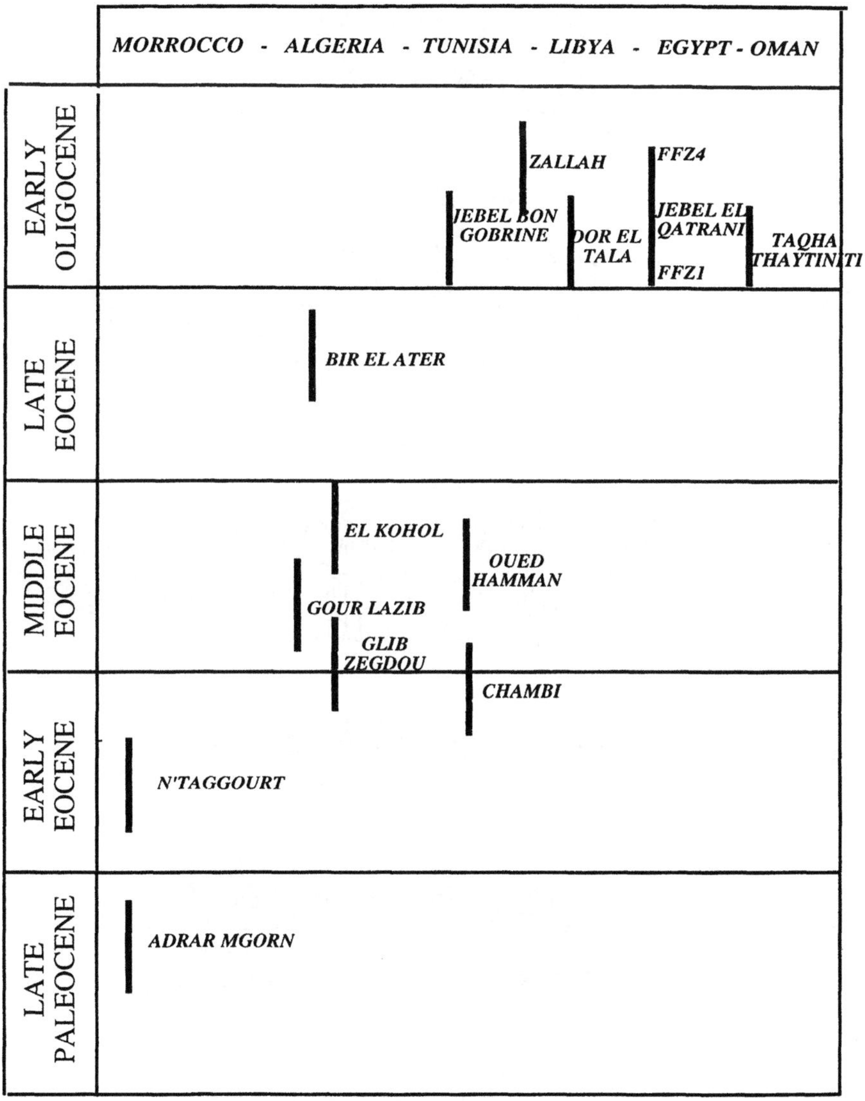

FIGURE 6. Stratigraphic distribution of main Paleogene mammalian localities in northern Africa and the Arabian Peninsula. (The late Paleocene proboscidean *Phosphatherium escuillei* [Gheerbrant *et al.*, 1998] comes from an unknown locality in the Ouled Abdoun Basin, Morocco and is not included here).

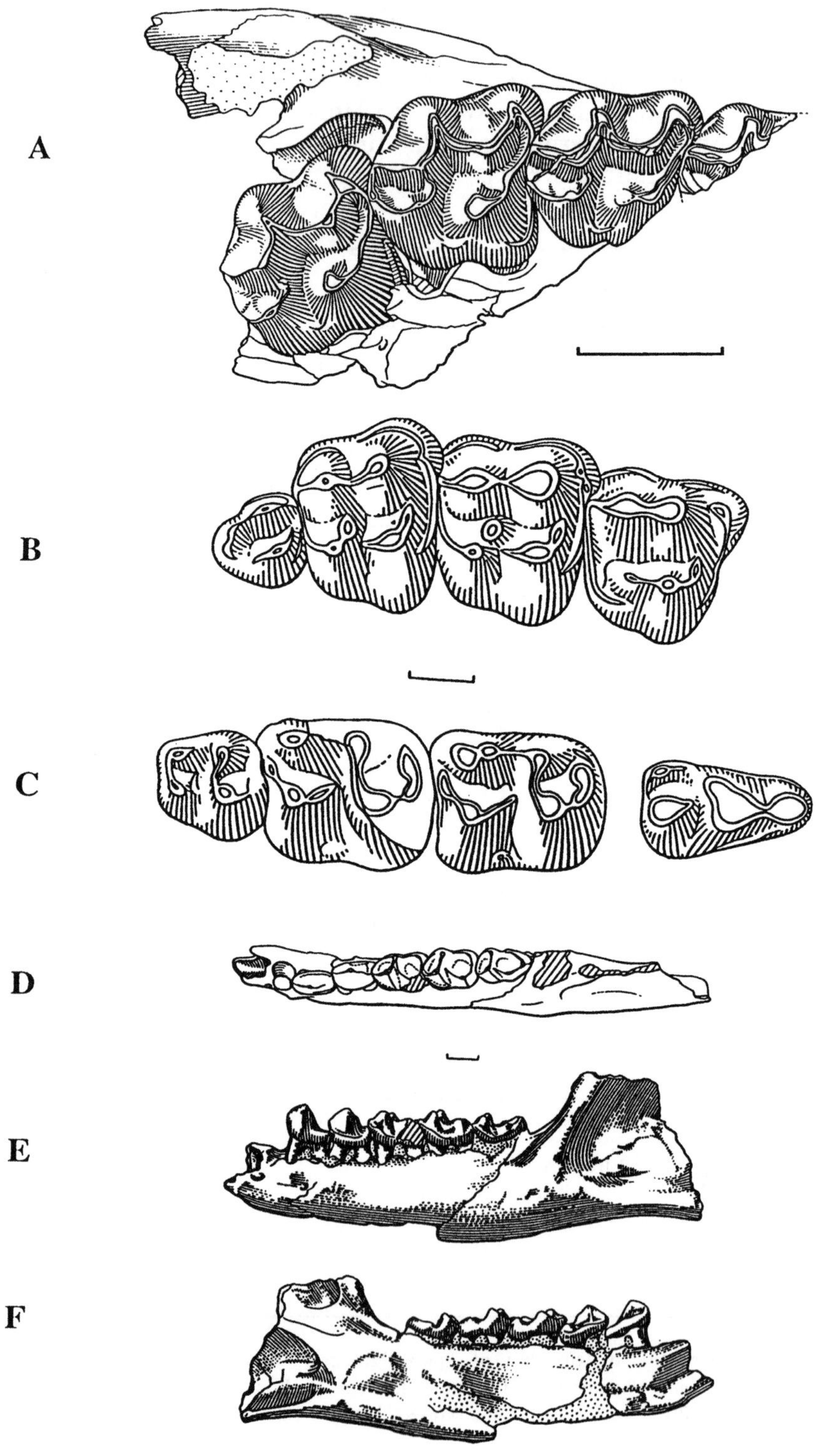
A
B
C
D
E
F

Table 1. List of biota collected from Chambi limestone, Chambi Massif, Tuisia

- Charophytes:
  - *Raskyella* cf. *R. sahariana* Mebrouk, 1993
- Vertebrata:
  - Amphibia
    - Batraciens indet.
  - Reptilia
    - Squamata indet.
  - Mammalia
    - Metatheria
      - Marsupialia
        - Peradectidae
          - *Kasserinotherium tunisiense* Crochet, 1986
    - Eutheria
      - ?Lipotyphla
      - ?Erinaceomorpha
        - *Ghambilentes foussanensis* (Gheerbrant and Hartenberger, 1999)
      - Chiroptera
        - Rhinolophoidea
          - Nycterididae
            - gen. et sp. indet.
        - Vespertilionoidea
          - Philisidae
            - *Dizzya exsultans* (Sigé, 1991)
      - Primates
        - Adapidae
          - *Djebelemur martinezi* Hartenberger and Marandat, 1992
      - Rodentia
        - Zegdoumyidae
          - *Zegdoumys sbetlai* Vianey-Liaud *et al.*, 1994
      - Macroscelidea
          - *Chambius kasserinensis* Hartenberger, 1986
      - Hyracoidea
        - Pliohyracidae
          - *Titanohyrax tantulus* Court and Hartenberger, 1993
      - Order indet.: in Court, 1993

◄

FIGURE 7. Best preserved specimens of mammals found at Chambi (1984 expedition). All specimens are in the collections of the Service de la Carte Géologique, Office National des Mines, Tunis. A. Hyracoidea—*Titanohyrax tantulus* Court and Hartenberger, 1992, right maxilla with dP4-M2, holotype. Scale bar = 10 cm. B and C. Macroscelidea—*Chambius kasserinensis* Hartenberger, 1986, right maxilla with P4-M3, holotype (B) and left dentary with m1–3 (C). Scale bar = 1 mm. D-F. Primates—*Djebelemur martinezi* Hartenberger and Marandat, 1992, left dentary with p3-m3, holotype in occlusal (D), lateral (E), and medial (F) views. Scale bar = 1 mm. Figure A is reproduced by permission of the Palaeontological Association. Figures B and C are from Hartenberger, 1986 and are reproduced by permission of the Institut de France, Academie des Sciences, Service des Comptes rendus. Figures D–F are reproduced by permission of the Instituto di Antropologia, Université di Firenze.

known that the end of the Eocene, marked by the *Grande Coupure* event, terminated this endemic period in Europe. A similar event probably occurred in North Africa but the appearance of anthracotheriids of Asian origin in the Oligocene of Fayum is, for the moment, the only supporting evidence.

The main interest of the Tunisian Chambi locality lies in the fact that the oldest occurrences on the African continent of representatives of Marsupialia and the eutherian orders Chiroptera, Macroscelidea, Hyracoidea, and Rodentia have been found there.

Crochet (1986) notes that the marsupials found at Chambi show affinities with European genera. He points out that Asian, European, and African marsupials otherwise known from the Eocene through Miocene, have restricted biogeographical ranges with only peradectines known from areas south of Tethys and herpetotheriids only known from areas north of Tethys. The fact that peradectine marsupials are present at Chambi suggests that some faunal interchange between the "Kasserine Island" and continental Europe may have been possible during the Eocene.

Among bats, the taxa represented at Chambi are members of the modern families Rhinolophoidea and Vespertilionidea. At the same time in Europe, only archaic groups of bats (archeonycterids, paleochiropterygids, hassianycterids) have been found. Based on this pattern of distribution, the African Paratethyean may be interpreted to have played an important role as a center of origin and subsequent dispersion of the modern bat clades. Recent discoveries in Europe may qualify this point of view (Sigé, 1997), but nevertheless, considering the age of Chambi with respect to that of European bat localities, it seems probable that transtethysan connections (from south to north), were still in existence at the end of the early Eocene (particularly for the paleochiropterygids).

The macroscelidid *Chambius* may share affinities with some condylarthrans (Louisininae) from the Paris Basin. Recently Sudre *et al.* (1993) have noted the occurrence of louisinine condylarths from the Paleocene of Morocco, confirming the notion of faunal interchange between North Africa and Europe at the time of the Paleocene/Eocene transition. Also the primate *Djebelemur* looks very similar to European adapids that have been recovered from early Eocene deposits in southern France and the Paris Basin (Hartenberger and Marandat, 1992).

All of the rodents from Chambi are included in the family Zegdoumyidae. Some of the representatives of this family (*Glibia*, *Glibemys*) are probable ancestors of Anomaluridae, a family that is frequently considered to be the sister group of European Theridomyidae. Also it should be noted that some of the molars attributed to *Zegdoumys* from Chambi have a phiomyid tooth pattern, emphasizing the "African style" of the rodent fauna.

Titanohyrax tantulus from Chambi is presently the oldest known hyracoid. Court and Hartenberger (1993) examined the occlusal relationships of *T. tantulus* molars and concluded that their pattern of attrition is of a hyracoid type, not at all like the condition observed in earliest perissodactyls. This contradicts evidence pointing to a phylogenetic link between hyraxes and perissodactyls as proposed by some authors (Fisher, 1986; Prothero *et al.*, 1988).

Overall the late-early to early-middle Eocene mammalian fauna from Chambi has a relatively strong African character. At this time North Africa was probably isolated from other areas as an island in the Tethys Seaway, but near the Paleocene/Eocene transition exchanges were possible, particularly with Europe.

ACKNOWLEDGMENTS. We are indebted to Dr. Ghozi and Dr. Ali Ben Hajl, Chiefs of Service de la Carte Géologique (Office National des Mines, Tunis) and to Professors Tlig and Zargouni (University of Tunis). Many thanks to M. Mansour and Taïeb Mohamed Tahar for facilities during field expeditions. Dr. G. F. Gunnell read and improved an earlier version of the paper. All type specimens are in the collections of the Office National des Mines (Tunis). N° I.S.E.M. 2001-016.

References

Biely, A., Rakus, M., Robinson, P., and Salaj, J., 1972, Essai de corrélation des formations miocènes au sud de la Dorsale Tunisienne, *Notes du Service Géologique, Tunis* **38**:73–92.

Burollet, P. F., 1956, Contribution à l'étude stratigraphique de la Tunisie Centrale, *Annales des Mines et de la Géologie, Tunis* **18**:350 pages.

Cavelier, C., 1984, Paléogène, in: *Synthése géologique du Sud-Est* (S. Debrand-Passard, ed.), Mém. Bureau de recherches géologiques et minières **125**:389–468.

Court, N., 1993, An enigmatic new mammal from the Eocene of North Africa, *Journal of Vertebrate Paleontology* **13**:267–269.

Court, N., and Hartenberger, J.-L., 1992, A new species of the hyracoid mammal *Titanohyrax* from the Eocene of Tunisia, *Palaeontology, London* **35**:309–317.

Crochet, J.-Y., 1986, *Kasserinotherium tunisiense* nov. gen., nov. sp., troisième marsupial découvert en Afrique (Eocène inférieur de Tunisie), *Comptes-rendus de l'Académie des Sciences, Paris, série 2* **302**:923–926.

Fejfar, O., 1987, Oligocene rodents from Zallah Oasis, *Libya, Münchener Geowissenschaftliche Abhandlungen, Reihe A* **10**:265–268.

Fischer, M., 1986, Die Stellung der Schliefer (Hyracoidea) im phylogenetischen System der Eutheria, *Courier Forschungsinstitut Senckenberg* **84**:1–132.

Flandrin, J., 1948, Contribution à l'étude stratigraphique du Nummulitique algérien, *Bulletin du Service de la Carte géologique d'Algérie, 2ème série, Stratigraphie, Alger* **19**:334 pages.

Freytet, P., 1973, Petrography and paleoenvironment of continental carbonate deposits with particular reference to the Upper Cretaceous and Lower Eocene of Languedoc (Southern France), *Sedimentary Geology* **10**:25–60.

Gheerbrant, E., Cappetta, H., Feist, M., Jaeger, J.-J., Sudre, J., Vianey-Liaud, M., and Sigé, B., 1993, La succession des faunes de vertébrés d'âge paléocène supérieur et éocène inférieur dans le bassin d'Ouarzazate, Maroc. Contexte géologique, portée stratigraphique et paléogéographique, *Newsletter in Stratigraphy* **28**:33–58.

Gheerbrant, E., and Haztenberger, J. L., 1999, Nouveou mammifèze insedivore (?Lipotypla ?Erinaceomorpha) de l' Éocène inférèuz de Chambi (Tamisie), *Paläontologische Zeitschrift* **73**:143–156.

Gheerbrant, E., Sudre, J., Cappetta, H., and Bignot, G., 1998, *Phosphatherium escuillei* du Thanétien du Bassindes Ouled Abdoun (Maroc), plus ancien proboscidien (Mammalia) d'Afrique, *Géobios* **30**:247–269.

Gingerich, P. D., 1993, Oligocene age for the Gebel Qatrani Formation, Fayum, Egypt, *Journal of Human Evolution, New York* **24:**207–218.

Godinot, M., 1994, Early North African primates and their significance for the origin of Simiiformes (= Anthropoidea), in: *Anthropoid Origins* (J. G. Fleagle, and R. F. Kay, eds.), Plenum Press, New York, pages 235–295.

Hartenberger, J.-L., 1963, Un gisement de la zone de Mammifères d'Issel (Éocène moyen) dans les calcaires lacustres d'Aumelas (Hérault), *Comptes-rendus som. Soc. géol. France* **1963:**321–322.

Hartenberger, J.-L., 1986, Hypothèse paléontologique sur l'origine des Macroscelidea, *Comptes-rendus de l'Académie des Sciences, Paris, série 2* **302:**247–249.

Hartenberger, J.-L., and Marandat, B., 1992, A new genus and species of an early Eocene primate from North Africa, *Journal of Human Evolution, Firenze* **7:**9–16.

Hartenberger, J.-L., Martinez, C., and Ben aid, A., 1985, Découverte de Mammifères d'âge Eocène inférieur en Tunisie Centrale, *Comptes-rendus de l'Académie des Sciences, Paris, série 2* **301:**649–652.

Hartenberger, J.-L., Crochet, J.-Y., Martinez, C., Feist, M., Godinot, M., Mannai Tayech, B., Marandat, B., and Sigé, B., 1997, Le gisement de Mammifères de Chambi (Éocène, Tunisie Centrale) dans son contexte géologique, in: *Apport à la connaissance de l'évolution en Afrique. Actes du Congrès Biochrom 97* (J.-P. Aguilar, S. Legendre, and J. Michaux, eds.), Mém. E. P. H. E., Montpellier **21:**263–274.

Mebrouk, F., 1993, *Charophytes du Tertaire continental de l'Algérie. Systématique et biostratigraphie*, Thèse, Université d'Oran, 173 pages.

Mebrouk, F., Bessedik, M., Mahboubi, M., and Feist, M., 1997, L'apport des Charophytes à la stratigraphie des formations continentales de l'Algérie, *Géobios* **30:**171–177.

Orth, B., 1987, The acid column process: an improvement in the concentration of carbonated fossiliferous sediment, *Curator* **30:**227–239.

Pickford, M., Thomas, H., Sen, S., Roger, J., Gheerbrant, E., and El-Sulaimani, Z., 1994, Early Oligocene Hyracoidea (Mammalia) from Thaytini and Taqah, Dhofar Province, Sultanate of Oman, *Comptes-rendus de l'Académie des Sciences, Paris, série 2* **318:**1395–1400.

Prothero, D. R., Manning, E. M., and Fischer, M., 1988, The phylogeny of ungulates, in: *The Phylogeny and Classification of the Tetrapods, Volume 2: Mammals* (M. J. Benton, ed.), Systematics Association Special Volume 35B, Clarendon Press, Oxford, 329 pages, pp. 201–204.

Riveline, J., 1986, *Les charophytes du Paléogène et du Miocène inférieur d'Europe occidentale*, Cahiers de Paléontologie (éditions du C. N. R. S., Paris).

Sassi, S., Triat, J.-M., Truc, G., and Millot, G., 1984, Découverte de l'Eocène continental en Tunisie Centrale: la formation du Jebel Chambi et ses encroutements carbonatès, *Comptes-rendus de l'Académie des Sciences, Paris, série 2* **299:**357–364.

Sigé, B., 1991, Rhinolophoidea et Vespertilionoidea (Chiroptera) du Chambi (Eocène inférieur de Tunisie). Aspects biostratigraphiques, biogéographiques et paléoécologiques de l'origine des chiroptères modernes, *Neues Jahrbuch Geologie und Palaontologie Abhandlugen* **182:** 355–376.

Sigé, B., 1997, Les remplissages karstiques polyphasés (Eocène, Oligocène, Pliocène) de St-Maximin (phosphorites du Gard) et leur apport à la connaissance des faunes européennes, notamment pour l'Eocène moyen (MP 13), in: *3 systématique: Euthériens entomophages. Actes du Congrès Biochrom 97* (J.-P. Aguilar, S. Legendre, and J. Michaux, eds.), Mém. E.P.H.E., Montpellier **21:**737–750.

Vianey-Liaud, M., Jaeger, J.-J., Hartenberger, J.-L., and Mahboubi, M., 1994, Les rongeurs de l'Éocène d'Afrique nord-occidentale (Glib Zegdou, Algérie; Chambi, Tunisie) et l'origine des Anomaluridae, *Palaeovertebrata* **23:**119–152.

Chapter 10

Gandhera Quarry, A Unique Mammalian Faunal Assemblage From the Early Eocene of Baluchistan (Pakistan)

PHILIP D. GINGERICH, MUHAMMAD ARIF, INTIZAR HUSSAIN KHAN, MUNIR UL-HAQ, JONATHAN I. BLOCH, WILLIAM C. CLYDE, and GREGG F. GUNNELL

1. Introduction

A newly discovered fossil locality, Gandhera Quarry, is yielding the most complete evidence to date on the structure of mammalian communities on the Indo-Pakistan subcontinent during the early Eocene. Gandhera Quarry is located in

PHILIP D. GINGERICH, JONATHAN I. BLOCH, and GREGG F. GUNNELL • Museum of Paleontology, University of Michigan, Ann Arbor, MI 48109-1079. MUHAMMAD ARIF, INTIZAR HUSSAIN KHAN, and MUNIR UL-HAQ • Geological Survey of Pakistan, Quetta, Pakistan. WILLIAM C. CLYDE • Department of Earth Sciences, University of New Hampshire, Durham, NH 03824-3589.

Eocene Biodiversity: Unusual Occurrences and Rarely Sampled Habitats, edited by Gregg F. Gunnell, Kluwer Academic/Plenum Publishers, New York, 2001.

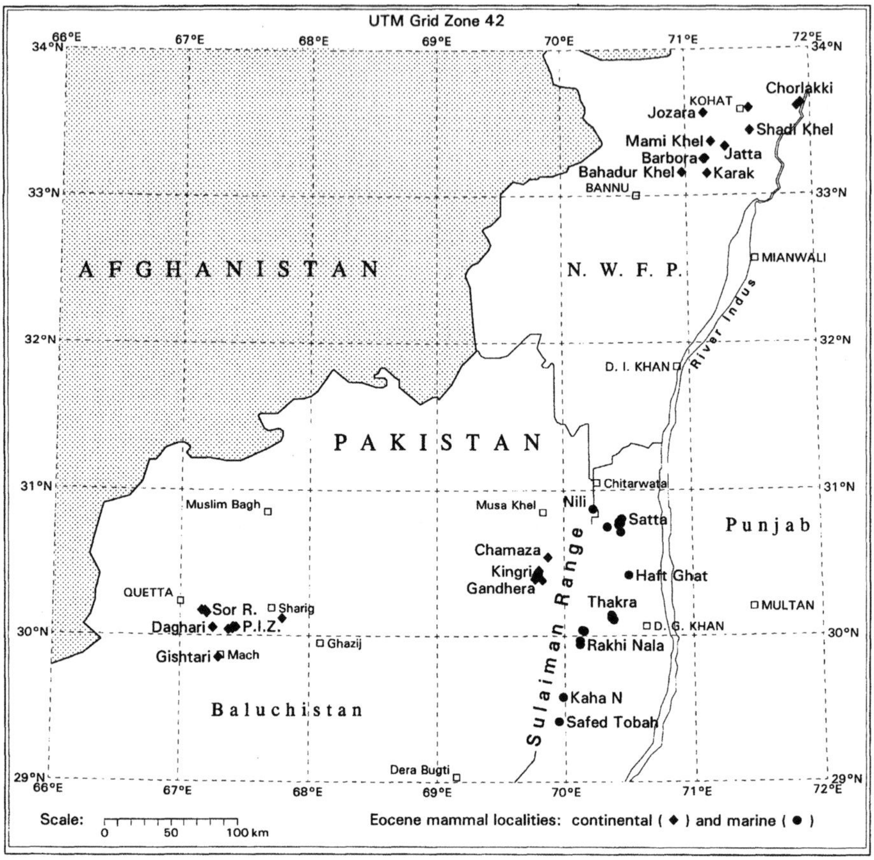

FIGURE 1. Map of northwestern Pakistan showing the location of Gandhera southeast of Kingri in eastern Baluchistan.

Musa Khel District, Baluchistan Province, Pakistan (Figures 1–3). The locality was discovered by Munir-ul-Haq in April, 1999, while prospecting for new localities with a Geological Survey of Pakistan-University of Michigan (GSP-UM) field party. The purpose of the present paper is to provide preliminary findings concerning the depositional environment, taphonomic characteristics, and taxonomic composition of Gandhera Quarry and its faunal sample. More complete taxonomic treatment is in progress and the paleobiogeographic implications of the Gandhera faunal assemblage will be discussed elsewhere (Gingerich *et al.*, in preparation).

2. Geologic Setting

Gandhera Quarry is located in the uppermost part of the mammal-bearing, fluvial, upper Ghazij Formation (Gingerich *et al.*, 1997, 1998, Figure 2). The

FIGURE 2. Photograph showing location of Gandhera Quarry in the upper part of the fluvial upper Ghazij Formation. View is to north. Collector standing in center of circle gives scale. Arrow points up-section. Dashed line approximates contact with overlying Drug Limestone forming bluff at left.

FIGURE 3. Gandhera Quarry. Fossil producing level is at the top of the slightly overturned flaggy sandstone (arrow). Hammer is shown for scale. Arrow points up-section. View is to north.

Ghazij Formation is a shallowing-upward sequence of: (1) marine shales in the lower part; (2) paludal coal-bearing sandstones and shales in the middle part; and (3) fluvial sandstones and floodplain mudstones with well developed paleosols in the upper part. It is on the order of 2600 m thick in the Kingri area (Warwick *et al.*, 1998). Marine shales in the lower Ghazij Formation yield planktonic foraminifera indicative of P6–7 age (Fritz and Khan, 1975), while the overlying Drug Limestone has zone P9 forams at the top (Afzal, 1996), making the Gandhera faunal assemblage a probable zone P7–9 equivalent (51–50 Ma) or late early Eocene (late Ypresian on the time scale of Berggren *et al.* 1995; Figure 4).

Gandhera Quarry is a 65 meter long sandstone-clay interface rich in small mammal jaws, teeth, and postcrania. The sequence is slightly overturned (Figure 3) so that the sandstone is now on top of the clay but the reverse was true during deposition. Well preserved dentitions and postcrania are cemented in a hard calcarenite matrix in the top 2–3 cm of the sandstone. This calcarenite facies is commonly preserved as a lag deposit in upper Ghazij sandstones and is largely composed of reworked carbonate nodules and less abundant lithic fragments cemented together with a carbonate/quartz cement. The calcarenite layer at Gandhera contains relatively coarse grains and larger bone elements at the northern end of the deposit and fines laterally to the south where grain size and bone elements are smaller. Above this hard layer, the shale is much softer and more friable and contains relatively more incomplete remains of small mammals. Above the shale is a greenish sandy clay that contains fewer bone elements.

3. Collecting Techniques

A variety of collecting techniques were undertaken at Gandhera Quarry in order to assess the best ways to approach future work. First, the small valley and drainage beneath the quarry layer were intensively surface prospected resulting in the collection of 187 maxillae and dentaries, 81 isolated whole teeth, and 1918 identifiable postcranial elements. Large blocks of the calcarenite layer were extracted and transported to the preparation laboratory at the University of Michigan where they are now being prepared. The shale layer above the calcarenite was quarried and dry screened resulting in collection of 41 maxillae and dentaries, 112 isolated whole teeth, and 1004 identifiable postcranial elements. Fresh samples of the shale layer were also collected ($\approx$ 5 kg) and transported to Michigan for wet screening. Nested screens of decreasing size (1000, 710, and 500 µm) were used, which resulted in the procurement of 13 additional maxillae and dentaries, 257 isolated whole teeth, and 918 identifiable postcranial elements. Quarrying produced some nearly complete bone elements but is very time consuming as, even when relatively intact, bones are microfractured making extraction difficult. Table 1 summarizes the identifiable elements collected at Gandhera Quarry in four days of field work and Figure 5 summarizes relative bone abundance.

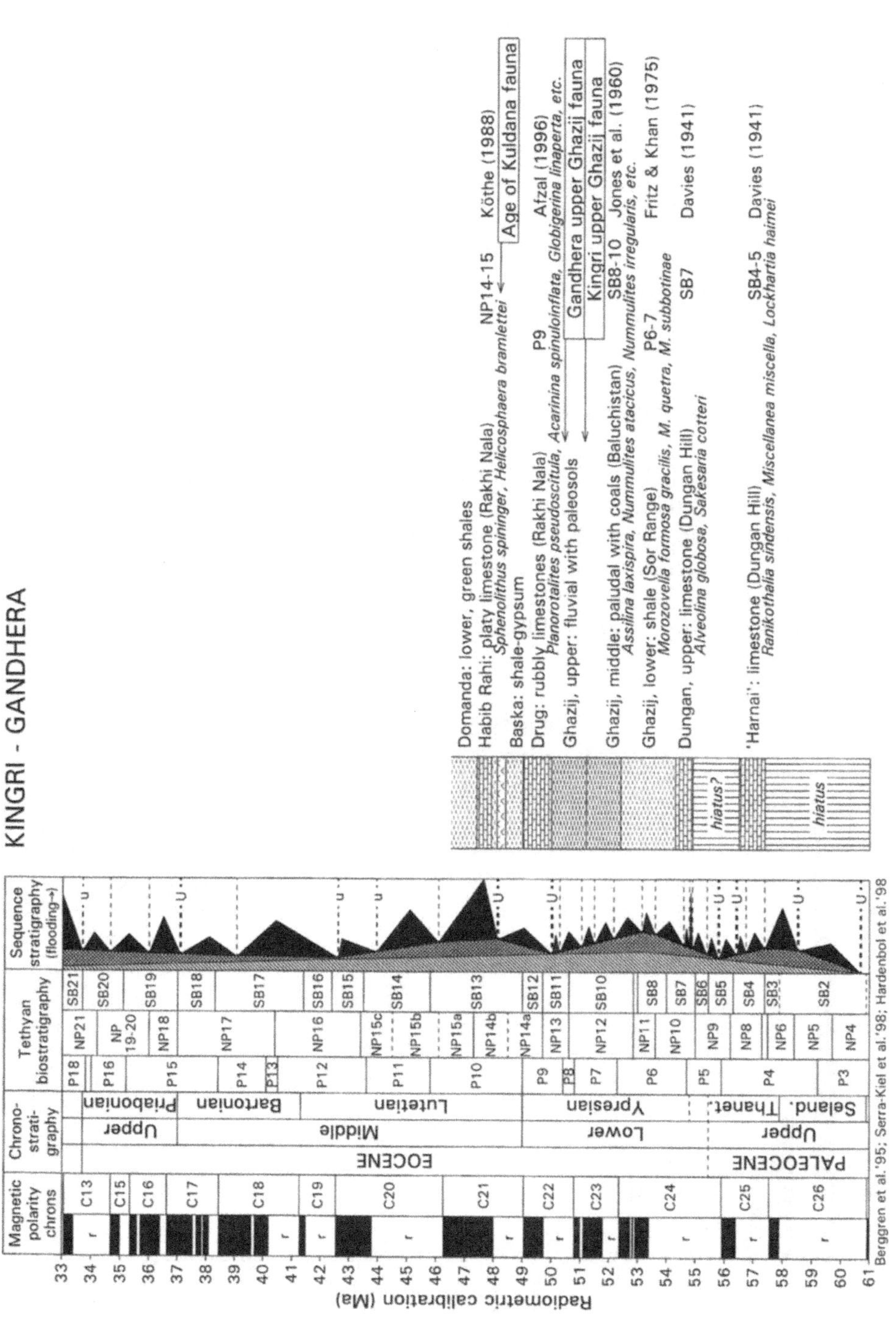

FIGURE 4. Correlation chart showing the stratigraphic position and age of Gandhera Quarry relative to earlier mammal-bearing units in the Ghazij Formation and the later Kuldana Formation faunas.

Table 1. Identifiable mammalian elements at Gandhera Quarry

Source	Maxillae/Dentaries	Whole Teeth	Tooth Fragments	Postcrania
Surface	187	81	98	1,198
Dry Screen	41	112	522	1,004
Wet Screen	13	257	2,387	918
Totals	241	450	3,007	3,840

Total of all identified elements = 7,538

4. Taxonomic Composition

Three intervals within the Ghazij Formation of Baluchistan have produced mammalian remains (Gingerich *et al.*, 1997, 1998). The lowest interval is in the middle part of the Ghazij Formation and has produced two taxa of quetta-cyonids, an archaic family of condylarths, from coal and overlying mudstone lithologies (Gingerich *et al.*, 1997, 1999), and a possible anthracobunid from the coal unit (Ginsburg *et al.*, 1999). A second interval is in the lower part of

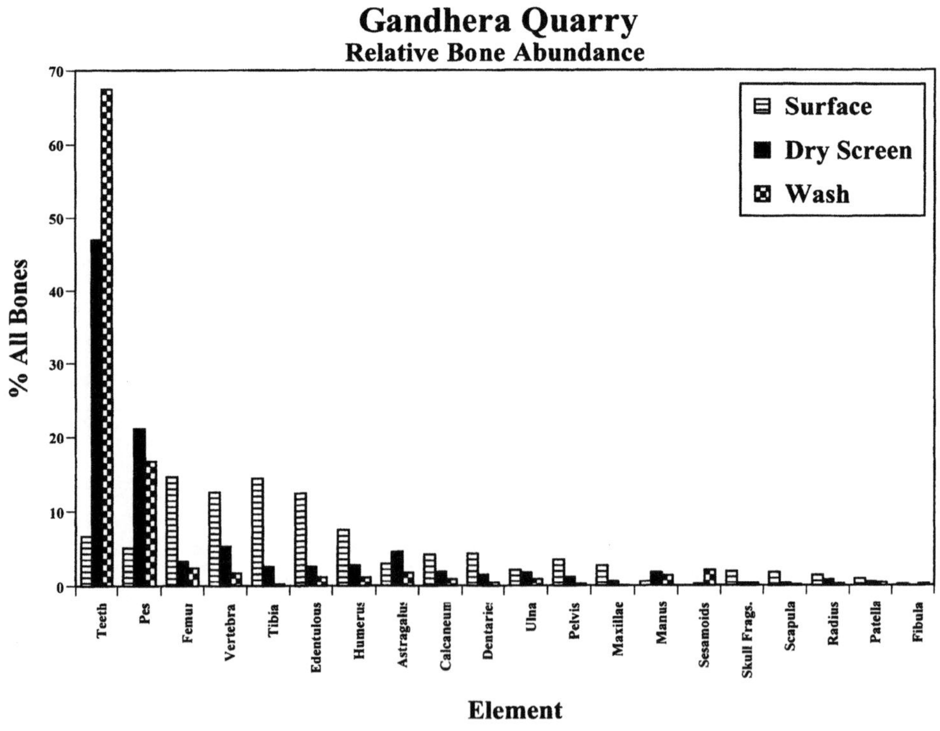

FIGURE 5. Relative bone abundance by collecting technique at Gandhera Quarry.

the upper Ghazij and has produced a more extensive faunal sample consisting of primitive perissodactyls, tillodonts, and quettacyonids, from calcarenite and mudstone. Gingerich, *et al.*, 1998. The upper part of the upper Ghazij Formation has produced the most diverse fauna and includes the Gandhera Quarry sample. The fauna from the upper part of the upper Ghazij is more typical of other holarctic early Eocene faunas known from North America, Europe, and Asia in that it includes primates, artiodactyls, perissodactyls, rodents, and hyaenodontid creodonts (Figure 6). Table 2 summarizes taxonomic composition through the Ghazij Formation.

5. Bones at Gandhera Quarry

5.1. Preservation

The calcarenite layer preserves relatively complete, solid elements, but otherwise bones tend to be broken and fractured in the rest of the sequence at Gandhera Quarry. Much of the breakage may be due to tectonic alteration as the sequence was overturned and compacted. Bones tend to be abraded and polished with few sharply preserved broken surfaces. Bones found in place in the calcarenite or in the shale directly above the calcarenite exhibit more freshly broken surfaces but even within these lithologies abraded and polished surfaces dominate. There is little evidence for any articulation or association of bones within the Gandhera sequence and there is no direct evidence of predator activity (bite marks, etching of teeth or bones due to gastrointestinal acids).

5.2. Distribution

Bones are scattered throughout the sequence with no dominant orientation. Within the calcarenite, bones are inter-laced and tightly juxtaposed with both horizontal (with respect to the bedding plane) and vertical orientations. The most substantial portion of the calcarenite layer is composed of bone, with only a small fraction made up of the silicified cementing matrix. Most bones found at Gandhera have nearly equivalent hydraulic properties (Voorhies Groups II and III, see Table 3). Somewhat larger elements are concentrated at the northern end of the quarry where sediments are coarser, and elements decrease in size southward as the sediments fine. Table 3 shows the distribution of bone elements by Voorhies group at Gandhera.

6. Summary

Much work remains to be done at Gandhera Quarry and any firm conclusions would be premature at this time. Table 4 summarizes sedimentary and taphonomic features at Gandhera as they are now understood. Based on these

Table 2. Ghazij Formation faunal list
(* - known from Gandhera Quarry; ** - only known from Gandhera Quarry)

<u>Upper part of upper Ghazij</u>
Primates
Adapiformes
1-2 taxa**
Perissodactyla
Brontotheriidae
3 taxa
Isectolophidae or Helaletidae
4 taxa*
Chalicotheriidae
1 taxon
New Family?
Large, bunodont taxon
Artiodactyla
Dichobunidae
3-4 taxa*
Rodentia
Ctenodactylidae
1-2 taxa**
Mesonychia
Mesonychidae
1 taxon
Creodonta
Hyaenodontidae
1-2 taxa**
Insectivora/Proteutheria
3-4 taxa**
<u>Lower part of upper Ghazij</u>
Condylarthra
Quettacyonidae
Sororocyon usmanii
Obashtakaia aeruginis
Perissodactyla
Families Uncertain
3-4 taxa (includes large, bunodont taxon)
Tillodontia
New Family
2-3 taxa
<u>Middle part of Ghazij Fm.</u>
Condylarthra
Quettacyonidae
Quettacyon parachai
Machocyon abbasi
Order Uncertain
Anthracobunidae?
Nakusia shahrigensis

FIGURE 6. Top—Artiodactyl dentitions, maxillae, and astragali of three sizes found at Gandhera Quarry; Bottom—Two perissodactyl dentitions and a pelvis in a block of Gandhera Quarry sandstone matrix.

features, a channel or splay depositional system is favored over an overbank floodplain interpretation. The fact that many of the bones at Gandhera are abraded and polished suggests that they were transported or reworked to some degree before being preserved. Their abundant preservation alongside reworked carbonate nodules suggests that they were hydraulically sorted and concentrated within an active (or reactivated) lag of a channel or splay. Once deposited, bones that were at the sand-shale contact were cemented and

Table 3. Gandhera quarry—voorhies dispersal groups (based on Voorhies, 1969; Dodson, 1973; Korth, 1979)

		GROUP (with element and number of specimens)		
Low Settling Velocity		⇔		High Settling Velocity
I	I/II	II	II/III	III
	Radius (56) Ulna (125) Scapula (59) Pelvis (128) Vertebrae (552)	Maxillae (104) Teeth (450) Humerus (312) Femur (555) Astragalus (208)	Tibia (467) Skull Frags. (65) Calcaneum (184)	Dentaries (579)
Totals By Group	920	1,629	716	579

hardened. Bones not encompassed by the calcarenite were cracked and broken further as strata were over-turned, producing the sequence now exposed at Gandhera.

The taxonomic composition of the mammalian assemblage at Gandhera will be important for interpreting the sequence of events that led to the diversification of many modern orders of mammals. The high diversity within

Table 4. Gandhera Quarry—sedimentary and taphonomic features (condition present at Gandhera in bold type; modified after Badgley, 1986)

	Expected	
Taphonomic/Sedimentary	Channel/Splay	Floodplain
Sediments coarsen laterally	**YES**	NO, OR LESS
Stratigraphic Distribution	**RESTRICTED HORIZONTALLY**	MORE HORIZONTAL DISPERSAL
Articulated skeletal elements	**NO**	OCCASIONALLY
Spatial Distribution	**SCATTERED**	SOME CLUSTERING
Dominant Voorhies Groups	**II-III**	I-III
Hydraulic Equivalence	**MOSTLY EQUIVALENT**	LESS EQUIVALENT
Tooth-Vertebra Ratio	**HIGH (Above 3.00 – Gandhera = 5.14)**	LOWER (Between 1-2)
Bone Damage	**ABRADED, POLISHED**	BROKEN, LESS ABRASION

families is surprising given the early Eocene age of the deposits. Much more study is required before these issues can be addressed in detail.

ACKNOWLEDGMENTS. We thank Mr. S. Hasan Ganhar, Director General of the Geological Survey of Pakistan, for encouraging and supporting the field program in the Ghazij Formation. Messrs. A. H. Usmani, Talib Hasan, and S. Ghazanfar Abbas were instrumental in the success of our field work and we gratefully acknowledge their support and help. Dr. William S. Sanders skillfully prepared all fossil material from Gandhera Quarry and Ms. Bonnie Miljour helped with preparation of figures. Field work in 1999 was funded by U. S. National Science Foundation grant EAR 9714923.

References

Afzal, J., 1996, Late Cretaceous to early Eocene foraminiferal biostratigraphy of the Rakhi Nala area, Sulaiman Range, Pakistan, *Pakistan Journal of Hydrocarbon Research, Islamabad* **8**:1–24.

Badgley, C., 1986, Counting individuals in mammalian fossil assemblages from fluvial environments, *Palaios* **1**:328–338.

Berggren, W. A., Kent, D. V., Swisher III, C. C., and Aubry, M.-P., 1995, A revised Cenozoic geochronology and chronostratigraphy, *SEPM Special Publication No. 54*, pp. 129–212.

Davies, L. M., 1941, The "Dunghan" limestone, and Ranikot beds in Baluchistan, *Geological Magazine* **78**:316–317.

Dodson, P., 1971, Sedimentology and taphonomy of the Oldman Formation (Campanian), Dinosaur Provincial Park, Alberta (Canada), *Palaeogeography, Palaeoclimatology, Palaeoecology* **10**:21–74.

Fritz, E. B., and Khan, M., 1975, Stratigraphy and paleontology of coal beds in the Ghazij Shale, Sor Range-Daghari coal field, Quetta Division, Pakistan, Project Report PK-15, *U. S. Geological Survey Open File Report* **75–274**:1–16.

Gingerich, P. D., Abbas, S. G., and Arif, M., 1997, Early Eocene *Quettacyon parachai* (Condylarthra), from the Ghazij Formation of Baluchistan (Pakistan): Oldest Cenozoic land mammal from South Asia, *Journal of Vertebrate Paleontology* **17**:629–637.

Gingerich, P. D., Arif, M., Hussain, I. H., and Abbas, S. G., 1998, First early Eocene land mammals from the upper Ghazij Formation of the Sor Range, Balochistan, in: *Siwaliks of South Asia* (M. I. Ghaznavi, S. M. Raza, and M. T. Hasan, eds.), Proceedings of the Third Geosas Workshop held at Islamabad, Pakistan, March 01-05, 1997, Geological Survey of Pakistan, Islamabad, pp. 1–17.

Gingerich, P. D., Arif, M., Khan, I. H., Clyde, W. C., and Bloch, J. I., 1999, *Machocyon abbasi*, a new early Eocene quettacyonid (Mammalia, Condylarthra) from the middle Ghazij Formation of Mach and Daghari coal fields, Baluchistan (Pakistan), Contributions from the Museum of Paleontology, University of Michigan **30**:233–250.

Ginsburg, L., Durrani, K. H., Kassi, A. M., and Welcomme, J.-L., 1999, Discovery of a new Anthracobunidae (Tethytheria, Mammalia) from the lower Eocene lignite of the Kach-Harnai Area in Baluchistan (Pakistan), *C. R. Acad. Sci., Paris* **328**:209–213.

Hardenbol, J. A., Thierry, J., Farley, M. B., Jacquin, T., de Graciansky, P.-C., and Vail, P. R., 1998, Mesozoic and Cenozoic sequence chronostratigraphic framework of European basins, in: *Mesozoic and Cenozoic Sequence Stratigraphy of European Basins* (P.-C. D. Graciansky, J. A. Hardenbol, T. Jacquin, and P. R. Vail, eds.), SEPM Society for Sedimentary Geology, Special Publication 60, pp. 3–13.

Jones, A. G., Manistre, B. E., Oliver, R. L., Willson, G. S., and Scott, H. S., 1960, *Reconnaissance Geology of part of West Pakistan* (Colombo Plan co-operative project conducted and compiled by Hunting Survey Corporation), Government of Canada, Toronto, 550 pp.

Korth, W. W., 1979, Taphonomy of microvertebrate fossil assemblages, *Annals of the Carnegie Museum* **48**:235–285.

Köthe, A., Khan, A. M., and Ashraf, M., 1988, Biostratigraphy of the Surghar Range, Salt Range, Sulaiman Range and the Kohat area, Pakistan, according to Jurassic through Paleogene calcareous nannofossils and Paleogene dinoflagellates, *Geologisches Jahrbuch, Hannover, Reihe B*, **71**:1–87.

Serra-Kiel, J., Hottinger, L., Caus, E., Drobne, K., Ferràndez, C., Jauhri, A. K., Less, G., Pavlovec, R., Pignatti, J., Samso, J. M., Schaub, H., Sirel, E., Strugo, A., Tambareau, Y., Tosquella, J., and Zakrevskaya, E., 1998, Larger foraminiferal biostratigraphy of the Tethyan Paleocene and Eocene, *Bulletin de la Société Géologique de France* **169**:281–299.

Voorhies, M. R., 1969, Taphonomy and population dynamics of an early Pliocene vertebrate fauna, Knox County, Nebraska, *University of Wyoming, Contributions to Geology, Special Paper No. 1*, pp. 1–69.

Warwick, P. D., Johnson, E. A., and Khan, I. H., 1998, Collision-induced tectonism along the northwestern margin of the Indian subcontinent as recorded in the upper Paleocene to middle Eocene strata of central Pakistan (Kirthar and Sulaiman Ranges), *Palaeogeography, Palaeoclimatology, and Palaeoecology* **142**:201–216.

Chapter 11

Paleoecology and Biostratigraphy of Marginal Marine Gulf Coast Eocene Vertebrate Localities

JAMES W. WESTGATE

1. Introduction

On-going vertebrate paleontologic research at Lamar University is focused on paleoecologic and biostratigraphic information provided by Eocene vertebrate remains preserved in marginal marine deposits across the Gulf Coastal Plain. Vertebrate faunas currently understudy were collected from outcrops in Texas, Arkansas, Alabama and Georgia.

Stratigraphic age assignments for Gulf Coast Paleogene stratigraphic units in this paper follow Dockery (1996). A latest Sabinian (latest early Eocene, early

JAMES W. WESTGATE • Department of Geology, Lamar University, Beaumont, TX 77710.

Eocene Biodiversity: Unusual Occurrences and Rarely Sampled Habitats, edited by Gregg F. Gunnell, Kluwer Academic/Plenum Publishers, New York, 2001.

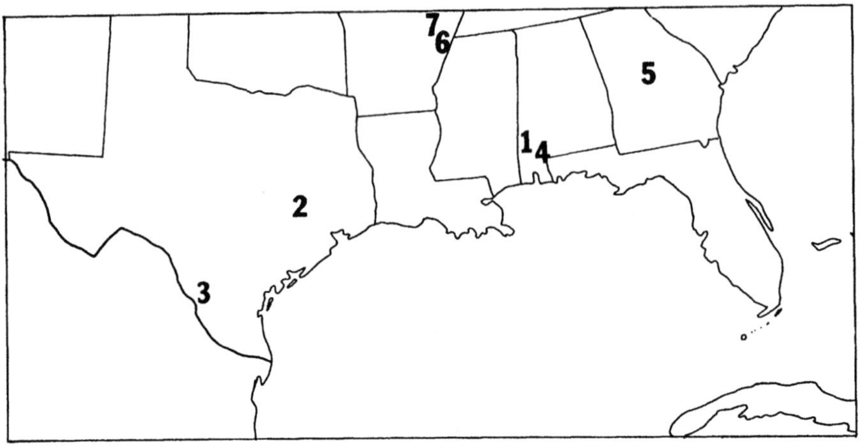

FIGURE 1. Eocene localities described in this paper: 1—Hatchetigbee Bluff, Washington County, Alabama; 2—Stone City Bluff, Burlson County, Texas; 3—Lake Casa Blanca, Webb County, Texas; 4—Little Stave Creek, Clarke County, Alabama; 5—Hardie Kaolin Mine, Wilkinson County, Georgia; 6—Crow Creek, St. Francis County, Arkansas; 7—Copperas Creek, Cross County, Arkansas.

Ypresian) fauna comes from the Hatchetigbee Formation, Wilcox Group. Specimens were collected from exposures at the formation's type locality at Hatchetigbee Bluff on the Tombigbee River, in Washington County, Alabama.

Two late Claibornian (late middle Eocene) faunas were collected from the Laredo and Stone City Formations, Claiborne Group, of Texas. The Laredo Formation specimens primarily come from a fossil quarry in estuarine deposits exposed in the spillway of Lake Casa Blanca near Laredo, in Webb County. The Whiskey Bridge site is at the type locality of the Stone City Formation where open marine deposits are exposed at Stone City Bluff on the Brazos River in Burlson County. A latest Claibornian (latest middle Eocene) fauna was collected from near-shore deposits of the Gosport Sand formation which crop-out on Little Stave Creek, in Clarke County, Alabama.

Sites in Arkansas and Georgia have yielded Jacksonian age (latest middle to late Eocene) vertebrates. Georgia specimens were recovered from near-shore deposits of the Clinchfield Formation (latest middle or earliest late Eocene, late Bartonian), Barnwell Group, through exposure of this deposit in the Hardie Kaolin Mine, near Gordon, in Wilkinson County. The Arkansas specimens were collected from estuarine Jackson Group (undifferentiated) outcrops on Crowley's Ridge, about 80 km west of Memphis, Tennessee. These exposures lie on Crow Creek near Forrest City, in St. Francis County, and along Copperas Creek near Wynne in Cross County.

Five of the seven faunas include at least one land mammal species. Land mammal specimens are extremely rare in Gulf Coast Eocene deposits. Only two early Eocene (Beard and Tabrum, 1991; Sheldon and Carpenter, 1997), four middle Eocene (Berry, 1924; Gazin and Sullivan, 1942; Thurmond and Jones,

1981; Westgate, 1988, 1990), and three late Eocene (Slaughter, 1978; Domning *et al.*, 1982; Westgate and Emry, 1985) Gulf Coast land mammal localities have been reported. Of these nine localities, six yielded a single land mammal specimen, while two produced only two specimens each. The middle Eocene Lake Casa Blanca locality is a spectacular exception having produced more than 1600 land mammal specimens which belong to at least 28 species.

North America's Eocene continental margin deposits are primarily dated using marine invertebrates. Eocene continental interior deposits are primarily dated using land mammal remains which are the basis for the North American Land Mammal Ages. Discoveries of land mammal remains in Gulf Coast marine deposits provide rare chances for direct biostratigraphic correlation of classic marine stratigraphic units with continental interior deposits bearing land mammals. Less than 20 years ago, it was thought that Uintan events in the North American interior were coincident with late Eocene Jackson Group deposition, while Chadronian faunas and Chadronian geologic events were thought to have occurred after Jacksonian time. Improved correlations have resulted in better synchronized time scales. It is now recognized that Uintan age volcanoes in Trans-Pecos Texas contributed volcaniclastic sediments to Claiborne sea deposits during middle Eocene time, while Chadronian land mammals lived concurrently with the giant archaeocete whales of the Jackson Sea during late Eocene time. These improved correlations allow us to better understand the timing of geologic processes in the North American interior and their role in the creation of the extensive deposits on the coastal plain. This knowledge will help us to better find and utilize the vast geologic resources stored in coastal plain deposits.

Although the remains of lower vertebrate species do not allow the precision in biostratigraphic correlations provided by mammals, they are an excellent source of paleoecologic data. The most common vertebrate remains from the fossil localities in this study belong to chondrichthyan and osteichthyan fishes. Most sites have also yielded reptilian remains. Aquatic species are in constant contact with the water they live in and many have a stenothermal and/or stenohaline physiology. Ichthyologists have established a detailed data base of the modern habitats of many living genera whose geologic ranges extend back to Eocene time. Many of these genera have a narrow range in habitat because of diet, behavioral, and physiological constraints. Fossil remains of these restricted habitat species serve as paleothermometers, paleosalinometers and paleobathymeters and allow detailed paleoecologic reconstructions of past environments to be made.

The following abbreviations are used: APL = antero-posterior length, AW = anterior width, dp = deciduous lower premolar, m = lower molar, M = upper molar, NALMA = North American Land Mammal Age, p = lower premolar, P = upper premolar, PW = posterior width, sec. = section, TMM 11111-1 = Texas Memorial Museum vertebrate fossil locality number-specimen number from that locality, USNM 1 = United States National Museum vertebrate fossil specimen number.

2. Wilcox Group

2.1. Hatchetigbee Formation

2.1.1. Hatchetigbee Bluff Local Fauna

Hatchetigbee Bluff on the Tombigbee River is the type locality of the Hatchetigbee Formation (Toulmin, 1977). The bluff lies on the west side of the river at navigational mile 107 in Washington County, Alabama. This is about 13 km downstream of the U.S. highway 84 bridge in the nw 1/4, nw 1/4, nw 1/4, of sec. 21, T. 8 N., R. 1 W. The Hatchetigbee Formation is the youngest formation in the Wilcox Group and is latest Sabinian (early Eocene, early Ypresian) in age. About 21 m of glauconitic, marly sands are exposed on the bluff. A ?*Crassostrea* sp. oyster reef and shells of *Venericardia hatcheplata* with articulated valves are exposed at the base of the bluff. One meter above the oyster bed is a mass mortality layer of articulated *V. hatcheplata* in life position. D. George, A. Sheldon and J. Westgate measured the stratigraphic section at the bluff and took bulk samples from four mollusc-rich horizons in 1997.

A partial left mandible bearing partial dp3 and 4 and a complete m1 from a small and strongly bilophodont species of *Coryphodon* was discovered just below the oyster bed by Mr. J. Gilley, owner of the bluff. The m1 displays only slight occlusal wear and dental x-rays indicate that the crown of m2 had not formed at the time of death. The specimen has been placed in the paleontologic collections of the University of Alabama and its species level identification is in progress. This is the only early Eocene land mammal known from Alabama and it is the easternmost occurrence of the genus in North America. Fragmentary remains of *Coryphodon* have been found in the late Paleocene, early Wasatchian age, Tuscahoma Formation at the Red Hot Truck Stop locality in Meridian, Mississippi (Beard *et al.*, 1995). The next closest occurrence of *Coryphodon* is *C. armatus* from the Hannold Hill Formation in Big Bend National Park, Texas (Lucas, 1989). Lucas (1989) assigned the Big Bend specimens to *C. molestus*, but Uhen and Gingerich agreed with Osborn (1898) and synonymized *C. molestus* with *C. armatus*.

Measurements of the Hatchetigbee *Coryphodon* m1 are APL 29 mm, AW 20 mm, and PW 18 mm, with a jaw depth of 52 mm below the m1. Sheldon and Carpenter (1997) reported that the size differential between *C. armatus* and the Hatchetigbee specimen is not significant. The m1 of *C. armatus* from the Hanold Hill Formation has an APL range of 23.7–24.5 mm, with a mean of 24.2 mm; AW ranging from 17.4–18.3 mm, with a mean of 17.8 mm; and a PW range of 17.8–18.6 mm, with a mean of 18.2 mm (individual measurements from Lucas, 1989 with recalculated means). In northwestern Wyoming, *C. armatus* m1's range in APL from 23.1–28.9 mm, while their widths range from 15.6–19.7 mm (Uhen and Gingerich, 1995). However, the m1 measurements of three other northwestern Wyoming *Coryphodon* species reported by Uhen and Gingerich (1995); *C. proterus* (APL 27.7–29.9 mm, W 19.5–23.1 mm), *C.*

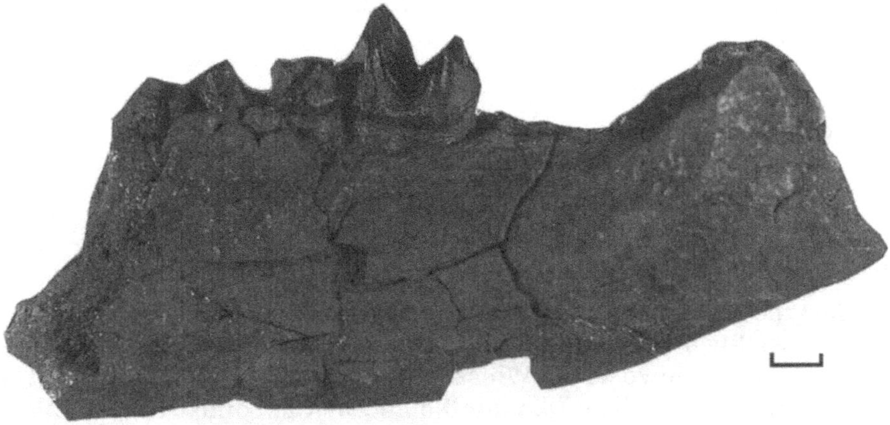

FIGURE 2. Partial left mandible of *Coryphodon* sp. with a partial dp4 and m1 from Hatchetigbee Bluff, Alabama. Bar scale = 1 cm.

radians, (APL 27.3–29.3 mm, W 16.6–21.3 mm), and *C. lobatus* (APL 29.1–32.8 mm, 20.9–22.8 mm) indicate that all are of comparable size with the Hatchetigbee specimen. Species level assignment of the Hatchetigbee Bluff *Coryphodon* must await detailed morphologic comparison with all of these species of *Coryphodon* (Westgate and Sheldon, in prep.). Even then, taxonomic assignment below the generic level may not be possible.

Sheldon and Carpenter (1997) incorrectly reported that the specimen was collected from the Bashi Marl Member of the Hatchetigbee Formation. There is no Bashi Marl exposed at Hatchetigbee Bluff, as all of the Hatchetigbee beds there belong to the upper part of the formation (Toulmin, 1977). Sheldon and Carpenter (1997) also incorrectly described the teeth in the Hatchetigbee specimen as being partial p4 and m1, with a complete m2.

The presence of *Coryphodon* in both the upper Tuscahoma Formation and the upper Hatchetigbee Formation indicates that these units in the Wilcox Group are correlative with strata in the western interior of North America which yield Clarkforkian and Wasatchian land mammal faunas. The strongly bilophodont cusps of the Hatchetigbee Bluff *Coryphodon* suggest a Wasatchian age for the upper Hatchetigbee Formation. Although it is possible that the chronologic range of *Coryphodon* in southeastern North America lasted longer than it did in the interior. The Alabama coastal setting could have served as a refugium from whatever environmental changes caused the extinction of *Coryphodon* in western North America. In the absence of associated land mammal taxa this cannot be determined.

The presence of the oyster reef and the low invertebrate species diversity in the beds at the base of Hatchetigbee Bluff may be indicative of estuarine depositional conditions. If so, the sediment-trap nature of the estuarine basin would be an excellent environment in which to bury additional land mammal specimens.

3. Claiborne Group

3.1. Stone City Formation

3.1.1. Whiskey Bridge Local Fauna

The type section of the Stone City Formation is exposed on the south bank of the Brazos River in Burlson County, Texas where Texas Highway 21 crosses the river. This is known as the Whiskey Bridge locality (TMM 42986). The formation lies in the middle of the middle Eocene Claiborne Group and laterally interfingers with the upper portion of the Sparta Sand formation and disconformably underlies the Cook Mountain Formation (Stenzel *et al.*, 1957). In 1996, about 100 kg of bulk sample was collected from the Main Glauconite bed, Bed "S," of Stenzel *et al.* (1957). The presence of solitary hexacorals, and the predominance of glauconite and other clays in this bed indicate it formed in a low energy, open marine environment. Remains from 10 vertebrate species were recovered by screen-washing (George and Westgate, 1998). Five shark species include *Striatolamia macrota* (sand tiger shark), TMM 42986-1; *Abdounia enniskilleni* (requiem shark), TMM 42986-2; *A. recticona* (= "*Galeorhinus recticonus claybornensis*" White, 1956), TMM 42986-3; *Galeocerdo latidens* (tiger shark), TMM 42986-4;, and *Negaprion gibbesi* (lemon shark), TMM 42986-5. Four teleosts include an ?ariid catfish, the drum *Jefitchia* sp., *Sphyreana* sp. (barracuda), and *Trichiuris* sp. (cutlassfish). This is a warm water fish assemblage. The reassignment by Cappetta (1987) of "*Galeorhinus recticonus*" to the carcharhinid, *Abdounia recticona*, could be environmentally significant, as the living *Galeorhinus galeus* inhabits cool waters in addition to tropical seas. *Abdounia* is an extinct genus and its paleoenvironmental restrictions are not clear. The clayey nature of the deposit and the lack of terrestrial species, may indicate that the deposit formed well offshore.

3.2. Laredo Formation

3.2.1. Lake Casa Blanca Local Fauna

The Spillway Quarry, TMM 42486, has yielded more than 1600 land mammal specimens and is the main source of specimens comprising the Casa Blanca local fauna. The quarry is located in the spillway of Lake Casa Blanca in Casa Blanca State Park, just east of the Laredo city limits in Webb County, Texas. This site lies stratigraphically in the middle of the Laredo Formation, Claiborne Group (Westgate, 1988a, 1988b). Fifteen tons of bulk sample were excavated from an up to 1 m thick, 14 m long and up to 6 m wide lens of sandy, oyster-shell hash. Samples were excavated using small hand tools and macro-

A

B

FIGURE 3. Middle Eocene Laredo Formation exposures in Lake Casa Blanca State Park. A. Outcrops in the Lake Casa Blanca spillway, view looking northeast. The "X" marks the location of the TMM 42486 vertebrate fossil locality. B. Close-up view of TMM 42486 with burlap bags being filled with bulk sample.

specimens were collected during this process. The sample was then triple-screenwashed with the finest screen having a 500 micron mesh. The small and medium screen fractions were then acidized with a solution of 10% acetic acid and re-screenwashed. Samples were picked for micro-vertebrates using binocular microscopes.

3.2.1.1. Fauna and Flora. Fifty-nine species of vertebrates, and more than 15 invertebrate and plant species comprise the Casa Blanca fossil community (Table 1, Westgate, 1994a). The marine invertebrates are late middle Eocene in age. Most of the larger land mammals closely resemble those in two Uinta C NALMA faunas in Trans-Pecos, Texas; the Serendipity local fauna from the Big Bend region and the Candelaria local fauna from the Vieja region (Westgate, 1990). At least eight Casa Blanca mammal species may have been endemic to the Laredo area, including two primates, one artiodactyl and five rodents. The apparent endemism of these new species is probably partly due to the paucity of middle Eocene coastal plain land mammal localities.

Nine mouse-sized mammal species include micro-opossums, shrews, and rodents. Walton (1993a, 1993b) described two new, small rodent species, *Pauromys simplex* and *Microeutypomys karenae*. One mouse-sized rodent, *Laredomys riograndensis*, is the sole species from a new family, the Laredomyidae (Wilson and Westgate, 1991). A single tooth (42486-1323) documents the oldest occurrence of cf. *Nonomys* sp.

Four large mammals include the amynodont rhinoceros, *Amynodon advenus*, a titanothere, a mesonychid and a sirenian. Although the sirenian specimen from the Spillway Quarry is a single vertebral fragment, more complete remains of at least four individuals were discovered in Laredo Formation deposits exposed south of Lake Casa Blanca at TMM localities 41843 and 42985 (Westgate, 1990). These are the only Tertiary-age Gulf Coastal Plain sirenian specimens known west of Alabama and north of the Yucatan Peninsula.

The majority of the mammalian species (16) fall in the size range between modern jack rabbits and wolves. The most common of these are specimens from the selenodont artiodactyls *Protoreodon petersoni*, *P. parvus*, *Leptoreodon pusillus* and *L. leptolophus*. Remains of *Epihippus* cf. *E. gracilis* are also common and include two nearly perfect mandibles (TMM 42486-220, and-1600) with left and right rami and nearly complete dentition. *Laredochoerus edwardsi* is a new genus and species of leptochoerid artiodactyl (Westgate, 1994b). Isolated teeth of *Omomys* and *Mytonius* (or its synonyms) appear to be from new primate species (Beard and Westgate, in prep.).

Only three bird specimens have been recovered. One is a metatarsus from a large crane-like species similar to *Paragrus*. Reptilian remains come from at least 11 species of snake, crocodiles, turtles and lizards. *Pterosphenus schucherti* specimens from the Casa Blanca quarry are the only middle Eocene record of this giant "sea snake" and its only record from Texas. Crocodilians are represented by isolated skeletal elements, teeth from *Allognathosuchus* sp., and teeth and a partial cranium from *Pristichampsus vorax*. Busbey (1989) identified another partial cranium (42486-625) as that of the oldest known true

Table 1. List of fossil species in the Casa Blanca local community (TMM 42486, 42582)

GROUP (# of species)	SCIENTIFIC NAME	COMMON NAME
PLANTAE		
	Spinizonocolpites prominatus	mangrove palm
	Selaginella sp.	epiphytic lycophyte
	Bombacacidites paulus	rain forest/swamp tree
	Lanagiopollis eocaenica	tropical tree
	Paraoreomunea sp.	tropical tree
	Leea sp.	tropical climbing shrub
	Lygodiumsporites adriennis	tropical fern
	Pteris dentata	tropical fern
INVERTEBRATA		
	?*Brachidontes* sp.	mussel
	Crassostrea amichel	oyster
	Turritella cortezi	snail
	naticid	carnivorous snail
	decapod sp. 1	mud lobster
	decapod sp. 2	crab
	Bythocypris sp.	ostracod
	ostracod spp.	8 ostracod spp.
CHONDRICHTHYES		
	Striatolamia macrota	sand shark
	Carcharhinus sp.	requiem shark
	Galeocerdo eaglesomei	tiger shark
	Pristis sp.	sawfish
	Rhinobatis sp.	guitarfish
	Dasyatis sp.	stingray
	Myliobatis sp.	eagle ray
	Rhinoptera sp.	cownose ray
OSTEICHTHYES		
	Lepisosteus sp.	gar
	megalopid	tarpon
	cf. *Paralbula marylandica*	bonefish
	cf. *Arius* sp.	sea catfish
	Diaphyodus wilsoni *	wrasse
	Jefitchia sp.	drum
	Cylindracanthus sp.	billfish
AMPHIBIA		
	anuran	frog
REPTILIA		
	cf. *Baptemys* sp.	river turtle
	cf. *Allaeochelys* sp.	river turtle
	Trionyx sp.	softshell turtle
	Xenochelys sp.	musk turtle
	emydid	pond turtle
	Hadrianus sp.	land tortoise
	Allognathosuchus sp.	crushing-toothed alligator
	Pristichampsus cf. *P. vorax*	semi-terrestrial crocodile
	crocodylid *	caiman
	glyptosaurine	lizard
	Pterosphenus schucherti	giant "sea" snake
AVES		
	gruiform	crane-like bird
	avian sp. 1	small ?water bird
	avian sp. 2	quail-sized bird

(*continued*)

Table 1. (Continued)

MAMMALIA		
	Peratherium comstocki	micro-opossum
	Peratherium marsupium	micro-opossum
	cf. *Simidecetes magnus*	insectivore
	Centetodon pulcher	insectivore
	Omomys or ?*Mytonius* new sp.*	tarsier-like primate
	?*Mytonius* new sp.*	tarsier-like primate
	Microsyops sp.	primate
	mesonychid	primitive carnivore
	Hyopsodus cf. *H. uintensis*	condylarth
	cf. *Uintacyon* sp.	carnivoran
	Procynodictis cf. *P. vulpiceps*	carnivoran
	sirenian	sea cow
	Epihippus cf. *E. gracilis*	four-toed horse
	Amynodon advenus	rhinoceros
	?*Triplopus* sp.	rhinoceros
	brontothere	titanothere
	*Laredochoerus edwardsi***	artiodactyl
	Protoreodon petersoni	artiodactyl
	Proreodon parvus	artiodactyl
	Leptoreodon pusillus	artiodactyl
	Leptoreodon leptolophus	artiodactyl
	Toromeryx marginensis	artiodactyl
	Mytonomys new sp.*	large rodent
	cylindrodontid**	meso- rodent
	Microparamys sp.	micro-rodent
	*Laredomys riograndensis****	micro-rodent
	*Pauromys simplex**	micro-rodent
	*Microeutypomys karenae**	micro-rodent
	cf. *Nonomys* sp.	micro-rodent

*= new species; **= new genus and species; ***= new family, genus and sp.

caiman from North America. Turtle shell fragments are common and indicate a high diversity of species in the community. Shell fragments from cf. *Baptemys, Xenochelys, Hadrianus,* cf. *Allaeochelys, Trionyx* and an emydid have been recognized (Westgate, 1989).

While shark teeth are abundant in the fossil quarry, the shark fauna was depauperate with only three species of sand tiger (*Striatolamia macrota*), tiger (*Galeocerdo eaglesomei*), and a requiem shark (*Carcharinus* sp.), present. It is notable that the sand tiger shark, *Striatolamia macrota,* which has large and distinctively grooved teeth, is not found in Gulf Coastal sediments younger than Claibornian in age. *Striatolamia* appears to have been replaced by an ecologically similar, but smooth-cusped, sand tiger species, *Carcharias hopei* by Jacksonian time, as indicated in the following sections on faunas from Arkansas and Georgia. The geographic distribution of living sharks indicates that the late Claiborne Sea probably had temperatures similar to today's Caribbean Sea, while the Jackson Sea's temperature range was more like that of the modern Gulf of Mexico. The disappearance of *Striatolamia macrota* from the Gulf of Mexico, and the appearance of large whale and cool-water shark populations indicates that Gulf water temperatures dropped as the Earth

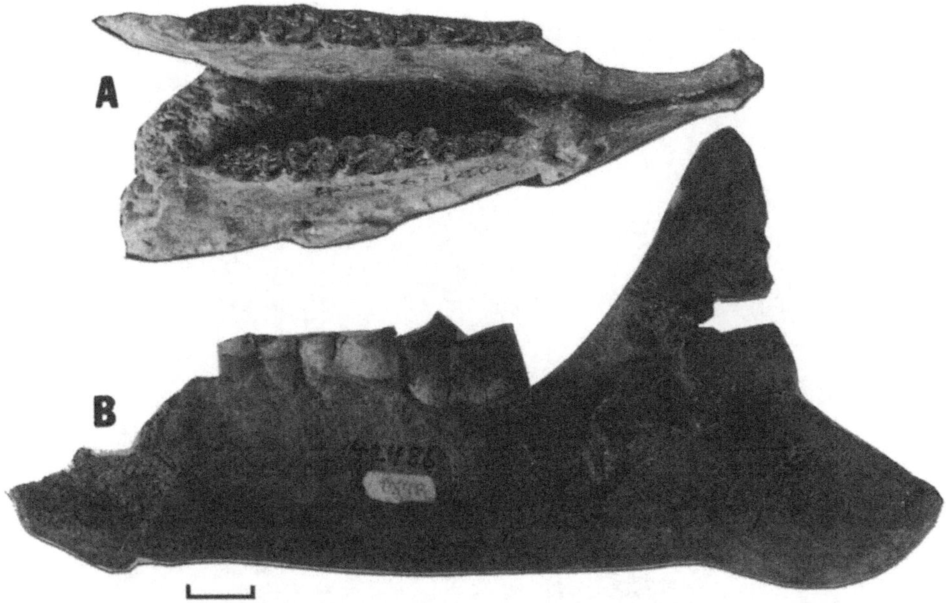

FIGURE 4. Mandibles from the Casa Blanca quarry. A. Nearly complete mandible (TMM 42486-1600) of *Epihippus* cf. *E. gracilis* with left p2–4, m1–3 and right p3–4. B. Left mandible (TMM 42486-1548) from an *Amynodon advenus* calf with dp3–4 and newly erupted m1, m2 crown formed in crypt, but unerupted. Bar scale = 1 cm for A., 2 cm for B.

shifted from its ice cap-free "Greenhouse" climate to the beginnings of today's "Icehouse" climate.

The Casa Blanca batoids include stingrays (*Dasyatis* sp.), eagle rays (*Myliobatis* sp.), cownose rays (*Rhinoptera* sp.), guitarfish (*Rhinobatis* sp.), and sawfish (*Pristis* sp.). There are seven species of osteichthyans. Gar (*Lepisosteus* sp.) scales, teeth and vertebrae are extremely common. Several dentaries (42486-620) represent the oldest record of the tarpon family in North America. A new wrasse species, *Diaphyodus wilsoni*, is the first reported occurrence of this well known European genus in North America. The drum fish, *Jefitchia* sp., is documented by a single otolith (ear stone).

A diverse flora is indicated by pollen, fruits, leaves and wood. The microflora is composed of pollen and spores from six species of tropical lowland forest trees, ferns, a climbing shrub, and an epiphytic club moss (Westgate and Gee, 1990; Gee and Westgate, 1990). The most paleoecologically significant of these is pollen of an ancient relative of the living mangrove-forming palm, *Nypa fructicans*. Living *Nypa* form a mangrove ecotone in the inter-tidal zone of Old World estuaries in wave-protected areas having brackish water with temperatures 24°C (75°F) or higher. The Casa Blanca deposits are unique in preserving evidence of the only known *Nypa* mangrove facies on this continent. Banana-sized fruits (seeds) found in mudstones lateral to the vertebrate

FIGURE 5. In-situ *Nypa* sp. fruit in mudstone lateral to the TMM 42486 quarry. Coin scale is 2.5 cm in diameter.

quarry resemble those of *Nypa burtini* which are well known from the early Eocene London Clay (England) and Brussels Sand (Belgium) deposits. These fruits and other macroflora are currently understudy (Dilcher *et al.*, in prep.).

Invertebrate remains locally include oyster reefs in life-position comprised of articulated and disarticulated valves of *Crassostrea amichel* at TMM localities 42582 and 42587. The marine guide fossil *Turritella cortezi*, was found in a sandstone bed exposed below the Lake Casa Blanca dam at TMM 42587. Giant mud lobster burrows underlie the reef at TMM 42582. External casts of the brackish water mussel ?*Brachidontes* are common in leaf-rich mudstones lateral to the vertebrate quarry.

Ostracods associated with the *Crassostrea amichel* reef were recovered from TMM 42582 (Borne, 1993). Nine species from six genera were identified in a sample of more than 500 complete carapaces. Three phractocytherid species, including *Phractocytheridea compressa*, make up 94% of the sample. *Neocyprideis* sp., *Bairdia* sp., and *Bythopypris* spp. and three haplocytherid species are also present (Borne, 1998, written comm.).

3.2.1.2. Geologic Age. The Uintan NALMA is partially defined by the evolutionary appearance of *Epihippus* and the invasion of North America by *Amynodon*. The presence of remains of these two genera in the Lake Casa Blanca beds indicate that the fauna is Uintan in age. The suite of selenodont artiodactyls with the rodent *Mytonomys* n. sp, allows refinement of the age assignment to Uinta C. The Casa Blanca fauna closely resembles the Serendipity local fauna of the Big Bend region of Texas. Biotite-bearing ashes above and below the Serendipity beds yielded K-Ar dates of 42.7 and 43.9 MA (Wilson, 1986).

Stenzel (1940) used the marine gastropod *Turritella cortezi* to correlate Laredo Formation beds of the Rio Grande Embayment with beds in the middle Eocene Landrum Member of the Cook Mountain Formation exposed at Alabama Ferry on the Trinity River of Texas. Ghosh (1972) reported a K-Ar average date of 42 MA for bentonites from that outcrop. Recalibration of that date is 42.2 MA (Westgate, 1988a). Nannoplankton from the Cook Mountain Formation are assigned to the upper half of Zone NP 16 (Dockery, 1996). This part of NP 16 has generally been correlated with the lower Bartonian Stage of Europe (Harland *et al.*, 1989; Dockery, 1996), however, Martini (1986) and Speigler (1986) correlated it with the upper part of the Lutetian Stage. The best age assignment for the Casa Blanca local fauna is middle Eocene, late Lutetian or early Bartonian, 42–43 MA.

3.2.1.3. Paleoecology. Comparison of modern geographic ranges and environmental requirements of taxa related to extinct Casa Blanca species provides a basis for the paleoecologic reconstruction of the coastal habitat in which the community lived. Three environmental factors; temperature, rainfall and salinity, are addressed.

Three of the turtle groups present have very restricted modern habitat distributions. Carrettochelyids today are found only in the coastal tropics of southern New Guinea and Northern Territory, Australia. Dermatemyid and staurotypine turtles range in the New World tropics from southern Mexico to Guatemala. The mangrove palm, *Nypa fruticans*, requires tropical waters 24°C (75°F) or warmer (Westgate and Gee, 1990). These modern geographic ranges of *Nypa* and the three turtle groups suggest that the coastal areas of southern New Guinea/Northern Territory, Australia and the Caribbean coasts of the Yucatan Peninsula, Mexico and northern Central America are realistic models

FIGURE 6. Artistic reconstruction of the middle Eocene Casa Blanca mangrove community by A. Salazar (from Westgate, 1999; reproduced by permission of Texas Parks and Wildlife Press).

for the climatic regime which prevailed in the Rio Grande Embayment when the Casa Blanca community inhabited it.

Although *Nypa* is an excellent paleothermometer, it is a poor paleo-rain gauge because its intertidal habitat is not dependent on precipitation rates. The presence of ferns and tropical epiphytes in the paleobotanical community suggests that rainfall amounts in the Laredo area were significantly higher than that which supports today's chaparral flora. Average annual rainfall at Laredo now is 55 cm (21 in).

Living reef-forming *Crassostrea virginica* prefer salinities averaging 20 parts per thousand. The low ostracod diversity at TMM 42482 and the occurrence of closely related living ostracods in brackish water habitats suggests hyposaline depositional conditions (P. Borne, 1998, written comm.). *N. fructicans* and the mussel *Brachidontes* today also prefer brackish, estuarine waters. The association of fossil *Crassostrea amichel* reefs, *Brachidontes* sp. and the ostracod fauna, in a *Nypa* mangrove habitat, indicates water salinity was brackish.

The Casa Blanca community has the greatest diversity of terrestrial and aquatic plants and animals of any Paleogene vertebrate fossil locality on the North American Coastal Plain. This complex community was comprised of species from freshwater, estuarine, mangrove, marine and lowland forest habitats. The climate in the Rio Grande Embayment was tropical and probably much wetter when the Casa Blanca community inhabited it, than it is today. Several species, including *Spinizonocolpites prominatus, Galeocerdo eaglesomei, Diaphyodus wilsoni* and cf. *Allaeochelys* sp., appear more closely related to Old World taxa than to fossil groups known from the Eocene of western North America. This indicates a zoogeographic dispersal route continued to exist along the coastal remnants of the Tethys Sea during middle Eocene time.

3.3. Gosport Sand

3.3.1. Little Stave Creek Local Fauna

Little Stave Creek is a deeply incised tributary of the Tombigbee River lying 5 km north of Jackson in sec. 19 and 20, T. 7 N., R. 2 E., in Clarke County, Alabama. The stratigraphic section was described by Gardner (1957). Extremely fossiliferous beds of the Gosport Sand, the youngest formation in the Claiborne Group, are exposed along the creek banks. Dockery (1996) correlated the Gosport Sand with medial Bartonian age beds in Great Britain. Locally the Gosport Sand is about 7 m thick and was deposited in a near-shore environment (Toulmin, 1977). Thurmond and Jones (1981) reported the discovery of a partial m1 from *Vulpavus* sp. at this site. In 1994, a Lamar University crew bulk sampled about 250 kg from the basal 0.5 m thick glauconitic sand (TMM 43412) and screen-washed it for vertebrate remains. Many of the recovered teeth are heavily worn from transportational abrasion and are not identifiable to generic level. Although this formation has been intensively sampled for several decades, there is no evidence of the presence of archaeocete whales in this community. Unworn *Striatolamia macrota* teeth clearly display striations, while those whose roots and cusps show abrasive wear have lost their striations and might be confused with teeth of *Carcharias hopei*. The taxa listed with the greatest to least number of specimens recovered are, *Negaprion gibbsi*, TMM 43412-4 (117); *Striatolamia macrota*, TMM 43412-1 (60); *Galeocerdo latidens*, TMM 43412-2 (13); *Hemipristis curvatus*, TMM 43412-3 (4); and *Abdounia recticona* (= "*Galeorhinus recticonus claybornensis*" White, 1956), TMM 43412-5 (4). This is a warm water shark assemblage. That interpretation agrees with the paleoecologic conclusions which Hansen (1987) made about the Gosport Sand molluscan community which he found to be dominated by subtropical genera. Although no land mammals were recovered in our sample, the sandy nature of the deposit indicates it formed in a near-shore or beach facies (Gardner, 1957), and there is a high potential for discovering additional land mammal remains at this site.

4. Barnwell Group

4.1. Clinchfield Formation

4.1.1. Hardie Kaolin Mine Local Fauna

The Hardie Mine local fauna consists of a diverse near-shore marine community predominated by chondrichthyans, with a few species of reptiles, cetaceans and osteichthyans. The coarse, well sorted sands, crocodilian bone

FIGURE 7. North wall of the Hardie Kaolin Mine in Wilkinson County, Georgia. The white kaolin at the base is unconformably overlain by the one meter thick coarse sand of the Clinchfield Formation at the bottom of the Barnwell Group exposures. The Twiggs Clay conformably overlies the Clinchfield Formation. CF—Clinchfield Fm., K—kaolin, TC—Twiggs Clay.

fragments and a land mammal tooth fragment, indicate that the shoreline was close by. The most recently excavated portion of the open pit kaolin mine (TMM 43390) lies 3.2 km northwest of the Gordon, Georgia post office in Wilkinson County. Kaolin mining here exposed a one meter thick layer of fossiliferous Clinchfield Formation sands. The Clinchfield Formation is the basal unit of the Barnwell Group and its coarse sands were deposited as the Jackson Sea transgressed across central Georgia either early in the late Eocene (Huddlestun and Hetrick, 1986; Huddlestun, 1993) or at the end of the middle Eocene (Dockery, 1996). Dockery (1996) indicated that the initial Jackson Sea transgression occurred in Mississippi during the late Bartonian. The Hardie mine lies about 15 km southeast of exposures of metamorphic crystalline rocks of Georgia's Piedmont Region where the Jackson Sea completed its transgression inland. In the mine walls, Clinchfield Formation sands unconformably overlay non-fossiliferous kaolin clay and underlie the locally non-fossiliferous Twiggs Clay. Mining procedures called for the Clinchfield Formation sands to be placed on top of the spoil piles, where through rapid weathering, vertebrate fossils were left on the spoil surfaces as lag deposits. Because the other formations exposed in the pit are non-fossiliferous, although the vertebrate remains are not in situ, it is possible to determine their original stratigraphic position to within the one meter locally occupied by the Clinchfield Formation. The kaolin excavation is complete and the mine site is awaiting reclamation.

4.1.1.1. Vertebrate Fauna. More than twenty-seven vertebrate species in the Hardie Mine local fauna have been identified from several thousand teeth and bone fragments. This mine is a favorite collecting locality for local scouting and school groups who have hauled away many pounds of specimens which undoubtedly could have added to the size of the faunal list. The fauna was first reported by Westgate, Gillette and Rolater (1994) and the paleocommunity structure was described by Riley and Westgate (1999). Eleven species of sharks, five batoids, five bony fish, three reptiles, and three mammalian species are preserved in the TMM collections (Table 2). Cappetta (1987) reassigned Case's (1981) "*Galeorhinus huberensis*" to *Physogaleus* sp. That assignment is used here.

It is noteworthy for biostratigraphic purposes that the sand tiger shark *Carcharias hopei* appears to have replaced *Striatolamia macrota* in the sand tiger shark niche with the transgression of the Jackson Sea across the Gulf Coastal Plain. We collected 812 *C. hopei* teeth and no *S. macrota* teeth. Only teeth displaying little transportational wear were counted to prevent contamination of the fauna through reworking of older fossils.

Table 2. The Hardie Mine local fauna (TMM 43390)

CLASS	SPECIES	COMMON NAME	SPECIMENS
CHONDRICHTHYES			
	Ginglymostoma serra	nurse shark	13 teeth
	Carcharias hopei	sand tiger shark	922 teeth
	Carcharodon auriculatus	great white shark	2 teeth
	Isurus sp.	mako shark	4 teeth
	Lamna sp.	porbeagle	124 teeth
	Abdounia enniskilleni	carcharhinid	492 teeth
	Galeocerdo latidens	tiger shark	219 teeth
	Hemipristis curvatus	snaggletooth shark	170 teeth
	Negaprion gibbesi	lemon shark	1538 teeth
	Physogaleus sp.	carcharhinid	19 teeth
	Squatina sp.	angel shark	1 tooth
	Propristis schweinfurthi	sawfish	238 rostral teeth
	Raja sp.	skate	4 scutes
	Dasyatis sp.	stingray	53 teeth
	Myliobatis sp.	eagle ray	8 dental battery, 284 teeth
	Rhinoptera sp.	cownose ray	67 teeth
OSTEICHTHYES			
	Lepisosteus sp.	gar	33 teeth
	Sphyraena sp.	barracuda	335 teeth
	Trichiuris sp.	cutlassfish	27 teeth
	Phyllodus sp.	none	1 dental battery
	teleost	bony fish	242 otoliths
REPTILIA			
	Trionyx sp.	softshell turtle	17 shell frags.
	crocodylid	crocodile	25 teeth, 10 bone frags.
	Pterosphenus schucherti	giant aquatic snake	14 vertebrae
MAMMALIA			
	Basilosaurus cetoides	ancient whale	18 vertebral frags.
	Zygorhiza kochii	ancient whale	4 teeth, 1 cervical
	brontotheriid	titanothere	1 tooth frag.

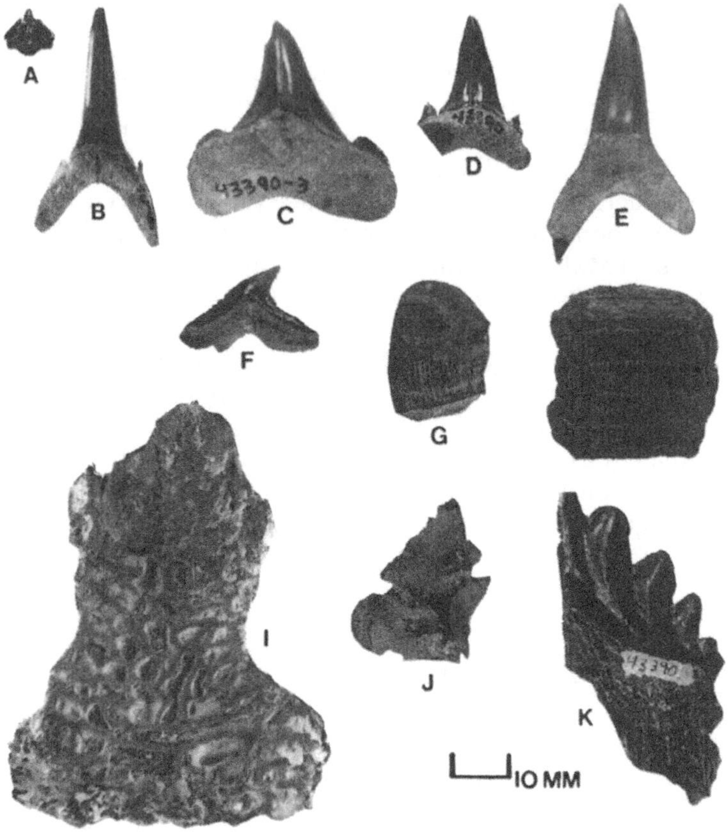

FIGURE 8. Vertebrate remains from TMM 43390, the Hardie Kaolin Mine. A. *Ginglymostoma serra*, tooth, 43390-1; B. *Carcharias hopei*, tooth, 43390-2; C. *Carcharodon auriculatus*, tooth, 43390-3; D. *Lamna* sp., tooth, 43390-4; E. *Isuris* sp., tooth, 43390-5; F. *Galeocerdo latidens*, tooth, 43390-6; G. *Propristis schweinfurthi*, rostral tooth, 43390-7; H. *Myliobatis* sp., dental battery, 43390-8; I. crocodylid, frontal from cranium, 43390-87; J. *Pterosphenus schucherti*, vertebra, 43390-9; K. *Zygorhiza kochii*, partial ?premolar tooth, 43390-10.

Several modern stenothermal chondrichthyan genera are represented by congeneric fossil species in the Hardie Mine. These sharks are of special interest because of their potential use as paleothermometers. It is paradoxical that both cold water-loving and warm water-loving sharks are present in the Hardie Mine fauna. The modern porbeagle, *Lamna nasus*, is restricted to cold waters of 18°C or cooler (Compagno, 1984). The living great white, *Carcharodon carcharias*, prefers cold and warm temperate seas (Compagno, 1984). Thus, it is occasionally found in the northern Gulf of Mexico in cool water months, but not in the tropical waters of the Caribbean Sea. Mako sharks, *Isurus oxyrhinchus*, prefer waters 16°C and warmer (Compagno, 1984). Tiger, *Galeocerdo cuvieri*, and nurse, *Ginglymostoma cirratum*, sharks are even more thermophilic, with tiger sharks inhabiting waters 21°C or warmer (Lineweaver and Backus, 1973).

The presence of teeth from both cold water porbeagles (124) and great whites (2), along with warm water tiger (219) and nurse sharks (13) indicates that the near-shore waters at the Hardie Mine site fluctuated from cold to warm. This could be the result of seasonal changes in solar warming as in the northern Gulf of Mexico which today ranges from about 10°C in the winter up to 32°C in the summer. Because the Florida Peninsula did not block access from the open North Atlantic to the northern Gulf of Mexico, cold water sharks would have had free access to the cool Gulf waters during winter months. The

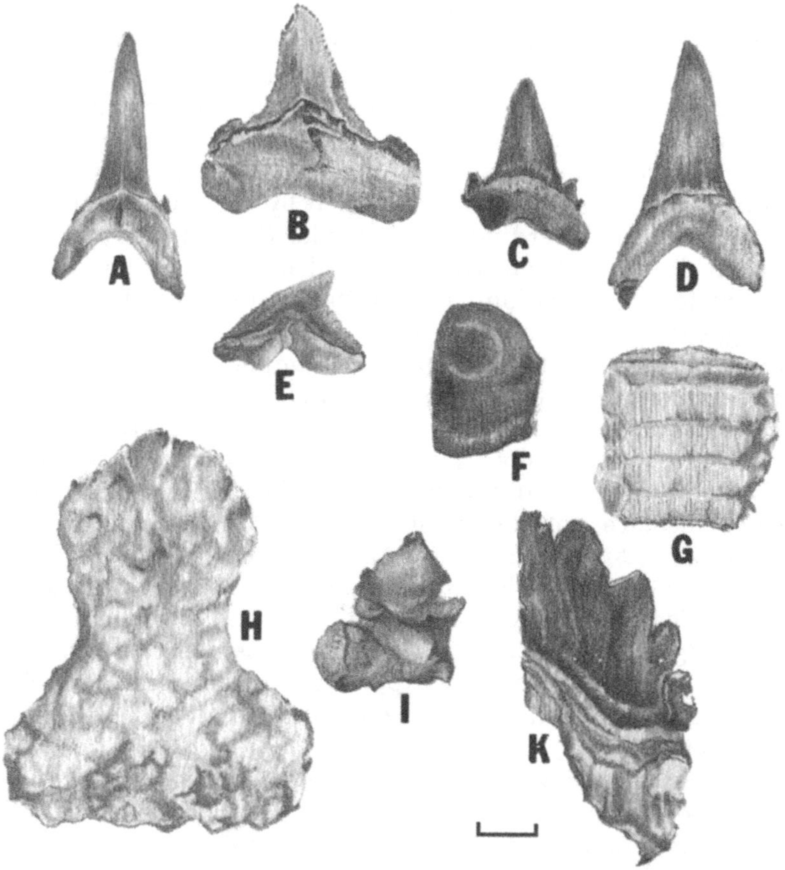

FIGURE 9. Drawings of vertebrate remains from TMM 43390, the Hardie Kaolin Mine. A. *Carcharias hopei*, tooth, 43390-2; B. *Carcharodon auriculatus*, tooth, 43390-3; C. *Lamna* sp., tooth, 43390-4; D. *Isuris* sp., tooth, 43390-5; E. *Galeocerdo latidens*, tooth, 43390-6; F._*Propristis schweinfurthi*, rostral tooth, 43390-7; G. *Myliobatis* sp., dental battery, 43390-8; H. crocodylid, frontal from cranium, 43390-87; I. *Pterosphenus schucherti*, vertebra, 43390-9; J. *Zygorhiza kochii*, partial ?premolar tooth, 43390-10. Bar scale = 1 cm.

presence of cold water sharks in the near-shore deposits of the Hardie Mine and their absence in the same facies recorded by the Gosport Sand indicates that the Jackson Sea's coastal waters were significantly cooler during parts of the year than those on the shoreline of the Claiborne Sea prior to its regression in late middle Eocene time. It is noteworthy that the Claiborne Sea regression appears to have occurred soon after the glaciation of Antarctica began. Oceanic drill core evidence indicates that a large glacier complex may have formed on the portion of Antarctica near Prydz Bay by 42.5 MA (Barrett, 1989).

Seasonal wind shifts could also have caused upwelling of cool waters from an offshore trench situated in the middle of the Suwanee Strait which separated coastal Georgia from the Florida banks in late Eocene time (Huddlestun, 1993). In the deepest part of the Suwanee Strait and parallel to the Georgia coast, lay the 150 m deep Gulf Trough. The Suwanee Current flowed from the Gulf of Mexico into the open North Atlantic along this current swept trench (Huddlestun, 1993). Upwelling of trench waters would have not only lowered water temperatures, but would also have brought nutrients into the photic zone to enhance the coastal ecosystems and perhaps establish a solid trophic base for a cetacean community.

Shifts in coastal currents also cause changes in coastal ocean temperature. Mid-Atlantic state coastal waters are warmed by the Gulf Stream in summer and cooled by the southward flowing Labrador Current in winter. The vertebrate remains do not indicate whether local temperature changes were the result of variation in insulation, wind shifts causing up-welling to come and go, or changes in strength and direction of the Suwanee Current. It is likely that all three of these agents interacted in a complex and dynamic manner.

Ribs, vertebrae, and teeth from the archaeocete whales *Basilosaurus cetoides* and *Zygorhiza kochii* are distinctive and relatively common in the Clinchfield Formation at the Hardie Mine. Archaeocete whales became common on the Gulf Coast for the first time in the Jackson Sea as indicated by numerous occurrences of the giant-sized *Basilosaurus cetoides*, and the smaller-sized, *Zygorhiza kochii*. These two archaeocetes are well known from Jacksonian deposits of equivalent and younger ages in every Gulf Coastal state, except Texas. They have also been found in the Mississippi Embayment of Arkansas (Westgate and Salazar, 1996) and a *B. cetoides* rib was found in a Jackson Group outcrop on the Mississippi River in Tennessee. This abundance of archaeocetes in Jacksonian deposits contrasts greatly to their rarity in Claibornian deposits.

In his extensive review of the Archaeoceti, Kellogg (1936) reported that only two Claibornian specimens were known from North America. One is a single *Protocetus* vertebra from the Cook Mountain Formation of Leon County, Texas in the TMM collections. The other, USNM 13856, is two archaeocete caudal vertebrae found with 17 mostly articulated "sirenian" vertebrae in the Yegua Formation of Burlson County, Texas (Kellogg, 1936). Domning *et al.* (1982) suggest that these "sirenian" vertebrae (USNM 13857) appear to be archaeocete vertebrae and noted that the remains actually came from the Cook Mountain Formation. These 19 vertebrae may have come from a single whale.

A third Claibornian archaeocete specimen in the TMM collection is a ?*Protocetus* rostrum which was collected from the Hurricane Lentil of the Cook Mountain Formation in Houston County, Texas. Thus, only three Claibornian archaeocete specimens are known from Texas and all three came from the Cook Mountain Formation in the east central portion of the state.

A single archaeocete specimen has been reported from the Claibornian of Louisiana. Uhen (1998) based *Natchitochia jonesi* on a partial vertebral column (USNM 16805), which, as with the Texas Claibornian whales, was collected from the Cook Mountain Formation.

Three fragmentary protocetid specimens are known from the Cook Mountain Formation correlate in Alabama, the Lisbon Formation. These include USNM 449546, CIS field # 98-30 and a University of Michigan specimen (Uhen, 1998, written comm.).

Hulbert *et al.* (1998) reported the discovery of an articulated protocetid archaeocete, *Georgiacetus vogtlensis*, from the middle Eocene Blue Bluff unit of the McBean Formation in Burke County, Georgia. This is the only fairly complete Claibornian age archaeocete known from the North American Coastal Plain. Two fragmentary specimens found nearby were referred to *G. vogtlensis*.

Albright (1996) described three isolated teeth from the middle Eocene-age upper Santee Limestone of South Carolina which resemble those of *G. vogtlensis*. Three partial protocetid specimens (USNM 310633, 449548 and 449549) are known from the Comfort Member of the Castle Hayne Formation, and a partial protocetid mandible (USNM 392014) was collected from the New Bern Formation, all in North Carolina (Uhen, 1998, written comm.).

It appears that although archaeocete whales lived along the coast of the Claiborne Sea, it was not suitable for large populations which would have left a more distinctive fossil record. However, archaeocete whale remains are common in deposits which accumulated on the continental shelf of the Jackson Sea, and even in the near-shore sands and estuarine deposits which accumulated as the shoreline transgressed inland (see the Crow Creek and Lover's Leap local faunas). Remains of *Basilosaurus* and *Zygorhiza* are common enough to make them useful guide fossils for differentiating Jacksonian deposits from Claibornian ones.

Coincident terrestrial vertebrate community changes on the Georgia coast are poorly documented. Crocodilian remains in the Hardie Mine fauna suggest that winters were relatively frost-free. A fragment of a deuterocone from a titanothere premolar was collected at the Hardie Mine by Mr. Bobby Strange and placed in his private museum in Sandersville, Georgia. The tooth is comparable in size to the P3 of the Uintan aged titanothere *Notiotitanops mississippiensis* collected from the Cook Mountain Formation in Clarke County, Mississippi (Gazin and Sullivan, 1942), although the cusps on that type specimen are too worn for detailed morphologic comparison. The relatively small size of the Hardie Mine titanothere suggests that either the deposition of the Clinchfield Formation predated the Chadronian NALMA, or ecological conditions on the Georgia coast did not allow a great increase in size to occur in titanothere species locally, while titanotheres in the continental interior displayed a massive increase in size from Uintan to Chadronian time.

Two probable entelodont teeth found in Twiggs County, Georgia (Domning *et al.*, 1982), and a partial *Leptotragulus* sp. jaw (USNM 244447) collected in Jefferson County, Georgia, all from Barnwell Group exposures, prove that there is high potential for finding additional late Eocene land mammal remains locally. Understanding of the changes which occurred in the terrestrial mammal community in southeastern North America during late Eocene time must await further discoveries.

5. Jackson Group

5.1. (Undifferentiated)

5.1.1. Crow Creek Local Fauna

A diverse estuarine vertebrate fauna has been collected from *Crassotrea* oyster-rich beds exposed on Crow Creek in St. Francis County, Arkansas (Figure 8A). Estuarine facies in Gulf Coast Eocene deposits are rare and the invertebrate and vertebrate suite of fossils in the Crow Creek beds is unique. The invertebrates are dominated by the oyster *Crassostrea alabamensis*, the mussel, *Mytilus hamatoides* and the clam *Venericardia planicosta*. A concentration of large freshwater snails, *Hemisinus jacksonius* (Figure 8B), was also found at one fossil quarry in association with two land mammal specimens (Westgate, 1982). Harris and Palmer (1947) interpreted the invertebrate assemblage to indicate that the deposit formed on an intertidal shoreline under the influence of coastal rivers. The U. S. Geological Survey does not differentiate the Jackson Group deposits in northeastern Arkansas into formations because of the estuarine facies present and the restricted nature of outcrops on Crowley's Ridge where these late Eocene beds are exposed. Jackson Group deposits in Mississippi range from late middle to latest Eocene (late Bartonian through Priabonian) in age (Dockery, 1996). The Crowley's Ridge deposits formed during the maximum extent of the Jacksonian transgression into the Mississippi Embayment.

5.1.1.1. Vertebrate Fauna. Several papers have reported on various aspects of the Crow Creek vertebrate fauna including the fishes (Westgate, 1984, 1985; Westgate and Salazar, 1996), reptiles (Westgate and Ward, 1981; Westgate, 1984) land mammals (Westgate and Emry, 1985, Westgate, 1986) and cetaceans (Palmer, 1939; Westgate and Salazar, 1996). Most of those papers display locality maps and stratigraphic sections. Revision of shark taxonomy by Cappetta (1987) and others has required changes in the previously reported faunal lists. Currently, "*Odontaspis*" *hopei* (Agassiz, 1843) = *Carcharias hopei*; "*Scyliorhinus*" *enniskilleni* White, 1956 = *Abdounia enniskilleni*; *Galeocerdo* "*alabamensis*" Leriche = *G. latidens*; and *Hemipristis* "*wyattdurhami*" White, 1956 = *H. curvatus*.

A

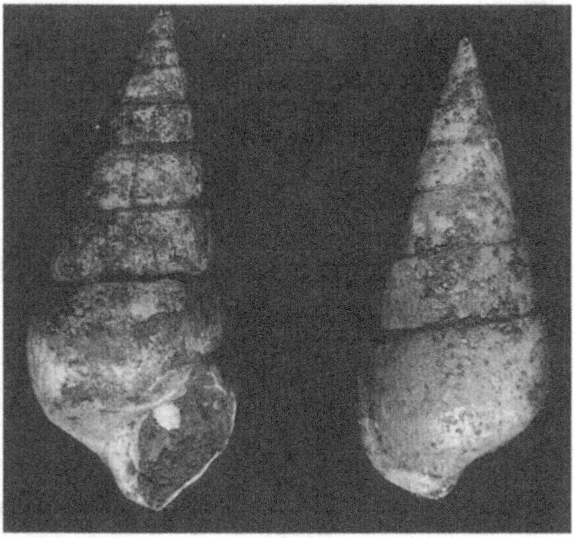

B

FIGURE 10. A. Fossil quarry locality TMM 43593 in an oyster reef facies of late Eocene Jackson Group outcrops on Crow Creek, St. Francis County, Arkansas, downstream from the U. S. Hwy. 70 bridge. B. Specimens of the freshwater snail *Hemisinus jacksonius* collected in association with land mammal remains at TMM 43593.

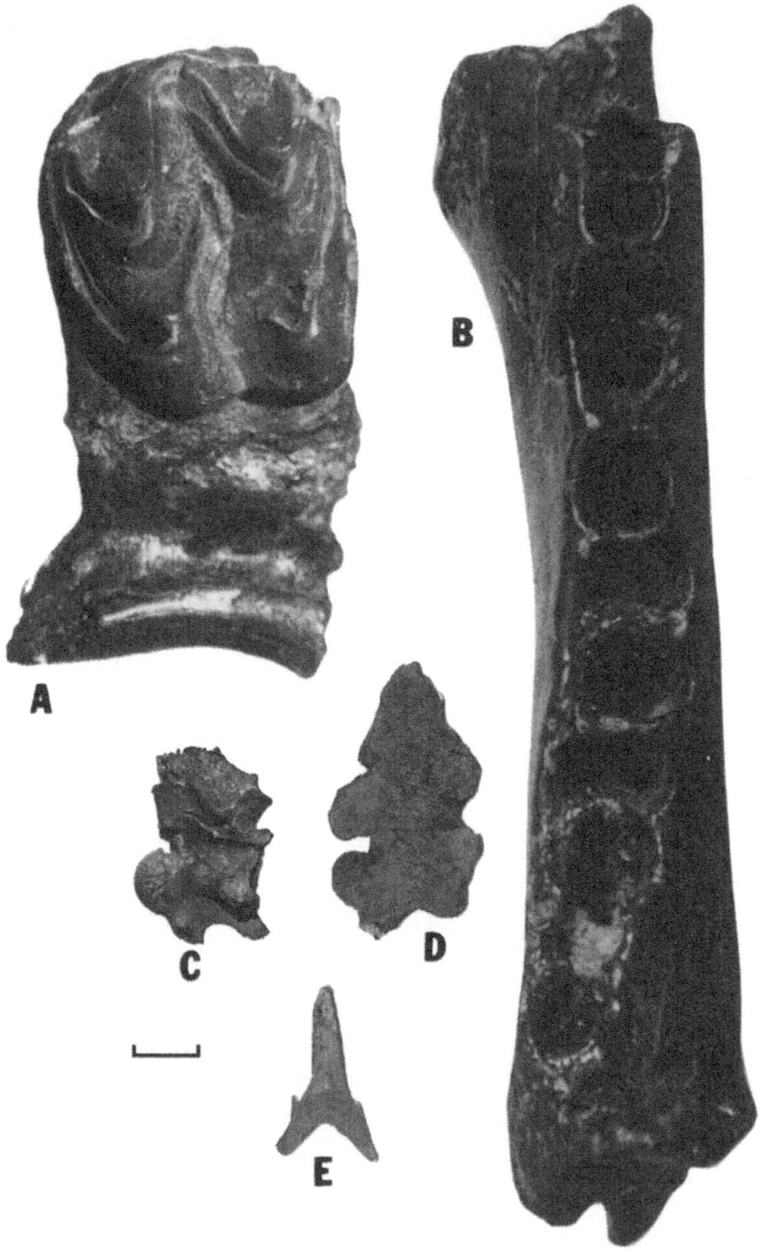

FIGURE 11. Jacksonian vertebrate remains from Arkansas; A–C from the Crow Creek outcrops in St. Francis County; D–E from Lover's Leap on Copperas Creek, Cross County. A. cf. *Diplobunops matthewi*, right M3 with maxillary fragment, USNM 299805; B. cf. *Pantolestes natans*, right mandibular ramus, USNM 299909; C. *Pterosphenus schucherti*, vertebra, USNM 299599: D. *P. schucherti*, vertebra, TMM 43596-3; E. *Carcharias hopei*, tooth, TMM 43596-1. Bar scale = 3 mm in A, 3.5 mm in B, 10 mm in C–E.

Table 3. The Crow Creek local fauna (TMM 43593).

CLASS	SPECIES	COMMON NAME	SPECIMENS
CHONDRICHTHYES	*Ginglymostoma serra*	nurse shark	5 teeth
	Carcharias hopei	sand tiger shark	37 teeth
	Galeorhinus sp.	tope shark	2 teeth
	Hemipristis curvatus	snaggletooth shark	1 tooth
	Abdounia enniskilleni	requiem shark	4 teeth
	Galeocerdo latidens	tiger shark	7 teeth
	Negaprion gibbesi	lemon shark	32 teeth
	Pristis sp	sawfish	1 rostral tooth
	Propristis schweinfurthi	sawfish	2 rostral teeth
	Dasyatis sp.	stingray (male & female)	16 teeth
	Myliobatis sp.	eagle ray	many tooth frags.
OSTEICHTHYES	*Lepisosteus* sp.	gar	3 teeth, 2 scales
	cf. *Arius* sp.	catfish	6 fin spines
	Sphyreana sp.	barracuda	28 teeth
	Trichiuris sp.	cutlassfish	6 teeth
REPTILIA	*Trionyx* sp.	softshell turtle	shell fragments
	crocodylid	crocodile	jaw frag., 4 teeth, 2 vertebrae, scute
	Pterosphenus schucherti	giant aquatic snake	6 vertebrae
MAMMALIA	cf. *Pantolestes natans*	aquatic carnivore	mandibular ramus
	cf. *Diplobunops matthewi*	primitive artiodactyl	M3
	Basilosaurus cetoides	primitive whale	vertebra, vert. epiphysis
	Zygorhiza kochii	primitive whale	deciduous P4 or p4, atlas

Taxa and specimen counts in Table 3 are based primarily on specimens which came from a now concrete rip-wrap covered vertebrate fossil quarry in an oyster reef facies just downstream of the U. S. Highway 70 bridge at TMM 43593. The specimens described by Westgate (1984) and Westgate and Emry (1985) are in the USNM collections (Figure 9). The record of four additional chondrichthyan species, *Ginglymostoma serra, Galeorhinus* sp., *Propristis schweinfurthi,* and *Dasyatis* sp. (Westgate and Salazar, 1996), is based on specimens obtained from more recent bulk sampling of a *Crassostrea-Venericardia* shell hash 300 m upstream of the bridge in the NE 1/4 of the SW 1/4, of sec. 25, T. 5 N., R. 3 E. (TMM 43594). That bed is currently inaccessible because of lateral stream shifting.

An archaeocete whale atlas from *Zygorhiza kochii* (TMM 43594-1, Figure 10A) was collected by P. Borne and J. Westgate in 1991 about 10 m upstream of the *Crassostrea-Venericardia* quarry (Westgate, 1993). A deciduous *Zygorhiza kochii* premolar (TMM 43595-1, Figure 10B) and a *Basilosaurus cetoides* vertebra (Figure 11) in the Paleontological Research Institute collections (Palmer, 1939) were both collected from outcrops of an unnamed western tributary which enters Crow Creek about 300 m downstream of the bridge in the se 1/4, se 1/4, SW 1/4, of sec. 25, T. 5 N., R. 3 E. A vertebra from cf. *Pontogeneus brachyspondylus* (USNM 299599) was also collected by J. Ward and J. Westgate in 1979 from an outcrop of the Pastoria Sand Member of the White Bluff Formation (*sensu* Wilbert, 1953) exposed at Red Bluff in Jefferson County, Arkansas. These archaeocete specimens indicate that at least three whale

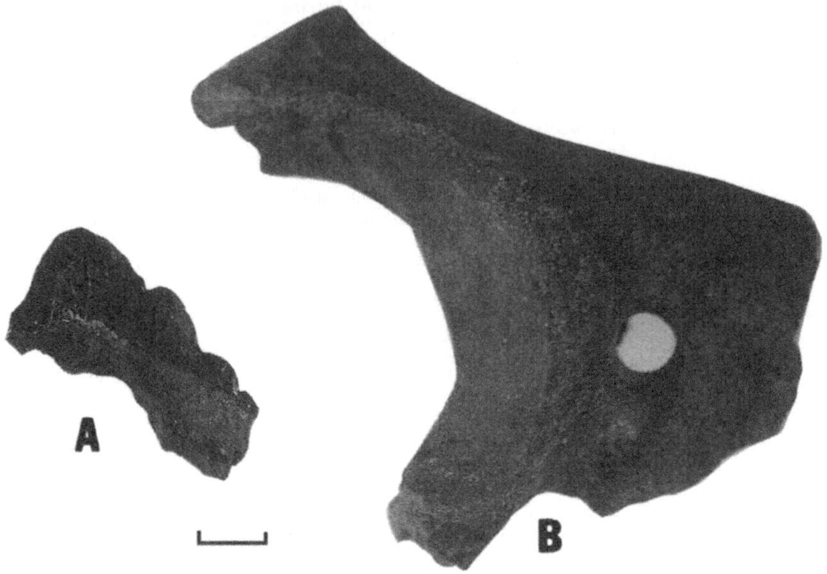

FIGURE 12. Remains of the archaeocete whale *Zygorhiza kochii* from Jacksonian outcrops on Crow Creek, St. Francis County, Arkansas. A. deciduous premolar, TMM 43595-1. Bar scale = 1 cm; B. left half of atlas vertebra, TMM 43594-1.

species inhabited the northern Mississippi Embayment of Arkansas (Westgate and Salazar, 1996).

The Crow Creek vertebrate species only indirectly verify that the facies is an estuarine and not an open marine near-shore (up to 10 m depth) deposit. The fish fauna is comprised of genera which today inhabit both near-shore and estuarine environments. Living *Trionyx* spp. and crocodiles are known to actively venture into open marine environments from protected coastal waters. The giant snake, *Pterosphenus schucherti*, is well known from late Eocene open marine deposits, including the Hardie Mine local fauna, but it also inhabited the middle Eocene Casa Blanca mangrove estuary. Both archaeocete whales, *B. cetoides* and *Z. kochii*, are normally found in open continental shelf deposits or in near-shore facies such as the Clinchfield Formation sands. However, it does not seem unlikely that the smaller-sized *Zygorhiza kochii* might also have actively foraged in food-rich estuaries having open access to the sea. The large size of up to 20 m (Kellogg, 1936) of *B. cetoides* may have prevented it from entering shallow estuaries. However, 12 m-long gray whales (*Eschrictius robustus*) today actively enter shallow Mexican coastal lagoons to mate and calve. Bloated carcasses of *B. cetoides* undoubtedly were transported through estuarine inlets occasionally. Fortunately, the vertebrate remains are associated with the *Crassostrea-Mytilus* invertebrate suite and the estuarine origin of the deposit is recognizable.

The presence of the oreodont cf. *Diplobunops matthewi* and the riparian mammal, cf. *Pantolestes natans*, implies that the shoreline was close by. However, land mammal remains on the Gulf Coast today are observed more often on the open marine beaches than in estuarine environments. The incomplete nature of the land mammal specimens allows only a broad correlation between the Crow Creek fauna and Uintan to Chadronian faunas in the continental interior (Westgate and Emry, 1985).

FIGURE 13. Vertebra from the giant archaeocete whale *Basilosaurus cetoides* found in Jacksonian deposits exposed in a tributary of Crow Creek, St. Francis County, Arkansas. Bar scale = 2.5 cm.

It must be concluded that the Crow Creek vertebrate fauna could be found in both estuarine and near-shore facies and that in the absence of an invertebrate assemblage, the two habitats could not be differentiated using these vertebrate taxa. Most of the Crow Creek species are also found in the near-shore facies which preserved the Hardie Mine fauna. It is noteworthy, however, that the lamnid sharks, *Lamna*, *Carcharodon*, and *Isurus*, which are present in the Hardie Mine's near-shore deposits are missing in the Crow Creek estuarine fauna.

5.1.2. Lover's Leap Local Fauna

The Lover's Leap local fauna comes from Jackson Group deposits exposed on Copperas Creek and its tributaries between Wynne and Levesque in Cross County, Arkansas. Vertebrate remains were collected by surface prospecting of two outcrops on the south side of Copperas Creek. The first outcrop, TMM 43596, is a 38 m tall bluff known locally as Lover's Leap, 0.5 km south of the town of Levesque where Arkansas highway 163 crosses the creek. The second outcrop, TMM 43597, is 1.2 km due southwest of Lover's Leap and directly southeast of the Crowley's Ridge Country Club in the NE 1/4, NE 1/4 of Sec. 13, T. 7 N., R. 3 E.

The bluff at Lover's Leap exposes 25 m of mudstones and sandy mudstones which are Eocene in age (Figure 12). The 1993 Geologic Map of Arkansas shows only Claiborne Group exposures in Cross County, and the stratigraphic section of the Lover's Leap outcrops described by Guccione and others (1988) states that all of the Eocene beds here belong to the Claiborne Group. However, the upper half of the Eocene deposits at Lover's Leap actually belong to the Jackson Group. This was determined by the discovery in 1992 by P. Borne and J. Westgate of two teeth from the late Eocene sand tiger shark *Carcharias hopei* (TMM 43596-1, Figure 9E; and 43596-2), in association with a complete vertebra from the aquatic snake *Pterosphenus schucherti* (TMM 43596-3, Figure 9D) and one tooth each from *Abdounia enniskilleni* (TMM 43596-4) and a small carcharhinid shark (?*Negaprion gibbsi*, TMM 43596-5) in a bed about 14 m above creek level which Guccione and others (1988) named Zone 4. The mudstones and sands below the vertebrate-bearing horizon (Zones 1–3 and the lower half of 4, Guccione *et al.*, 1988) are dark brown, bear lignite and resemble Claiborne Group deltaic deposits. However, a lignitic log 1 m above creek level displays borings from the shipworm *Teredo* sp., indicating that at least hyposaline conditions existed here when the wood was buried. This may actually be a deltaic facies of the Jackson Group. If so, the entire 25 m of Eocene beds are actually Jacksonian in age and none are Claibornian. Two meters of sand and gravel of the Lafayette Formation of Plio-Pleistocene age unconformably overlie the Jackson Group beds. Four meters of Pleistocene loess caps the section.

The bluff southeast of Crowley's Ridge country club exposes a 15 m thick section of Eocene beds called the Country Club locality. A concentration of vertebrate remains were discovered by P. Borne and J. Westgate in 1992 in a

FIGURE 14. Lover's Leap on Copperas Creek, near Levesque, in Cross County, Arkansas. The basal 25 m of mudstones and sands are Eocene marginal marine deposits. The "X" indicates the horizon with yielded a suite of Jacksonian vertebrates (TMM 43596) indicating that at least the upper half of these Eocene beds are Jacksonian in age. Six meters of Plio-Pleistocene Lafayetee Formation (LF) sands and gravel and Pleistocene loess (L) cap the bluff.

gypsum-rich layer about 6 m above creek level. Taxa present include *Carcharias hopei* (TMM 43597-1, 30 teeth), teleosts (TMM 53597-2, 3 vertebrae), a turtle (TMM 43597-4, shell fragment), and a crocodilian (TMM 43597-4, dermal scute). This suite of vertebrates closely resembles the estuarine community preserved in the Crow Creek local fauna 25 km to the south in St. Francis County, and indicates that the Jackson Group lies no higher than 6 m above the bed of Copperas Creek at this point.

Additional Jackson Group outcrops of glauconitic clayey sands are exposed 1.5 km southwest of the Country Club locality on Browner's Lake Creek which drains the south-central portion of Sec. 13, T. 7 N., R. 3 E.

Remains of additional Jacksonian species have been collected in Cross County, but they are in private collections. In 1992, Mr. Paul Sides of Levesque, Arkansas collected two articulated *Basilosaurus cetoides* vertebrae from the bed of a small unnamed stream near where it joins Copperas Creek 2.7 km due west of the Lover's Leap locality. Mr. Larry Lowman, of Wynne, Arkansas and former park naturalist at Village Creek State Park, collected several hundred Jacksonian vertebrates, including teeth of *Carcharias hopei* and a partial vertebra of *Pterosphenus schucherti*, from beds exposed during the construction of the Lake Austell dam in the State Park, about 0.7 km north of the Cross/St. Francis County line.

The Lover's Leap local fauna indicates that Jackson Group deposits extend in the subsurface from the exposures on Crow Creek in St. Francis County, northward into Cross County at least as far as Copperas Creek. The 1993 Geologic Map of Arkansas shows the northernmost Jacksonian outcrops ending about three km south of the Cross/St. Francis County line. This map should be amended by extending the Jackson Group outcrops 13 km to the north to Levesque, Arkansas and westward along Copperas Creek at least 2.7 km (Westgate and Salazar, 1996) to indicate the northernmost transgression of the Jackson Sea into Arkansas.

6. Discussion

The fossil localities in this study were chosen because of their potential for providing new biostratigraphic and paleoecologic information. Marginal marine deposits have a high potential for yielding land mammal remains. In addition, several localities are of similar age, but provide fossils from distinctly different depositional facies. The three late Claibornian localities sampled estuarine mangrove swamp (Lake Casa Blanca), sandy near-shore (Little Stave Creek), and muddy continental shelf (Whiskey Bridge) depositional environments. The three Jacksonian faunas were buried in estuarine oyster-reef (Crow Creek and Lover's Leap) and sandy near-shore (Hardie Mine) habitats. Analysis of faunas of similar age allows the reconstruction of co-existing Gulf Coastal paleocommunities in estuarine and near-shore (less than 10 m) environments. Differences in faunas of different age, but from the same facies, such as the latest Claibornian Little Stave Creek Gosport Sand fauna and the early Jacksonian Hardie Mine fauna, reflect environmental and evolutionary changes affecting Gulf Coast communities.

The Little Stave Creek and Hardie Mine faunas both come from sandy near-shore deposits. The Gosport Sand has been dated at 40.7 Ma and the Clinchfield Formation's western correlate, the Moodys Branch Formation, has yielded dates of 40.35 and 38.3 Ma. (Harland *et al.*, 1989). The Gosport Sand was deposited along the shoreline during a regressive sea level phase at the end of Claibornian time. During the subsequent one-half to two million years, the Jackson Sea transgressed across central Georgia to the edge of the Piedmont region. The coarse sands of the Clinchfield Formation are the near-shore remnants of this shifting shoreline. The shark faunas in these two near-shore deposits (Little Stave Creek and Hardie Mine) show distinct differences (Westgate and Gillette, 1995). The Little Stave Creek sharks are genera which today only inhabit warm waters. Some of the genera such as *Galeocerdo* (tiger sharks), and *Ginglymostoma* (nurse sharks) are also in the Hardie Mine fauna. However, cool water sharks such as *Lamna* (porbeagle) and *Carcharodon* (great white) are common in the Hardie Mine fauna as well. The appearance of remains of these cool water sharks in the Jacksonian near-shore facies, and their absence in the late Claibornian near-shore deposits, may indicate that a significant drop in water temperature occurred in the Gulf of Mexico between the times of the Little Stave Creek and Hardie Mine faunas. The mixture of warm and cool water species in the Hardie Mine fauna may indicate that water temperatures fluctuated seasonally, as they do in the Gulf of Mexico today, with modern surf temperatures ranging from 10–32°C (50–89°F) along the northern shore. Seasonal drops in Jacksonian Sea water temperature may have resulted from shifting currents, upwelling and/or winter cooling.

Whether these cooler oceanic conditions were the result of world-wide climatic change, as the co-incident initial growth of the Antarctic ice cap might suggest (Barrett, 1989), or the result of up-welling, or both, is not clear. However, the transgression of the Jackson Sea across the Gulf Coastal Plain brought with it a cetophilic environment and allowed the establishment of a large whale community dominated by *Zygorhiza kochii* and *Basilosaurus cetoides*, for the first time. The synchronous appearance of the porbeagle *Lamna* sp. and an unprecedented abundance of great white shark remains (*Carcharodon auriculatus*) in Jackson Sea deposits (Dockery and Manning, 1986) suggests that the new whale community may have helped to support a new Gulf Coastal shark community. The dietary predilection for marine mammals by modern great white sharks living in cool ocean waters is well documented (Compagno, 1984). It is reasonable to assume that preying on marine mammals by *C. auriculatus* was a common occurrence in late Eocene seas.

Some of the most common vertebrate fossils in late Claibornian and Jacksonian beds are the teeth of sand tiger sharks. Unworn teeth from the Claibornian shark, *Striatolamia macrota*, have distinctive striations on the main cusp and are present in the Whiskey Bridge, Casa Blanca and Little Stave Creek local faunas. Unworn teeth of the Jacksonian sand tiger, *Carcharias hopei*, do not display these striations and are present in the Hardie Mine, Crow Creek and Lover's Leap local faunas. Eocene deposits often have lost their invertebrate faunal assemblages through weathering or other post-depositional destruction of $CaCO_3$ minerals. Often the vertebrate assemblages survive this post-depositional alteration. In such cases, the common occurrence of *S. macrota* and *C. hopei* teeth on outcrop surfaces, their large size and distinctive morphology, and their presence in both open marine and estuarine facies, make them excellent guide fossils in differentiating late Claibornian from Jacksonian deposits in the field.

ACKNOWLEDGMENTS. Consultation with numerous paleontologists has been necessary to identify the diverse assortment of fossil remains discussed above. I am indebted to experts in the following areas: plants—D. Dilcher (Florida State Museum), C. Gee (University of Bonn, Germany) and V. Krassilov (Palaeontological Institute, Moscow, Russia); mammals—C. Beard and M. Dawson (Carnegie Museum of Natural History), R. Emry (U. S. National Museum of Natural History), A. Walton (Austin Community College), E. Lundelius, Jr. and J. Wilson (University of Texas), and F. C. Whitmore, Jr. (U. S. Geological Survey); birds—S. Olsen (U. S. National Museum of Natural History); crocodilians—A. Busbey (Texas Christian University) and W. Langston (University of Texas); turtles—H. Hutchison (University of California-Berkeley); fishes—R. Estes (San Diego State University), P. Forey (British Museum of Natural History), R. Purdy (U. S. National Museum of Natural History) and G. Stringer (Northeast Louisiana University); decapods—A. Williams (U. S. National Museum of Natural History); ostracods—P. Borne (Louisiana State University), and oysters—R. Perrault.

Abby Salazar painted the reconstruction of the Casa Blanca mangrove community, drew Figure 9, and assisted in the preparation of black and white photographs. Lake Casa Blanca specimens found during the period 1990–1997 were collected under Texas Parks and Wildlife Department permits. Superintendent Dan Olivas and the staff at Lake Casa Blanca State Park supported our fossil excavation program in many ways. Field and research expenses were underwritten by grants from the National Geographic Society; Donors of The Petroleum Research Fund, administered by the American Chemical Society; Office of Research, Lamar University; Geology Foundation, University of Texas-Austin; Geological Society of America; American Association of Petroleum Geologists; Gulf Coast Section, Society of Economic Paleontologists and Mineralogists; Gulf Coast Association of Geological Societies; and Sigma Xi.

References

Barrett, P. J., (ed.), 1989, Antarctic Cenozoic history from the CIROS-1 drill-hole, *New Zealand Department of Scientific Research Bulletin* **245**:1–254.

Beard, C., and Tabrum, A., 1991, The first early Eocene mammal from eastern North America: an omomyid primate from the Bashi Formation, Lauderdale County, Mississippi, *Mississippi Geology* **11**:1–6.

Beard, C., Dawson, M., and Tabrum, A., 1995, First diverse land mammal fauna from the early Cenozoic of the southeastern United States: the early Wasatchian Red Hot local fauna, Lauderdale County, Mississippi, *Geologic Society of America Abstracts with programs*, A-453.

Berry, E. W., 1924, The middle and upper Eocene floras of southeastern North America, *U. S. Geological Survey Professional Paper* **92**:1–206.

Borne, P., 1993, Casa Blanca ostracod assemblage from the Laredo Formation, Webb County, Texas, *Texas Academy of Sciences Abstracts with Program.*

Busby, A. B., 1989, An Eocene caiman from the Texas Gulf Coast, *Journal of Vertebrate Paleontology* **9**:15A.

Cappetta, H., 1987, Chondrichthyes II, Mesozoic and Cenozoic Elasmobranchii, in: *Handbook of Paleoichthyology* (H. Schultze, and O. Kuhn, eds.), Gustav Fischer Verlag, New York, pp. 1–193.

Case, G., 1981, Late Eocene selachians from south central Georgia, *Palaeontographica Abt. A* **176**:52–79.

Compagno, L. J. V., 1984, Sharks of the world: An annotated and illustrated catalog of shark species known to date, *Food and Agriculture Organization of the United Nations, Fisheries Synopsis* **125(4)**:1–655.

Dockery, D., III., 1996, Toward a revision of the generalized stratigraphic column of Mississippi, *Mississippi Geology* **17**:1–9.

Dockery, D., III, and Manning, E., 1986, Teeth of the giant shark *Carcharodon auriculatus* from the Eocene and Oligocene of Mississippi, *Mississippi Geology* **7**:7–19.

Domning, D., Morgan, G., and Ray, C., 1982, North American Eocene sea cows (Mammalia, Sirenia), *Smithsonian Contributions to Paleobiology* **52**:1–69.

Gardner, J., 1957, Little Stave Creek, Alabama-Paleoecologic study, in: Treatise on marine ecology and paleoecology, Memoir 67 (H. S. Ladd, ed.), *Geological Society of America* **(2)**:573–588.

Gazin, C. L., and Sullivan, J. M., 1942, A new titanothere from the Eocene of Mississippi, with notes on the correlation between the marine Eocene of the Gulf Coastal Plain and continental Eocene of the Rocky Mountain region, *Smithsonian Miscellaneous Collections* **101(13)**:1–13.

Gee, C., and Westgate, J. W., 1990, Paleoecological reconstruction of a middle Eocene mangrove community from southwest Texas using palynology, *Palynology* **14**:212.

George, D., and Westgate, J. W., 1998, The Whiskey Bridge local fauna: Fossil fishes from the type locality of the middle Eocene Stone City Formation, Claiborne Group, *Texas Academy of Science 101st Annual Meeting, Program and Abstracts*, p. 157.

Ghosh, P. K., 1972, *Use of bentonites and glauconites in potassium-40/argon-40 dating in Gulf Coast stratigraphy*, Ph.D. dissertation, Rice University, Houston (unpublished).

Guccione, M. J., Prior, W. L., and Rutledge, E. M., 1988, Crowley's Ridge, Arkansas, in: *South-central Section Centennial Field Guide*, Volume 4 (O. T. Hayward, ed.), Geological Society of America, pp. 225–230.

Hansen, T., 1987, Extinction of late Eocene to Oligocene molluscs: relationship to shelf area, temperature changes and impact events, *Palaios* **2**:69–75.

Harland, W. B., Armstrong, R., Cox, A., Craig, L., Smith, A., and Smith, D., 1989, *A geologic time scale*, Cambridge University Press, Cambridge, 263 p.

Harris, G. D., and Palmer, K., 1947, The mollusca of the Jackson Eocene of the Mississippi Embayment (Sabine River to the Alabama River): Part II, Univalves and index, *Bulletins of American Paleontology, Paleontologic Research Institute* **30(117)**:207–563.

Huddlestun, P., 1993, A revision of the lithostratigraphic units of the coastal plain of Georgia: the Oligocene, *Georgia Geologic Survey Bulletin* **105**:1–152.

Huddlestun, P., and Hetrick, J., 1986, Upper Eocene stratigraphy of central and eastern Georgia, *Georgia Geologic Survey Bulletin* **95**:1–78.

Hulbert, R., Petkewich, R., Bishop, G., Bukry, D., and Aleshire, D. P., 1998, A new middle Eocene protocetid whale (Mammalia: Cetacea: Archaeoceti) and associated biota from Georgia, *Journal of Paleontology* **72**:907–927.

Kellogg, R., 1936, A review of the Archaeoceti, *Carnegie Institution of Washington* **482**:1–366.

Lineweaver, T., and Backus, R., 1973, *The natural history of sharks*, Doubleday and Co., New York, 239 p.

Lucas, S., 1989, *Coryphodon* (Mammalia, Pantodonta) from the Hannold Hill Formation, Eocene of Trans-Pecos, Texas, *Pearce-Sellards Series, Texas Memorial Museum* **46:**1–16.

Martini, E., 1986, The northwest German Tertiary basin Eocene, in: *Northwest Germany during the Tertiary*, Volume 18 (H. Tobien, ed.), Beitrage zur regionalen Geologie der Erde, Gebruder Borntraeger, Berlin, pp. 660–668.

Osborn, H. F., 1898, Evolution of the Amblypoda. Part 1. Taligrada and Pantodonta, *Bulletin, American Museum Natural History* **10:**169–218.

Palmer, K. V., 1939, *Basilosaurus* in Arkansas, *American Association of Petroleum Geologists Bulletin* **23:**1228–1229.

Riley, K. S., and Westgate, J. W., 1999, Community structure of the Jacksonian age (late Eocene) Hardie Mine local fauna, Wilkinson County, Georgia, *Texas Academy of Science Annual Meeting Program with Abstracts.*

Sheldon, A., and Carpenter, M., 1997, First occurrence of early Eocene mammal in Alabama is the easternmost occurrence of *Coryphodon* from the Bashi formation, Wasatchian mammal age, *Geological Society of America Annual Meeting Abstracts with Programs*, A-105.

Slaughter, B. H., 1978, Occurrences of didelphine marsupials from the Eocene and Miocene of the Texas Gulf Coastal Plain, *Journal of Paleontology* **52:**744–746.

Speigler, D., 1986, Subdivision of the northwest German Tertiary (Paleogene and Neogene) by means of planktic foraminifera, in: *Northwest Germany during the Tertiary*, Volume 18 (H. Tobias, ed.), Beitrage zur regionalen Geologie der Erde, Gebruder Borntraeger, Berlin, pp. 213–299.

Stenzel, H. B., 1940, New zone in the Cook Mountain Formation, the *Crassatella texalta* Harris-*Turritella cortezi* Bolwes zone, *American Association of Petroleum Geologists Bulletin* **24:**1633–1675.

Stenzel, H. B., Krause, E., and Twining, J., 1957, Pelecypoda from the type locality of the Stone City beds (middle Eocene) of Texas, *Bureau of Economic Geology, University of Texas, Pub.* **5704:**1–237.

Thurmond, J. T., and Jones, D. E., 1981, *Fossil vertebrates of Alabama*, University of Alabama Press, Tuscaloosa, 244 p.

Toulmin, L. D., 1977, Stratigraphic distribution of Paleocene and Eocene fossils in the eastern Gulf Coast region, *Geological Survey of Alabama Monograph* **13:**1–602.

Uhen, M. D., 1998, New protocetid (Mammalia, Cetacea) from the late middle Eocene Cook Mountain Formation of Louisiana, *Journal of Vertebrate Paleontology* **18:**664–668.

Uhen, M., and Gingerich, P. D., 1995, Evolution of *Coryphodon* (Mammalia, Pantodonta) in the late Paleocene and early Eocene of northwestern Wyoming, *University of Michigan, Contributions from the Museum of Paleontology* **29:**259–289.

Walton, A. H., 1993a, *Pauromys* and other small Sciuravidae (Mammalia, Rodentia) from the middle Eocene of Texas, *Journal of Vertebrate Paleontology* **13:**243–261.

Walton, A. H., 1993b, A new genus of eutypomyid (Mammalia, Rodentia) from the middle Eocene of the Texas Gulf Coast, *Journal of Vertebrate Paleontology* **13:**262–266.

Westgate, J. W., 1982, *Hemisinus jacksonius* (Gastropoda) as an aid in determining potential of marine deposits for yielding remains of terrestrial mammals, *Missouri Academy of Science Transactions* **16:**49–53.

Westgate, J. W., 1984, Lower vertebrates from the late Eocene Crow Creek local fauna, St. Francis County, Arkansas, *Journal of Vertebrate Paleontology* **4:**536–546.

Westgate, J. W., 1985, Shark hunting on the Arkansas ocean deep, *Sea Frontiers* **31:**241–246.

Westgate, J. W., 1986, Late Eocene land mammals from the Gulf Coastal Plain, in: *Montgomery Landing Site, Marine Eocene (Jacksonian) of central Louisiana* (J. A. Schiebout, J. A. and W. van den Bold, eds.), Gulf Coast Association of Geological Societies Symposium Proceedings, pp. 223–229.

Westgate, J. W., 1988a, Biostratigraphic implications of the first Eocene land-mammal fauna from the North American coastal plain, *Geology* **16:**995–998.

Westgate, J. W., 1988b, *Biostratigraphic and paleoecologic implications of the first Eocene land mammal fauna from the North American Coastal Plain*, University of Texas-Austin, Ph.D. Dissertation, 181 p. (unpublished).
Westgate, J. W., 1989, Lower vertebrates from an estuarine facies of the middle Eocene Laredo Formation (Claiborne Group) of Webb County, Texas, *Journal of Vertebrate Paleontology* **9**:282–294.
Westgate, J. W., 1990, Uintan land mammals (excluding rodents) from an estuarine facies of the Laredo Formation (middle Eocene, Claiborne Group) of Webb County, Texas, *Journal of Paleontology* **64**:454–468.
Westgate, J. W., 1993, A stratigraphic and paleoecologic analysis of land mammal-bearing Eocene strata on the Gulf Coastal Plain, *Annual Report on Research, Petroleum Research Fund, American Chemical Society* **37**:418.
Westgate, J. W., 1994a, Eocene forest-swamp, *Research & Exploration, National Geographic Society* **10**:78–89.
Westgate, J. W., 1994b, A new leptochoerid from middle Eocene (Uintan) deposits of the Gulf Coastal Plain, *Journal of Vertebrate Paleontology* **14**:296–299.
Westgate, J. W., 1999, *After the dinosaurs: A Texas tropical paradise recovered at Lake Casa Blanca*, Texas Parks and Wildlife Press, Austin, 69 p.
Westgate, J. W., and Emry, R. J., 1985, Land mammals of the Crow Creek local fauna, late Eocene, Jackson Group, St. Francis County, Arkansas. Journal of Paleontology **59**:242–248.
Westgate, J. W., and Gee, C., 1990, Paleoecology of a middle Eocene mangrove biota (vertebrates, plants, and invertebrates) from southwest Texas, *Palaeogeography, Palaeoclimatology, Palaeoecology* **78**:163–177.
Westgate, J. W., and Gillette, M., 1995, Comparison of Eocene Clinchfield Formation (Barnwell Group) and Gosport Sand (Claiborne Group) Coastal vertebrate faunas and their stratigraphic implications, *Geological Society of America Annual Meeting Abstracts with Programs*, p. 453.
Westgate, J. W., and Salazar, A., 1996, Additions to the late Eocene (Jacksonian) cetacean and chondrichthyan faunas of Arkansas, *Geological Society of America, South-Central Section Annual Meeting, Abstracts with Programs*, p. 68.
Westgate, J. W., and Ward, J. F., 1981, The giant aquatic snake *Pterosphenus schucherti* (Palaeophidae) in Arkansas and Mississippi, *Journal of Vertebrate Paleontology* **1**:161–164.
Westgate, J. W., Gillette, M., and Rolater, E., 1994, Paleoecology of an Eocene coastal community from Georgia, *Journal of Vertebrate Paleontology* **14(3)**:52A.
White, E. I., 1956, The Eocene fishes of Alabama, *Bulletins of American Paleontology* **36(156)**:123–153.
Wilbert, L. J., 1953, The Jacksonian Stage in southeastern Arkansas, *Arkansas Division of Geology Bulletin* **19**:1–125.
Wilson, J. A., 1986, Stratigraphic occurrence and correlation of early Tertiary vertebrate faunas, Trans-Pecos, Texas: Agua Fria-Green Valley areas, *Journal of Vertebrate Paleontology* **6**:350–373.
Wilson, J. A., and Westgate, J. W., 1991, A lophodont rodent from the middle Eocene of the Gulf Coastal Plain, Texas, *Journal of Vertebrate Paleontology* **11**:257–260.

Chapter 12

Taphonomic Interpretation of Gnat-Out-of-Hell, an Early Uintan Small Mammal Locality in the Uinta Formation, Utah

MEG L. THORNTON and D. TAB RASMUSSEN

1. Introduction

Fossil mammals from the Uinta Formation of northeastern Utah play a central role in our understanding of the Uintan "North American Land Mammal Age" (Scott and Osborn, 1890; Black and Dawson, 1966; Walsh, 1996; Rasmussen

MEG L. THORNTON and D. TAB RASMUSSEN • Department of Anthropology, Washington University, St. Louis, MO 63130.

Eocene Biodiversity: Unusual Occurrences and Rarely Sampled Habitats, edited by Gregg F. Gunnell, Kluwer Academic/Plenum Publishers, New York, 2001.

et al., 1999). However, the mammal collections from the Uinta Formation are notably short on small mammals. Historically, most of the work on the Uinta Formation mammals, and consequently our understanding of mammalian evolution in the Rocky Mountain region during the Uintan, has been based on mammals from about the size of *Leptotomus* (a large ischyromyid rodent), *Leptotragulus* (a selenodont artiodactyl), and *Epihippus* (a horse), on up to brontotheres and amynodontid rhinos. Smaller taxa, such as marsupials, insectivores, apatemyids, primates, and the smaller rodents, have rarely been recovered from the Uinta Formation.

The paucity of small mammal specimens provides an incomplete view of the Uintan faunas, which consequently clouds and distorts our view of mammalian evolution during the middle Eocene. Powder Wash, a very productive locality in the Bridgerian aged beds underlying the Uinta Formation, yields practically nothing but small mammals (Dawson, 1968). The best Rocky Mountain assemblage younger than the Uinta Formation, the mammals of Badwater Creek, Wyoming, is also primarily a small mammal assemblage (Black, 1970; Setoguchi, 1975; Krishtalka and Setoguchi, 1977). Therefore, Uinta Formation samples are bracketed by younger and older faunas that have not been directly comparable because the size biases inevitably lead to unevenness of taxonomic representation. Similarly the most important assemblages of Uintan mammals outside the Rocky Mountain region, those of San Diego and Ventura Counties in southern California, include rich assemblages of small-bodied mammals (Kelly, 1990; Walsh, 1996). To meaningfully compare the Uinta Formation mammals to other assemblages that are most relevant to understanding mammalian evolution and biogeography during the Middle Eocene, small mammals of the Uinta Formation must be found. A new locality representing an unusual occurrence of small mammals allows us to take steps in this direction.

Teams from Washington University and the Utah Field House have collection fossil vertebrates from the Uinta Formation for several years (Rasmussen *et al.*, 1999). Most of the localities encountered are similar geologically to Myton Pocket, a region of badlands of several square kilometers, and the best-studied site in the Uinta formation (Hamblin, 1987). Myton Pocket consists primarily of overbank siltstones and mudstones along with sandstone paleochannels that were deposited on the delta plain of an ancient lake. Hamblin (1987:57) interpreted the habitat of Myton Pocket as being savanna-like with "streamside forest, some swampy or marshy areas, and forested highlands". The broader chronological, geological, and paleoclimatic context of the Uinta Formation has been reviewed in Prothero (1996) and Rasmussen *et al.* (1999).

Within the Uinta Formation, there are a variety of taphonomic contexts. Fossils of large mammals such as brontotheres and amynodontids are often found in paleochannels and coarse sheet sandstones (Douglass, 1910). Medium-sized and small mammals are commonly encountered in overbank mudstones and siltstones, some of which show paleosol development. The Uinta Formation is unusual among major fossil-bearing western Eocene formations in having relatively poor preservation of teeth; these often shatter when exposed on the surface. At many sites, fossils are too fragile to be successfully excavated or wet screened. Typically small mammals are uncommon at most localities. When small mammal remains are recovered from sites, they are often edentulous jaws. The insectivore record from the Uinta Formation, for example, consists of dozens of toothless dentaries, but very few identifiable teeth (Krishtalka, 1976; Rasmussen *et al.*, 1999). This contrasts sharply with sites from the underlying Green River formation such as Powder Wash, which have abundant small mammal teeth and jaws (Gazin, 1958; Dawson, 1968).

A new small mammal locality in the Uinta Formation is exceptional for the abundance and preservation of small mammals and delicate bird remains. This locality (WU-18, or Gnat-Out-of-Hell) is located in Coyote Basin in Uinta B rocks and in its broadest dimensions measures approximately 70 × 180 m (Figure 1). Microfossils at this locality come from a dark maroon-brown mudstone exposed on low-relief slopes and pan-like flats at the base of a moderate bluff (Figures 2–3). The fossil-producing bed is dissected by a small arroyo in one area. Fossils are recovered by surface collecting and limited screen washing. Literally thousands of specimens have been recovered from Gnat-Out-of-Hell yielding 20 genera of mammals (Table 1). The site also contains unstudied birds and reptiles. Among the unique or unusual occurrences at Gnat-Out-of-Hell are the following: the primate *Chipetaia* (Rasmussen, 1996);

FIGURE 1. Photo of the northeast portion of WU-18 (Gnat-Out-Of-Hell) showing the sparsely vegetated, exposed mudstones cut by arroyos and bounded by low bluffs. The microfossil producing layer is the darker mudstones exposed on the lowest flats and cut by gullies.

FIGURE 2. Two collectors work the productive flats of WU-18. Fossil have been recovered by surface collecting and a limited amount of screen washing.

FIGURE 3. Dr. Ellen Miller demonstrates surface collecting technique on Gnat-Out-Of-Hell. As recommended by 19th Century paleontologist O. C. Marsh, Miller keeps her nose about 6 inches above the ground, pointed down not up.

Table 1. Mammals recovered from Gnat-Out-Of-Hell (WU-18).

Insectivora	*Ischyrotomus compressidens*
?Leptictidae	*Ischyrotomus*, new species
Genus and species indeterminate	*Microparamys dubius*
Dormaaliidae	Sciuravidae
New genus, new species #1	*Sciuravus altidens*
New genus, new species #2	Cylindrodontidae
Proteutheria	*Pareumys grangeri*
Pantolestidae	*Pareumys milleri*
Pantolestes sp.	Family indet.
Apatotheria	*Natrona* sp.
Apatemyidae	Condylarthra
Apatemys uintensis	Hyopsodontidae
Primates	*Hyopsodus uintensis*
Omomyidae	Perissodactyla
Chipetaia lamporea	Brontotheriidae
Creodonta	Genus and species indet.
Hyaenodontidae	Equidae
Limnocyon sp.	*Epihippus uintensis*
Carnivora	Artiodactyla
Miacidae	Homacodontidae
Miacis sp., cf. *M. gracilis*	*Mesomeryx grangeri*
?New genus, very small	Agriochoeridae
Miocyon sp.	*Protoreodon* sp.
Rodentia	Protoceratidae
Ischyromyidae	*Leptotragulus proavus*

two species of a new genus of large dormaaliid (yet undescribed); specimens of the diminutive artiodactyl, *Mesomeryx* (Townsend and Rasmussen, 1995); the only specimens of the rodent *Natrona* outside Badwater (M. D. Dawson, Carnegie Museum, pers. comm.); a new species of the rodent *Ischyrotomus*; the largest Uinta Formation sample of *Microparamys*; the best upper and lower rodent dentitions from the Uintan Formation (Dawson, 1966); and the most abundant bird fauna from the Uinta Formation.

Although most of the fossils occur as isolated, broken elements, we have recovered a few small concretions containing masses of crushed mammal remains. These concretions, and the abundance of small bone at the site, were hypothesized to represent owl pellets or mammalian predator coprolites (Rasmussen, 1996). Knowing how the assemblage accumulated is important for interpreting paleocommunities and early Uintan ecosystems. The purpose of this paper is to evaluate the hypothesis that the Gnat-Out-of-Hell fossil assemblage represents accumulation by owls or other predators, particularly diurnal raptors or small mammalian predators. Data were collected on the size, anatomical representation, patterns of breakage, and the condition of the fossils from the site. These were then analyzed in a number of ways that facilitated comparison to modern assemblages created by living owls and other predators.

2. Taphonomy of Predator Assemblages

Modern investigations of faunal assemblages, whether archaeological or paleontological, must consider multiple taphonomic processes. Most assemblages are affected by numerous damaging processes that can affect the makeup and character of the bone assemblage. These processes include predation by many different animals, weathering, trampling, water transport, and the process of burial. Sometimes these processes leave marks that can be used to attribute the damage to a specific taphonomic agent. However many times there are few clues as to what caused the damage or what processes produced an assemblage. A general review of predator assemblages is provided here to illustrate which variables were evaluated.

Predators can have a highly varied affect on bones. For comparison to Gnat-Out-of-Hell, it seemed most prudent to focus on small mammals bone assemblages accumulated by raptors and small carnivores, due to the size of the bones preserved. Raptors regurgitate pellets containing the bones and other inedible portions of their prey (Reed and Reed, 1928; Chitty, 1938; Crass, 1944; de Graaf, 1960; Carnegie, 1961; Benson, 1962; Seigfried, 1965; Brooke, 1967, 1973; Dodson and Wexlar, 1979; Brain, 1981; Bramwell *et al.*, 1987; Hoffman, 1988; Andrews, 1990). In contrast small mammalian carnivores accumulate bones in either their feces or in their food caches (Dodson, 1973; Andrews and Evans, 1983; Hoffman, 1988; Horwitz and Goldberg, 1989; Andrews, 1990). Both raptors and mammalian carnivores can produce similar results that vary slightly in the amount of damage done to the bones of their prey (breakage, teeth marks, and corrosion) and bone representation (element and prey size). Much of the data on bird and mammal predators is difficult to use in a quantitative analysis, since the data are usually presented in summary tables that are not standardized across studies. Comparisons among studies are therefore qualitative and descriptive. Below we outline what a typical assemblage would be for each possible agent.

Owls usually swallow small prey whole (Andrews, 1990), but larger animals are often ripped into pieces. Because they process prey minimally, owls do not typically cause a lot of bone breakage. However, breakage may occur after the pellet has been deposited on the ground or in the nest by trampling by either the owl or a larger animal (Andrews, 1990). Therefore an owl pellet typically should contain nearly complete skeletons with little or no breakage and no other damage such as gastric etching or tooth marks.

Diurnal raptors, such as hawks, falcons, and eagles ingest their food by holding it in their talons and ripping pieces from the prey. These activities, as well as the strong neck muscles and large beaks, cause more bone damage than that observed in owls, yet not as much damage as is inflicted by mammalian carnivores (Hoffman, 1981). Raptors tend to cause more gastric damage than owls, both etching and weakening bones, which can cause more breakage or completely destroy them. However, research on Peregrine Falcon pellets suggests that there can also be a considerable amount of rounding caused by gastric juices and no clear examples of gastric etching (Thornton, unpublished). The assemblage of a diurnal raptor would consist of some complete but fragile bones with no teeth marks, considerable rounding of the bones, and maybe some gastric etching.

Andrews and Evans (1983) compared various groups of mammalian carnivores in terms of breakage and representation. Carnivore species within families tend to show less variation than across families. Viverrids tend to cause moderate to great bone breakage in prey because they rip the carcass into small pieces and chew a considerable amount during ingestion. This can completely destroy fragile bones like the cranium. These carnivores usually cause some "rounding and polishing" of the bone during digestion (Andrews and Evans, 1983). However, viverrids rarely leave tooth marks on their bones of their prey.

Canids cause considerable bone breakage during mastication. Prey skulls and occasionally distal limbs of small- to medium-sized mammals are not ingested, resulting in their absence. Canids also damage bone during digestion, including rounding and etching of both dental and bone elements. One species of canid, the bat-eared fox, has proved the exception because it does not cause much gastric digestion nor does it leave teeth marks like other canids in this study (Andrew, 1991).

The larger species of mustelids occasionally consume bone. Typically the bone found in their scat are flakes and unrecognizable bone fragments. The prey's head is rarely eaten. There is some rounding and polishing but little corrosion, except for some severe dentin corrosion that can occur. There are also few to no tooth marks. Andrews and Evans (1983) had very little information on Felidae. The information they did have showed extreme damage, breakage, and corrosion.

Some other animals can produce faunal assemblages as a result of scavenging. Packrats (*Neotoma floridana*) have been observed caching bones. Bone in such an assemblage tends to be highly gnawed. New and Old World porcupines also collect bones but they tend to be those of medium and large animals of the region (Brain, 1981). Harvester ants can also carry large numbers of bones to the close vicinity of an anthill. An anthill studied by Shipman and Walker (1980) showed that all skeletal elements are represented in the anthill, but the smaller elements such as carpals and tarsals were underrepresented.

3. Methods

3.1. Fieldwork

Fossils were recovered by surface collecting and by screen washing of bagged matrix from Gnat-Out-of-Hell (WU-18). Two of the richest pockets of small fossil remains (the "Main Quarry" and "Tony's Pocket") were sampled intensively by randomly collecting a large sample of broken bone fragments even if they were not identifiable. On other parts of the quarry, only fragments that were anatomically or taxonomically useful were typically collected. The matrix of the Main Quarry would not screen wash well; most of the screen wash sample comes from Tony's Pocket. Specimens were assigned field numbers, and then sorted and analyzed in St. Louis. Specimens are catalogued into the Carnegie Museum, the permanent repository, but they remain for now in the study collections at Washington University.

3.2. Analyses

Data were collected on 4152 fossil fragments. Data included taxonomic affiliation (typically, only to classes Mammalia, Reptilia, or Aves), and when possible, the fragments were identified by anatomical element, portion of element, relative completeness (25, 50, 75, 100%), length and width, and any type of damage (tooth marks, gastric etching, etc.).

Data from Gnat-Out-of-Hell were compared to data published in the literature on accumulations of animal remains made by a variety of owls, diurnal raptors, and small mammals. The publications utilized were the following: Reed and Reed, 1928; Chitty, 1938; Crass, 1944; de Graaf, 1960; Carnegie, 1961; Benson, 1962; Seigfried, 1965; Brooke, 1967, 1973; Dean, 1973, 1974; Dodson and Wexlar, 1979; Brain, 1981; Andrews and Evans, 1983; Bramwell *et al.*, 1987; Hoffman, 1988; Andrews, 1990.

Data were analyzed in four ways. Because publications differed in the exact way that the quantitative data were presented, it was difficult to always make precise quantitative comparisons. Therefore, the first analysis was a qualitative comparison among data sets using simple descriptive statistics (mean, range, proportions, etc.). This was the only analysis to take advantage of the entire Gnat-Out-of-Hell data set.

The second analysis was a cluster analysis of a portion of the Gnat-Out-of-Hell data set compared to assemblages made by 18 predators as reported by Andrews (1990). Sixteen quantitative variables were used for clustering. The eighteen predators and the data are shown in Table 2. Clustering was performed on SAS using two clustering criteria: average linkage and centroid linkage. The purpose of this analysis was to determine multivariate pattern of associations among data sets based on linear distance measures.

The third analysis was a SAS discriminant function applied to the same data as the cluster analysis. In this analysis, the 18 modern analogs were categorized into three groups: owls, diurnal raptors, and mammalian predators (Table 1). The purpose of this analysis was to see if modern assemblages made by the three categories of predators could be statistically distinguished from each other, and if so, with which group, if any, the Gnat-Out-of-Hell assemblage was statistically associated.

In the fourth analysis, principal components (using SAS) were derived for the multivariate data for two purposes: (1) to evaluate which skeletal variables were most important in distinguishing among groups; and (2) to allow us to visually evaluate the patterns of clustering and discrimination revealed in the previous two analyses.

4. Results

4.1. Pellet Descriptions

Only four concretions with bone have been recovered from Gnat-Out-of-Hell, but they show dense packing of small rodent bones and jaws (Figure 4). Concretions like this have not been found at other Uinta Formation sites. Densely concentrated small bone in concretions only from this site are unlikely to be the result of depositional processes not involving an animal agent. The packing of bone fragments closely resembles that seen in predator feces or pellets, and suggests that the fragmentation of elements is predepositional.

4.2. Qualitative Comparisons

A summary of the reduced data set from Gnat-Out-of-Hell appears in Table 2. Owls show a very obvious bias towards leaving complete bones in their pellets, directly opposed to the pattern observed in the Gnat-Out-of-Hell assemblage, which has no complete bones, even within the few concretions. The carnivore and raptor assemblages have half as many complete bones as the owl assemblages. Note that proximal and distal ends are more common than the more fragile shafts (Table 2).

The high quantity of bone breakage in the Gnat-Out-of-Hell assemblage leads us to dismiss owls as the bone-accumulating agent. Diurnal raptors may be a likely accumulator given the amount of breakage and rounding. Small carnivores also cause severe breakage to bones of their prey, but they produce rounding, polishing, teeth marks, and some gastric digestion. Among carnivores, viverrids show few tooth marks on the bones of their prey. Diagenetic processes may also be responsible, except that there is high breakage also seen within the concretions.

The Gnat-Out-Of-Hell assemblage most closely resembles those of raptors and perhaps small carnivores except for the lack of surface damage. There is a possibility that such damage has been abraded off over time by wind, water, shift in soil, and various other taphonomic processes, although these fossils exhibit

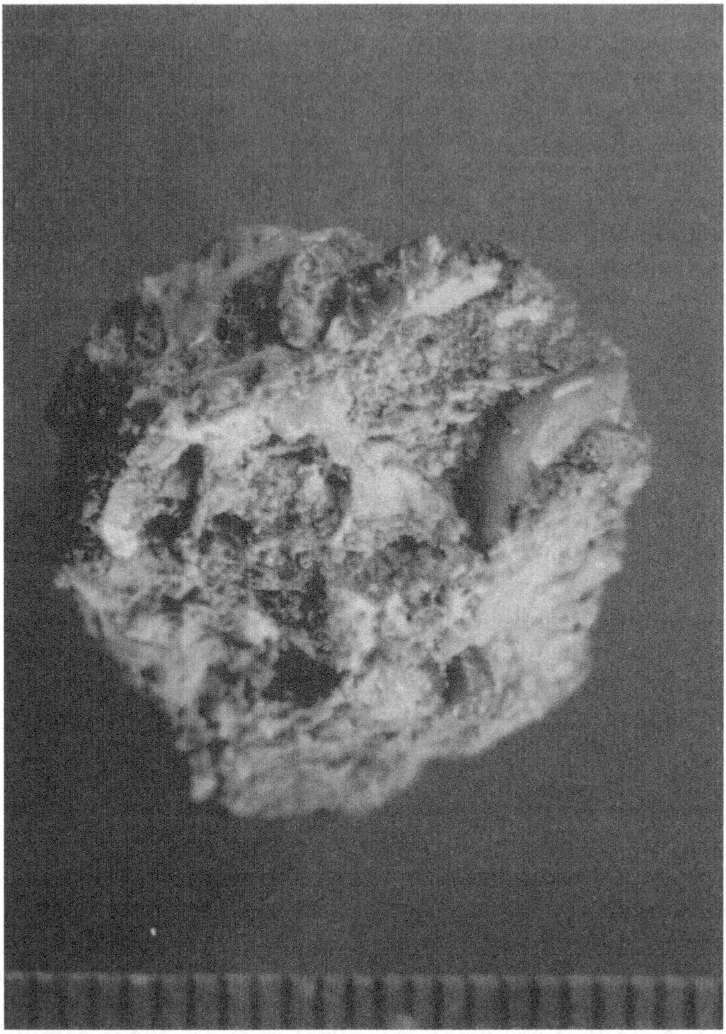

FIGURE 4. A concretion containing multiple crushed specimens of *Pareumys grangeri*, the most common mammal at WU-18. A cheek dentition is visible towards the top, and an incisor is visible on the right. The concretion is composed primarily of fossil bone. Scale at the bottom is in 1 mm units.

no signs of abrasion. High breakage could be due to diagenetic processes rather than predator behavior, and this will be evaluated in the discussion section.

4.3. Cluster Analysis

The cluster analysis placed WU-18 near a combined cluster of raptors and small carnivores and distinctly separate from the owls (with the exception of the Little Owl; Figure 5).

4.4. Discriminant Function

Each of the 18 species in the discriminant function was correctly placed in its predator group. The greatest generalized squared distance among groups was between carnivores and diurnal raptors, while the closest was between carnivores and owls. The Gnat-Out-of-Hell assemblage was classified with diurnal raptors. However, as there were only two species within the diurnal raptor group (Table 2) there was poor statistical resolution of the "group". The two were a small falcon and a harrier.

4.5. Principal Components Analysis

The plots of principal components show that Gnat-Out-of-Hell is, in reality, an outlier relative to the other predator accumulations included in the quantitative comparisons. Plots of the first three principal components are shown in Figures 6a-b. The first component accounts for about 56% of the

Table 2. Data used in cluster and discriminant analyses.

Species	Predator	Humerus				Ulna				Femur				Tibia			
	Group	C	P	S	D	C	P	S	D	C	P	S	D	C	P	S	D
Fossil	Unknown	0	48	8	43	0	55	24	21	0	39	5	56	0	20	2	77
Barn owl	Owl	99	0	0	1	97	3	0	0	97	1	2	0	98	1	1	0
Snowy owl	Owl	75	4	8	12	76	24	0	0	88	4	0	8	88	8	4	0
Long-eared owl	Owl	96	0	1	3	95	4	1	0	96	3	1	0	93	6	1	0
Short-eared owl	Owl	88	3	2	7	92	8	0	0	97	2	1	0	99	1	0	0
Verreaux eagle owl	Owl	96	0	2	2	98	2	0	0	97	2	1	0	99	1	0	0
Spotted owl	Owl	4	47	11	38	85	12	3	0	66	32	2	0	71	0	29	0
European eagle owl	Owl	82	7	0	11	0	73	0	0	83	12	3	2	86	9	0	5
Great grey owl	Owl	89	4	4	4	96	4	0	0	90	8	2	0	93	7	0	0
Tawny owl	Owl	53	7	12	28	69	31	0	0	52	22	6	20	85	7	4	4
Little owl	Owl	33	33	16	16	100	0	0	0	12	64	12	12	33	8	50	8
Mongoose	Carnivore	30	29	9	32	8	92	0	0	12	52	13	23	37	25	0	38
Genet	Carnivore	33	13	10	44	54	46	0	0	12	51	20	17	57	27	0	16
Bat-eared Fox	Carnivore	26	7	15	52	57	43	0	0	3	87	3	7	10	80	0	10
Coyote	Carnivore	7	38	17	38	25	75	0	0	0	42	28	30	0	90	0	10
Red Fox	Carnivore	0	8	9	83	0	67	33	0	0	53	21	26	0	67	0	33
Pine Martin	Carnivore	0	30	19	51	25	75	0	0	0	50	50	0	0	82	0	18
Kestrel	Raptor	44	4	27	25	32	52	8	8	20	48	24	7	31	27	25	14
Hen harrier	Raptor	22	7	39	32	60	40	0	0	20	40	20	20	22	22	33	22

C = complete
P = proximal end
S = shaft
D = distal end
Numbers represent percentage of specimens per bone in each breakage category

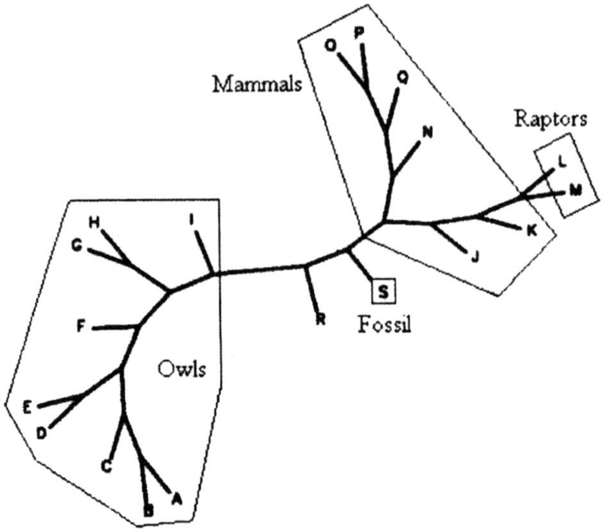

FIGURE 5. Tree diagram with taxonomic groupings resulting from the cluster analysis. The branch lengths are proportional to distance. The two raptors cluster closely together (L and M) and are attached to a cluster of mammalian carnivores (J, K, N-Q). The Gnat-Out-Of-Hell assemblage (S) falls closest to these groups, although it is intermediate between them and the owls. Most of the owls form a cohesive group (A-I) with the little owl (R) falling equally remotely between the owls and the carnivores.

variance, and it separates owls (low scores) from a combined group of carnivores and diurnal raptors (high scores). Gnat-Out-of-Hell scores slightly higher than the highest scoring carnivore. On this component, nothing is further from the owls than Gnat-Out-of-Hell. The heaviest loading variables for the first component are the presence of complete humerus, femur, and tibia.

The second component accounts for about 16% of the variance and does not distinguish among the three groups of the extant predators. Individual species of Carnivora score from very high to very low on the second component, with owls tending towards the positive scores and diurnal raptors towards the negative scores. Gnat-Out-of-Hell is a far outlier on this component with a very high score. The highest loadings on this component were the distal end and shaft scores, particularly for the ulna.

The third component accounts for about 10% of the variance. Cumulatively, the first three components account for 82% of the variance; no further component accounts for more than 5.5% of the total variance. The third component separates the diurnal raptors (high scores) from the carnivores (low scores, with very little variance). Owls are spread broadly on the third component. High loadings for this component are tibia shaft and humerus shaft. Taken in combination, the first and third components provide clean separation among the three predator groups and the fossil assemblage (Figure 6b), whereas the second component mainly separates the fossil assemblage from everything else.

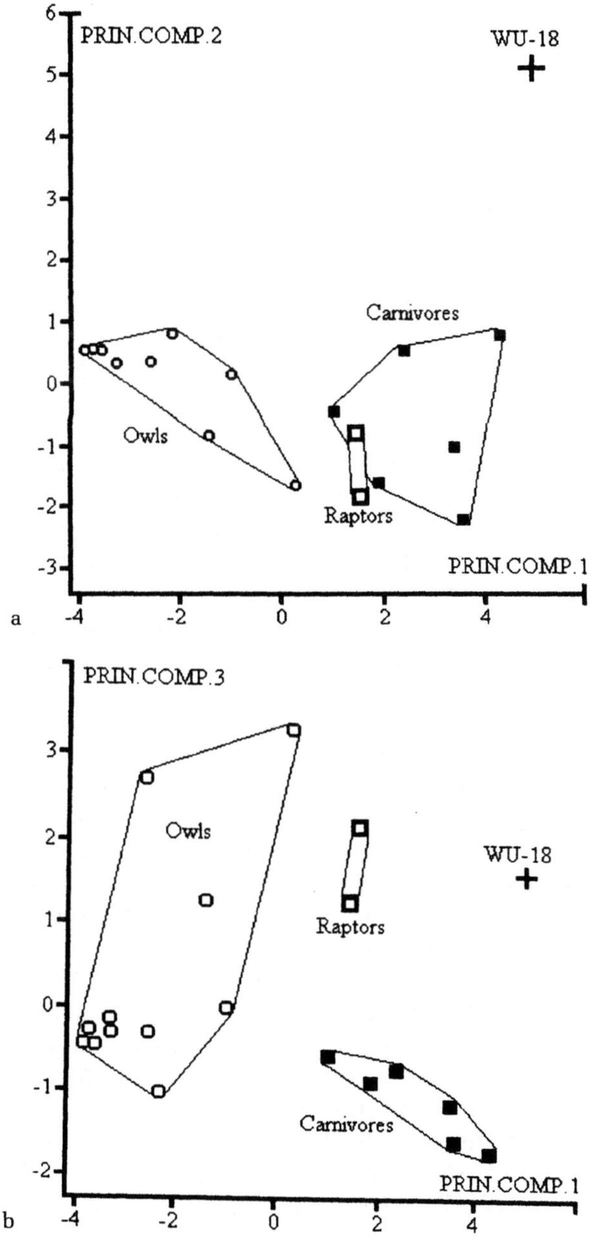

FIGURE 6. Bivariate plots of principal component scores for the extant taxa and the fossil assemblages in the multivariate analyses. (a) Principal Components 1 and 2 show a wide separation of the fossil assemblage from any of the extant taxa. (b) Principal Components 1 and 3 allow resolution of the extant taxa into distinct groups. Component 1 separates the owls from the other groups, Component 3 partially separates the avian predators from the mammalian carnivores, and Component 2 primarily separates the fossil from the extant taxa (this component represents an unknown "paleo factor").

5. Discussion

That the bones from Gnat-Out-of-Hell accumulated largely by the action of predators seems probable. The small size of the prey remains, their consistent patterns of breakage, and most importantly, the preservation of a few concretions showing the broken up and packed together bones from several individual animals, are solid evidence for the action of predators (Mellet, 1974). The null hypothesis, that the assemblage accumulated without predator behavior, can be reasonably rejected.

Although isolated elements of large animals up to the size of brontotheres are found at Gnat-Out-of-Hell (presumably components of the depositional background), the largest mammal found commonly in the hot spots for the prey remains is *Mesomeryx grangeri*, a diminutive artiodactyl the size of a modern cottontail rabbit (*Sylvilagus*). The habitat preferences of the putative prey species are not known well, but limited information currently supports the interpretation that predators were working forested areas. The primate *Chipetaia* is represented by limb and foot bones indicating it was an arboreal leaper (G. C. Conroy and D. T. Rasmussen, unpublished). Apatemyids are also arboreal (Koenigswald, 1987). *Mesomeryx* was a forest floor omnivore (Townsend and Rasmussen, 1995), as were probably the dormaaliids. The rodents would be most interesting in terms of assessing the habitat hunting preferences of the predator, but their postcrania have not yet been analyzed. Fish and turtle remains, generally the most common elements found at small vertebrate localities in the Uinta Formation, are nearly absent.

All of our analyses suggest that this assemblage is unlike those generated by living owls. Owls are among the most common avian predators known from the Eocene; their earliest fossil record is from the North American Paleocene, and they are widespread in the Eocene (Olson, 1985). The bird remains from Gnat-Out-of-Hell have yet to be studied in detail, but preliminary inspection reveals nothing reminiscent of owls or other raptors.

There are some similarities between the Gnat-Out-of-Hell assemblage and those made by viverrids. The modern family Viverridae used for our statistical analogies was not present in the Eocene. The mammalian predators of the Uinta Formation belong to the carnivore family Miacidae and the creodont family Hyaenodontidae, overlapping the viverrids in size, inferred locomotion, and diets. At Gnat-Out-of-Hell, there are at least three carnivores, a large raccoon-sized species of *Miocyon*, a skunk-sized species of *Miacis* resembling and possibly conspecific with *M. gracilis*, and a very small weasel-sized species of indeterminate genus (Table 1). Based on size, the most likely miacid suspect as the accumulator of the bones would be the skunk-sized *Miacis* (the tiny miacid, possibly the smallest species known for the family, is small enough to have been a victim). Also in the fauna is one medium-sized creodont, *Limnocyon*, which presumably had behavior generally similar to that of viverrids and miacids; this animal is also of a size that would be expected to generate an assemblage similar to that at Gnat-Out-of-Hell. Interestingly, the six species of modern carnivores used in the quantitative comparisons often clustered relatively close together, and so statistically, it was easier to characterize them than it was to understand the diversity within diurnal raptor assemblages.

Perhaps the most promising predator category to account for the Gnat-Out-of-Hell assemblage is that of diurnal raptors. The similarities to raptor assemblages include the breakage patterns and prey composition. Our quantitative comparisons contained only two diurnal raptors, the Kestrel (*Falco*) and the Hen Harrier (*Circus*). Both of these favor small rodents as prey (plus many invertebrates for the Kestrel). Comparative data were not available for accipitrids, or forest falcons, most likely responsible for an assemblage like this. With a larger sample and broader range of assemblage profiles, it might be that Gnat-Out-of-Hell would overlap in some ways with diurnal raptors.

The ambiguous aspects of our results suggested to us that the approach of using living groups to assess the fossil assemblage may be flawed even with a more complete sample of living taxa for comparison, especially when extrapolating back in time as far as the Eocene, when taxonomic groups and ecological relationships may have differed substantially from the modern analogs. One problem is that when studying modern groups it is clear which bones truly belong to the predator pellet or feces, whereas with the fossils it is difficult to separate predator food items from later intrusive rodent remains and from other bones washed in with the sediment. The latter problem is probably minimal, because the standard Uinta Formation background of turtles, fish and large mammal bones are not a substantive part of the Gnat-Out-of-Hell sample, but some of the rodents may be intrusive burrowers or foragers entering the assemblage subsequent to the action of the predator. A second problem is that diagenetic or erosional fragmentation of the material cannot always be reliably separated from damage caused by the predators. One of the major ways in which the Gnat-Out-of-Hell assemblage differs from owl assemblages is in the high degree of fragmentation among the fossils. While it may be that the fragmentation was generated as a result of post-predator processes, the cemented concretions provide some evidence that the remains were already badly fragmented at the time they were packed into pellets or feces.

A final problem has to do with the logic of using statistical analogy between the fossil assemblage and those of modern animals. For a relatively simple analogy from present to past (e.g. estimating mammal body size from tooth size) there are reasons to assume that the cause of the statistical relationships now observed held true in the past. In the case of predator assemblages, however, there are numerous important variables that may not relate to each other in the same way today as they did in the Eocene. We do not know much about the diversity of predator behavior in the Eocene, and one of the most important groups, the creodonts, is extinct. Even less is known of Eocene avian predators. Also, given the behavior of a predator, its assemblages are going to reflect, in part, the nature of the prey community, how it is structured taxonomically and ecologically (by habitat, size, activity cycle, etc.). Finally, we know nothing of the post-predator biotic factors that influenced Eocene predator assemblages: for example, were there mammals specialized to utilize resources left over in predator dens or bird nests? In short, the more simple and constant the underlying reason for a correlation between two variables, the more likely the statistical approach will yield satisfactory results. We suggest that the intervening variables are too complex to expect precise, certain answers when looking back at predator-prey relationships in the Eocene. The quantitative approaches used in this report may be of greatest value for paleontological sites of Plio-Pleistocene age or even younger archaeological sites in which there is greater continuity of taxonomic and ecological representation across samples.

Gnat-Out-of-Hell remains a unique occurrence in the Uinta Formation. Small mammals have been recovered at several other sites, but none are taphonomically like Gnat-Out-of-Hell. The small mammals of Gnat-Out-of-Hell have been, and will continue to be useful in tracing patterns of mammalian evolution and biogeography in the Middle Eocene. For example, it is now clear that the small mammals from this site show greater similarity and evolutionary continuity with earlier and younger sites within the Rocky Mountain region than they do with contemporaneous but geographically distant sites in California and Texas, reflecting patterns of faunal endemism and provincialism during the Eocene (Rasmussen *et al.*, 1999). The dentitions and postcranial adaptations of rare mammalian taxa are being illuminated (e.g., *Mesomeryx* and *Chipetaia*, Townsend and Rasmussen, 1995) and more taxa await detailed study. Gnat-Out-of-Hell is a good example of how one or a few unusual occurrences in an area outweigh dozens or hundreds of "typical" localities in terms of their potential value. Understanding as much as possible about the processes leading to the formation of unusual occurrences will maximize our abilities to find these sites and to fully utilize the information contained within them.

ACKNOWLDGMENTS. We wish to thank the Utah Bureau of Land Management (who issued the collecting permit) and especially Blaine Phillips of the Vernal office; the staff of the Utah Field House in Vernal; the employees of Hatch River Expeditions; and Chevron employees at Redwash. All of these individuals and groups contributed to the success of the field operations. Many crew members have worked on Gnat-Out-of-Hell but we wish to thank especially A. R. Friscia, G. C. Conroy, and K. E. Townsend. We thank Gregg Gunnell for helpful comments and critique of the manuscript.

References

Andrews, P., 1990, *Owls, Caves, and Fossils*, University of Chicago Press, Chicago.

Andrews, P., and Evans, E. M. N., 1983, Small mammal bone accumulations produced by mammalian carnivores, *Paleobiology* **9**:289–307.

Benson, C. W., 1962, The food of the spotted eagle-owl *Bubo africanus*, *Ostrich* **33**:35.

Black, C. C., 1970, Paleontology and geology of the Badwater Creek area, central Wyoming. Part 5. The cylindrodont rodents, *Annals of Carnegie Museum* **41**:201–214.

Black, C. C., and Dawson, M. R., 1966, A review of late Eocene mammalian faunas from North America, *American Journal of Science* **264**:321–349.

Brain, C. K., 1981, *The Hunters or the Hunted?*, University of Chicago Press, Chicago.
Bramwell, D., Yalden, D., and Yalden, P., 1987, Black Grouse as the prey of the Golden Eagle at an archaeological site, *Journal of Archaeological Science* **14**:195–200.
Brooke, R. K., 1973, Notes on the distribution and food of the Cape Eagle Owl in Rhodesia, *Ostrich* **44**:137–39.
Brooke, R. K., 1967, On the food of the Wood Owl, *Ostrich* **38**:55.
Carnegie, A. J. M., 1961, The stomach contents of a spotted eagle owl (*Bubo africanus*), *Ostrich* **32**:97.
Chitty, D., 1938, A laboratory study of pellet formation in the short-eared owl, *Proceedings of the Zoological Society of London, Series A*, **108**:2.
Crass, R. S., 1944, The birds of Sulenkama, Qumbu District Cape Province, *Ostrich* **15**:10–20.
Dawson, M. R., 1968, Middle Eocene rodents (Mammalia) from northeastern Utah, *Annals of the Carnegie Museum* **39**:327–370.
Dawson, M. R., 1966, Additional late Eocene rodents (Mammalia) from the Uinta Basin, Utah, *Annals of the Carnegie Museum* **38**:97–114.
Dean, W. R. J., 1974, Analysis of *Tyto alba* pellets from Angola, *Zoologica Africana* **9**:89–92.
Dean, W. R. J., 1973, Analysis of a collection of Barn Owl, *Tyto alba* pellets from Warmbaths, Transvaal, *Zoologica Africana* **8**:75–81.
Dodson, P., 1973, The significance of small bones in paleoecological interpretation, *University of Wyoming Contributions in Geology* **12**:15–19.
Dodson, P., and Wexlar, D., 1979, Taphonomic investigations of owl pellets, *Paleobiology* **5**:275–284.
Douglass, E., 1910, Preliminary description of some new titanotheres from the Uinta deposits, *Annals of the Carnegie Museum* **6**:304–313.
Gazin, C. L., 1958, A review of the middle and upper Eocene primates of North America, *Smithsonian Miscellaneous Collections* **136**:1–112.
Graaff, G. de, 1960, A preliminary investigation of the mammalian microfauna in Pleistocene deposits of caves in the Transvaal System, *Palaeontographica Africana* **7**:59–118.
Hamblin, A., 1987, Paleoeography and Paleoecology of the Myton Pocket, Uinta Basin, Utah (Uintan Formation-Upper Eocene), *Brigham Young University Geology Studies* **34**(1):33–60.
Hoffman, R., 1988, The contribution of raptorial birds to patterning in small mammal assemblages, *Paleobiology* **14**:81–90.
Horwitz, L. K., and Goldberg, P., 1989, A study of Pleistocene and Holocene hyaena coprolites, *Journal of Archaeological Science* **16**:71–94.
Kelly, T. S., 1990, Biostratigraphy of the Uintan and Duchesnean land mammal assemblages from the Middle Member of the Sespe Formation, Simi Valley, California, *Natural History Museum of Los Angeles County, Contributions in Science* **49**:1–42.
Koeningswald, W. von, 1987, The Eocene niche of early Tertiary apatemyids, an extinct group of mammals, *Nature* **326**:595–597.
Koeningswald, W. von, Storch, G., and Rishter, G., 1992, Primitive insectivores, extraordinary hedgehogs, and long fingers, in: *Messel: An Insight into the History of Life and of the Earth* (S. Schaal, and W. Ziegler, eds.), Clarendon Press, Oxford, pp. 161–177.
Krishtalka, L., 1976, Early Tertiary Adapisoricidae and Erinaceidae (Mammalia, Insectivora) of North America, *Bulletin of the Carnegie Museum of Natural History* **1**:1–40.
Krishtalka, L., and Setoguchi, T., 1977, Paleontology and geology of the Badwater Creek area, central Wyoming. Part 13. The late Eocene Insectivora and Dermoptera, *Annals of the Carnegie Museum* **46**:71–99.
Mellet, J. S., 1974, Scatalogical origin of microvertebrate fossil accumulations, *Science* **185**:349–350.
Olson, S. L., 1985, The fossil record of birds, in: *Avian Biology*, Volume VIII (D. S. Farner, J. R. King, and K. C. Parkes, eds.), Academic Press, New York, pp. 79–238.
Prothero, D. R., 1996, Magnetic stratigraphy and biostratigraphy of the Middle Eocene Uinta Formation, Uinta Basin, Utah, in: *The Terrestrial Eocene-Oligocene Transition in North America* (D. R. Prothero, and R. J. Emry, eds.), Cambridge University Press, Cambridge, pp. 3–24.

Rasmussen, D. T., 1996, A New Middle Eocene omomyine primate from the Uinta Basin, Utah, *Journal of Human Evolution* **31**:75–87.

Rasmussen, D. T., Conroy, G. C., Friscia, A. R., Townsend, K. E., and Kinkel, M. D., 1999, Mammals of the Middle Eocene Uinta Formation, in: *Fossil Vertebrates of Utah* (D. D. Gillette, ed.), Special Publication of the Utah Geological Survey, Salt Lake City, pp. 401–420.

Reed, C., and Reed, B., 1928, The mechanism of pellet formation in the great horned owl (*Bubo virginianus*), *Science* **68**:359–360.

Scott, W. B., and Osborn, H. F., 1890, The Mammalia of the Uinta Formation, *Transactions of the American Philosophical Society* **1887**:255–264.

Seigfried, W. R., 1965, On the food habits if the spotted eagle owl, *Ostrich* **36**:146.

Setoguchi, T., 1975, Paleontology and geology of the Badwater Creek area, central Wyoming. Part 11. Late Eocene marsupials, *Annals of the Carnegie Museum* **45**:496–502.

Shipman, P., and Walker, A., 1980, Bone-collecting by harvesting ants, *Paleobiology* **6**:496–502.

Townsend, K. E., and Rasmussen, D. T., 1995, Skeletal remains of the small artiodactyl, *Mesomeryx grangeri*, from the Uinta Formation, Uinta Basin, Utah, *Journal of Vertebrate Paleontology* **15**(3):57A.

Walsh, S. L., 1996, Middle Eocene mammalian faunas of San Diego County, California, in: *The Terrestrial Eocene-Oligocene Transition in North America* (D. R. Prothero, and R. J. Emry, eds.), Cambridge University Press, Cambridge, pp. 75–119.

Chapter 13

A Mass Death Accumulation of *Coryphodon anthracoideus* (Mammalia: Pantodonta) at Roehler's *Coryphodon* Catastrophe Quarry (Lower Eocene, Wasatch Formation), Washakie Basin, Wyoming

ELIZABETH MCGEE

1. Introduction

As a richly fossiliferous bone bed with over 350 bones from a single taxon, Roehler's *Coryphodon* Catastrophe Quarry (RCCQ) in the Washakie Basin of

ELIZABETH MCGEE • Department of Biological Sciences, San José State University, San José, CA 95192-0100.

Eocene Biodiversity: Unusual Occurrences and Rarely Sampled Habitats, edited by Gregg F. Gunnell, Kluwer Academic/Plenum Publishers, New York, 2001.

Wyoming is an intriguing paleontological phenomenon. *Coryphodon* was a large bodied, subdigitigrade, graviportal herbivorous browser that first appeared in North America in the latest Paleocene (Clarkforkian Land Mammal Age) and was ubiquitous in early Eocene terrestrial faunas (Lucas, 1984). It is suggested that RCCQ is a mass death accumulation, but that evidence for such an event is obscured by the overprint of at least one additional stage which intervened between the time of death and the time of final deposition. This paper draws on faunal, taphonomic, and sedimentologic evidence to reconstruct the sequence of events leading to formation of the RCCQ bone bed, and to ascertain the extent and degree of bias caused by successive modifications.

2. Regional Geology and Paleontology

The Washakie Basin is an intermontaine desert basin located approximately 30 miles southeast of the city of Rock Springs, Wyoming (Figure 1). The basin

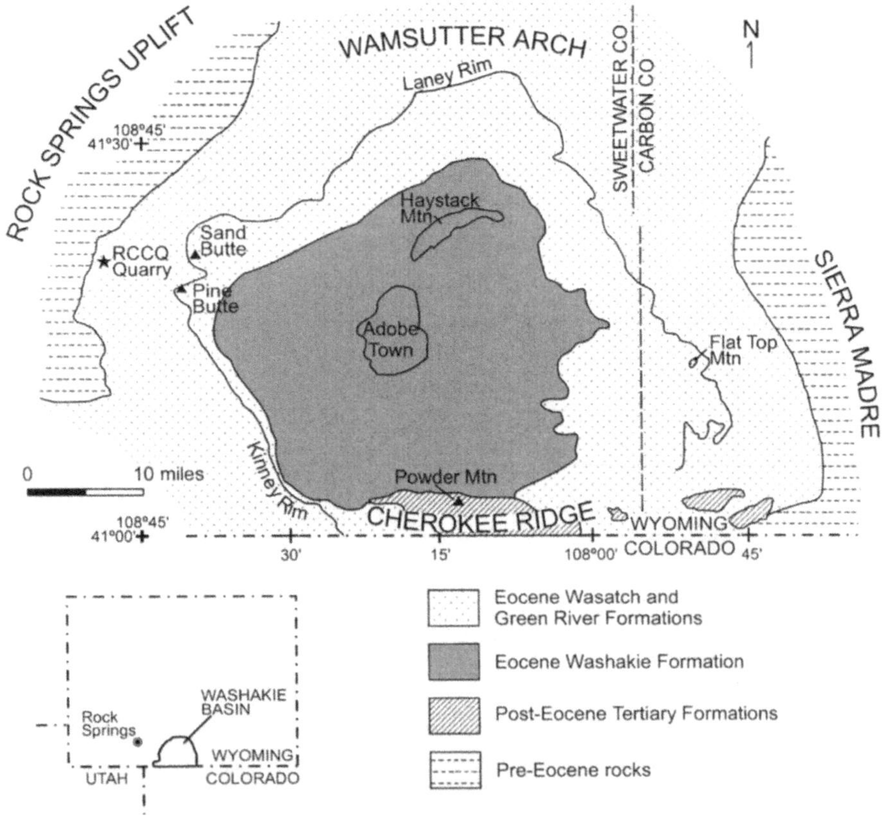

FIGURE 1. Location of Roehler's *Coryphodon* Catastrophe Quarry along the western rim of the Washakie Basin in south central Wyoming.

spans approximately 6475 km^2. With 80 percent of the outcrops in the basin consisting of lacustrine and fluvial rocks of Eocene age, the Washakie Basin contains one of the more complete sequences of continental Eocene deposits known in North America. Rocks from the Fort Union Formation (Paleocene) and of the Browns Park Formation (Miocene), as well as surficial Quaternary deposits make up the other 20 percent of the exposures (Roehler, 1969; Turnbull, 1972, 1978).

In 1970, D.E. Savage and W.A. Clemens of the University of California Museum of Paleontology (UCMP) led a class field expedition to an area near the Bitter Creek Railroad Station on the northern rim of the basin. Discoveries there prompted D.E. Savage, J.H. Hutchison, and B.T. Waters to begin a long-term exploration of the area that yielded over 86 localities of Wasatchian vertebrates (Savage, pers. comm.). Additionally, interest by the United States Geological Survey (USGS) in oil shale deposits in the Washakie Basin resulted in extensive mapping and study of the deposits and, coincidentally, in the discovery of several vertebrate localities (Roehler, 1969, 1973).

In the early 1970s, USGS geologist Henry Roehler discovered a localized bone bed, which was later named "Roehler's *Coryphodon* Catastrophe Quarry" (UCMP V72193), in the Main Body of the Wasatch Formation. Approximately 31 m^2 of the quarry was excavated in the 1970s and during one brief field season in 1987; overall the quarry is approximately 7 m^2 (E–W) by 15 m^2 (N–S) in extent. The bone bed is a laterally extensive deposit and consists almost entirely of disarticulated *Coryphodon anthracoideus* bones in a silty sandstone unit. In addition to *Coryphodon*, the RCCQ assemblage contains a small number of *Hyracotherium* (Perissodactyla), ?*Pachyaena* (Cete), Creodonta indet., turtle, bird, and fish bones (2.8 percent of the total assemblage).

2.1. Depositional Environment

The area around RCCQ is dominated by sandstone and siltstone facies. The sandstone facies is characterized by crossbedding and planar laminations; a few conglomeratic layers are present which fine upward and are demarcated from previous levels by scour marks. The clasts of these conglomerates are typically intraformational sandstone (as noted by comparison with nearby sediments elsewhere in the section), but there are some fossil bone clasts. The siltstone facies is fine grained, greenish grey in color with some organic matter (as evidenced by the dark color and by carbon stains). Gypsum occurs in several places interbedded with sandy silt and silty sand layers.

The relatively high frequency of crossbedding suggests a fluvial system. The siltstone facies could possibly represent an abandoned channel or a crevasse splay. The presence of ripple marks and sand-size sediment argues against the possibility of overbank deposits. The association of the siltstone facies with sandstone facies below and above, and the absence of large scale crossbedding and large sandstone bodies indicate that these deposits probably accumulated on a floodplain. The sandstone facies is much more common than the siltstone facies, indicating either a river of low sinuosity in which overbank flooding was rare, or a situation in which channel migration resulted in reworking of the floodplain deposits, thus decreasing the likelihood of their preservation.

The dominant bedforms at RCCQ are graded bedding, and bedding with either diagonal or horizontal lamination. While crossbedding is a product of the downcurrent migration of sand waves, graded beds are indicative of a decaying current (Pettijohn *et al.*, 1987). Ripple marks suggest a lower flow regime (where the Froude number is less than 1) was operating at least part of the time (Simons and Richardson, 1961; Pettijohn *et al.*, 1987). Because planar bedding structures have been observed in both upper and lower flow regimes, it is not possible to comment on flow regime using this type of bedform. However, a lower flow regime is also consistent with observed grain diameters (0.16–0.18 mm).

The bones at RCCQ are predominantly entombed in loosely lithified sandstone with localized patches of concretionary sandstone. The deposit is more laterally than vertically extensive, and is rectangular with the long axis running parallel to what is believed to be the direction of the current based on observations on the orientation of bedforms. The deposit is topped by overburden 0.5 to 1 m in thickness. Extant vegetation covers the deposit and most of the fossil material has sustained some damage by roots. Directly below the bone bed is a sandy siltstone unit of the siltstone facies. The quarry sample consists of more than 350 bones of *Coryphodon anthracoideus*, ranging in size from phalanges and isolated teeth a few centimeters long to elements as large as pelves (Figure 2). In addition to whole or nearly whole *Coryphodon* bones, over 300 *Coryphodon* bone fragments have been recovered.

3. Taphonomy

Broadly speaking, vertebrate remains in the fossil record can accumulate either in a gradual (attritional) manner, or, less commonly, as the result of single, temporally restricted events ("catastrophes"). Faunal criteria are often used to initially distinguish between these two modes of accumulation. A catastrophic, or mass death, accumulation formed in a fluvial environment is characterized by the predominance of a single taxon which shows an age profile that is representative of a living herd (Kurtén, 1953; Voorhies, 1969; Saunders, 1977; Turnbull and Martill, 1988; but see Lyman, 1994). Bones in mass death accumulations are also relatively more complete, and are frequently found in association with one another, reflecting localized conditions of death and/or transport (Behrensmeyer, 1975, 1991). Assemblages formed over a gradual period of time usually sample many taxa, and have an age profile that reflects greater mortality in juveniles and old adults (Klein and Cruz-Uribe, 1984; Haynes, 1985, 1987). Additionally, attritional assemblages tend to consist of fragmented and dispersed bones.

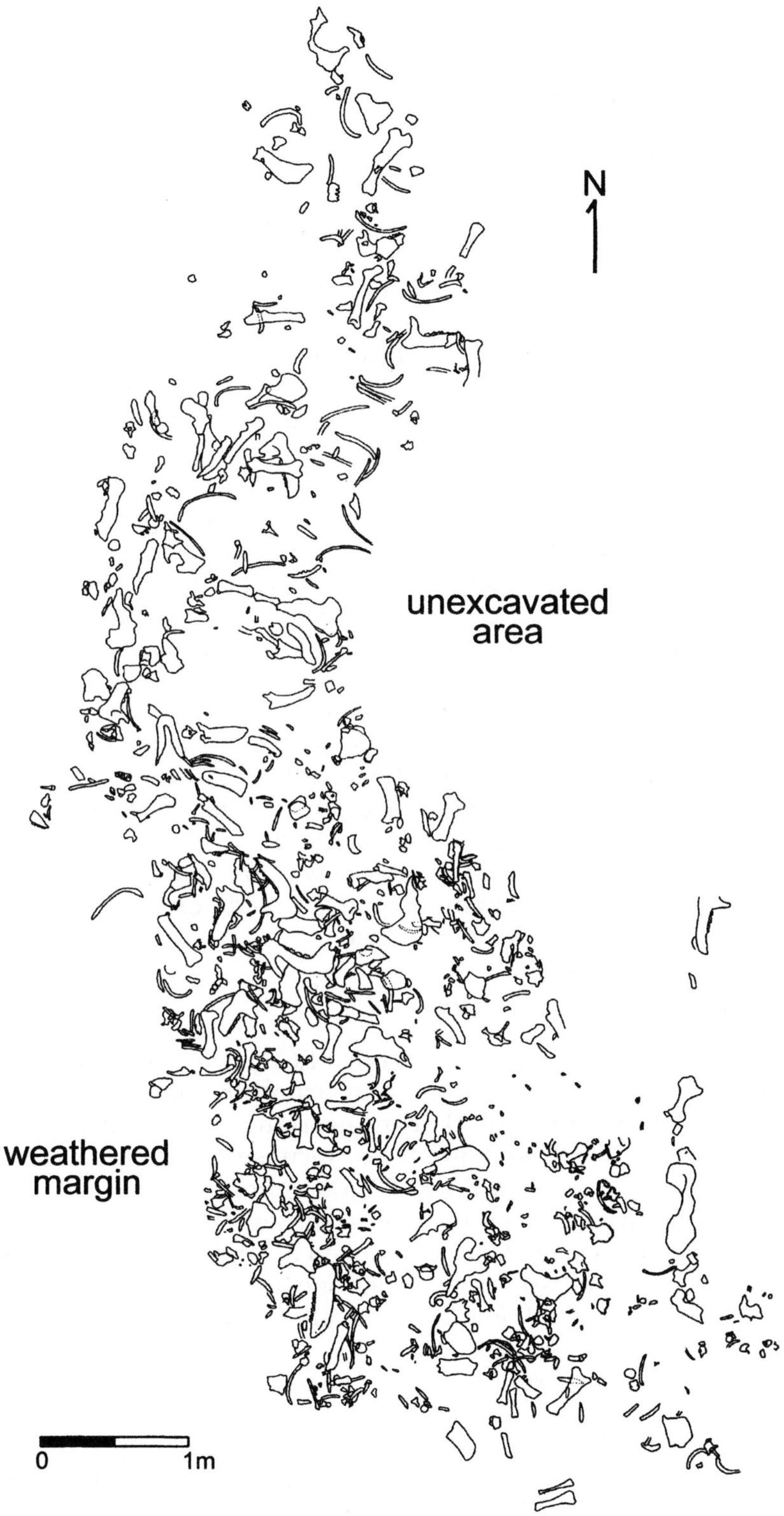

FIGURE 2. Plan view of excavated portions of the *Coryphodon anthracoideus* bone bed at RCCQ. Quarry dimensions run 7 m^2 (E–W) by 15 m^2 (N–S).

3.1. The Mosaic Nature of RCCQ

RCCQ is problematic in that it appears to be neither a strictly attritional assemblage nor a catastrophic one. RCCQ resembles a catastrophic assemblage in that the Quarry is dominated by a single species (McGee, in prep.) of a reputed herding mammal (Hébert, 1856; Osborn, 1898; Simons, 1960; Lucas, 1984). Herding presumably makes animals more susceptible to mass deaths (Boaz, 1982; Behrensmeyer and Hook, 1992).

RCCQ is unlike a mass mortality assemblage in several ways. First, the age profile at RCCQ consists almost entirely of adults of different ages. Catastrophic assemblages should be, in principle, a "snapshot" of the population's age profile at the time of death. The teeth in all 10 dentaries recovered are fully erupted and are at various stages of wear. Only one juvenile bone (a tibia) has been recovered, so that juveniles account for less than 1 percent of the assemblage. Furthermore, unlike other large mammal mass mortality assemblages (e.g., *Mesatirhinus sp.* from the Washakie Basin [Turnbull and Martill, 1988] and *Coryphodon molestus* from AMNH Quarry 242 in the San Juan Basin [Lucas, 1984]), the bone assemblage at RCCQ is highly fragmented (nearly 50% of the total assemblage) with no association between skeletal elements. Lastly, there is no indication of rapid deposition of strata. Aside from the fossil bones at RCCQ, there is very little clastic material either in the quarry itself or in the vicinity. Furthermore, bedforms identified in the sandstone facies are suggestive of a low energy depositional environment. Therefore, while the monospecific nature of the RCCQ assemblage suggests a catastrophic event, it is also evident that as the *Coryphodon* assemblage was transformed from a life assemblage to a fossil sample, a certain amount of modification took place which now obscures the history of the bone bed.

3.1.1. Formational Stages of RCCQ

As a working model, it is proposed that RCCQ was formed over the course of several stages: 1) death; 2) simultaneous deposition of a number of carcasses with immediate interim burial; 3) later reworking with limited transport; and 4) final deposition. Of these four stages, reconstructing the event or events which resulted in the death of the RCCQ *Coryphodon* is the most speculative. Although in rare instances the mode of death leaves an imprint in the fossil record, e.g., unequivocal evidence of tooth marks indicating predation (Martill, 1991), post-mortem processes can smudge or completely obliterate that imprint (Andrews and Ersoy, 1990). For example, although death by predation could initially create tooth marks or characteristic bone breakage patterns, destruction during transport and subsequent diagenetic modifications would result in abrasion and fragmentation, making identification of carnivore patterns highly tenuous. It is useful, however, to consider how different causes of death could potentially effect different taphonomic patterns. Weigelt's (1927/1989) list is helpful, for example, in distinguishing between internal sources of death (disease and old age) and external sources of death (accidents). Accidents can affect single individuals, e.g., death due to intraspecific strife or predation, or multiple individuals, e.g., volcanic eruptions (Voorhies and Thomasson, 1979; Lyman, 1989), fire or freezing (Oliver and Graham, 1994), drowning and/or flooding (Boaz, 1982; Agenbroad, 1984; Capaldo and Peters, 1995), or starvation and/or drought (Conybeare and Haynes, 1984; Haynes, 1985, 1987, 1988; Rogers, 1990; Varricchio, 1995; Coombs and Coombs, 1997). Although accidental death events affecting multiple individuals are often referred to as "natural catastrophic mortality events" (Lyman, 1994), crossover between these categories occurs (e.g., individuals can drown as well as a herd), nor are catastrophic or mass death events limited to accidents, as in the case of disease epidemics. Capaldo and Peters (1995) analyzed modern bone deposits resulting from mass wildebeest drownings in Tanzania and identified signature criteria for these events. These criteria include the presence of adults from a single species, the absence of immature individuals, the predominance of axial elements and long bones, and the general absence of compact bones (Capaldo and Peters, 1995). As discussed below, RCCQ appears to meet these criteria, suggesting that *Coryphodon anthracoideus* at RCCQ may have died as a result of a mass drowning.

Reconstructing the events which occurred at stages 2 and 3 is facilitated by studying the composition of the assemblage. The relative abundance profile as defined by Voorhies (1969) is a convenient tool that uses assemblage composition to evaluate biases which affected stages 2 (interim burial) and 3 (subsequent reworking). Relative abundance is determined by first estimating the minimum number of individuals for the assemblage, which at RCCQ is 10 (based on dentaries), and then comparing the number of each element type present with the number expected if all 10 individuals had been preserved. For example, at RCCQ, 9 ulnae have been recovered, and based on an MNI of 10, 20 ulnae would be expected. The relative representation of this element is then 9/20 or 0.45. In the profile presented in Figure 3, it is readily apparent that a distinct bias exists in the number of postcranial elements present in the assemblage, with whole dentaries, scapulae and pelves being among the better represented elements, and carpals, tarsals, metapodia, and phalanges being among the more poorly represented. The RCCQ bone bed clearly does not represent a random sample of bones from a minimum of 10 individuals of *Coryphodon anthracoideus.*

3.2. Information Loss in Fluvial Environments

3.2.1. Vector Summation

Differences in relative abundance such as those seen in Figure 3 are sometimes attributed to the effects of sorting in a fluvial environment (Voorhies, 1969). Before considering sorting *per se*, however, the orientation of the bones was studied to determine if the bone bed was modified by fluvial activity. Typically, orientation data are analyzed graphically in the form of rose diagrams

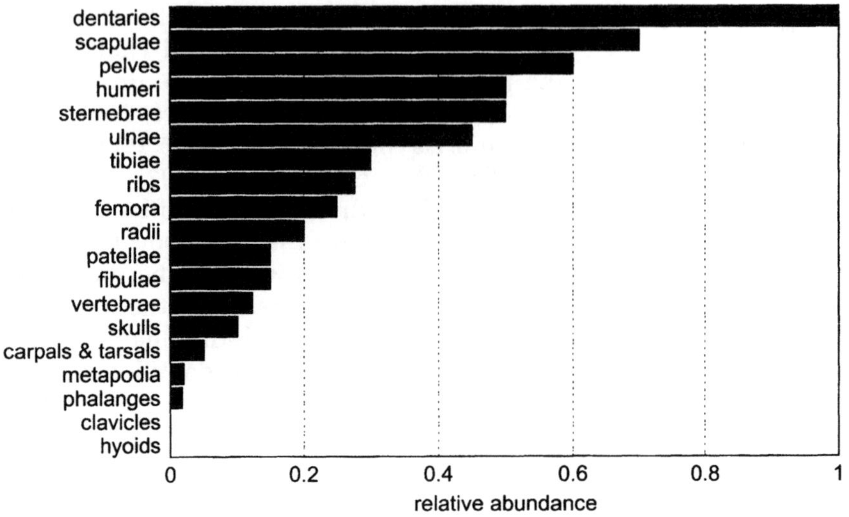

FIGURE 3. Relative abundance profile for *Coryphodon anthracoideus* elements at RCCQ, based on a minimum number of 10 individuals. Dentaries, scapulae, and pelves are well-represented, while carpals, tarsals, metapodia, and phalanges are not. The RCCQ assemblage does not represent a random sample of bones from a minimum number of 10 individuals of *Coryphodon anthracoideus.*

if the bone bed is laterally extensive (Lawton, 1977; Hunt, 1978; Shipman, 1981; Shipman *et al.*, 1981), or stereographic projections if the deposit has a substantial vertical component as well (Toots, 1965b; Hunt, 1978; Fiorillo, 1988a, 1988b). In addition, orientation analysis can be enhanced by using a quantitative method termed vector summation.

Vector summation compares the variance of vectors in a uniform circular distribution (i.e., a distribution in which all vectors have equal probability of representation) to the variance observed in the orientation of bones (the measured circular distribution). The null hypothesis is that variance of the *measured* circular distribution does not vary significantly from the variance of a *uniform* circular distribution. In other words, the bones at RCCQ are randomly oriented. The alternate hypothesis is the variance of the measured circular distribution is less than the variance of the uniform distribution, i.e., the bones at RCCQ are oriented in a non-random fashion.

Although Voorhies' flume studies (1969) demonstrated that pelves are the most sensitive indicators of current direction, only long bones were considered here since they are (as a group) more abundant at RCCQ, and are better preserved than are the pelves. The variance in the orientation of the long bones was estimated by first measuring the angle formed between the long axis of each long bone as plotted on a quarry map against a N–S standard. No distinction was made as to distal vs. proximal ends of the bones. Bones were treated as unipolar directional indicators and were assumed to be positioned "downstream". Only two quadrants are of relevance then, and measurements range from 0° (oriented to N–S vector) to (but not including) 180°.

To estimate the angular deviation, the azimuth and magnitude of the mean vector (in degrees) are first calculated. The angular variance is then the angular deviation squared. The azimuth of the mean vector (in degrees) is obtained by

$$\tan\Theta = \frac{\sum \sin\Theta}{\sum \cos\Theta} = \frac{16.25}{9.99} = 1.63$$

Θ is then 58°. It should be noted that although Θ is 58°, this does not imply current direction. As mentioned earlier, distal vs. proximal ends were not distinguished. Hence, current direction could either be 58° SW or 238° NE.

The magnitude of the mean vector, *r*, is obtained by:

$$r = \sqrt{\left(\frac{\sum \sin\Theta}{n}\right)^2 + \left(\frac{\sum \cos\Theta}{n}\right)^2} = \sqrt{.26 + .10} = .60$$

where $n = 32$, the number of data points in the sample. *r* is also a measure of consistency or dispersion of the data and can be used to estimate the angular deviation, *s* (Batschelet, 1981) by the following equation:

$$s = \sqrt{2(1 - r)}$$

For the RCCQ data, $s = .89$ radians or 51°. The variance of the measured circular distribution (in degrees) is then s^2 or 2601.

To determine if the variance of the measured circular distribution is significantly different from the variance of the uniform circular distribution, a variance ratio is calculated and compared with the F distribution. The variance of a uniform circular distribution is 6561 ($s = \pm 81$ degrees) (Dott, pers. comm.). The alternate hypothesis leads to a one-tailed test and F is calculated as:

$$F = \frac{s_1^2}{s_2^2} = \frac{6561}{2601} = 2.52$$

with n-1, or 31, degrees of freedom for both the variance of the measured circular distribution and the variance of the uniform circular distribution. For $F_{.05[31,31]}$, the critical value is 1.82. Because the observed variance ratio is greater than the critical value, the null hypothesis is rejected. The measured variance is significantly different from the uniform variance, i.e., the bones exhibit directional orientation. Given that the deposit underwent some modification by water, the question is then whether certain elements are underrepresented due to hydraulic sorting.

3.2.2. Voorhies Groups

Voorhies (1969) suggested that the relative frequencies of certain bones in fossil deposits is a function of the hydraulic behavior in bones of varying shape and/or size. Through a series of flume experiments using bones from sheep and coyotes, Voorhies (1969) noted that certain elements (e.g., skulls and rami) occurred as lag, while other elements (e.g., vertebrae, ribs, sacra, phalanges and carpals) remained in suspension. From his observations, Voorhies (1969) identified a three step sequence of removal under "typical" fluvial conditions. Group I is defined as elements that tend to be affected immediately by slight currents; transportation occurs mainly by saltation or even flotation. Group II consists of bones that are moved from their original position later than Group I; the primary mode of transportation is by traction. Group III elements are those that tend to resist transport and remain behind as a lag deposit (Behrensmeyer, 1975). These now-familiar "Voorhies Groups" are the basis for a model that predicts assemblage composition based on the dispersal potential of different groups of bones. For example, Voorhies noted that the sacrum and sternum float until waterlogged; therefore, their final deposition may be far removed from other elements from the same carcass. Similarly, some elements have a tendency to travel in suspension over greater distances (e.g., phalanges) and as a result may be consistently underrepresented in an assemblage.

Behrensmeyer (1975) expanded on Voorhies' stream table experiments, and showed that Voorhies Groups are generally size dependent. Figure 4 shows the distribution of bones from a sheep, a suid, and a hippo in Group I, a transitional I/II group, and Group III. In general, the number of elements in Group I decreases as the size of the animal increases. For an animal as large as the hippo, only phalanges and sesamoids will be sorted; all other elements will remain as lag. *Coryphodon*, with an estimated weight of 150 to 300 kg (Lucas, 1984), is comparable in body mass to the suid (*Hylochoerus*; Nowak, 1991) represented in Figure 4. All *Coryphodon* bones, then, except for ribs, patellae,

	Ovis	suid	hippo
GROUP I	phalanges sacrum sternum vertebrae (except atlas) patella proximal humerus distal metacarpal rib sesamoids	podials phalanges patella sesamoids	terminal phalanges sesamoids
GROUP I/II	astragalus podial atlas proximal metapodials phalanges	rib astragalus podial phalanges	terminal phalanges phalanges sesamoids
GROUP III	ALL OTHER BONES (= LAG)		

FIGURE 4. Voorhies Groups in animals of different sizes. The number of elements in Group I decreases as the size of the animal increases. *Coryphodon* is comparable in body mass to the suid.

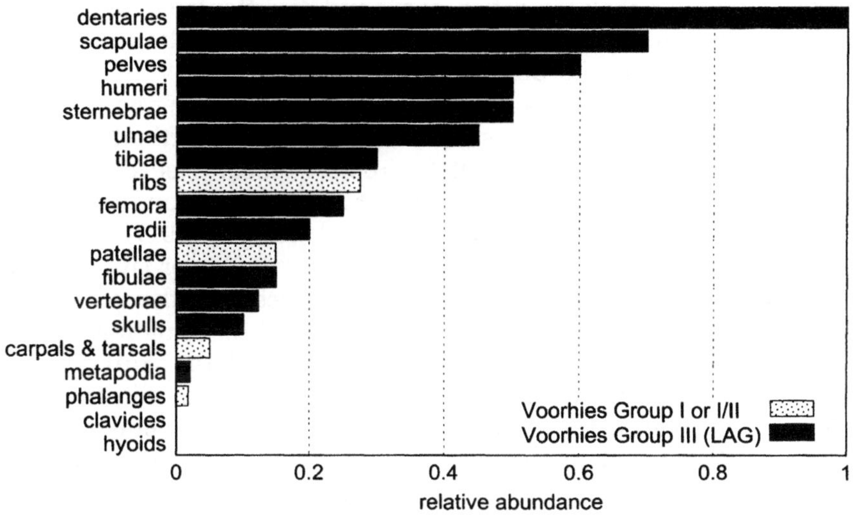

FIGURE 5. Relative abundance of *Coryphodon anthracoideus* skeletal elements shown according to Voorhies Groups. All *Coryphodon* bones except for ribs, patellae, carpals, tarsals, and phalanges will occur as lag elements.

carpals, tarsals, and phalanges, will occur as lag elements. Under the premise of Voorhies Groups, it is predicted that these elements will be underrepresented in a *Coryphodon* assemblage. Furthermore, the remaining Group III elements should be present in numbers proportionate to their representation in the skeleton.

Figure 5 shows the relative abundance profile for RCCQ with elements shaded according to Voorhies group number. While float elements (i.e., Group I or I/II) tend to be underrepresented, several lag elements, such as the skull and many limb bones, are underrepresented as well. Furthermore, Group III elements are not present in numbers proportionate to their representation in the skeleton (as indicated by relative abundance indices < 1.0). There is, then, no obvious association between relative representation and dispersal potential of individual bones.

Bones may be underrepresented for other reasons, however: lack of preservation, or simply chance (McGee and Martin, 1995). Removal by scavengers is another source of bias, although there is no indication of scavenging or carnivore activity at RCCQ. Alternatively, assemblage composition may be influenced by decomposition and disarticulation patterns.

3.3. Disarticulation Sequences

It is suggested that some of the bias seen in the relative abundance profile for RCCQ likely reflects information loss through disarticulation prior to burial.

The order in which a skeleton decomposes and disarticulates can influence the damage, destruction, transport, association, and spatial patterning of bones in fossil assemblages (Hill and Behrensmeyer, 1984).

Hill (1979) proposed considering animal remains in terms of disarticulation units that are or have been subject to particular disarticulation patterns. Several authors (Weigelt, 1927/1989; Toots, 1965a; Dodson, 1973) have observed sequences of disarticulation under a number of conditions (water vs. dry land; small animals vs. large animals, etc.). Weigelt (1927/1989) observed that in cows trapped in a deep water salt marsh, the lower jaw disarticulates first, followed by the phalanges, other limbs bones, and then the rest of the skeleton. Toots (1965a) in contrast observed the disarticulation sequence of coyotes on dry land and suggested a sequence in which the skull disarticulates first, followed by the entire forelimb and/or hindlimb as a unit, the atlas, ribs, individual limb bones, the dentary, and finally the vertebral column. Hill (1979) and Hill and Behrensmeyer (1984) noted that this disarticulation sequence is consistent across several groups of modern land mammals of medium to large size, except for the domestic cow. In the case of the cow, the terminal phalanges tend to stay bound up longer. In Hill's (1979) study, the disarticulation sequence of the modern topi *Damaliscus korrigum* Ogilby agreed more closely with the disarticulation sequence of the coyote reported by Toots (1965a) than with the sequence of the cow reported by Weigelt (1927/1989). Hill (1979) attributed these differences to the environment, particularly the presence or absence of water. In dry areas where mummification occurs due to the quick evaporation of body fluids, skeletons resist disarticulation. The amount of body fluid also affects the rate of decomposition. Decay and disarticulation proceed more rapidly in animals with a large amount of body fluids. The elephant, for example, is completely disarticulated by the time the skin and other tissues have disappeared. Likewise, areas of the body with less body fluid (tarsals and carpals) will remain as a unit longer than other parts of the body (Hill, 1979). Hence, water is a major factor in determining the disarticulation sequence. The same skeleton in dry conditions would disarticulate differently in humid conditions. In humid conditions, a cow will disarticulate from the extremities inward; whereas in drier conditions, disarticulation proceeds outward from the body to the extremities.

Coryphodon at RCCQ most likely decomposed in water or very near a water source, since burial by sediments carried by water is essential to the process of preservation and fossilization. If this is accurate, these animals would have disarticulated from the extremities inward, with the result that in the fossil sample, distal limb bones will be less abundant relative to proximal limb bones. This interpretation is supported by Figure 6, where the relative abundance of RCCQ elements is redrawn to show complementing pairs of forelimb and hind limb elements. Relative abundance for all elements except for the femora decreases proximally to distally down the limbs, i.e., pelves ($n = 12$) and scapulae ($n = 14$) are quite abundant, whereas elements occurring more distally in the limb are less abundant (e.g., tibiae ($n = 6$) and radii

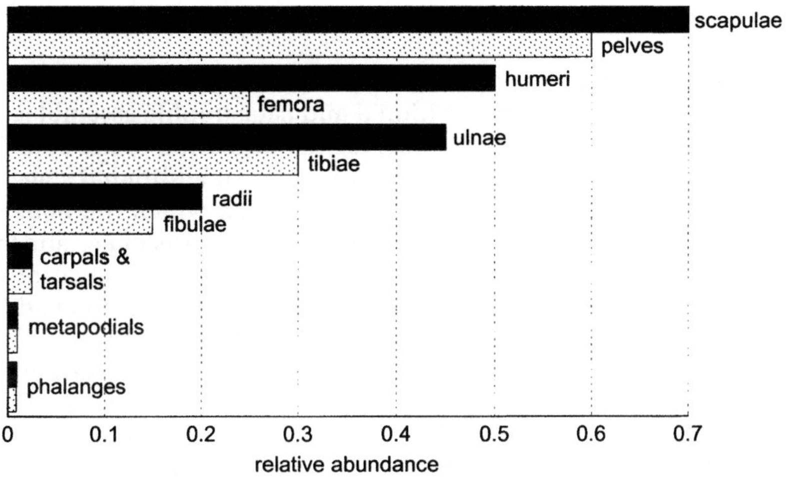

FIGURE 6. Comparison of relative abundance of forelimb and hind limb elements of *Coryphodon anthracoideus* at RCCQ. Proximal limb elements are more common than distal limb elements.

($n = 4$). Therefore, disarticulation patterns could have contributed to at least some of the bias observed at RCCQ.

3.4. Scenario of Events at RCCQ

Bias in the RCCQ assemblage can be attributed to information loss that occurred during several separate events. First, a number of *Coryphodon* carcasses were deposited simultaneously. These underwent partial decay—particularly in the limbs—and then were buried. Upon interim burial, the carcasses were at a later stage of disarticulation, having had distal elements removed earlier. The absence of clastic material at RCCQ may reflect the deposition of bones via a carcass rather than as individual particles. Second, the bone bed was reworked, and a limited amount of transport occurred. Reworking also disassociated and fragmented skeletal elements. During this phase, the majority of elements were not affected as much by sorting as by reorientation. Given this second stage, it is not surprising that juveniles are underrepresented. Juvenile skeletons are more susceptible to breakage and their transport potential as carcass floats is different from adults (Capaldo and Peters, 1995). The fact that the composition of the bone bed remained remarkably homogeneous suggests that this reworking, while intense, was brief. Third, the bones at RCCQ were deposited for the final time. During this stage, the primary agents modifying the assemblage were diagenetic in nature—structural deformation (e.g., faulting, uplifting), bioturbation (notably, destruction cause by roots), and erosion of the bone bed. Aside from the destruction caused by roots, the degree of alteration during this phase appears to be minimal.

4. Conclusion

Roehler's *Coryphodon* Catastrophe Quarry is characterized by a mosaic of features that are indicative of both attritional and catastrophic assemblages. These features are best interpreted by considering the formation of the bone bed over a series of stages, from death to final deposition. An additional stage between interim burial (stage 2) and stage 4 (final burial) in which the bone bed was reworked and the elements transported over a short distance, appears to be most responsible for giving the bone bed the appearance of an attritional assemblage, i.e., an fragmented assemblage of adult individuals entombed in a silty sandstone. RCCQ is best categorized as a catastrophic assemblage dominated by a single species (presumably of a herding animal) deposited in a fluvial environment in which the transport and disarticulation potential of carcasses exerted the greatest influence on the composition of the assemblage.

The study of "catastrophes" in the fossil record raises interesting issues concerning the applicability of certain models in the literature. Hill's (1979) study of topi disarticulation shows that bones that would be easily transported as single units (Voorhies Group I) were bound up in carcasses for long periods of time and thus impervious to transportation (de facto Voorhies Group III). This effectively reverses the order bones are introduced to the fluvial system. Hill (1979) noted that bones such as phalanges, carpals, tarsals and sesamoids initially remain bound up in units that are more resistant to decomposition, while as isolated bones in flume experiments, these bones are transported immediately. The carcass model presents a more complicated scenario of a series of successive winnowing stages based not on the physical properties (density, shape, etc.) of each bone type, but rather on the effects of differential decay on releasing bones. Because each process has potentially opposite effects on spatial patterning, models should be used with a clear understanding of what stage is being treated. In this study, Voorhies' model is preferable in analyzing processes following decay but before final burial, whereas a carcass disarticulation model is appropriate to interpret patterns produced by death and decay, but not necessarily transport.

Mass death assemblages represent geologically instantaneous events that essentially "freeze frame" life in the fossil record. These assemblages afford a unique opportunity to study certain paleobiological and ecological features of the assemblage that are ordinarily not preserved. However, even mass death assemblages have sustained a certain amount of information loss or bias, and it is therefore important to understand how this bias has affected our interpretation of the history of the assemblage.

ACKNOWLEDGMENTS. I would like to thank Drs. J. Howard Hutchison, Donald E. Savage, and W.B.N. Berry (UCMP) for their sponsorship of this project. Drs. Jennifer Hogler and Donald Lofgren (Raymond Alf Museum) provided many good ideas and criticisms, and were invaluable in the field. Kyoko Kishi (University of Texas at Austin) and Dr. Howard Hutchison prepared the illustrations. I thank Kyoko Kishi and Mark Goodwin (UCMP) for patiently teaching me how to resurrect smashed *Coryphodon* specimens. I also thank Drs. Gregg Gunnell (University of Michigan) and David Krause (SUNY Stony Brook) and an anonymous reviewer for critiques of earlier versions of this paper. Special thanks to Susan Mitnick Falsetti and Dr. Tony Falsetti (University of Florida) for their assistance and support during this project.

References

Agenbroad, L. D., 1984, Hot Springs, South Dakota: Entrapment and taphonomy of Columbian Mammoth, in: *Quaternary Extinctions: a Prehistoric Revolution* (P. S. Martin and R. G. Klein, eds.), University of Arizona Press, Tucson, pp. 113–127.

Andrews, P., and Ersoy, A., 1990, Taphonomy of the Miocene bone accumulations at Pasalar, Turkey, *J. Hum. Evol.* **19**:379–396.

Batschelet, E., 1981, *Circular Statistics in Biology*, Academic Press, New York.

Behrensmeyer, A. K., 1975, The taphonomy and paleoecology of Plio-Pleistocene vertebrate assemblages east of Lake Rudolf, Kenya, *Bull. Mus. Comp. Zool.* **146**:473–578.

Behrensmeyer, A. K., 1991, Terrestrial vertebrate accumulations, in: Taphonomy: *Releasing the Data Locked in the Fossil Record* (P. A. Allison, and D. E. G. Briggs, eds.), Plenum Press, New York, pp. 291–335.

Behrensmeyer, A. K., and Hook, R. W., 1992, Paleoenvironmental contexts and taphonomic modes, in: *Terrestrial Ecosystems Through Time: Evolutionary Paleoecology of Terrestrial Plants and Animals* (A. K. Behrensmeyer, J. D. Damuth, W. A. DiMichele, R. Pots, H.-D. Sues, and S. L. Wing, eds.), University of Chicago Press, Chicago, pp. 15–136.

Boaz, D., 1982, Modern riverine taphonomy: Its relevance to the interpretation of Plio-Pleistocene hominid paleoecology in the Omo Basin, Ethiopia, Ph.D. thesis, University of California, Berkeley.

Capaldo, S. D., and Peters, C. R., 1995, Skeletal inventories from wildebeest drownings at lakes Masek and Ndutu in the Serengeti ecosystem of Tanzania, *J. Arch. Sci.* **22**:385–408.

Conybeare, A., and Haynes, G., 1984, Observations on elephant mortality and bones in water holes, *Quat. Res.* **22**:189–200.

Coombs, M. C., and Coombs, W. P., Jr., 1997, Analysis of the geology, fauna, and taphonomy of Morava Ranch Quarry, Early Miocene of Northwest Nebraska, *Palaios* **12**:165–187.

Dodson, P., 1973, The significance of small bones in paleoecological interpretation, *Univ. Wyo. Contrib. Geol.* **12**:15–19.

Fiorillo, A. R., 1988a, A proposal for graphic presentation of orientation data from fossils, *Univ. Wyo. Contrib. Geol.* **26**:1–4.

Fiorillo, A. R., 1988b, Taphonomy of Hazard Homestead Quarry (Ogallala Group), Hitchcock County, Nebraska, *Univ. Wyo. Contrib. Geol.* **26**:57–97.

Haynes, G., 1985, Age profiles in elephant and mammoth bone assemblages, *Quat. Res.* **24**:333–345.

Haynes, G., 1987, Proboscidean die-offs and die-outs: Age profiles in fossil collections, *J. Arch. Sci.* **14**:659–668.

Haynes, G., 1988, Mass deaths and serial predation: comparative taphonomic studies of modern large mammal death sites, *J. Arch. Sci.* **15**:219–235.

Hébert, E., 1856, Recherches sur la faune des premiers sédiments tertiaires Parisiens (mammifères pachydermes du genre *Coryphodon*), *Ann. Sci. Nat.* **6**:87–136.

Hill, A., 1979, Disarticulation and scattering of mammal skeletons, *Paleobiology* **5**:261–274.

Hill, A., and Behrensmeyer, A. K., 1984, Disarticulation patterns of some modern East African mammals, *Paleobiology* **10**:366–376.

Hunt, R. M., Jr., 1978, Depositional Setting of a Miocene Mammal Assemblage, Sioux County, Nebraska (U.S.A.), *Palaegeogr. Palaeoclimatol. Palaeoecol.*, **24**:1–52.

Klein, R. G., and Cruz-Uribe, K., 1984, *The analysis of animal bones from archeological sites*, University of Chicago Press, Chicago.

Kurtén, B., 1953, On the variation and population dynamics of fossil and recent mammal populations, *Acta Zool. Fenn.*, **76**:1–122.

Lawton, R., 1977, Taphonomy of the dinosaur quarry, Dinosaur National Monument, *Univ. Wyo. Contrib. Geol.* **15**:119–126.

Lucas, S. G., 1984, Systematics, Biostratigraphy, and Evolution of Early Cenozoic *Coryphodon* (Mammalia, Pantodonta), Ph.D. thesis, Yale University, New Haven.

Lyman, R. L., 1989, Taphonomy of cervids killed by the 18 May 1980 volcanic eruption of Mount St. Helens, Washington in: *Bone Modification* (R. Bonnichsen and M. Sorg, eds.), Center for the Study of the First Americans, University of Maine, Orono, pp. 149–167.

Lyman, R. L., 1994, *Vertebrate Taphonomy*, Cambridge University Press, Cambridge.

Martill, D. M., 1991, Bones as stones: the contribution of vertebrate remains to the lithologic record, in: *Processes of Fossilization* (S. K. Donovan, ed.), Columbia University Press, New York, pp. 270–292.

McGee, E. M., Intraspecific variation in *Coryphodon anthracoideus* from Roehler's *Coryphodon* Catastrophe Quarry. In preparation.

McGee, E. M., and Martin, L., 1995, Chance and the taphonomy of the hominoid assemblage from the middle Miocene site at Pasalar, Turkey, *J. Hum. Evol.* **28**:325–341.

Nowak, R. M., 1991, *Walker's Mammals of the World*, Johns Hopkins University Press, Baltimore.

Oliver, J. S., and Graham, R. W., 1994, A catastrophic kill of ice-trapped coots: time-averaged versus scavenger-specific disarticulation patterns, *Paleobiology* **20**:229–244.

Osborn, H. F., 1898, A complete skeleton of *Coryphodon radians*. Notes upon the locomotion of this animal, *Bull. Am. Mus. Nat. Hist.* **10**:81–91.

Pettijohn, F. J., Potter, P. E., and Siever, R., 1987, *Sand and Sandstone*, 2nd Edition, Springer-Verlag, New York.

Rogers, R. R., 1990, Taphonomy of three dinosaur bone beds in the Upper Cretaceous Two Medicine Formation of northwestern Montana: evidence for drought-related mortality, *Palaios* **5**:394–413.

Roehler, H. W., 1969, Stratigraphy and oil-shale deposits of Eocene rocks in the Washakie Basin, Wyoming, *Twenty-first Annu. Field Conf. Wyo. Geol. Assoc. Guideb.*, pp. 197–206.

Roehler, H. W., 1973, Stratigraphy of the Washakie Formation in the Washakie Basin, Wyoming, *Geol. Surv. Bull.* 1369.

Saunders, J. J., 1977, Late Pleistocene vertebrates of the western Ozark Highland, Missouri, *Ill. State Mus. Rep. Invest. No. 33*, Springfield, Illinois.

Schäfer, W., 1972, *Ecology and Palaeoecology of Marine Environments*, Irmgard Oertel, trans., Oliver and Boyd, Edinburgh.

Shipman, P., 1981, *Life History of a Fossil*, Harvard University Press.

Shipman, P., Walker, A., Van Couvering, J. A., Hooker, P. J., and Miller, J. A., 1981, The Fort Ternan hominoid site, Kenya: geology, age, taphonomy and paleoecology, *J. Hum. Evol.* **10**:49–72.

Simons, D. B., and Richardson, E. V., 1961. Forms of bed roughness in alluvial channels, *Am. Soc. Civil Eng. Proc., Jour. Hydraulics Div. 87 (HY-3)*, pp. 87–105.

Simons, E. L., 1960, The Paleocene Pantodonta, *Tran. Am. Philos. Soc.* **50**:1–99.

Toots, H., 1965a, Sequence of disarticulation in mammalian skeletons, *Univ. Wyo. Contrib. Geol.* **4**:37–39.

Toots, H., 1965b, Random orientation of fossils and its significance, *Univ. Wyo. Contrib. Geol.* **4**:59–62.

Turnbull, W. D., 1972, The Washakie Formation of Bridgerian-Uintan ages, and the related faunas, in: *Guideb. Field Conf. on Tertiary Biostratigraphy of Southern and Western Wyoming*, (R. M. West, ed.), privately printed, pp. 3–19.

Turnbull, W. D., 1978, The mammalian faunas of the Washakie Formation, Eocene age, of southern Wyoming, Part I Introduction: The geology, history, and setting, *Fieldiana, Geol.* **33**:569–601.

Turnbull, W. D., and Martill, D. M., 1988, Taphonomy and preservation of a monospecific titanothere assemblage from the Washakie Formation (Late Eocene), Southern Wyoming: an ecological accident in the fossil record, *Palaeogeogr., Palaeoclimatol. Palaeoecol.* **63**: 91–108.

Varricchio, D. J., 1995, Taphonomy of Jack's Birthday site, a diverse dinosaur bonebed from the Upper Cretaceous Two Medicine Formation of Montana, *Palaeogeogr. Palaeoclimatol. Palaeoecol.* **114**:297–323.

Voorhies, M. R., 1969, Taphonomy and population dynamics of an early Pliocene vertebrate fauna, Knox County, Nebraska, *Univ. Wyo. Contrib. Geol. Spec. Pap.* **1**:1–69.

Voorhies, M. R., and Thomasson, J. R., 1979, Fossil grass anthoecia within Miocene rhinoceros skeletons: Diet in an extinct species, *Science* **206**:331–333.

Weigelt, J., 1927, Rezente wirbeltierleichen und ihre paläobiologische bedeutung, Max Weg, Leipzig.

Weigelt, J., 1989, Recent Vertebrate Carcasses and their Paleobiological Implications, Judith Schaefer, trans., University of Chicago Press, Chicago.

Chapter 14

Meniscotherium Mass-Death Assemblages

THOMAS E. WILLIAMSON

1. Introduction

Meniscotherium is a relatively small condylarth that is considered unusual in having a specialized dentition consisting of molarized posterior premolars and selenodont-lophodont molars (Figure 1). It unites this specialized dentition with a very generalized skeleton (Williamson and Lucas, 1992). Two species of *Meniscotherium* are recognized, *M. chamense* and *M. tapiacitum* (Williamson, 1989; Williamson and Lucas, 1992). *M. chamense* from the San Jose Formation of the San Juan Basin, New Mexico probably was about 58 cm in length and ranged from between 4.88–9.76 kg in body weight. *M. tapiaci-*

THOMAS E. WILLIAMSON • New Mexico Museum of Natural History and Science, Albuquerque, NM 87104-1375.

Eocene Biodiversity: Unusual Occurrences and Rarely Sampled Habitats, edited by Gregg F. Gunnell, Kluwer Academic/Plenum Publishing, New York, 2001.

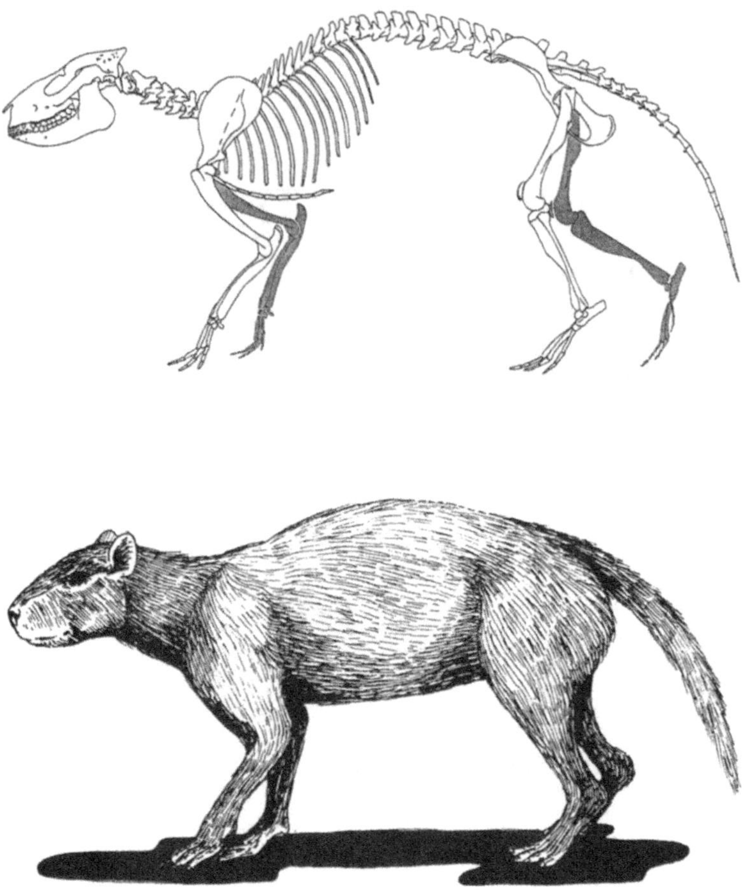

FIGURE 1. Reconstruction of the early Eocene (Wasatchian) condylarth *Meniscotherium chamense* (modified from Williamson, 1989 and Williamson and Lucas, 1992).

tum was approximately 70% the size of San Juan Basin *M. chamense* in linear dimensions (about 41 cm in length) and probably weighed between 1.7 and 3.3 kg (Williamson and Lucas, 1992). The apparently erratic occurrence of *Meniscotherium* has long been noted (Van Houten, 1945) and the presence/absence of *Meniscotherium* in the San Jose Formation was believed to be controlled by the depositional environment as reflected in particular facies association (Granger, 1914; Simpson, 1948b; Gazin, 1952, 1962, 1965a, b). *Meniscotherium* is often found in large concentrations in the San Jose Formation, New Mexico and as a result, it is one of the best known early Paleogene condylarths.

Three unusual occurrences of *Meniscotherium* are analyzed in this paper. Simpson (1948b) initially reported on a large concentration of *M. chamense* in the Tapicitos Member of the San Jose Formation of northwestern New Mexico, the AMNH 150 Quarry (referred to as locality NMMNH-L201 later in this paper). The University of California, Berkeley recovered a significant sample of *M. tapiacitum* from the Wasatch Formation of southwestern Wyoming from UCMP locality V71237. An additional large sample of *M. chamense* was recovered from the Regina Member of the San Jose Formation of northwestern New Mexico by the New Mexico Museum of Natural History and Science from NMMNH locality L-203. These three concentrations of *Meniscotherium* are especially significant because mass-death assemblages are generally rare for early Paleogene mammals and especially for condylarths. Therefore, these localities are analyzed below in an attempt to ascertain their probable cause. Also, the fossils these localities offer potential community population samples for *Meniscotherium* allowing for examination of its community structure and analysis of its intraspecific variation.

Abbreviations used in this paper are AMNH = American Museum of Natural History; NMMNH = New Mexico Museum of Natural History and Science; UALP = University of Arizona Museum of Paleontology; and UCMP = University of California Museum of Paleontology.

2. Ucmp Locality V71237

UCMP locality V71237 quarry is located in the Bitter Creek area of southwestern Wyoming in the Wasatch Formation (Washakie Basin). It was discovered in 1974 by J. H. Huchison of the University of California at Berkeley from the "Lower *Meniscotherium* Zone" (Savage *et al.*, 1972). The fossil remains of at least twelve individuals of *Meniscotherium tapiacitum* have been recovered from this site so far.

A map of a portion of the V71237 quarry (Figure 2) shows the elements of *Meniscotherium* and a small associated fauna to be largely disarticulated. Many of the bones are fragmented but all show an apparently unweathered surface. The equal area plot made from the quarry map (Figure 2) shows that the long-bones are weakly oriented bimodally (see Appendix 3 for statistics) but the two modes are not perpendicular suggesting that the flow was not unidirectional over the entire area or shifted direction during deposition of the bone-laden sediment.

A frequency plot of the various elements of *M. tapiacitum* from locality UCMP V71237 (Figure 3) as well as a Voorhies diagram (Figure 4) shows that it is depleted in smaller elements making it a winnowed assemblage.

3. Nmmnh Locality L-201

The most famous of *Meniscotherium* mass-death assemblages is from NMMNH L-201 (also known as the AMNH 150 quarry; Figure 5). This locality was discovered in the Tapicitos Member of the San Jose Formation, San Juan Basin, northwestern New Mexico by George Gaylord Simpson and William Fish in 1947 (Simpson, 1948b, 1981, and unpublished field notes at the AMNH) and has been collected by AMNH, UCMP, UALP, NMMNH and probably by several

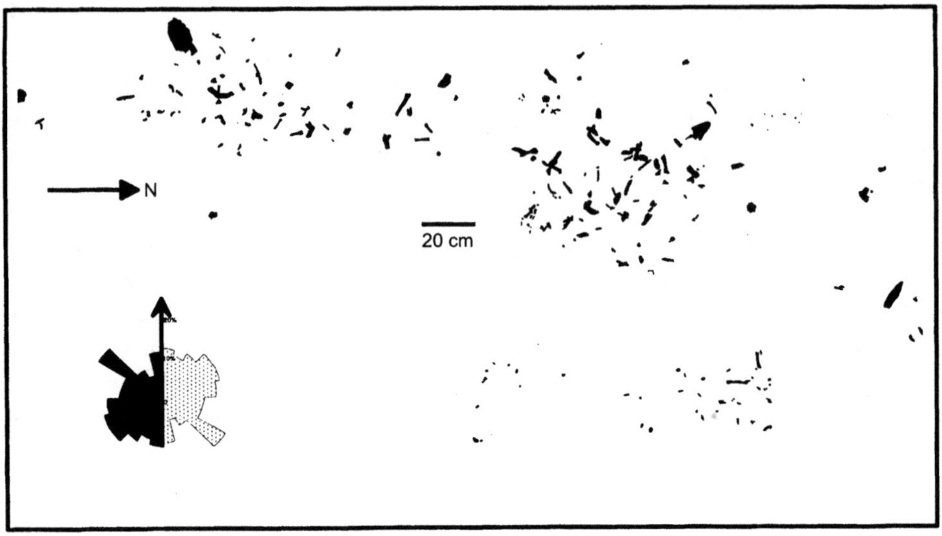

FIGURE 2. Quarry map showing the distribution of skeletal elements of *M. tapiacitum* from UCMP locality V71237 with the accompanying Rose diagram showing long-bone orientation (see Appendix 2 for statistical information) (unpublished data courtesy of Dr. J. H. Hutchison).

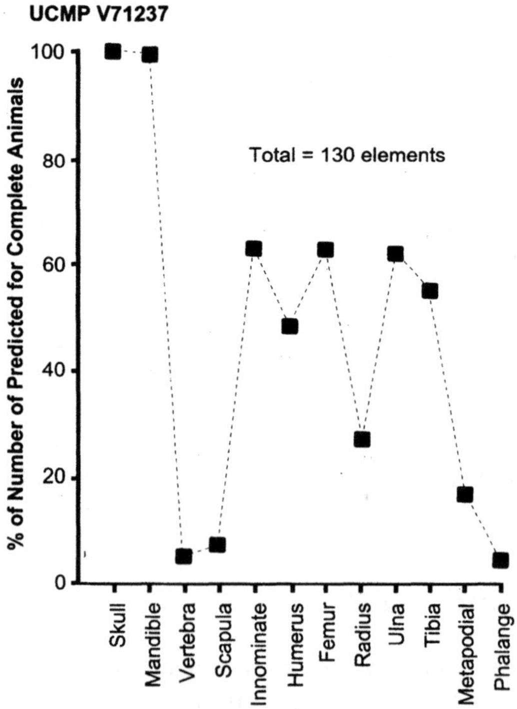

FIGURE 3. Relative bone frequency plot of 130 elements from locality UCMP V71237. Seven skulls were collected from this locality so the relative numbers of other elements were compared to the number expected for seven individuals (unpublished data courtesy of Dr. J. H. Hutchison).

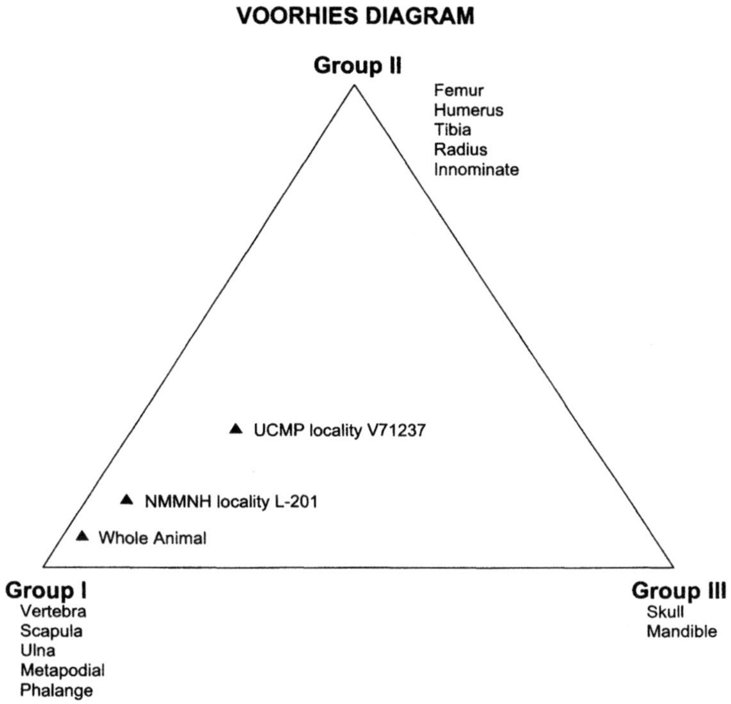

FIGURE 4. Voorhies diagram of localities NMMNH L-201, L-203 and UCMP V71237. A few items from V71237, for example a partially articulated manus, suggest that remains *of M. tapiacitum* were buried very shortly after death. Studies of natural decay of modern animals of similar size to *M. tapiacitum* (about 2 kg.) suggests that the remains in this quarry were buried in less than three weeks of death if exposed to the air and insects under warm, humid conditions (Payne, 1965).

other institutions in subsequent years. Based on the number of M^1s, a minimum of 46 individuals have been recovered from this locality. This locality is far from exhausted and the total numbers of individuals still remaining buried within the bonebed are probably far greater. The remains of *M. chamense* as well as other animals are found in a 7-to-13-cm-thick interval of poorly sorted muddy siltstone in a generally coarsening upward sequence of muddy siltstone alternating with relatively thin beds of sandstone (Figure 6). The sequence of strata that contains L-201 is in the Fbssc facies described by Smith (1988) and the Fbsc + Fl lithofacies described by Smith and Lucas (1991). These strata are interpreted to represent splay and overbank sedimentation relatively proximal to trunk streams (Smith, 1988; Smith and Lucas, 1991). These overall fining-upward sequences of sand and mud represent deposition of waning flows of relatively high-energy overbank events.

The NMMNH L-201 quarry was opened and the bone-producing interval worked for several weeks during the summer of 1987 (Williamson and McCarroll, 1988) and again in 1997 by workers of the NMMNH. Several square meters of the bone bed were quarried and mapped (Figure 7). Matrix collected

FIGURE 5. Photograph of locality NMMNH L-201 in the Tapicitos Member of the San Jose Formation, northwestern New Mexico.

from selected portions of the bonebed was then screenwashed in order to collect smaller elements that might be missed by quarrying alone. During this operation, the bones of *Meniscotherium* were found to be, as Simpson (1948b) had described, largely disarticulated. In addition, skulls are often found missing anterior teeth and the alveoli are filled with sediment. Isolated teeth are also found throughout the bone bed. However, at least one skull was recovered from this locality articulated with the mandible (AMNH 48125). Also, limb bones and partial manus and pes are occasionally found in articulation. Although Simpson (1948b) observed that even the most delicate bones are found intact, it is not uncommon to find fragmented bones including even larger elements such as tibiae and femora. However, all bone surfaces do appear fresh and lack any sign of subaerial weathering (e.g., Behrensmeyer, 1978; Bown, 1980).

A map of a portion of the L201 quarry (Figure 7) shows that skeletal elements tend to occur in patches (see also Simpson, 1948b). Lenses of coarser sediment largely devoid of fossils separate these patches of bone accumulation. Skeletal elements also occur above and below the coarser lenses. An equal area rose diagram of long-bone orientation (Figure 7) show a weakly bimodal distribution, with the two modes roughly perpendicular (see Appendix 3 for statistics).

Two types of ichnofossils were observed within or near the main bone bed. The first and most abundant is a smooth-walled, tubular burrow about 1 cm in diameter that is usually sinuous and similar to the "Type 5" burrow described by Bown and Krause (1983) except these are usually oriented nearly vertically and are also similar to the "Type 1" burrows described by Smith (1988) in the Regina and Tapicitos members of the San Jose Formation. Longitudinal cross

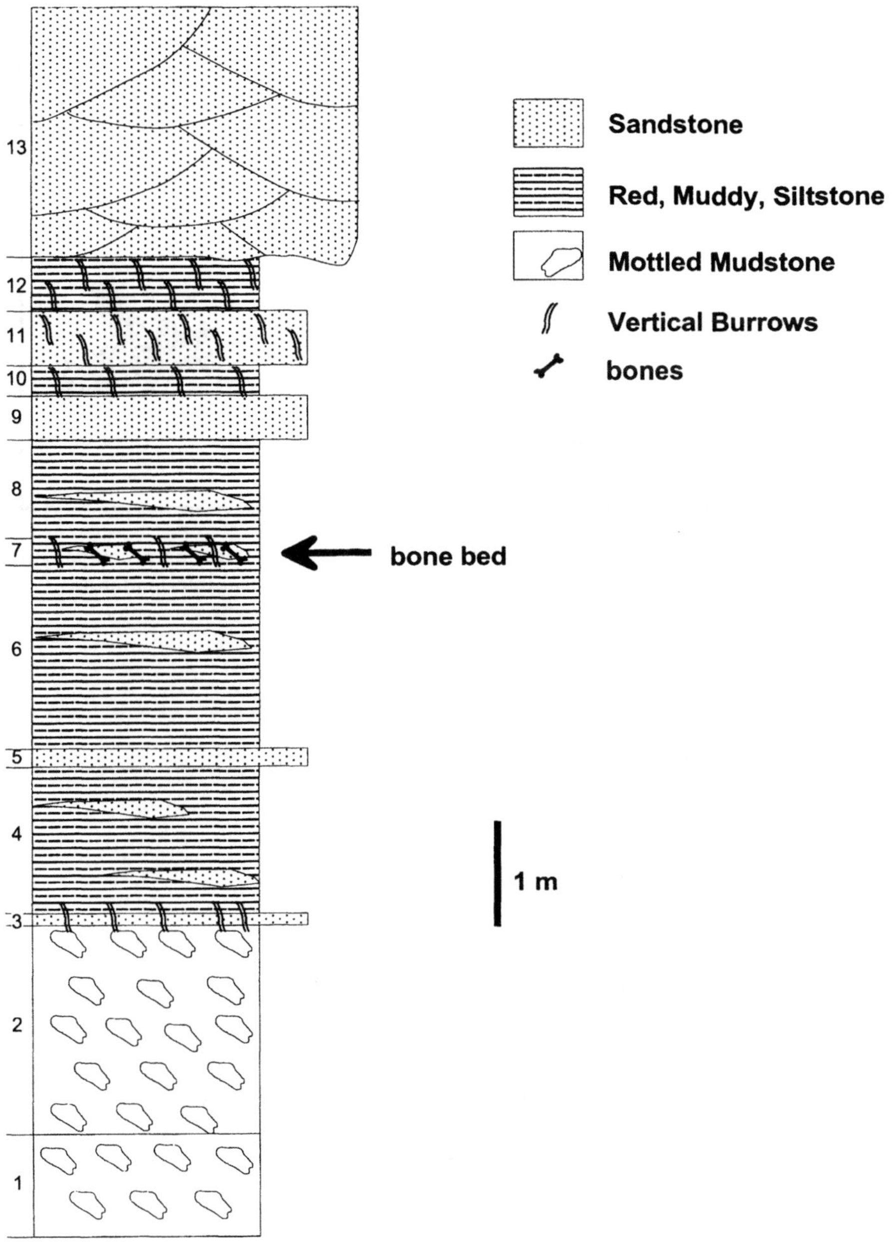

FIGURE 6. Measured section through locality NMMNH L-201. The numbers on the measured section correspond to the lithologic descriptions in Appendix 1.

sections indicate that the infilling of at least some of these burrows is meniscate. The second type of burrow is cylindrical and about 2 cm in diameter. These burrows are also sinuous and nearly vertical oriented. These burrows are generally filled with a very coarse matrix and their walls are rough,

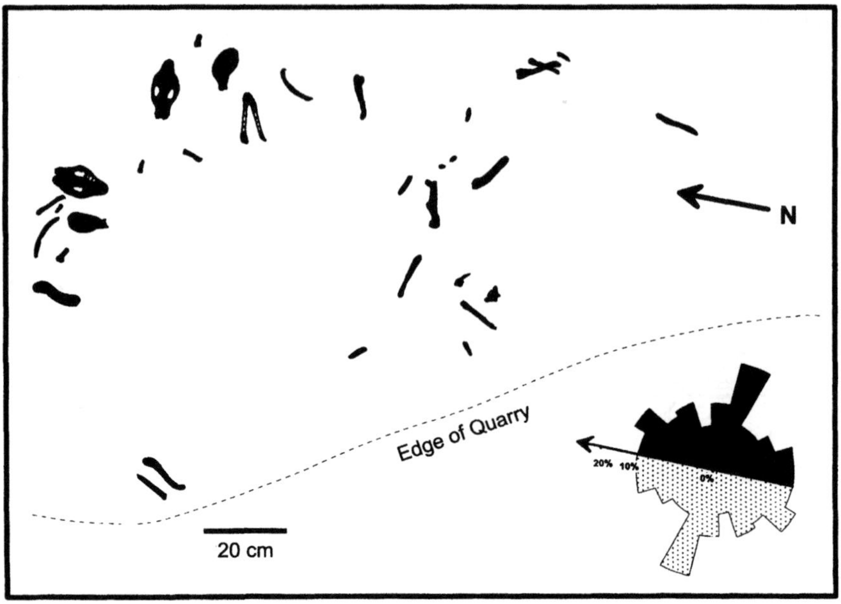

FIGURE 7. Map of a portion of locality NMMNH L-201 quarry with the accompanying Rose diagram showing long-bone orientation (see Appendix 2 for statistical information).

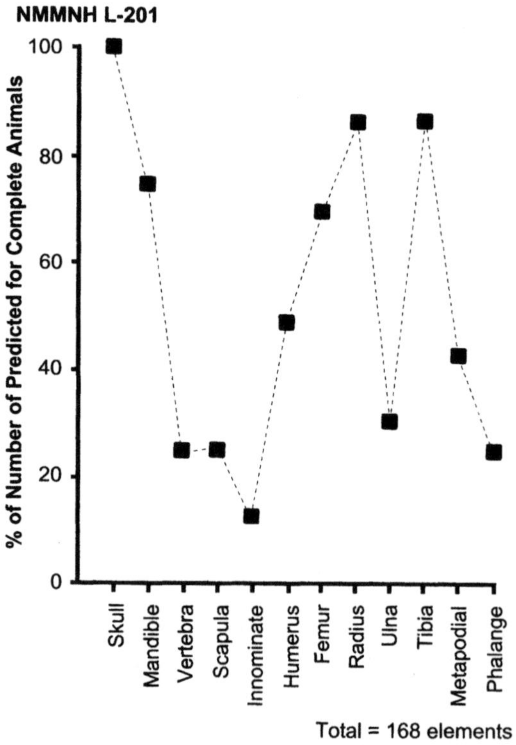

FIGURE 8. Relative bone frequency plot of 168 elements collected from locality NMMNH L-201 by the NMMNH during 1987. Four skulls were obtained so the relative numbers of other elements were compared to that expected for four individuals.

FIGURE 9. Photograph of locality NMMNH L-203.

possibly indicating that they were lined with fecal pellets. These burrows are similar to ones described from the alluvial Willwood Formation (Bown and Krause, 1983) and can probably be attributed to arthropods.

Plots of relative frequency of various elements recovered from the L-201 quarry during the summer of 1987 by the NMMNH (Figure 8) as well as a Voorhies diagram (Figure 4) indicate that L-201 quarry is depleted in those bones (Voorhies Group I) that tend to be removed from an assemblage by saltation or suspension in running water.

4. Nmmnh Locality L-203

The NMMNH locality L-203 was discovered during the Fall of 1986 in the Regina Member, San Jose Formation by Ruby Williamson of the University of New Mexico. Cramial and postcramial fragments were initially found eroding from the sides of a small ridge laterally adjacent to a channel-sand body (Figures 9–12). A minimum of 46 individuals of *Meniscotherium chamense* have thus far been recovered from this locality, based on the number of upper left M^1's.

The remains of *Meniscotherium* at the L-203 locality originate primarily from a laterally restricted area (Figures 9–12) as a bone bed in a thin, muddy siltstone that describes a widely u-shaped cross-section. This siltstone overlies a surface that closely resembles the "paleogully system" depicted by Bown and Krause (1987, figure 3) in the upper Paleocene/lower Eocene Willwood

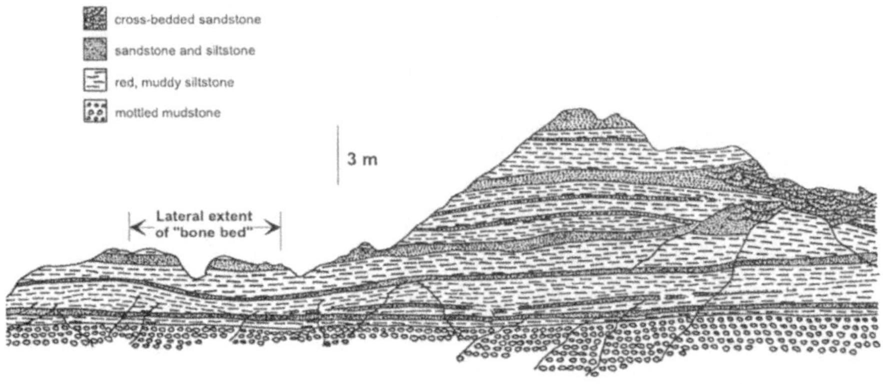

FIGURE 10. A line drawing based on Figure 9 showing distribution of bone bed and nearby facies relationships.

Formation, Bighorn Basin, Wyoming. Bone elements are also found in the red mudstone that immediately overlies the siltstone of the main bone bed but these are scanty and tend to be very fragmentary. In addition to the bone bed and overlying mudstone, remains of *M. chamense* were also found from the siltstone capping the ridge containing the bone bed of L-203. A string of articulated vertebrae and apparently associated skull and mandible fragments *of M. chamense* were found eroding from the top of this siltstone. The bone bed siltstone and the capping siltstone both contain short, vertical burrows. In general, the strata

FIGURE 11. A close-up photograph of locality NMMNH L-203 bone bed.

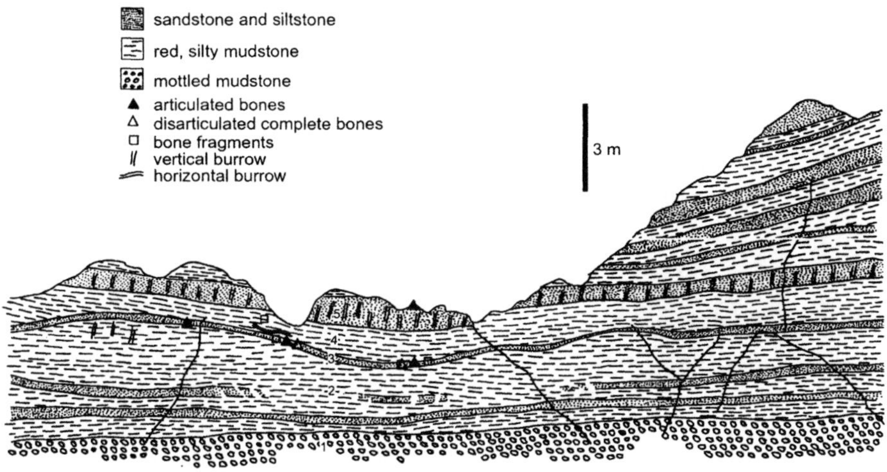

FIGURE 12. A line drawing based on Figure 11 showing bone and facies relationships. Numbers on line drawings correspond to lithologic descriptions given in Appendix 1.

containing L-203 (Appendix 2) resemble the Fbssc facies described by Smith (1988) and the Fbsc + Fl lithofacies described by Smith and Lucas (1991) in the Regina and Tapicitos members of the San Jose Formation. These deposits, like those that contain L-201, are interpreted to represent deposition of fining-upward sequences of sandstone, siltstone, and claystone by waning flows following high-energy overbank events. Bioturbation of the claystone and siltstone occurred between overbank events indicating significant hiatuses typically separated these events. The channel sandstone laterally adjacent to the bone bed (Figure 13) probably represents the margin of the associated trunk channel.

The main bone bed is a poorly sorted muddy siltstone about 10 cm thick. The bone bed contains articulated, nearly complete skeletons, disarticulated bones, and fragmented bones (Figure 12). Two blocks containing articulated remains of *M. chamense* were removed from opposite sides of the ridge containing the bone bed (Figure 13). Block 1 contained two partial skeletons; NMMNH P-3441 (on the left) and NMMNH P-3442 (on the right). NMMNH P-3442 is missing the right innominate and both hind limbs. A possibly associated partial and fragmented pes articulated to a fibula was found less than 10 cm from the partial pelvis of this individual. Block 2 contained a nearly complete but badly concreted skeleton, NMMNH P-3438. Immediately beneath the pelvis of this specimen were two fragmented femora. These femora are both missing their distal articular ends, terminating in jagged breaks just above the expansion of the distal condyles. Although no large-scale quarry operations were undertaken at this locality, the exposed margins of the bone bed were uncovered and additional remains removed from the bone bed include complete skulls, partial skulls, complete mandibles, isolated teeth, and complete limb elements.

The various stages of completeness of articulated skeletons occurring with disarticulated and also partial skeletal remains within the main bone bed

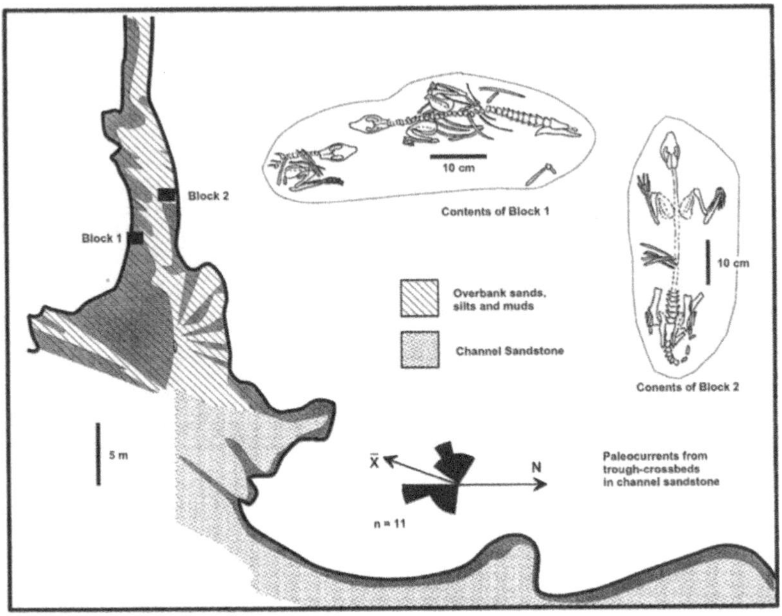

FIGURE 13. Map of locality NMMNH L-203 showing contents of two quarried blocks removed from the main bone-producing interval. Paleocurrents taken from trough cross-bed-sets of a nearby channel sandstone are also indicated.

suggests that animals were buried after spending different lengths of time exposed to weathering and possibly also scavenging on the surface. Hill (1976) and Toots (1965), in studies of the natural decay of animal carcasses, found that the hind limbs were often the first part to separate from the rest of the skeleton. This suggests that the partial skeleton of block 1 was in the early stages of decay when it was buried. Other bones, such as femora, are missing articular ends. Binford (1981), in studies of carnivore damage to various bones, found that dogs and wolves often remove the proximal articular end of femora. The medial ridge of the trochlea of the femur is also sometimes furrowed and removed (Binford, 1981). This type of damage is similar to the damage observed on the two femora removed from block 2. Unfortunately, the presence of tooth punctures and furrowing that often result from scavenger activity can not be verified due to the concreted and obscured surfaces of the bones recovered.

Abundant ichnofosssils similar to the "Type 5" (Bown and Krause, 1983) burrows observed at locality L-201 were observed within the main bone bed. Because dense concentrations of these burrows occur in the main bone bed at locality L-203 and do not extend into the overlying mudstones, there was probably a significant hiatus between deposition of the bone bed and the overlying mudstone.

The presence of very fragmented skeletal elements in the mudstones overlying the main bone bed of L-201 is consistent with highly weathered and possibly scavenged material being reworked from the main bone accumulation into overlying units during subsequent fluvial activity. The presence of articulated skeletal material of *Meniscotherium* in the siltstone cap above the main bone bed suggests a later burial in a second, unrelated flow event.

5. Discussion

The three localities discussed above, UCMP V71237 and NMMNH localities L201 and L203 are each dominated by the remains of a single species of *Meniscotherium*. Catastrophic events are usually invoked to explain the accumulations of a single taxon of terrestrial vertebrate. Many different mechanisms have been suggested to explain similar accumulations of other vertebrates and it is possible that a different cause of accumulation may have been responsible for each of the above assemblages.

The sedimentological evidence of UCMP locality V71237 and NMMNH locality L201 are each consistent with burial of large numbers of *Meniscotherium* relatively soon after death by a single fluvial event. Both localities contain largely disarticulated but relatively unaltered skeletal elements that show a weakly bimodal orientation and are depleted in smaller skeletal elements consistent with winnowed assemblages indicating that animals were buried by fluvial processes soon after death.

The main bone bed of NMMNH locality L203 is also consistent with a large number of *Meniscotherium* being buried soon after death by a single fluvial event. NMMNH locality L203 is the only one of the three localities that contains largely articulated skeletons. However, disarticulated and fragmentary elements are also abundant from the main bone bed of L203. NMMNH-L203 is also very unusual in that *Meniscotherium* is very rare from the Regina Member of the San Jose Formation yet at this site, *Meniscotherium* is found in great abundance. Indeed, the absence of *Meniscotherium* from the Regina Member was the primary way of defining the "Almagre fauna" (Granger, 1914). At this site there also is evidence that there may have been subsequent burial events following deposition of the main bone bed after a substantial hiatus as evidenced by partially articulated remains of *Meniscotherium* from a siltstone lense that is significantly above the main bone-producing interval. This suggests that there might have been a repeating mechanism for killing large numbers of *Meniscotherium* at or near the site.

The analysis of the three localities discussed above indicates that each is the result of large numbers of *Meniscotherium* being killed by a catastrophic event rather than by an attritional accumulation of animals over an extended period. In all three localities, the carcasses of *Meniscotherium* were probably produced within several weeks and buried by a relatively typical proximal floodplain flooding event soon after this.

All three *Meniscotherium*-dominated assemblages described above are interpreted to represent catastrophic death assemblages. The presence of at least three mass-death assemblages representing both species of

Meniscotherium suggests that *Meniscotherium* was a gregarious animal that made it susceptible to death and burial en-masse. Also, all three assemblages appear to be mixed with relatively typical overbank sediments of a proximal floodplain environment. It also appears that the flooding events that buried the assemblages postdated the event that killed *Meniscotherium*.

Whatever the cause of the mass-death of *Meniscotherium*, it appears clear that *Meniscotherium* must have been at least occasionally present in large congregations. Therefore, the large accumulation of dead *Meniscotherium* may have resulted from the attritional death of a large localized colony. However, the pristine surfaces of bones from the NMMNH L-201 and UCMP V71237 assemblages argue against a relatively slow accumulation of animals and instead indicate that the animals all died within a relatively short interval, that is, within several weeks. Alternatively, a large community of *Meniscotherium* may have been struck by disease. For example, Kingdon (1971) reported that colonies of the hyrax *Procavia* that contained populations of 25–60 individuals occasionally are decimated by epidemics. However, the presence of a relatively diverse associated fauna (see below) argues against this as few diseases can infect such a large suite of distantly related animals. Drought is also an agent of death often invoked to explain large mass-death assemblages of terrestrial vertebrates (Shipman, 1981). Unfortunately, the three sites analyzed above do not offer any obvious clues to point to the agent of death.

6. Associated Faunas

Meniscotherium has long been noted for its unusual distribution marked by great abundance in some strata (e.g., Tapicitos Member, San Jose Formation, San Juan Basin, New Mexico) contrasted with near absence in others (e.g., the Willwood Formation, Bighorn Basin, Wyoming; see Williamson and Lucas, 1989). Much of the speculation surrounding the distribution of *Meniscotherium* stems from its occurrence within the San Jose Formation of New Mexico. There, *Meniscotherium* is typically very abundant in the reddish sediments of the Tapicitos Member but generally very rare in the drab sediments of the Regina Member. This distribution was therefore, usually attributed to environmental controls where *Meniscotherium* was restricted from more paludal environments and largely confined to drier upland environments (Simpson, 1948a; Webb, 1977). Gazin (1965a) gave a brief account of the faunal association of *Meniscotherium* in order to compare faunas in which *Meniscotherium* was abundant with those in which *Meniscotherium* was rare or absent. Kihm (1984) also argued that the distribution of *Meniscotherium* in the DeBeque Formation of the Piceance Creek basin of northwestern Colorado reflected environmental control that should be reflected by its co-occurrence with other taxa.

The NMMNH L201, L203 and UCMP V71237 localities are believed to represent catastrophic death assemblages. Therefore, they potentially preserve a fauna that lived in sympatry with *Meniscotherium*. All three assemblages are dominated by *Meniscotherium*, but also have small associated faunas that include other mammals, reptiles and fish. Williamson (1989, table 4) and Williamson and Lucas (1992, table 22) published a brief faunal list associated with NMMNH localities L201 and L203 and UCMP V71237. A revised list is indicated in Table 1.

All three localities contain relatively diverse assemblages that are fairly representative of Wasatchian assemblages as a whole (see Savage and Russell, 1983). However, much of the associated faunas recovered from localities L201 and L203 is represented by relatively fragmentary material such as isolated teeth and postcranial fragments that may represent transported and/or reworked material or accidental associations rather than the remains of animals killed by the same event that decimated *Meniscotherium*. The possible exception to this are the rodents from NMMNH locality L201. Several fairly complete skulls and disarticulated postcranial remains of rodents, especially of *Paramys copei*, are represented in the AMNH, NMMNH, and UALP collections. Also, a complete skull of *Knightomys depressus* (NMMNH P-5552) was recently recovered from the main bone bed. The significance of this association is not clear.

7. Population Samples of *Meniscotherium*

Fossil localities that yield large numbers of a species are very useful in paleontology for establishing intraspecific variability, sexual dimorphism and demography (Gingerich, 1981). Unfortunately, large samples of homogeneous groups or populations of early Eocene mammals are extremely rare (Gingerich, 1981; Kurten, 1969). Small potential population samples have been reported for the perrisodactyl *Hyracotherium* (Gingerich, 1981) and for the pantodont *Coryphodon* (Lucas, 1984 and this volume; McGee, 1993).

7.1. Age Structure

Catastrophic death assemblages potentially preserve an unbiased sample of a living population of animals. The age structure of populations of *Meniscotherium* (Figure 14) was estimated by using samples from the three localities discussed above. All three assemblages contain a wide range of age classes ranging from very young individuals with relatively unworn deciduous molars to those with extremely worn permanent cheek teeth.

All three localities are assumed to yield unaltered samples of the original living population. However, it is likely that they actually represent biased samples. For example, small, young animals are often found to be depleted in fossil assemblages because they are easily removed or destroyed by scavengers. Also, smaller and more fragile bones may be easily destroyed or removed from an assemblage before burial, or they may become fragmented and unrecognizable before recovery because of other processes. The relatively undamaged

Table 1. Associated vertebrate faunas of the three localities discussed in the text. All three localities are believed to represent mass-death assemblages. NMMNH localities L201 and L203 both yield large samples of *Meniscotherium chamense* from the San Jose Formation, Tapicitos and Regina Members, respectively, of the San Juan Basin, New Mexico. UCMP locality V71237 from the Wasatch Formation, Washakie Basin, Wyoming yielded a sample of *M. tapiacitum*.

UCMP V71237

Taxa	Elements represented
Ostracoda unident.	Tests
Gastropoda unident.	Tests
cf. *Lepisosteus* sp.	Scales
Amiidae	vertebrae and skull fragments
Echmatemys sp.	Shell fragments
cf. *Glyptosaurus* sp.	Scutes
Crocodylia indet.	Teeth and scutes
Multituberculata unident.	Isolated teeth
"Insectivora" unident.	Isolated teeth
Tetonius sp.	Dentary fragment with teeth
Coryphodon sp.	Dentary with teeth
Esthonyx sp.	Dentary with teeth
Hyracotherium sp.	Isolated teeth
Rodentia unident.	Isolated teeth

NMMNH L201

Taxa	Elements represented
Squamata indet.	Limb fragment
Saniwa ensidens	maxillary fragment and vertebrae
Chelonia unident.	Shell fragments
Crocodylia indet.	Isolated teeth and scutes
Coryphodon sp.	Tooth and bone fragments
Prototomus viverrinus	Dentary with teeth
Oxyaena sp.	Tooth fragment
Artiodactyla unident.	Limb fragments
Hyracotherium sp.	Isolated teeth
Hyracotherium vasacciense	Dentary with teeth
Cantius sp.	Dentary with teeth
Knightomys depressus	Skull
Paramys copei	Skulls, isolated teeth and disarticulated postcrania
Paramys excavatus	Isolated teeth

NMMNH L203

Taxa	Elements represented
Anguidae unident.	Vertebra
Saniwa ensidens	Vertebra
Coryphodon sp.	Tooth and postcranial fragments
Creodonta indet.	Isolated tooth
Esthonyx sp.	Isolated tooth
Didymictis protenus	Tooth fragment
Hyracotherium sp.	Isolated teeth
Cantius frugivorus	Dentary fragment with teeth

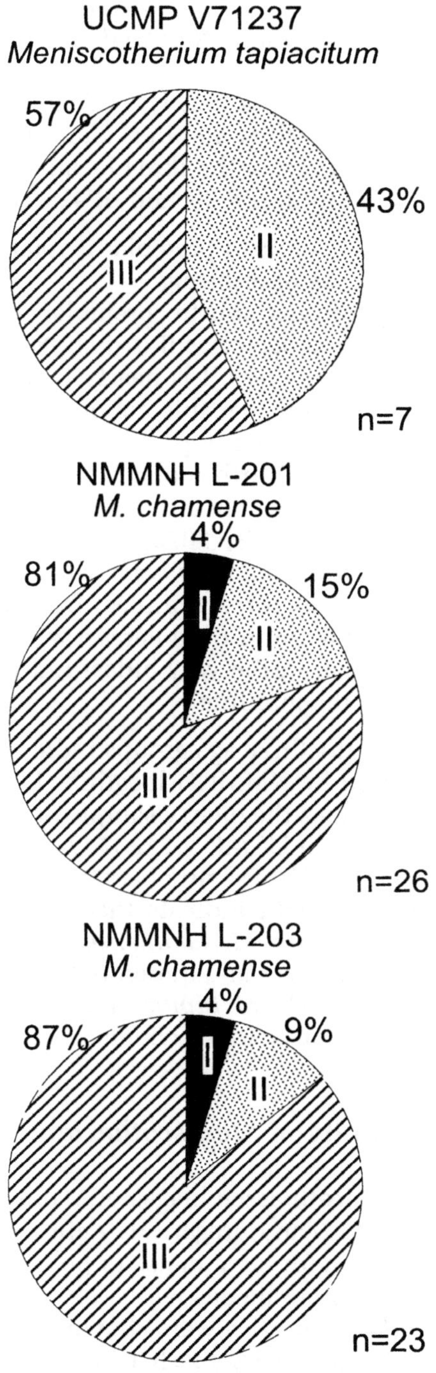

FIGURE 14. Age structure of the NMMNH L-201, L-203, and UCMP V71237 *Meniscotherium* assemblages showing the relative percentages of the three age classes described in the text. Class I represents infants, class II represents juveniles, and class III represents adults. Only those specimens from each locality that preserved a portion of the occlusal surface of P_4 were assigned to an age class.

condition of most of the bone seen in the NMMNH locality L201 and UCMP V71237 at least indicates that burial followed quickly after death, perhaps minimizing these effects.

The fossil skulls and mandibles recovered from the three localities were often found fragmented, and teeth were often found disassociated. Therefore, individuals of *Meniscotherium* from the three localities were divided into age categories based on the condition of P4's and dP4's. The age classes are:

I. Infant—The deciduous fourth premolar is relatively unworn. The first molar is in the process of erupting.
II. Juvenile—The deciduous fourth premolar is worn. The division between age class I and age class II is distinct and the amount of wear on all specimens of age class II is nearly identical. This suggests that births were seasonal and that this cohort represents a single generation among all three quarry samples.
III. Adult—The permanent fourth molar is fully erupted. Further breakdown of this group is possible although it would be completely arbitrary and subjective (Morris, 1978) and was not attempted for this study. However, a wide range of wear was observed among the permanent dentition of all three samples, from virtually unworn third molars to extremely worn third molars in which dentine was exposed over the entire crown. In the very oldest individuals, the third molar was worn down to the gumline.

The relative abundances of the three age groups indicates that the two samples of *M. chamense* represented by NMMNH localities L201 and L203 have relatively fewer juveniles than does the quarry sample of *M. tapiacitum*. However, the very small sample size from UCMP V-71237 Quarry (n = 7) makes this statistically somewhat uncertain. If these numbers truly do reflect the relative age group abundance, then this might indicate that *M. tapiacitum* had a higher recruitment rate than the larger *M. chamense*. In living mammals, small mammals tend to have shorter lives and therefore higher recruitment rates than larger mammals (Eisenberg, 1981). *M. tapiacitum* might have had larger litters, given birth to young more often than *M. chamense*, or both.

7.2 Intraspecific Variability of *Meniscotherium*

The presence of mass-death assemblages of *Meniscotherium* suggests that it was a gregarious animal. Many gregarious animals exhibit sexual dimorphism due to sexual selection for large males in polygynous breeding systems. At least two mammals contemporaneous with *Meniscotherium*, *Hyracotherium tapirinum* and *Coryphodon molestus*, were shown to have sexual dimorphism as expressed in both canine and body size (Gingerich, 1981; Lucas, 1984). Generally, in modern dimorphic mammal species, central cheek teeth have a unimodal distribution while males have larger canines than females (Gingerich, 1981). Male skull size is often larger in dimorphic species (Gingerich, 1981).

For example, Gingerich (1981) showed that within a sample of 24 individuals of *Hyracotherium tapirinum*, the coefficient of variation (CV) of cheek teeth (P4-M3) was between 3.5–6.9. This is considered to be typical for a mammal species (Simpson *et al.*, 1960). However, upper and lower canines of this sample had a CV of 19.4–28.4 in measures of length and width. Various measures of several skulls from this sample also indicated a relatively high amount of variation (CV = 9.3–19.1; Gingerich, 1981).

In order to test for sexual dimorphism in *Meniscotherium chamense*, measurements of the dentition including several canines (Table 2) and seven skulls (Table 3) were analyzed from NMMNH L-201 (all metric information is available in Williamson and Lucas, 1992 under AMNH 150). The upper and lower molars show relatively low variability with a CV of 3.48–6.99. The first molar is the least variable with a CV of 3.48–4.29. Canine size shows only moderate variability with a CV of 7.3–9.2. This value is consistent with canine variablility "within sex" of dimorphic primate taxa (Gingerich and Schoeninger, 1979) or "across sexes" of monomorphic species of mammals (e.g., Gingerich and Winkler, 1979).

Skull measures also show a generally low to moderate variablity (Table 3). Skull length shows a very low CV of only 1.5 but some measures show higher variability, especially the maximum transverse width across the zygomatic arches (CV = 19.1) and the width of the foramen magnum (CV = 29.2). These high measures of variability are attributed to postmortem distortion of the skulls. More difficult to explain is the relatively moderate variability of the minimum transverse width across the supraorbital crests (V = 10.3). However, it is possible that some of this high variability is also due to some postmortem deformation of the specimens and possibly also to the small sample sizes available. *Meniscotherium chamense* does not exhibit marked sexual dimorphism in skull size or tooth size.

Although the presence of sexual dimorphism in mammals can indicate that there is sexual selection for larger males in a polygynous breeding systems, lack of such sexual dimorphism in mammals does not exclude gregarious behavior or even a polygynous breeding system. For example, permanently territorial or harem-holding males may be monomorphic or smaller than females (Raedeke, 1979; Eisenberg, 1981).

8. Summary

The examination of three fossil accumulations containing the abundant remains of the early Eocene (Wasatchian) condylarth *Meniscotherium* indicate that they represent catastrophic mass-death assemblages. All three are found in relatively typical proximal floodplain fluvial deposits and fail to reveal any particularly unusual killing mechanism for *Meniscotherium* or the associated fauna. However, the relatively large number of mass-accumulations of *Meniscotherium* suggest that it has a predisposition for burial en-masse probably due to its living in large groupings.

Table 2. Metrical data of *Meniscotherium chamense* teeth from NMMNH L-203. Abbreviations are n, number; OR, observed range; x, sample mean; s, standard deviation; and V, coefficient of variation. Measurements are in mm (teeth measured to nearest 0.1 mm).

Tooth		n	OR	X	SE	s	CV
I1	L	2	2.6-2.7	2.65	----	----	----
	W	2	2.0-2.2	2.10	----	----	----
I2	L	3	2.3-2.7	2.53	0.10	0.17	6.71
	W	3	1.6-2.4	2.10	0.21	0.36	17.41
I3	L	4	2.5-2.8	2.63	0.06	0.11	4.51
	W	4	2.0-2.4	2.20	0.08	0.16	7.27
C1	L	8	2.8-3.5	3.09	0.08	0.23	7.32
	W	8	2.6-3.0	2.64	0.08	0.23	8.71
P1	L	1	3.6	----	----	----	----
	W	1	2.8	----	----	----	----
P2	L	9	3.8-5.1	4.24	0.12	0.35	8.18
	W	9	2.7-3.6	3.33	0.09	0.28	8.41
P3	L	23	5.4-6.8	6.04	0.08	0.39	6.45
	W	24	5.7-7.0	6.43	0.07	0.36	5.60
P4	L	29	7.0-8.1	7.59	0.06	0.31	4.13
	A	28	8.2-9.9	8.95	0.07	0.36	4.02
	P	18	4.4-5.8	5.29	0.08	0.36	6.81
M1	L	33	8.2-9.7	8.95	0.06	0.37	4.15
	A	28	9.1-11.5	10.56	0.08	0.44	4.13
	P	32	9.6-11.9	10.25	0.08	0.44	4.29
M2	L	26	9.2-11.2	10.27	0.10	0.50	4.91
	A	27	11.2-13.2	12.29	0.10	0.51	4.15
	P	27	10.4-12.2	11.25	0.10	0.50	4.44
M3	L	24	8.3-9.5	9.03	0.07	0.34	3.79
	A	23	10.7-12.7	11.53	0.10	0.47	4.07
	P	25	9.1-11.3	10.44	0.10	0.48	4.62
DP3	L	1	5.9	----	----	----	----
	W	2	5.5-5.7	5.60	----	----	----
DP4	L	2	7.0-7.7	7.35	----	----	----
	W	2	6.6-8.2	7.40	----	----	----
i1	L	3	2.0-2.1	2.03	0.03	0.05	2.31

(*continued*)

Table 2. (continued)

	W	3	1.4-1.9	1.67	0.12	0.21	12.60
i2	L	3	2.2-2.6	2.40	0.09	0.16	6.79
	W	3	1.9-2.3	2.07	0.10	0.17	8.21
i3	L	3	2.3-2.7	2.50	0.09	0.16	6.52
	W	3	1.6-2.3	2.10	0.09	0.16	7.60
c1	L	11	2.4-3.3	2.93	0.08	0.27	9.22
	W	11	2.6-3.1	2.90	0.06	0.20	6.90
p1	L	3	2.2-3.2	2.77	0.13	0.42	14.78
	W	3	2.0-2.7	2.43	0.09	0.31	12.80
p2	L	5	3.2-3.6	3.40	0.06	0.13	3.71
	W	3	2.4-2.5	2.43	0.03	0.05	2.06
p3	L	11	5.0-5.7	5.45	0.08	0.25	4.52
	W	11	3.2-3.7	3.45	0.04	0.14	4.06
p4	L	17	7.2-8.6	7.79	0.11	0.47	6.07
	A	18	4.5-5.2	4.83	0.05	0.20	4.14
	P	18	4.4-5.8	5.29	0.08	0.36	6.81
m1	L	23	7.5-8.4	7.86	0.06	0.30	3.80
	A	19	5.4-6.0	5.74	0.05	0.20	3.48
	P	20	5.8-6.4	5.99	0.05	0.22	3.67
m2	L	23	8.3-9.8	8.95	0.07	0.32	3.57
	A	22	6.1-7.0	6.37	0.06	0.27	4.24
	P	26	5.8-6.7	6.17	0.06	0.29	4.70
m3	L	15	9.0-11.0	9.82	0.14	0.54	5.51
	A	17	5.3-6.6	6.08	0.08	0.35	5.76
	P	17	4.7-6.1	5.29	0.09	0.37	6.99
Dp3	L	5	4.9-5.9	5.38	0.20	0.45	8.35
	W	5	3.2-3.9	3.34	0.13	0.28	8.38
Dp4	L	5	8.6-8.8	8.72	0.04	0.08	0.86
	A	5	4.6	4.60	----	----	----
	P	5	5.2-5.6	5.40	0.08	0.18	3.33

Each locality contains a potential population sample of *Meniscotherium* that may allow the age structure of a community of *Meniscotherium* to be determined as well as means for gauging intraspecific variability. Based on an analysis of tooth and skull size of these samples, there is no indication of sexual dimorphism for *M. chamense*.

ACKNOWLEDGMENTS. Thanks are extended to Michael J. Novacek, Malcolm C. McKenna and John Alexander of the American Museum of Natural History, Everett Lindsay of the University of Arizona, and J. Howard Hutchison and

Table 3. A statistical summary of skull measurements of specimens of *Meniscotherium chamense* from the NMMNH L-201 quarry. Abbreviations are n, number; OR, observed range; x, sample mean; s, standard deviation; and V, coefficient of variation. Measurements are in mm (teeth measured to nearest 0.1 mm)

Measurement	n	OR	X	SE	s	CV
1	7	116.0-121.5	119.64	0.64	1.80	1.05
2	7	43.5-60.0	49.60	2.06	5.44	10.96
3	5	24.0-30.4	25.72	1.00	2.66	10.32
4	6	47.0-78.0	64.88	4.68	12.38	19.08
5	5	42.0-57.2	48.04	2.15	5.69	11.85
6	3	22.0-28.0	25.33	1.15	3.06	12.06
7	2	24.0-26.5	25.25	----	----	----
8	3	8.5-9.2	8.90	0.14	0.36	4.05
9	3	10.7-18.0	13.50	1.49	3.94	29.15
10	2	18.0-21.0	19.50	----	----	----
11	7	58.5-64.0	61.26	0.78	2.08	3.39

Measurements: 1 = Maximum Length
2 = Length from tip of snout to anterior orbits
3 = Minimum transverse width across supraorbital crests
4 = Maximum transverse width across zygomatic arches
5 = Length of nasals
6 = Skull width across occiput
7 = Maximum transverse width across mastoid processes
8 = Height of foramen magnum
9 = Width of foramen magnum
10 = Height of occiput (measured from the superior border of the foramen magnum)
11 = Maximum length of tooth row (measured from the labial surface of I1 to the posterior face of M3)

Donald E. Savage of the University of California Museum of Paleontology for allowing access to their collections and offering assistance. I especially offer thanks to J. Howard Hutchison for the permission to use unpublished field notes (J. H. Hutchison also graciously allowed the reproduction of the quarry map of UCMP V71237 based on his unpublished field notes). I thank numerous volunteers for their assistance in the field including Ron Armor, Lloyd and Michelle Chino, Charles Jaynes, Gwen Lindquist, Steve McCarroll, Tracy O'Kelly, Warren Slade, Scott Sucher, and Ruby Williamson. I thank the Santa Fe National Forest for granting permission to excavate and collect vertebrate fossils from NMMNH locality L-201. The study was made possible through grants and funding from the University of New Mexico, the New Mexico Geological Society, the University of New Mexico Student Research Allocations Committee, and the New Mexico Friends of Paleontology.

References

Behrensmeyer, A. K., 1978, Taphonomic and ecologic information from bone weathering, *Paleobiology* **4**:150–162.

Binford, L. R., 1981, *Bones: ancient men and modern myths*, Academic Press, New York.

Bown, T. M., 1980, The Willwood Formation (lower Eocene) of the southern Bighorn basin, Wyoming, *Univ. Mich. Pap. Paleontol.* **24**:127–133.

Bown, T. M., and Krause, M. J., 1983, Ichnofossils of the alluvial Willwood Formation (Lower Eocene), Bighorn Basin, northwestern Wyoming, U.S.A., *Palaeogeogr. Palaeoclim. Palaeoecol.* **43**:95–128.

Bown, T. M., and Krause, M. J., 1987, Integration of channel and floodplain suites, I. Developmental sequence and lateral relations of alluvial paleosols, *J. Sed. Petr.* **57**:587–601.

Eisenberg, J. F., 1981, *The Mammalian Radiations: An Analysis of Trends in Evolution, Adaptation, and Behavior*, The University of Chicago Press, Chicago.

Gazin, C. L., 1952, The lower Eocene Knight Formation of western Wyoming and its mammalian faunas, *Smithson. Misc. Collect.* **117**:1–82.

Gazin, C. L., 1962, A further study of the lower Eocene mammalian faunas of southwestern Wyoming, *Smithson. Misc. Coll.* **144**:1–82.

Gazin, C. L., 1965a, A study of the early Tertiary condylarthran mammal *Meniscotherium*, *Smithson. Misc. Coll.* **149**:1–111.

Gazin, C. L., 1965b, *Early Eocene mammalian faunas and their environment in the vicinity of the Rock Springs Uplift, Wyoming*, in: Wyoming Geol. Assoc. Guidebook, 19th Field Conf., Sedimentation of Late Cretaceous and Tertiary outcrops, Rock Springs Uplift, pp. 171–180.

Gingerich, P. D., 1981, Variation, sexual dimorphism, and social structure in the early Eocene horse *Hyracotherium* (Mammalia, Perissodactyla), *Paleobiology* **7**:443–455.

Gingerich, P. D., and Schoeninger, M. J., 1979, Patterns of tooth size variability in the dentition of primates, *Am. J. Phys. Anthropol.* **51**:457–466.

Gingerich, P. D., and Winkler, D. A., 1979, Patterns of variation and correlation in the dentition of the red fox, *Vulpes vulpes*, *J. Mamm.*, **60**:691–704.

Goddard, E. N., Trask, P. D., Deford, R. K., Rove, O. N., Singewald, J. T., Jr., and Overbeck, R. M., 1979, *Rock-color chart*, Geological Society of America, Boulder.

Granger, W., 1914, On the names of lower Eocene faunal horizons of Wyoming and New Mexico, *Bulletin of the American Museum of Natural History* **33**:201–207.

Hill, A. P., 1976, On carnivore and weathering damage to bone, *Curr. Anthrop.*, **17**(3), 335–336.

Kihm, A. J., 1984, *Early Eocene mammalian faunas of the Piceance Creek Basin, northwestern Colorado*, Unpublished PhD disseration, University of Colorado, Boulder.

Kingdon, J., 1971, *East African Mammals: An Atlas of Evolution. Vol. 1*, Academic Press, New York.

Kurten, B., 1969, Sexual dimorphism in fossil mammals, in: *Sexual dimorphism in fossil Metazoa and taxonomic implications*, (G. E. G. Westerman, ed.), Int. Union Geol. Scie. Series A, Schwizerbart'sche Verlagsbuchhandlung; Stuttgart, West Germany, pp. 226–233.

Lucas, S. G., 1984, *Systematics, biostratigraphy and evolution of early Cenozoic Coryphodon (Mammalia, Pantodonta)*, Unpublished doctoral dissertation, Yale University, New Haven.

McGee, E. M., 1993, The taphonomy of Roehler's *Coryphodon* catastrophe quarry (lower Eocene, Wasatch Formation, Washakie Basin, Wyoming, *Journal of Vertebrate Paleontology* **13**:49A.

Morris, P., 1978, The use of teeth for estimating the age of wild mammals, in: *Development, Function and Evolution of Teeth*, (P. M. Butler and K. A. Joysey, eds.), Academic Press, London, pp. 341–353.

Payne, J. W., 1965, Summer carrion study of the baby pig *Sus scrofa*, *Ecology* **46**:592–602.

Raedeke, K. J., 1979, *Sexual selection and dimorphism in the guanaco* (*Lama guanicoe*), Abstract 73 in Abstracts of Technical Papers, 59th Annual Meeting, Amer. Soc. Mammal.

Savage, D. E., Waters, B. G., and Hutchison, J. H., 1972, *Northwestern border of the Washakie Basin, Wyoming*, in: Field Conference on Tertiary Biostratigraphy of Southern and Western Wyoming, pp. 32–39. Privately printed and distributed.

Savage, D. E., and Russell, D. E., 1983, *Mammalian paleofaunas of the world*, Addison-Wesley, London.

Shipman, P., 1981, *Life History of a Fossil: An Introduction to Taphonomy and Paleoecology*, Harvard University Press, Cambridge.

Simpson, G. G., 1948a, The Eocene of the San Juan Basin, New Mexico, *American Journal of Science* **246**:157–282, 363–385.

Simpson, G. G., 1948b, A fossil-collecting campaign in New Mexico, *Science* **107**:207–212.

Simpson, G. G., 1981, History of vertebrate paleontology in the San Juan Basin, in: *Advances in San Juan Basin Paleontology* (S. G. Lucas, K. Rigby, Jr., B. S., and Kues, eds.), University of New Mexico Press, Albuquerque, pp. 3–25.

Simpson, G. G., Roe, A., and Lewontin, R. C., 1960, *Quantitative Zoology*, Harcourt, Brace and Col., New York.

Smith, L. N., 1988, *Basin analysis of the lower Eocene San Jose Formation, San Juan Basin, New Mexico and Colorado*, Unpublished PhD dissertation, University of New Mexico, Albuquerque.

Toots, H., 1965, Sequence of disarticulation in mammalian skeletons, *Universtiy of Wyoming Contributions to Geology* **4**:37–39.

Van Houten, F. B., 1945, Review of latest Paleocene and early Eocene mammalian faunas, *Journal of Paleontology* **19**:421–461.

Webb, S. D., 1977, A history of savanna vertebrates in the New World. Part 1: North America, *Ann. Rev. Ecol. Syst.* **8**:355–380.

Williamson, T. E., 1989, *The osteology, functional morphology, taphonomy and paleobiology of early Cenozoic Meniscotherium (Mammalia, "Condylarthra")*, Unpublished M.S. thesis, University of New Mexico, Albuquerque.

Williamson, T. E., and Lucas, S. G., 1989, Systematics and paleobiogeography of late Paleocene-early Eocene *Meniscotherium* (Mammalia, "Condylarthra") from western North America, *Journal of Vertebrate Paleontology* **9**:44A.

Williamson, T. E., and Lucas, S. G., 1992, *Meniscotherium* (Mammalia, "Condylarthra") from the Paleocene-Eocene of western North America, *New Mexico Museum of Natural History Bulletin* **1**, 75 p.

Williamson, T. E., and McCarroll, S., 1988, Taphonomy of two *Meniscotherium* assemblages from the Eocene San Jose Formation, San Juan Basin, New Mexico, *New Mexico Geology* **10**:66.

Appendix 1

Measured section through the NMMNH L-201 quarry. The stratigraphic section was measured with a 1.5-m-long staff and Brunton compass. Colors are those of Goddard *et al.* (1979).

unit	lithology	thickness (m)
13	Sandstone; light gray (N7); quartzose; coarse to very coarse; subangular to subrounded; poorly sorted; calcareous, clay rip-ups; trough-cross-bedded, basal scour; forms channel trending 260–280N.	2.5+
12	Muddy siltstone; pale brown (5 YR 5/2) with light gray (N7) blotches and burrowfill; poorly sorted; calcareous; burrowed with sinuous 0.5–1-cm-thick burrows seemingly randomly oriented; non-calcareous.	0.65
11	Sandstone; light gray (N7); quartzose, very fine—find grained; angular; massive; nodular weathering; calcareous.	0.6
10	Muddy siltstone; pale brown (5 YR 5/2) with light gray (N7) blotches and burrow fill; poorly sorted; burrowed with predominantly vertical, slightly sinuous burrows about 1 cm in diameter; calcareous.	0.35

9	Sandstone; light gray (N7); quartzose, very fine—fine grained; angular; moderately sorted; massive; nodular weathering; calcareous.	0.4
8	Same as 10.	1.0
7	Same as 10 except for the presence of largely disarticulated bone in a layer about 7–13 cm thick. The bone lies between light gray (N7), poorly sorted, quartzose sandy lenses. Elongate carbonate-rich concretions about 3 cm in diamter and 10 cm in length with their long-axes oriented vertically are common. Some concretions are also formed around bone.	0.7–0.13
6	Same as 10.	2.1
5	Sandstone; pale yellowish brown (10 YR 6/2); quartzose; fine-medium grained, moderately sorted; subangular; massive; calcareous.	0.13
4	Muddy silstone; moderate brown (5 YR 3/4); poorly sorted containing several thin layers of light gray (N7), quarzose, poorly sorted, very fine grained and angular sandstone; calcareous.	1.5
3	Same as 5 except for trough-cross-bedding. thickens to North.	0.1
2	Silty mudstone; medium bluish gray (5 B 5/1), moderate yellowish brown (10 YR 5/4) and pale brown (5 YR 5/2) mottled on fresh surface; weathers pale brown (5 YR 5/2); calcareous.	2.25
1	Silty mudstone; moderate brown (5 YR 4/4), medium grey (N5), and moderate yellowish brown (10 YR 5/4) mottled on fresh surface; weathers pale yellowish brown (10 YR 6/2); non-calcareous.	1.0

Appendix 2

Measured Section through the NMMNH L-203 quarry. The stratigraphic section was measured with a 1.5-m-long staff and Brunton compass. Colors are those of Goddard *et al.* (1979).

unit	lithology	thickness (m)
6	Muddy silststone; light brown (5 YR 5/6) alternating with sandstone; pale brown (5 YR 5/2); interfingers to the east with trough-cross-bedded channel sandstone up to about 4 m thick. Channel forms bench trending N-S. Articulated vertebrae, skull and mandible fragments found in situ immediately above sand that is capping the ridge over the main bone-producing interval.	5.7+

5	Sandstone; pale brown (5 YR 5/2); quartzose; fine-medium grained moderately; sorted; contains verteical burrows about 1 cm in diameter; massive; well indurated in nodular patches; calcareous.	0.75
4	Silty mudstone; greyish red (10 R 4/2); highly friable with many "slicks"' upon quarrying, found to be devoid of bone except for one dentary fragment found in situ.	0.3
3	Sandy and muddy siltstone; light gray (N7), greyish red (10 R 4/2); and moderate yellowish brown (10 YR 4/2) mottled on fresh surface, weathers pale brown (5 YR 5/2); poorly sorted; calcareous; contains abundaant articulated, disarticulated, and fragmented bone, devoid of burrows except for rare horizontal burrows filled with light gray (N7) sand and about 1 cm in diameter found at top; bones are restricted to about 9 m wide area filling a paleo-depression about 0.35 m deep.	0.5–0.75
2	Muddy siltstone; light brown (5 YR 5/6) with thin bands of pale brown (5 YR 5/2) silty mudstone; poorly sorted; slightly calcareous, contains carbonized plant fragments.	1.8
1	Silty mudstone; dark yellowish orange (10 YR 6/6) and light bluish gray (5 B 7/1) mottled.	1.6

Appendix 3

Paleocurrent data was derived from long-bone orientation from localities NMMNH L-201 and UC V71237 and represented by equal area rose diagrams in Figures 2 and 5 respectively. Here, this data is analyzed for statistical significance (after Carver, 1971).

Interval	NMMNH L-201 Interval Size		UCMP V71237 Interval Size	
	15°	30°	15°	30°
0–14°	7		4	
15–29°	7	14	3	7
30–44°	4		4	
45–59°	3	7	6	10
60–74°	3		6	
75–89°	12*	15	5	11*
90–104°	5		3	
105–119°	2	7	3	6
120–134°	5		9*	
135–149°	4	9	2	11*
150–164°	9*		1	
165–179°	7	16*	4	5
Total	68		50	
Avg. (Ne)	5.67	11.33	4.17	8.33
Std. Dev. (s)	2.87	4.13	2.12	2.66
Significant concentrations *		(n > n + s)		

Chapter 15

Taphonomy, Fauna, and Depositional Environment of the *Omomys* Quarry, an Unusual Accumulation From the Bridger Formation (Middle Eocene) of Southwestern Wyoming (USA)

PAUL C. MURPHEY, LISA L. TORICK, EMILY S. BRAY, ROBERT CHANDLER, and EMMETT EVANOFF

PAUL C. MURPHEY, LISA L. TORICK, EMILY S. BRAY, and EMMETT EVANOFF • University of Colorado Museum, Boulder, CO 80309-0315. ROBERT CHANDLER • Department of Biology, Georgia State College and University, Milledgeville, GA 31061.

Eocene Biodiversity: Unusual Occurrences and Rarely Sampled Habitats, edited by Gregg F. Gunnell, Kluwer Academic/Plenum Publishers, New York, 2001.

Abstract

Taphonomic, faunal, and lithologic analysis of an unusual fossil accumulation from the upper Bridger Formation is undertaken to infer its origin and depositional environment. The sample includes over 2300 vertebrate, gastropod, and plant specimens. What makes this accumulation so unusual is the high concentration of mandibular, maxillary, and postcranial remains of *Omomys* (Primates, Omomyidae), avian skeletal remains, and eggshell fragments. These "unusual" components of the assemblage are superimposed on a more "typical" Bridger fauna, which occurs at the quarry and lateral to it in the same stratigraphic interval. Bones of *Omomys* and birds are mostly disarticulated and fragmentary, oriented horizontally to sub-horizontally, and exhibit minimal modifications by primary weathering and hydraulic sorting, suggesting minimal trampling and transport, and rapid burial. The *Omomys* Quarry assemblage was initially hypothesized to represent an owl pellet accumulation formed in close proximity to a nesting site, but further analysis indicates that the taphonomy of the accumulation is more complex. Bone breakage patterns of *Omomys* and the absence of digestive corrosion of tooth enamel are consistent with mammalian remains in modern owl pellets, however the dominant type of eggshell is not owl or falconiform. The avian skeletal remains are dominated by wading and shore-edge birds, but forest and ground birds are also represented. In addition to environmental conditions which may have favored localized preservation, four taphonomic agents are postulated for the formation of the accumulation: 1) an attritional accumulation of aquatic taxa in lacustrine sediments, 2) an attritional accumulation of both aquatic and terrestrial taxa in shoreline sediments, 3) an attritional accumulation consisting primarily of bird bones and eggshell formed in close proximity to a nesting area, and 4) a predator accumulation dominated by *Omomys* but probably including other vertebrates formed by owls in close proximity to a nest, day roost or night feeding station. The study concludes that, like other lacustrine-associated vertebrate accumulations in the upper Bridger Formation, the *Omomys* Quarry assemblage was formed in a combination of distal shore, littoral, and sub-aerial muddy lake margin environments.

1. Introduction

One of the world's most abundant and diverse middle Eocene vertebrate faunas is preserved in the Bridger Formation of southwestern Wyoming. Over 86 species representing 67 genera, 30 families, and 13 orders of fossil mammals are recognized (Gazin, 1976). Joseph Leidy's 1869 description of *Omomys carteri* was the first scientific description of a fossil from the Bridger Formation. Subsequently, Bridger fossils have been the subject of numerous publications including many classic papers by pioneers of American vertebrate paleontology (Cope, 1872, 1873a, 1873b; Granger, 1908; Leidy, 1869, 1871, 1872a, 1872b; Marsh, 1871, 1886; Matthew, 1909; Osborn, 1929). Like many other highly fossiliferous formations, the Bridger contains an abundance and diversity of fossils that make it well suited for taphonomic and paleoecologic studies. Preserved in a variety of sedimentary environments, preservational states, assemblages, and in locally varying abundances, these fossils include primarily vertebrates and molluscs, with less common plants and ichnofossils. Despite the relative ease with which diverse and statistically significant samples can be collected, and the large historical collections of Bridger vertebrates available in many museums, paleoecologic studies of Bridger faunas are few (Gunnell, 1997; Gunnell and Bartels, 1994). This is the first detailed taphonomic study of a microvertebrate accumulation from the Bridger Formation.

It has long been recognized that taphonomic processes have created discrepancies between ancient communities and fossil assemblages, and made the fossil record difficult to interpret (Efremov, 1940; Kidwell, 1985; Weigelt, 1989; Wing *et al.*, 1992). Dodson (1980) considered taphonomy to be the study of the processes between death and burial of an organism, including cause of death, decay, transportation, and burial. Efremov's (1940) original and broader definition of taphonomy as the "science of the laws of burial" included diagenesis and large scale biases in fossil preservation as well as pre-burial processes. Behrensmeyer and Kidwell's (1985) all-encompassing definition of taphonomy as "the study of processes of preservation and how they affect information in the fossil record" most closely reflects the nature of recent taphonomic research. The recognition of taphonomic biases inherent in the fossil record is critical to paleoecological interpretations (Badgley, 1986; Behrensmeyer, 1978; Vrba, 1980; Winkler, 1983).

In the last thirty years, many taphonomic studies have been based on empirical observations of modern taphonomic processes with the goal of more accurately interpreting fossil accumulations (Andrews, 1990; Behrensmeyer, 1975; Dodson, 1973; Dodson and Wexlar, 1979; Haynes, 1983; Hoffman, 1988; Kusmer, 1990; Mayhew, 1977; Voorhies, 1969). Most of the earlier and frequently cited actualistic studies were based on medium and large sized mammals (Behrensmeyer, 1975, 1978; Shotwell, 1955; Voorhies, 1969). Although the usefulness of microvertebrates to paleocommunity reconstructions has been questioned because of their susceptibility to transport (Dodson, 1973; Korth, 1979), they have been the subject of increasing study over the last twenty years by both paleontologists and archaeologists (Andrews, 1990; Blob and Fiorillo, 1996; Cassiliano, 1997; Fernandez-Jalvo, 1995; Fernandez-Jalvo *et al.*, 1998; Khajuria and Prasad, 1998; Korth, 1979; Kusmer, 1990; Lyman, 1994; Winkler, 1983). Mammalian carnivores have been cited as accumulators of microvertebrates (Andrews and Evans, 1983; Haynes, 1983; Maas, 1985;

Mellett, 1974), and the importance of avian predators to the formation of such assemblages is widely recognized (Andrews, 1990; Dodson and Wexlar, 1979; Hoffman, 1988; Kusmer, 1990; Mayhew, 1977).

This paper presents a faunal and taphonomic analysis of the *Omomys* Quarry, a microfossil assemblage from the Bridger Formation. The assemblage is unusual because it contains a large concentration of skeletal remains of the primate *Omomys* (Omomyidae), avian skeletal remains, and eggshell fragments, all of which are relatively rare in the Bridger. The skeleton of north American middle Eocene omomyids is poorly known, and the *Omomys* Quarry sample represents the largest collection of skeletal specimens known for a single species of omomyid to date (Covert, 1995). Hirsch and Kohring (1992) published the only other report to date of fossil eggshell from the Bridger. Also unusual are plant fossils which include fragments of fern petioles and at least two kinds of dicotyledenous wood. In addition to these unusual components of the *Omomys* Quarry assemblage, the typical, or "background" fauna for lacustrine and associated deposits in the Bridger is present, although remains of are less numerous than would normally be expected. Typically, Bridger fossil faunas include a mixed aquatic and terrestrial assemblage of gastropods, fish, amphibians, reptiles, and a mammalian fauna which is dominated by *Hyopsodus* (Condylarthra, Hyopsodontidae). Since *Omomys* is thought to have been nocturnal (Covert, 1995), and its remains at the *Omomys* Quarry are generally consistent with mammalian remains in modern owl accumulations (Andrews, 1990; Dodson and Wexlar, 1979; Hoffman, 1988; Korth, 1979; Kusmer, 1990; Mayhew, 1977), the hypothesis that this assemblage represents an owl accumulation was initially proposed (Murphey and Burger, 1996). Subsequent analysis, however, indicates that the taphonomic history of the *Omomys* Quarry assemblage is more complex.

The *Omomys* Quarry assemblage provides a unique opportunity to study a fossil accumulation which is apparently the result of multiple taphonomic processes. In an attempt to analyze these processes, the *Omomys*, birds, and other vertebrates in the sample were analyzed separately for comparative purposes whenever appropriate. In this paper, we attempt to: 1) describe the *Omomys* Quarry fauna with particular emphasis on its unusual elements: *Omomys*, birds, and eggshell; 2) identify the taphonomic processes involved, 3) ascertain the causes of mortality, and 4) propose a depositional model for the *Omomys* Quarry and other fossiliferous lacustrine and lacustrine-associated deposits in the upper Bridger Formation.

2. Materials and Methods

The *Omomys* Quarry is situated in a small valley at the base of a south facing hill on the northeastern flank of Hickey Mountain, Uinta County, Wyoming. The fossils were quarried from two distinct lithologic units: the Hickey Mountain limestone (an informal lithostratigraphic unit within the Bridger C), and a thin overlying mudstone.

Surface collection and sampling of unconsolidated "popcorn" matrix occurred in June and July, 1993, and approximately 1 m^2 of sediment was screen washed. All screen washing was done using a 530 micron, 33 mesh plastic screen backed by a $\frac{1}{4}$ inch metal screen for support. During May, 1994, the remainder of the unconsolidated matrix from the site was bagged for screen-washing, and approximately 4 m^2 of the 8–10 cm thick Hickey Mountain limestone (HMLs) and 2–3 cm thick fossiliferous mudstone overlying it was excavated. The excavation employed a wooden 1 m^2 frame which was subdivided into 20 × 20 cm squares constructed of string. Prior to excavation, the wooden frame was placed over each square meter of the area to be excavated and the grid was traced onto the rock surface using a water-proof marker. The frame was moved across the quarry as the excavation proceeded. Quarry rock was oriented, carefully removed, labeled with quarry coordinates, and transported to the lab for processing. The orientation (dip and strike) of long bones and shaft fragments was measured by re-orienting quarry pieces to their original compass direction.

Fossils were removed from the quarry rock samples using two methods. The first was used for only four samples, each representing one of the 20 × 20 cm squares of the excavation grid, and each being approximately 4,600 cm^3 in size. These samples were randomly selected in order to study the general fossil distribution pattern that was observed in cross sections of quarry rock samples. Because the HMLs and overlying mudstone are much more indurated than the bones preserved within them, attempts to mechanically or "hand" prepare specimens from the rock while chemically stabilizing proved damaging to the fossils and extremely time intensive because of their small size and fragility. Fossils were hand prepared from these samples so that their distribution within the rock could be studied in detail.

For the other 65 quarry rock samples, an alternative method to remove fossils was used. These were broken down individually in a 30%–40% hydrogen peroxide solution. Various approaches, all attempting to salvage orientation data in different ways, were attempted. Regrettably, these met with little success, and useable orientation data for these samples was lost during fossil removal. Because each sample was processed separately, however, it was possible to place all fossils from each quarry sample into their respective squares on a quarry map so that the general lateral distribution of fossils could be examined. This also assisted in the repair of specimens which were broken and separated during the excavation. It is important to note that as a result of the use of this preparation technique for the majority of quarry samples, the aquatic fossils from the limestone were mixed with the aquatic fossils from the overlying mudstone. This did not have a significant affect on the terrestrial vertebrates, because they were contained only in the mudstone.

After the matrix was broken down, the samples were screen washed, dried and picked. All fossils, including unidentifiable bone fragments, were saved for analysis. Taxonomic and morphologic identifications of fossils were made with a stereoscopic binocular microscope, and further microscopic analyses were done with scanning electron microscopy. Vertebrate specimens were examined with regard to the following criteria: age at time of mortality (for *Omomys* only), weathering, surface modification, distribution, orientation, breakage, and transport. Rock samples from the *Omomys* Quarry, other localities within the HMLs, and other limestones were analyzed using thin sections and x-ray diffraction.

All eggshell was cleaned ultrasonically in deionized, distilled water. Eggshell thicknesses were measured with digital calipers and a micrometer. After an initial hand sort, two stages of sorting followed. The first was based on general eggshell morphology such as shell thickness, pattern, shape and number of the pore openings, and sculpturing of the shells' outer surface. The second was based on the microstructural and ultrastructural characteristics of the eggshell as determined by microscopic analysis. Radial sections of each specimen were embedded in epoxy and thin sectioned for histostructural analysis under transmitted and polarized light microscopy. Fragments of each type were also mounted for analysis by high magnification SEM to distinguish the ultrastructural characteristics, recrystallization, and to view details of the inner and outer shell surfaces. Analyses of radial sections are preferred as this permits the observation of the progression of eggshell growth, from the inner shell membrane up through the outer shell surface. The analytical techniques for eggshell utilized in this study are described by Erben (1970), Pooley (1979), Packard (1980) and Hirsch (1983). The structural and parataxonomic terminology follows that of Mikhailov (1991, 1997b), Hirsch (1994), and Mikhailov *et al.* (1996).

In addition to field and laboratory studies of the *Omomys* Quarry, the distribution and lithofacies of the HMLs were mapped, and other HMLs fossil localities documented. Seven detailed stratigraphic sections through the HMLs and its overlying and underlying units were described, and thicknesses between the HMLs and more widespread stratigraphic markers were measured (Murphey, 1995). All of the fossils from the *Omomys* Quarry (UCM Locality 93026-OQ) are housed in the collections of the University of Colorado Museum.

Abbreviations: CL, continuous layer; H:W, height to width ratio; HMLs, Hickey Mountain limestone; MNI, minimum number of individuals; ML, mammillary layer; NISP, number of identified specimens; OC, organic core; SD, secondary deposit; SEM, scanning electron microscopy; SU, shell unit; UCM, University of Colorado Museum, Boulder, Colorado, USA.

3. Geology

The Bridger Formation is a sequence of green and brown mudstones and claystones, scattered ribbon and sheet sandstones, widespread micritic and silicified limestones, and rare, thin but widespread ash-fall tuffs (Bradley, 1964; Evanoff *et al.*, 1998; Murphey, 2001; Roehler, 1992). The first stratigraphy of the Bridger Formation was proposed by Matthew (1909), who divided the entire formation into five informal units: Bridger A–E (oldest to youngest).

Matthew's (1909) subdivisions were based on five areally extensive "white layers": the Cottonwood, Sage Creek, Burnt Fork, Lonetree, and upper white layers, from lowest to highest. These "white layers" are tan to white, resistant limestones that were deposited in broad shallow lakes, and are readily distinguishable from the fluvial mudstones and sandstones which comprise most of the upper Bridger Formation.

Recent work has led to further subdivisions and a clarification of Matthew's (1909) stratigraphy with the recognition of 25 marker beds in the Bridger B–D (Evanoff *et al.*, 1998; Murphey, 1995, 2001). Nine of these are widespread enough and at the appropriate position to subdivide Bridger submembers B–D into lower, middle, and upper parts, and many occur across the entire southern Green River Basin, a distance of 63 km (Murphey, 2001). In this paper the term "upper Bridger" is used to refer to Bridger C, D, and E, with the C and D equivalent to the Twin Buttes Member and the E equivalent to the Turtle Bluff Member of the Bridger Formation (Evanoff *et al.*, 1998). Our stratigraphic nomenclature (Figure 1) maintains much of Matthew's (1909) terminology, however, we have replaced the term "white layer" has been replaced with the lithology of each unit.

3.1. Middle Eocene Lakes in the Green River Basin

Lake Gosiute, which occupied a maximum of 15,400 km^2 of the Green River Basin during the middle Eocene (Roehler, 1993), was deposited over approximately 3 m.y. (Krishtalka *et al.*, 1987). This lake is represented by the Green River Formation, an approximately 840 m thick lacustrine sequence (Roehler, 1992) well known for its economic geology and excellent fossil preservation. The lower Bridger Formation (Bridger A–B) was deposited lateral to the Laney Shale member of the Green River Formation (Bradley, 1964; Roehler, 1993), and many of the widespread limestones in Bridger A and B are actually intertonguing lacustrine rocks of the Green River Formation. The Sage Creek limestone forms the boundary between the lower and upper Bridger, the base of the Bridger C (Bradley, 1964; Evanoff *et al.*, 1998; Matthew, 1909), and apparently represents the last extension of Lake Gosiute in the southern Green River Basin (Murphey, 2001).

After the deposition of Lake Gosiute, lacustrine sedimentation continued during deposition of the Bridger C, D, and E, but on a much reduced scale as smaller and shallower lakes and ponds. These lakes and ponds are represented by relatively thin lacustrine limestones (including Matthew's "white layers") within the dominantly fluvial upper Bridger sequence (Evanoff *et al.*, 1998; Murphey, 1995, 2001). The limestones range from 9 km across to more than 60, and in thickness from 4 cm for a portion of the Henrys Fork limestone (Murphey, 1995) to 4.1 m at the type locality of the Sage Creek limestone (Murphey, 2001; Sinclair, 1906). The thickest portions of most upper Bridger lacustrine units are usually less than 1 m (Roehler, 1993; Murphey, 2001).

While sands are common in fluvial facies of the upper Bridger, the

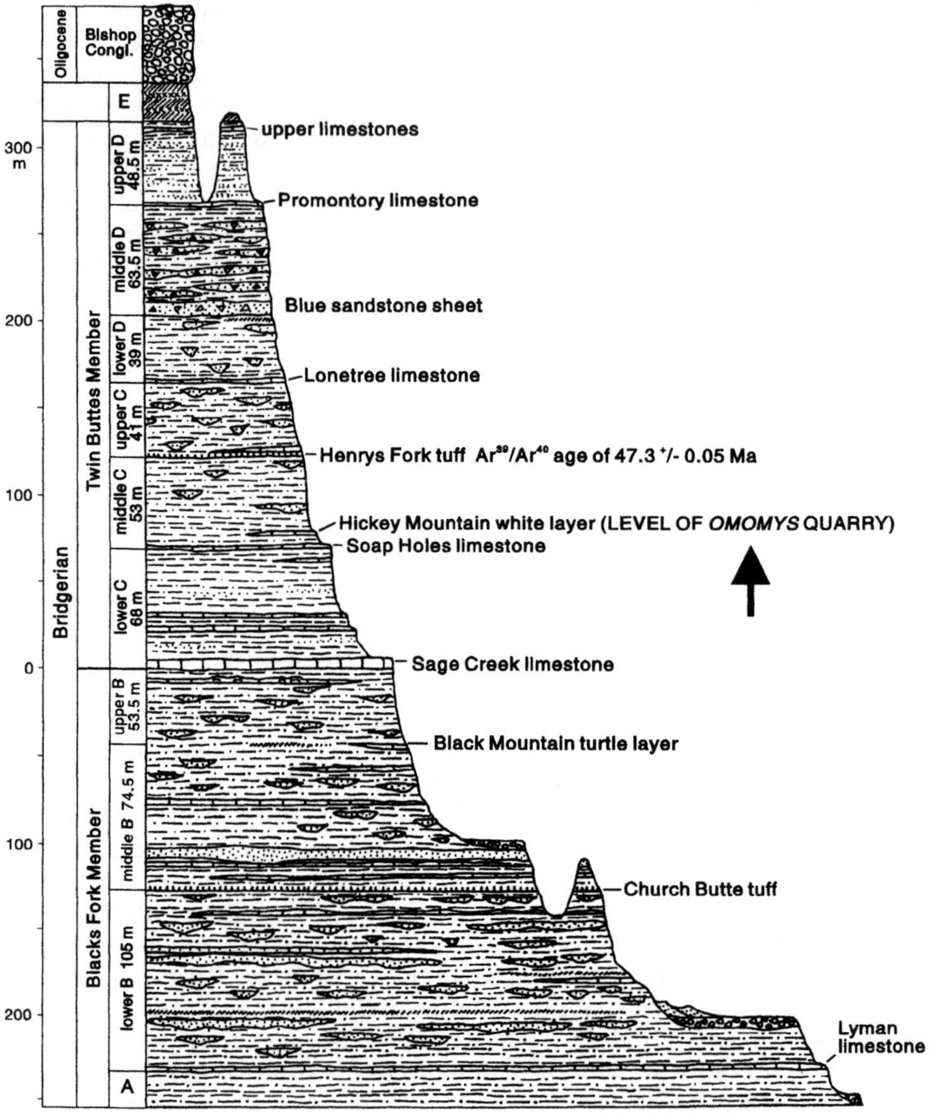

FIGURE 1. Widespread marker beds of the Bridger Formation, southwestern Wyoming, showing the stratigraphic position of the Hickey Mountain limestone and the *Omomys* Quarry.

coarsest lacustrine and lacustrine-associated sediments in this interval are typically fine-grained sandy muds. Most of the lakes appear to have been perennial, with periodic fluctuations in water level, and were probably sustained by groundwater or small, localized streams. Lacustrine deposition in the Bridger may represent periods of decreased regional sedimentation, and associated fossil accumulations may be concentrated in lacustrine rocks relative to rocks with higher sedimentation rates (see Kidwell, 1985).

3.2. Geology of the *Omomys* Quarry

Two distinct fossil-bearing lithologic units occur at the *Omomys* Quarry. The lower is the Hickey Mountain limestone, which contains mostly an aquatic invertebrate and vertebrate fauna. The upper is a calcareous sandy mudstone, which contains a mixed aquatic and terrestrial invertebrate and vertebrate fauna. Descriptions of these units, as well as a list of the fossils which were found in each unit for four quarry samples, are provided in Table I.

Although the maximum size of the original lake is unknown due to limited modern exposures of the HMLs, the bed is partially exposed over a distance of approximately 9 km. Lithologies of the HMLs vary from tan marlstones and micrites, to well-indurated brown cherty limestone, to pale green calcareous sandy mudstone, to well indurated tan sparry limestone which weathers to rusty brown. The overlying highly fossiliferous mudstone occurs over most of the HMLs's distribution, especially where the HMLs is micritic. Most fossils occur in localized concentrations.

The HMLs occurs 70 m above the base of the Bridger C and 48 m below the Henrys Fork tuff. The Henrys Fork tuff has recently been re-dated at 46.92 ± 0.17 Ma using Ar40/Ar39 techniques (Murphey *et al.*, 1999). The HMLs and *Omomys* Quarry are interpreted to be slightly older than the Henrys Fork tuff.

4. The *Omomys* Quarry Fossil Assemblage

Shotwell (1955) pointed out that quarry samples do not produce complete faunal lists of any of the communities represented. This is certainly the case for the *Omomys* Quarry in relation to the known fauna of lacustrine and associated rocks in the upper Bridger, and for the formation in general (see Gazin, 1976). The estimation of numbers of individuals in fossil assemblages forms the basis for paleocommunity reconstruction in many paleoecological analyses, and is an important factor in determining the taphonomic history of fossil assemblages. The difficulty of estimating numbers of individuals represented in an assemblage based on complete bones is greatly exaggerated by the high degree of bone fragmentation typical of vertebrate fossils. Various methods to count individuals, each with their own merits and disadvantages, are used (see Badgley, 1986; Gilinsky and Bennington, 1994; Holtzman, 1979; Lyman, 1994).

Table I. Geologic units of the *Omomys* Quarry, and the fossil content of four typical quarry samples illustrating the general pattern of fossil distribution within the Hickey Mountain limestone and overlying mudstone

Geology		Samples and their fossil content			
Unit	Description	E3-14	E3-6	D3-4	D4-8
sandy mudstone (2-3 cm thick)	Pale yellowish brown 5YR 7/2, calcareous, ~60% micrite and sparite, also contains smectites and illitic clays, angular to subrounded glass shards, quartz grains, and other detrital grains. Overlain by unfossiliferous non-calcareous pale olive claystone. Bottom contact diffuse	Snails, coprolite, gar scale, turtle shell fragment, alligator tooth, bird longbone shaft & talon, 22 eggshell fragments, 2 mammal teeth, 18 undet. bone fragments	Snails, bird femur & tibia, 46 eggshell fragments. Mammal vertebra, incisor, scapular fragment. 24 undet. bone fragments (mostly bird long bone shafts)	8 wood fragments, fish vertebra, 2 lizard dentaries, 2 alligator teeth. 3 bird phalanges, 1 talon, 1 tarsometatarsus. 5 mammal teeth, 2 vertebrae, 1 jaw, partial tibia and femur. 6 undet. bone fragments	Lizard osteoderm, bird vertebra, 2 bird talons, 13 eggshell fragments, mammal jaw and incisor. 11 undet. bone fragments
Hickey Mountain limestone (8-10 cm thick)	Grayish orange pink (5YR 7/2), micrite (~80%), also contains smectites, illitic clays, rare glass shards, and other detrital grains. Contains grayish yellow green clay balls in lower part, abundant snail shells and fish pellets. Underlain by unfossiliferous non-calcareous light olive gray sandy mudstone	Snails, fish vertebra, 3 gar scales	Snails	Snails, fish vertebra, palatal tooth plate, 11 gar scales	Snails, fish palatal tooth plate, mammal vertebra, 6 undet. bone fragments

Making the distinction between "specimens" and "skeletal elements" is critical to estimating the number of individuals and the degree of bone fragmentation in a fossil assemblage. We follow Badgley (1986) and Lyman (1994) in defining a "specimen" as a discrete bone or tooth, or fragment of any degree of completeness in a fossil assemblage. A "skeletal element" is defined as a discrete anatomical unit such as single complete bone or tooth. Thus specimens can be, but are not necessarily, skeletal elements.

The *Omomys* Quarry sample includes 1183 identified vertebrate specimens consisting of mostly fragmentary bones and teeth, over 1000 eggshell fragments, approximately 100 gastropod shells, and 40 poorly preserved fragments of silicified wood (Table II). The *Omomys* Quarry assemblage is highly unusual for the Bridger Formation because of its relatively abundant *Omomys*, avian, and eggshell (primarily avian) specimens. These fossils are relatively uncommon in the upper Bridger, and are typically disassociated and widely dispersed. The total number of *Omomys*, avian, and eggshell specimens from the quarry is greater than the total number from all of UCM's 530 other Bridger fossil localities. In fact, the total number of *Omomys* specimens in the UCM collections from the entire upper Bridger Formation (not including the *Omomys* Quarry) is 35, and these consist of mostly isolated teeth. One way to quantitatively illustrate the unusual nature of the *Omomys* Quarry assemblage is to compare the number of identified specimens (NISP) and minimum number of individuals (MNI) of the most common orders of mammals in the *Omomys* Quarry assemblage with other upper Bridger assemblages.

The MNI method estimates the number of individuals present in an assemblage as the minimum number necessary to account for all of the faunal remains, and tends to underestimate the actual number of individuals. The NISP method assumes that each specimen represents an individual. Since this approach tends to maximize the number of possible individuals, it has also been referred to as the maximum number of individuals or XNI (Gilinsky and Bennington, 1994). One of the problems with the NISP approach is sample inflation (Badgley, 1986), since some individuals in any given sample contribute more than a single bone. These two basic approaches are discussed in detail by Badgley (1986), Holtzman (1979), and Lyman (1994). For taphonomic circumstances which involve the accumulation of partial remains of prey animals where the probability of association is intermediate to high, Badgley (1986) has advocated the use of the MNI approach over NISP. Both the NISP and MNI were found to be useful for different aspects of this study.

Three of the localities (UCM L. 93078, L. 92180, L. 93027) to which we compare the *Omomys* Quarry assemblage were selected because, like the quarry, they occur in calcareous mudstones overlying micritic limestones, represent similar depositional environments, and are also relatively rich fossil accumulations within the Bridger C. Locality 93078 occurs in the Henrys Fork limestone (43 m above the HMLs) on the south side of Cedar Mountain, and locality 92180 occurs in the Henrys Fork limestone on the south side of Sage Creek Mountain. Locality 93027 occurs in the Hickey Mountain limestone approximately 0.5 km to the southwest of the *Omomys* Quarry. The fourth

Table II. Fauna and flora of the *Omomys* Quarry with number of identified specimens for vertebrates shown in right column. Eggshell is not included

TAXA			NISP
PLANTAE			
	Division Chlorophyta		
		Chara sp.	
	Division Tracheophyta		
		Dennstaedtiopsis aerenchymata (fern)	
		2 types of dicotyledenous wood	
ANIMALIA			
Phylum Mollusca			
Class Gastropoda			
	Order Lymnophila (freshwater pulmonates)		
		Biomphalaria sp.	
		Omalodiscus sp.	
		Stagnicola? spp.	
		Physa spp.	
	Order Geophilia (land pulmonates)		
		Gastrocopta spp.	
		Oreoconus? sp.	
Phylum Chordata			
		Coprolites	4
Class Osteichthyes			
		Osteichthyes undet.	200
	Order Amiiformes		
		Amia sp.	8
	Order Lepisosteiformes		
		Lepisosteus sp.	98
	Order Siluriformes		14
Class Amphibia			
		Amphibia undet.	4
	Order Anura		
		Anura undet.	4
Class Reptilia			
	Order Chelonia		
		Chelonia undet.	5
		Trionychidae undet.	1
		Echmatemys sp.	1
		Palaeotheca sp.	2
	Order Squamata		
		Lacertilia undet.	117
		Iguanidae undet.	3
		Tinosaurus sp.	1
		Saniwa sp.	1
		Serpentes undet.	2
	Order Crocodilia		
		Allognathosuchus sp.	34
		Crocodilia undet.	4

continued

locality, 92183, is a highly fossiliferous point bar deposit exposed to the south of Sage Creek Mountain, 74 m above the HMLs.

The faunal comparison (Table III) is intended only to illustrate the unusually high number of primate specimens in the *Omomys* Quarry sample and the quarry's relative richness in terms of NISP. Combined, the relative abundances

Table II. (Continued)

TAXA			NISP
Class Aves			
	Order Ciconiiformes		
		Juncitarsus gracillimus	4
	Order Falconiformes		
		Accipitridae (2 species)	7
	Order Charadriiformes		
		Burhinidae	5
	Order Gruiformes		
		Rallidae	5
		Geranoididae	8
	Total Aves (includes undet. specimens not listed above)		(191)
Class Mammalia			
		Mammalia undet.	150
	Order Marsupialia		
		Peratherium sp.	13
		Peradectes sp.	1
	Order Rodentia		
		Rodentia undet.	13
		Paramys sp.	1
		Thisbemys sp.	1
		Microparamys sp.	1
		Ischyromyidae undet.	1
		Sciuravus sp.	9
		Pauromys sp.	2
		Sciuravidae undet.	6
	Order Apatemyida		
		Apatemys sp.	1
	Order Insectivora		
		Insectivora undet.	28
		Scenopagus sp.	2
		Entomolestes sp.	7
		Centetodon sp.	6
		Apternodontidae undet.	1
		Nyctitherium sp.	4
	Order Dermoptera		
		Uintasorex sp.	1
	Order Primates		
		Notharctus sp.	2
		Omomys sp. nov.	214
	Order Condylarthra		
		Hyopsodus sp.	22
	Order Artiodactyla		
		Artiodactyla undet.	1
		Homacodon sp.	1
	Order Perissodactyla		
		Hyrachyus sp.	1
		TOTAL	1,183

of mammalian groups based on the NISP from the four localities compared to the *Omomys* Quarry are typical of most upper Bridger assemblages. Since these four localities are geographically and/or stratigraphically separated from one another, the MNI takes into account the impossibility that any two localities could contain fragments of the same individual.

Table III. Comparison of the most common orders of mammals from the *Omomys* Quarry with other more typical UCM localities in the upper Bridger Formation based on NISP and MNI (Localities 92183, 93078, 92180, 93027)

	Omomys Quarry				Combined total from 4 other upper Bridger localities			
Taxonomic Group	NISP	% of total	MNI	% of total	NISP	% of total	MNI	% of total
Marsupials	14	4	4	8	12	4	3	4
Rodents	23	7	11	22	32	11	7	8
Insectivores	48	15	15	31	20	7	4	5
Primates	216	66	12	24	19	6	6	7
Condylarths	22	7	5	10	163	54	44	52
Artiodactyls	1	<1	1	2	8	3	4	5
Perissodactyls	1	<1	1	2	38	13	11	13
Carnivores	0	0	0	0	10	3	5	6
TOTALS	325		49		302		84	

The NISP (216) and MNI (12) of primates from the *Omomys* Quarry is much larger than the NISP (19) and MNI (6) of primates from the combined totals of the four other localities shown in Table III. Also, the proportion of primate specimens to other mammalian specimens is much larger in the *Omomys* Quarry assemblage (66%) than the other assemblages (6%). Note that the 216 NISP of primates from the *Omomys* Quarry shown in Table III includes two specimens of *Notharctus* (=2 MNI), and that the MNI of *Omomys* is ten. Although not shown in Table III, avian specimens on average comprise less than 1% of specimens collected from UCM upper Bridger localities, but comprise 16% of the total NISP of vertebrates from the *Omomys* Quarry.

The large number of skeletal specimens of *Omomys*, birds and eggshell fragments at the *Omomys* Quarry occur together with a more typical Bridger vertebrate assemblage consisting primarily of isolated teeth, jaw fragments with teeth, and isolated and usually fragmentary postcrania representing 778 specimens of at least 40 taxa (Tables II and IV). Typical Bridger mammalian samples are similar to this portion of the *Omomys* Quarry fauna, except that jaw fragments and isolated teeth of the condylarth *Hyopsodus* typically constitute about 50% of mammalian specimens as is shown by Table III.

The relatively high number of *Omomys* and avian specimens in the *Omomys* Quarry sample suggests that they have been concentrated as the result of some taphonomic process or combination of processes. The diversity of skeletal elements present and the lower variation in their relative representation in comparison to other vertebrate groups (Table IV) suggests that individuals of these taxa are represented by more complete and/or less disassociated skeletons than other vertebrate taxa. In non-*Omomys* mammals, for example, 18 bone types are present, however only four of these types (isolated teeth, metapodials/phalanges, vertebrae, and dentaries) account for 87% of the total number of specimens, and isolated teeth alone constitute 51% of specimens.

In *Omomys*, 18 bone types are present, and the 11 most numerous types (all with ten or more specimens per type) account for 91% of total *Omomys* specimens. Isolated teeth constitute only 21% of *Omomys* specimens. In birds, ten bone types are present, and eight of these (all with 12 or more specimens per type) constitute 98% of total avian specimens. Note that Table IV is based on NISP, and that many of the tabulated specimens are fragmentary portions of bones, such as proximal or distal ends of long bones.

4.1. Flora

Fossil plants are rare in the Bridger Formation. Although poorly preserved, several varieties are present in the *Omomys* Quarry assemblage. These include petiole fragments of the fern *Dennstaedtiopsis aerenchymata* and numerous fragments of at least two kinds of dicotyledenous wood (Steven Manchester, written communication, 1997). Like the terrestrial vertebrates, these fossils were preserved in the mudstone overlying the HMLs, and, along with the presence of arboreal mammals such as *Omomys*, indicate a wooded environment with conditions suitable for ferns. Charophytes are present in the HMLs at the *Omomys* Quarry, and locally throughout the HMLs's distribution. *Chara*, or Stonewort, is an algae that grows today submerged in calcium carbonate-rich standing waters on muddy or sandy substrates, and is common in alkaline lakes and ponds (pH 7.5–8.0) (Wetzel, 1983).

4.2. *Omomys*

As noted above, the *Omomys* sample from the quarry (214 NISP, 10 MNI) is relatively well-preserved and elementally diverse (Table IV). Jaw fragments with teeth and isolated teeth constitute 32% of the total number of *Omomys* specimens in the sample, while postcranial specimens both complete and fragmentary constitute 68%. The MNI of 10 for *Omomys* is indicated by the presence of 10 right distal femora.

Skeletal morphology of *Omomys* indicates that it was an active arboreal runner, leaper, and climber (Covert and Murphey, 1994; Covert, 1995). In tarsal morphology, it most closely resembles the extant mouse lemur, while in femoral and tibial morphology it resembles the acrobatic leaping galagos and tarsiers (Covert, 1995). Tarsal and limb bones indicate that *Omomys* had a body mass of approximately 250–350 g, and maxillary fragments indicate that it had relatively enlarged orbits indicating a nocturnal activity pattern, and fairly well developed shear on its molars indicating a diet of insects and fruit (Covert, 1995). Preliminary studies of *Omomys* specimens from the quarry suggest that they exhibit a significantly derived dental morphology in comparison to the Bridger B species *Omomys carteri*, and represent an undescribed species of this genus.

Table IV. Elemental representation (NISP) of vertebrates from the *Omomys* Quarry listed in order of abundance within each group. Eggshell fragments are not included

Specimen type	Number	Specimen type	Number
FISH		*MAMMALS* (non-*Omomys*)	
vertebrae	103	isolated teeth	143
scales	97	metapodials/phalanges	44
isolated teeth	36	vertebrae	31
palatal teeth plates	27	dentaries	25
miscellaneous undet. bones	26	undet. long bone shaft frags.	10
dentaries	21	claws	5
spines	10	astragali	4
TOTAL	**320**	maxillae	4
		calcanei	3
AMPHIBIANS		radii	2
humeri	4	misc. podials	1
vertebrae	3	bullae	1
dentaries	1	femora	1
TOTAL	**8**	fibulae	1
		humeri	1
CROCODILIANS		premaxillae	1
isolated teeth	36	scapulae	1
osteoderms	2	ulnae	1
TOTAL	**38**	TOTAL	**279**
TURTLES		*OMOMYS*	
carapace/plastron fragments	8	isolated teeth	46
phalanges	1	femora	24
TOTAL	**9**	tibiae	21
		misc. podials	16
SQUAMATES		humeri	16
dentaries	87	pelvic/sacral frags.	16
vertebrae	17	astragali	12
osteoderms	11	maxillae	12
phalanges	4	dentaries	11
undet. bones	3	calcanei	10
maxillae	1	metapodials/phalanges	10
isolated teeth	1	fibulae	5
TOTAL	**124**	ulnae	5
		radii	3
AVES		vertebrae	3
phalanges	50	scapulae	2
talons	31	patellae	1
tarsometatarsi	24	ribs	1
femora	20	TOTAL	**214**
humeri	18		
vertebrae	18	TOTAL (all groups combined)	**1,183**
tibiotarsi	14		
undet. bones	12		
synsacral fragments	3		
beaks	1		
TOTAL	**191**		

The age at death of individuals in a population can be compared on the basis of the stage of tooth wear (Hillson, 1986). Simple tooth wear states of individual *Omomys* upper and lower first and second molars from the quarry estimate the general population structure of the death assemblage. Juveniles are inferred to be represented by deciduous teeth, and young adults are inferred to be represented by "unworn" teeth, defined as exhibiting no visible wear facets. Adults are inferred to be represented by "slightly worn" teeth, which exhibit small circular wear facets on the tips of cusps, and older adults are inferred to be represented by "worn to very worn" teeth, which exhibit larger wear facets in which cusps have been partially or fully joined, and crown morphology has been partially to totally obliterated.

No deciduous teeth occur in the *Omomys* sample. Of ten first upper molars present in the sample, five (50%) are unworn, four (40%) exhibit slight wear, and one (10%) is worn to very worn. This pattern is similar to that which occurs in the ten first lower molars in the sample. Among nine second upper molars, six (67%) are unworn (Figure 2A), and three (33%) exhibit slight wear. Twelve second lower molars are present, of which six (50%) are unworn, four (33%) exhibit slight wear, and two (17%) are worn to very worn. The general wear patterns suggest, based on the tooth wear states defined above, that the *Omomys* death assemblage is dominated by young adults and adults, with no juveniles and relatively few older adults. This agrees with the general extent of epiphyseal fusion observed on proximal and distal femora, tibiae, and humeri. Of 41 specimens in which the epiphyseal area is well preserved, none exhibit unattached epiphyses, 16 (39%) exhibit partially visible epiphyseal sutures and are inferred to represent young adults, and 25 (61%) exhibit completely fused epiphyses and are inferred to represent adults. Together with dental wear, these results suggest a sample bias in favor of young adults and adults.

4.3. Fossil Eggshell

The analysis of eggshell microstructure has shown that it can be distinct at the ordinal and, in some extant avians, to the familial level (Mikhailov, 1992, 1995, 1997a; Tyler, 1964, 1969a,b; Tyler and Simkiss, 1959). Though eggshells of recent avian taxa and dinosaurian taxa have been studied in some detail (Bray and Hirsch, in press; Hirsch, 1994; Hirsch and Quinn, 1990; Mikhailov, 1994a, 1994b; Mikhailov *et al.*, 1994; Vianey-Liaud *et al.*, 1994; Zhao, 1975, 1979; Zhao and Jiang, 1974; Zhao and Li, 1993), eggshell of Eocene age has not (Hirsch, 1985; Hirsch and Kohring, 1992). Therefore Eocene-aged avian eggshell is difficult to identify. Eggshells are classified by the fundamental structural organization of the organic and inorganic shell components. The varieties of fundamental eggshell structures are referred to as "basic types" of eggshell organization.

The fossil eggshell from the *Omomys* Quarry consists of more than 1000 fragments which were collected from approximately 75% of the quarry area.

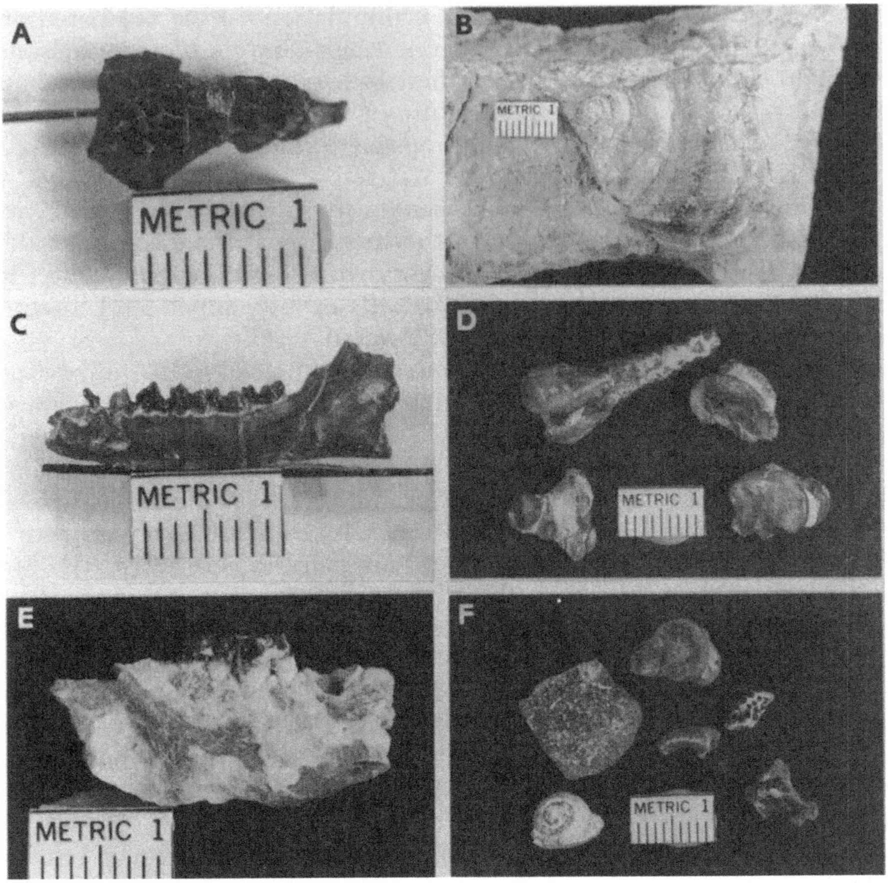

FIGURE 2. a, UCM 67864, right maxillary fragment of *Omomys* sp. with unworn RP3/-RM3/; b, UCM 30017, partial shell of *Oreoconus* (?) sp. with color bands; c, UCM 69052, left dentary of *Omomys* sp. with P/2-M/3, showing minimal primary weathering/corrosion of bone and teeth; d, UCM unnumbered, proximal and distal femora of *Omomys* sp. with slight exfoliation of outer layer of bone in places, but with generally minimal modifications by primary weathering/corrosion; e, UCM 84301, right dentary of *Paramys* sp. with extreme weathering and/or corrosion; f, specimens from *Omomys* Quarry sample E3–25, showing typical breakage (clockwise from top: distal femur, palatal tooth fragment, vertebra, partial snail, eggshell fragment; partial talon in center). Scale bar in all photographs = 1 cm.

No complete eggs were found, and all eggs appear to have been broken prior to burial. The eggshell sample is dominated by one variety of ornithoid eggshell. In hand specimens this shell appears well preserved, with discrete pore openings on the outer surface and preserved mammillae on the inner shell surface. However, when analyzed at higher magnifications by SEM, the absence of detailed preservation of the fine arrangement of biocrystalline material (ultrastructure) is apparent. This is also true for most of the other eggshell varieties from the *Omomys* Quarry. Due to the lack of fine ultrastructural preservation by which eggshell is generally classified, the eggshells from the quarry assemblage are distinguished by their histostructure; the organization of the individual shell units, the size and shape of the mammillae, and the characteristics of the pore system (microstructure).

Other eggshell varieties from the *Omomys* Quarry include ornithoid, crocodiloid and probable squamate basic types. Numerous very small (>2 mm) fragments with sculpturing different from each other and from the aforementioned varieties, do not possess distinct microstructures when viewed by SEM and therefore are not included in the eggshell descriptions below.

The dominant eggshell type from the quarry is of the ornithoid basic type/prismatic (neognathe) morphotype (UCM 82001). The outer surface is smooth to crenulate, light brown in color, and pore canal openings are round and occur singularly. In radial section (Figure 3A) the individual shell units height to width ratio (H:W) range from 1:6 to 1:3.6. The ratio of the mammillary layer to continuous layer (ML:CL) approaches 1:3, well within the typical range of avian eggshell (Mikhailov, 1997a). The mammillae do not show preservation of fine structure (Figure 3B) but some have retained the organic cores (Figure 3C). Most *Omomys* Quarry shells have a thin layer of calcite, a secondary deposit (SD), on the outer surface of the eggshell (Figure 3A). Under SEM a thin external zone is observed (Figure 3D). Under polarized light (Figure 3E), the extinction patterns show the crystallographic orientation of calcite roughly outlining the shell unit prisms and boundaries but is not reflective of the fine structural arrangement of biocrystalline material (Golubev, 1988; Mikhailov, 1997b). The pore system is angusticanaliculate consisting of straight, narrow pores with circular openings on the outer surface, but the accurate estimation of eggshell porosity is not easily determined due to the small size of shell fragments. Shell thickness ranges from 0.44–0.60 mm. The difficulty of determining the taxonomic affinity of this eggshell is compounded by the recrystallization of its original shell structure. Without preservation of the fine ultrastructure and ability to distinguish the ratios of the structural portions of the shell, accurate taxonomic assignment is not possible. In general, the morphology of this shell (Figure 3D) more closely resembles that found in Anseriformes than it does of Falconiformes (differing distinctly from Accipitridae and Pandionidae) and the thickness of the shell is twice of that found in extant Strigiformes.

The best preserved type of eggshell from the quarry is an avian prismatic eggshell known from only four small fragments excavated from the outer edge of the quarry area (UCM 82002). This shell averages 0.35 mm in thickness. The H:W ratio of shell units average 3.3:1 (Figure 4A). The mammillae are well preserved and under polarized light the shell units are well defined (Figure 4B). The outer surface is smooth and white. The pore system is angusticanaliculate and pore openings on the outer surface are round and singular. Because of the current lack of knowledge of Eocene-aged avian eggshell, these specimens were not identifiable.

The third type of eggshell is of the crocodiloid morphotype (UCM 82006). The shell is textured on the outer surface with enlarged pore openings. The wedge shaped shell units typical of the crocodiloid morphotype are relatively well preserved in radial section (Figure 4, C, D, E). Eggs of this type are known from other localities in the Bridger Formation (Hirsch and Kohring, 1992). Comparing this eggshell to that of extant crocodiles, the similarities are apparent (Figure 4F).

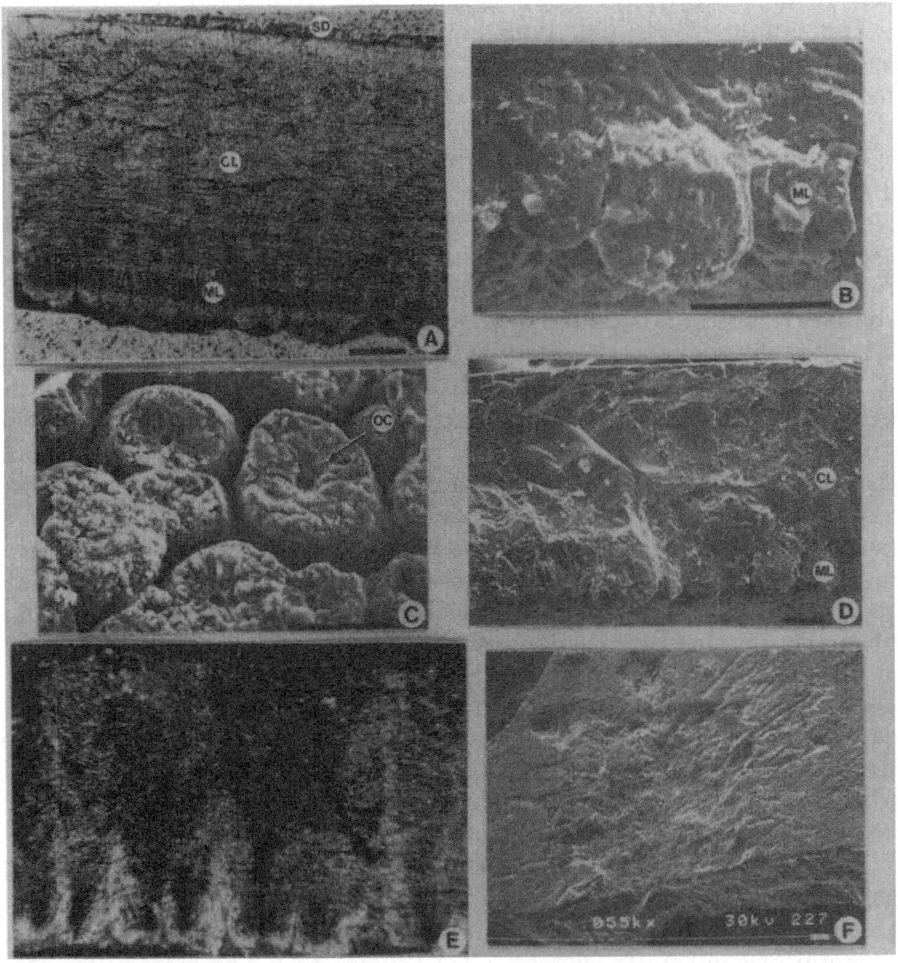

FIGURE 3. a–e, UCM 82001. a, radial thin section of eggshell with mammillary layer (ML), continuous layer (CL), and secondary layer (SL) of calcite on outer surface. Outer surface of eggshell is at top of photo in all radial views; b, radial section of mammillae (ML); c, view of inner surface of eggshell with organic cores (OC), the origin of growth of the mammillae; d, radial section with external zone; e, micrograph under polarized light with extinction patterns and vertical orientation of calcite growth; f, outer surface of eggshell with pitting and minute grooves possibly attributable to insects. Scale bar in all photographs = 100 μ.

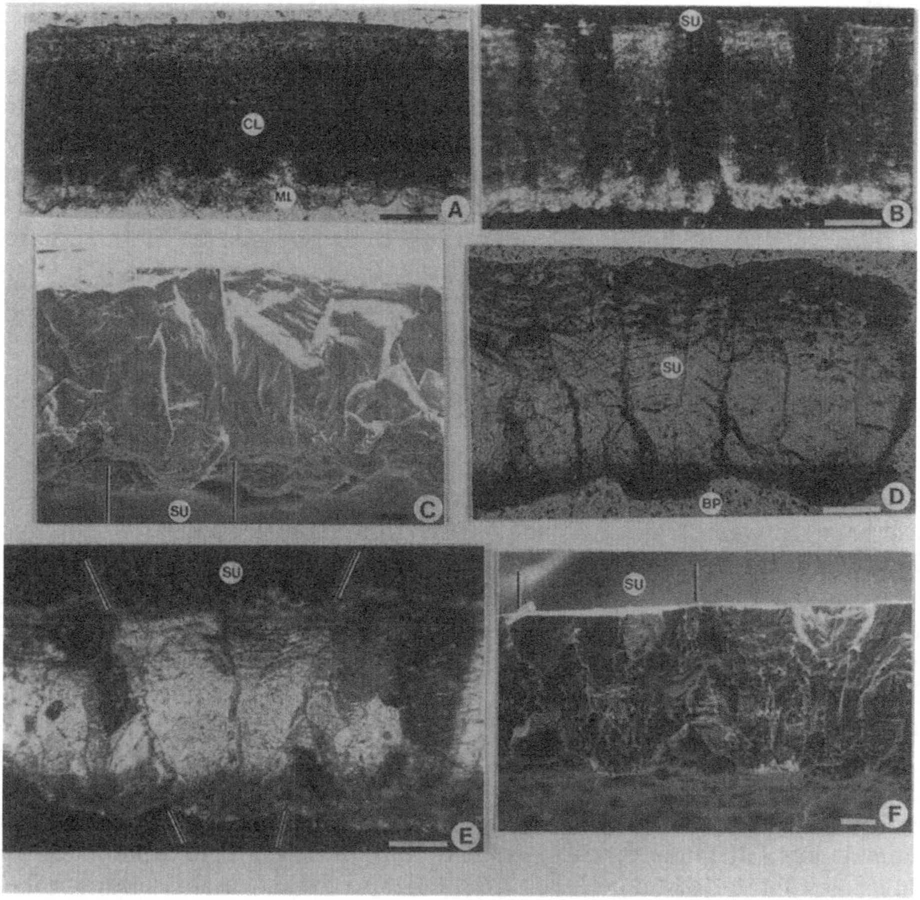

FIGURE 4. a and b, UCM 82002. a, radial thin section of ornithoid prismatic shell, mammillary layer (ML) and continuous layer (CL); b, polarized light radial view with sharp vertical extinction lines within shell units (SU). c–e, UCM 82006. c, wedge shaped shell units (SU) of tabular ultrastructure representative of the crocodiloid morphotype; d, in radial thin section, shell units (SU) and basal plate groups (BP); e, polarized light view of shell units (SU); f, for comparison, UCM OS-1314, *Crocodylus johnstoni* (fresh water crocodile), shell thickness 0.30–0.38 mm. Scale bar in all photographs = 100 μ.

4.4. Fossil Birds

The *Omomys* Quarry avifauna is an interesting and unusually diverse sample composed of predominantly wading and shore-edge birds. Forest and ground birds also are represented, and hawks are the only avian predators in the assemblage. Specifically, the avifauna includes a small flamingo (Ciconiiformes, Phoenicopteridae), two species of hawks (Falconiformes, Accipitridae), two species of rails and other gruiform birds (Gruiformes, Rallidae and Geranoididae), and a thick-knee (Charadriiformes, Burhinidae).

The small flamingo, *Juncitarsus gracillimus* Olson and Feduccia (1980), is identified based on three tarsometatarsi (left distal end, UCM 68688; right distal end fragment, UCM 68779; left proximal end, UCM 68756) and a right distal humerus (UCM 69924). The tarsometatarsi and the humerus are virtually identical to specimens of the type series figured by Olson and Feduccia (1980). The type locality for *J. gracillimus* is in the Bridger B, northeast of Twin Buttes, Sweetwater County, Wyoming.

Two species of hawks are identified from the quarry. Both appear to have an affinity with the modern genus *Buteo* (Accipitrinae). The larger species is represented by talons (UCM 69099, 69100, and 69301). Each exhibits the typical profile and cross section of this subfamily in comparison to falcons, osprey, and owls. The talons are comparable in size to the modern Red-tailed Hawk (*B. jamaicensis*) and Ferruginous Hawk (*B. regalis*).

The second hawk species is smaller in size than the above species and comparable to the modern Broad-winged Hawk (*Buteo platypterus*). The size difference between the hawk above and this smaller species is not sexual dimorphism but is a species level character state. This species is represented by two humeri (right distal end, UCM 68752; left distal end, UCM 68754), a fragmentary proximal sternum (UCM 68771), and distal left tibiotarsus (UCM 68778). The fossils are all very similar in comparative osteology and size to *B. platypterus* and represent an early record for the Accipitridae in North America.

There are at least two species of rails from the quarry. A diminutive left tarsometatarsus (UCM 68696) is comparable in size and osteology to the living Virginia rail (*Rallus limicola*). A second species is represented by a left tarsometatarsus (distal end, UCM 68684), left humerus (proximal end, UCM 68699), left femur (distal end, UCM 68765), and a right coracoid (humeral end, UCM 68765). This species also shows an affinity to the modern rails (*Rallus*) in comparative osteology. The fossils are larger than modern *R. longirostris* or *R. elegans*. The Lower Eocene species *Palaeorallus troxelli* Wetmore and *P. brodkorbi* Cracraft are only known from their holotype tibiotarsi and are not directly comparable to the rails from the *Omomys* Quarry (Cracraft, 1973; Olson, 1977). *Palaeorallus brodkorbi* is approximately the size of the larger species reported on here.

The Geranoididae is one of the more puzzling gruiform groups (Cracraft, 1973; Olson, 1985). The following fossils are referred to this group: six tibiotarsi (left proximal end, UCM 68761; left distal ends, UCM 69699, 68758, 69790; right distal ends, UCM 68757, 68760) and two distal tarsometatarsi (left distal end, UCM 68687; right distal end, UCM 68685). Although they have been placed in their own family (Wetmore, 1931), then in their own subfamily within the Gruidae (Brodkorb, 1967), and then back in their own family (Cracraft, 1969, 1973), this group has its affinities with modern South American seriemas (Gruiformes, Cariamidae), and the middle Tertiary Idiornithinae and Bathornithinae (Chandler, 1997).

The only shorebird (Charadriiformes) to be recovered from the Quarry is a thick-knee (Burhinidae) and is represented by a left tarsometatarsus, (distal end, UCM 68686), three humeri (right proximal, UCM 68697; right proximal fragment, UCM 68698; left proximal fragment, UCM 68700), and a left coracoid (humeral end, UCM 68780). The fossils are very similar osteologically to the modern genera of thick-knees or stone curlews (*Burhinus* and *Esacus*). Thick-knees are found in the temperate and tropical parts of the world except for North America, New Zealand, and the tropical Pacific. Thick-knees have been reported from the Lower Eocene London Clay, the Naze Formation (Daniels private collection, personal observation), the Miocene of Nebraska (Bickart, 1981), and the Pleistocene of Kansas (Feduccia, 1980).

Modern birds related to those found at the *Omomys* Quarry inhabit a variety of environments. With a preference for muddy, alkaline, or brackish lakes, modern flamingoes build their pedestal-like nests out of mud, usually in shallow water near lake margins or on mudflats adjacent to lakes. Adapted for running through thick grass and underbrush, modern rails are typically marsh inhabitants, although some live on open ponds and lakes, and a few species live in woodlands and even on dry plains. Rails build bulky nests out of vegetation which are well hidden in grasses on or near the ground. The two modern species of seriamas live in grasslands or sparse brushy forests, nesting either on or close to the ground. Although clearly related to "wading birds," only one of the nine modern species of thick-knees lives near water. Most inhabit dry uplands and prefer rough, pebbly areas. With an almost cosmopolitan distribution, modern members of the subfamily Accipitrinae are hunting hawks and eagles that prey largely on small mammals and birds, and occupy a wide variety of environments (Austin, 1977).

4.5. Gastropods

Gastropod shells are scattered throughout the *Omomys* Quarry matrix, and their preservation is typically poor, permitting identification only to the generic level. Shell material is present in some of the shells, and one shell of a large land snail (*Oreoconus*? sp.) retains color bands (Figure 2B). Five genera of aquatic snails and two of land snails occur in the quarry assemblage (Table II). Shells of aquatic snails are approximately four times as abundant as land snail shells.

The freshwater gastropods are all lung-bearing pulmonates. The most abundant shells are those of the large planorbid *Biomphalaria* sp., which is the most abundant gastropod in the Bridger Formation. The other freshwater taxa include the planorbid *Omalodiscus* sp., and the lymnaeids *Stagnicola* (?) spp. and *Physa* spp. The freshwater taxa suggest that the waters of Hickey Mountain lake: 1) had reduced oxygen content, 2) were turbid, and/or 3) were intermittent. Because modern planorbids are most common in water that is less than 1.8 m deep (Hanley, 1974), the abundance of planorbids in the assemblage (approximately 70% of all shells) indicates that the water was shallow. The land snails *Oreoconus* (?) sp. and two species of *Gastrocopta* are also common fossil snails of the Bridger.

5. Taphonomy of the *Omomys* Quarry

The *Omomys* Quarry provides an opportunity to apply actualistic taphonomic data to an unusual fossil accumulation. Bone weathering, hydraulic sorting, mechanical abrasion, distribution and orientation, digestive corrosion, and breakage, can be evaluated. The taphonomic interpretations made in this study are based on these physical aspects of the bones and teeth collected from the quarry, the geology of the Hickey Mountain limestone, modern predator-prey interactions, and habitat preferences.

5.1. Weathering

Behrensmeyer (1978) proposed five weathering stages for bones of medium and large sized mammals from various habitats in Amboseli National Park, Kenya. These stages correlate the condition of bones with the length of exposure to physical and chemical agents on the surface and in the soil zone. If primary weathering is distinguishable from transport, abrasion and diagenetic effects in fossil bones, it can provide information on sub-aerial exposure of bone prior to burial and the amount of time over which bones accumulated (Behrensmeyer, 1978). Experimental results reported by Andrews (1990) for the weathering of small mammal bones in a wet temperate climate indicate that small bones weather differently and at different rates than medium and large-sized bones from Amboseli.

Using either the criteria of Behrensmeyer (1978) or Andrews (1990), the weathering of bones from the *Omomys* Quarry suggests relatively rapid burial. None of the mammalian teeth from the quarry show any visible modification due to weathering. Of the 112 mammalian bones or bone fragments (excluding *Omomys*) for which a determination of weathering could be made, 65% exhibit no modifications due to weathering, suggesting a surface exposure time of 0–2 years (Andrews, 1990). The other 35% are only slightly weathered, equivalent to Behrensmeyer's stage 1 (0–3 years) and Andrews' range of 1–5 years. Of the 144 *Omomys* bones or bone fragments for which a determination of weathering could be made, 92% exhibit no modification due to weathering, suggesting a surface exposure time of 0–2 years (Andrews, 1990) (Figure 2, C, D). The other 8% are slightly weathered, equivalent to Behrensmeyer's stage 1 (0–3 years) and Andrews' range of 1–5 years. These data imply that all mammalian specimens were buried in less than five years, with most of them being buried in less than two. The variation in weathering states observed in the mammalian sample implies an attritional accumulation rather than a catastrophic assemblage. The *Omomys* specimens, however, are less weathered (92% no modification) than other mammals (65% no modification). This suggests that most of the *Omomys* remains endured a shorter period of exposure than the other mammalian remains.

5.2. Hydraulic Sorting

In general, the lithologies of the *Omomys* Quarry limestone and mudstone suggest quiet water and little or no transport, however, transport of small bones has been shown to occur even at low current velocities (Dodson, 1973). In addition, small bones that have dried on land may float for considerable distances, and then be moved along the bottom after sinking (Dodson, 1973). The possibility of bone sorting by transport of the quarry assemblage should therefore be considered.

Voorhies (1969) assigned skeletal elements of medium and large mammals to three groups based on their susceptibility to hydraulic sorting, or transport. These groups were later placed into five groups and named "Voorhies Dispersal Groups" by Behrensmeyer (1975). Dodson (1973) performed a similar experiment with a skeleton of a mouse, a frog, and a toad, and reported results generally similar to those of Voorhies (1969). These studies indicate that lighter bones with relatively large surface areas (group I) are more subject to transport than denser bone (group III). Dodson (1973) found that very low current velocities were sufficient to disperse small bones, and their transport produced high levels of breakage. It was concluded that because small bones are so easily dispersed, it might be difficult to tell if a small bone assemblage had not been heavily transported or at least modified. For this reason, Dodson considered small bones to be problematic for paleoecological reconstruction. Dodson (1973) concluded that a high representation of vertebrae, which are easily dispersed, to jaws, which are more difficult to move, would be one indication that "selective" transport had not occurred. These studies are based on experiments with complete bones, and not broken bones and partial skeletons which are so often characteristic of natural accumulations even prior to hydraulic transport. One might intuitively expect that fragmented bones would generally be more readily dispersed by currents because they are smaller and lighter, so it seems prudent to use caution when applying these "dispersal group" data to fossil assemblages dominated by bone fragments where primary breakage is apparent. Was the *Omomys* Quarry assemblage or any part of it modified by transport? Because of the small size and fragmentation of these remains, it could be difficult to tell using dispersal group data as Dodson's (1973) findings suggest. Mammalian skeletal specimens from the quarry represented by five or more specimens, both fragmentary and complete, were assigned to both Voorhies' (1969) and Dodson's (1973) dispersal groups (Table V). Terrestrial vertebrates other than mammals were not considered because to our knowledge there have been no studies of their bone dispersal patterns.

Isolated teeth are the most abundant mammalian element in the quarry sample. Of 248 mammalian specimens (excluding *Omomys*) listed in Table V,

Table V. Frequencies and transport groups of identified mammal bones represented by 5 or more specimens. "?" indicates elements that were not assigned to bone dispersal groups by Voorhies (1969) or Dodson (1973)

Taxonomic group and bone type	NISP	% of group	Voorhies 1969	Dodson 1973
MAMMALS (excluding *Omomys*) (MNI = 42)				
isolated teeth	143	58	?	?
metapodials/phalanges	44	18	I/II and II	?
vertebrae	31	12	I	I and I/II
dentaries	25	10	II/III	III
claws	5	2	?	?
TOTAL of 5 bone types	248			
OMOMYS (MNI = 10)				
isolated teeth	46	22	?	?
femora (8 proximal, 15 distal, 1 almost complete)	24	12	II	II
tibiae (13 proximal, 8 distal)	21	10	II	II
misc. podials	16	8	I	?
humeri (7 proximal, 9 distal)	16	8	II	II
astragali	12	6	I	?
maxillae	12	6	III	I
dentaries	11	5	II/III	III
pelvic frags. (5 ischium, 6 ilium and acetabulum)	11	5	II	I/II
calcanei	10	5	I	II/III
metapodials	10	5	II	?
fibulae (all distal)	5	2.5	?	II
sacral fragments	5	2.5	I	?
ulnae (3 proximal, 2 distal)	5	2.5	I/II	II/III
TOTAL of 13 bone types	204			

58% are isolated teeth, and of the 204 *Omomys* specimens, 22% are isolated teeth. Neither Voorhies (1969) nor Dodson (1973) included isolated teeth in their dispersal groups, although Dodson indicates elsewhere in his paper that mouse incisors were the last element to be moved. In Pleistocene-aged bones from California, Wolff (1973) found that isolated teeth were the most abundant element in transported assemblages (suggestive of Group 1-Early), but that larger rodent incisors were less susceptible to transport. These findings make it difficult to evaluate isolated teeth from the *Omomys* Quarry with regard to transport, particularly since only nine of the 189 isolated teeth from the quarry are rodent incisors.

Besides isolated teeth, the non-*Omomys* mammalian sample includes only four types of bones which are represented by five or more specimens (Table V). These include metapodials/phalanges, vertebrae, dentaries, and claws, and are representative of dispersal groups I, I/II, II, and II/III of Voorhies (1969), and groups I, I/II and III of Dodson (1973). According to Dodson's (1973) findings, the presence of phalanges, vertebrae (both group I) and dentaries (group III), and the approximately equal representation of vertebrae to dentaries (31 : 25), suggests that the non-*Omomys* mammalian assemblage has not been significantly modified by hydraulic transport.

In comparison with the non-*Omomys* mammalian remains, the *Omomys* sample includes a greater diversity of elements (13 types with five or more specimens) with more specimens per type, and a lower proportion of isolated teeth as discussed above. Although most of the *Omomys* specimens fall into either dispersal groups I or II, group III is also represented (Table V), which suggests that selective transport has not occurred. This is supported by the diverse and unusually high concentration of *Omomys* specimens from the quarry.

Mechanical abrasion of bone is known to produce rounded edges, breakage, and even the total destruction of bones and teeth (Andrews, 1990; Korth, 1979). Experiments have shown, however, that it may take several hundred hours of tumbling with silt, sand, and gravel to produce pronounced rounding of bones (Andrews, 1990). Rounded fossil bones are not uncommon in the Bridger Formation, especially when the bones occur in stream channel deposits. Considering the apparently relatively low hydraulic energy interpreted for the upper Bridger lakes and ponds and their associated deposits, it is not surprising that only 5% of mammalian specimens from the quarry exhibit any evidence of possible mechanical abrasion.

According to dispersal group data and overall lack of abrasion, hydraulic transport has not significantly modified the mammalian assemblage. Some other taphonomic agent(s) must therefore be assumed to be responsible for the differential representation of skeletal elements observed in the *Omomys* versus the non-*Omomys* mammalian sample.

5.3. Distribution and Orientation

Prior to their removal from the rock, the location and orientation of every fossil that was visible as the result of being exposed on the surfaces of quarry rock samples was recorded (see Materials and Methods). Both the Hickey Mountain limestone and overlying bone-rich mudstone were distinguishable in cross section by both color and lithology. Most bones, including the largest, best preserved, and most complete, were found in the mudstone, and these were from both aquatic and terrestrial vertebrates. Smaller and less complete bones, mostly from aquatic vertebrates, were found in the limestone. Numerous, mostly unidentifiable tiny bone chips were also present throughout the limestone and mudstone, although they were far more numerous in the mudstone. This pattern of fossil distribution was also present in the quarry samples that were hand prepared (Table I). The eggshell and fossil plant specimens (wood and fern stems) were preserved in the mudstone.

The orientation of long bones has been shown to produce reliable information on the direction of current flow (Voorhies, 1969). Forty one long bones or pieces thereof (range of 2–8 cm long) were found to exhibit essentially random directions, indicating that current flow was not a significant factor in their orientation (Murphey and Burger, 1996), and supporting the conclusion that significant hydraulic sorting did not occur. These bones were also mostly horizontal to subhorizontal (actual range 0° to 20° dip).

Despite their unusually high density, none of the *Omomys* bones showed evidence of articulation prior to burial. Some of the avian bones, however, appear to have been articulated at or shortly before burial, because limb elements which were articulated in life were found in close proximity to one another. For example, a femur, tibiotarsus, and tarsometatarsus were found together in one quarry sample with no other avian bones, and this pattern of distribution occurred in several other quarry samples.

The thinness of the bone-rich mudstone (2–3 cm) allows no stratigraphic patterns of fossil distribution, and thus it was impossible to tell if all or part of the *Omomys* sample was accumulated prior or subsequent to the avian sample, or if the background fauna was deposited throughout. Likewise, no stratigraphic patterns of fossil distribution were observed in the limestone. Either the sediments were too soft to permit such layering, or any stratigraphic sequence that may have originally existed was obliterated during compaction and lithification of the fine-grained, lime-rich sediments.

In the field, other observations regarding fossil distribution were made. The high concentration of *Omomys*, avian, and eggshell specimens occur only at the quarry. Lateral to the quarry at the same stratigraphic level are commonly occurring fossils of the typical, or background, upper Bridger fauna (mammals shown in Table III). Where *Omomys* or avian specimens (including eggshell) have been found in the HMLs, they are disassociated and widely dispersed, as they typically are throughout the upper Bridger Formation.

The main conclusions regarding fossil distribution at the *Omomys* Quarry are that the *Omomys*, birds, and most other non-aquatic vertebrate remains were deposited in muddy sediments overlying the marl that contains the aquatic vertebrates. Most of the *Omomys* and avian remains, including the avian eggshell, are probably associated (come from fewer individual animals) because of their unusually high concentration, elemental diversity, and relative completeness. The conclusion of association is supported by the uniform absence of significant primary weathering of *Omomys* bones, and the apparent absence of hydraulic sorting.

5.4. Corrosion

The corrosive effects of digestion of the surfaces of bones and teeth can be quite significant (Andrews, 1990; Kusmer, 1990) and sometimes dramatic (D'Errico and Villa, 1997). Because these corrosion patterns are not duplicated by any other taphonomic process, they may be used to identify bone assemblages derived from predators. Furthermore, because predators show differences in the pattern and extent of digestion of prey bone, it is possible to identify species-specific patterns of bone digestion for modern predators (Andrews, 1990). Based on extent of bone and teeth breakage and digestive corrosion, Andrews (1990) was able to establish five modification categories for modern predators. Mammalian carnivores are classified as category 3, 4 and 5 predators, because bones and teeth recovered from their scats generally exhibit the highest degree of digestive corrosion and fragmentation. Members of the

Order Falconiformes are also associated with moderate to extreme corrosion of bones and teeth (categories 3–5). Owls are associated with a range of bone and tooth corrosion from minimal or absent to extreme, depending on the species (categories 1–4). Several modern species of owls, however, are the only predators that produce minimal or no corrosion of bone and/or tooth enamel (category 1 of Andrews, 1990).

If the *Omomys* accumulation was predator mediated, a category 1 predator is most likely because there is no evidence of digestion on any *Omomys* teeth, and only eight postcranial specimens (5% of postcrania) show evidence of possible digestion (Figure 2, A, C, D). The absence of primary weathering damage on *Omomys* specimens greatly eased the difficulty of distinguishing between the results of weathering and digestive corrosion. On a severely weathered specimen, it might very well be impossible to identify the prior effects of digestion.

Among other mammals from the quarry, digestive corrosion where preserved ranges from absent or minimal (category 1) to extreme (category 5), suggesting a wider variety of predators (Figure 2E). Crocodilians, known to produce extreme digestive corrosion sometimes resulting in the total digestion of enamel and bone (Fisher, 1981), are a possible predator candidate for some of the most corroded and unidentifiable quarry bones.

The effects of digestive corrosion versus weathering in avian bones has been studied (Bochenski and Tomek, 1997), and it was found that modifications to bird bones as a result of these processes are generally similar to those in mammalian remains. Most broken avian bones from owl pellets have rounded and often thinned edges, and articular and breakage surfaces were found to be more susceptible to digestion damage than the surfaces of shafts. Similar rounding, however, is caused by soil corrosion. In contrast, weathering of bird bones produces sharp breakage surfaces with holes and depressions. Most avian long bone shafts from the *Omomys* Quarry are crushed, presumably as the result of sediment compaction, and this makes it extremely difficult to identify rounding and thinning, and to distinguish primary from post-burial breakage. Articular ends and the breakage surfaces on long bones, however, do not appear to be corroded.

In addition to modification by digestion, bones and teeth are subject to a variety of pre- and post-burial modifications by the prolonged activity by insects (Behrensmeyer, 1978), algae growth (Fernandez-Jalvo, 1995), etching and corrosion by plant roots, acid etching in highly acidic soils, and corrosion caused by organic acids (Andrews, 1990). A number of bones and eggshell from the *Omomys* Quarry show minute scratches, grooves, and pitting of as yet undetermined origin. Modifications to the external surface of an avian eggshell fragment in the form of pitting and minute grooves are possibly attributable to insects (Figure 3F).

5.5. Breakage

Assessments of the timing and degree of bone breakage, as well as associated modifications to skeletal element frequencies, are important for estimations of numbers of individuals present in vertebrate fossil assemblages and interpretations of taphonomic history. One of the drawbacks of using breakage data, however, is that bones can be broken by a number of different pre and post-depositional processes, and it is difficult to distinguish between them (Andrews, 1990).

Most bones from the *Omomys* Quarry, with the exception of some smaller skeletal elements such as podials and claws, are broken, even if the damage consists only of the removal of delicate processes (Figure 2F). This breakage may have occurred prior to burial, during compaction, during excavation, and during laboratory processing. Causes of primary breakage (that which occurred prior to burial) include any combination of predator/scavenger activity, organic decay, sub-aerial weathering, and transport. Primary versus "later" breakage was distinguishable in many quarry specimens, because during excavation and processing, fresh breaks were created. These were darker in color than unbroken surfaces, with sharper edges, and with no matrix accreted to the break surface. Older surfaces tended to be coarser, exhibit the same color as unbroken surfaces, and were more abraded in microscopic view with matrix accreted to their surfaces. Amongst specimens that were broken prior to excavation, it was possible to distinguish primary (pre-depositional) breaks from post-depositional breaks, because those bone fragments that were physically separated from their counterpart(s) within the rock, or that had no counterpart in the quarry sample, must have been broken prior to deposition. There is no evidence of soft sediment flow or other process that might have broken and dispersed bones after burial. Approximately 70% of vertebrate specimens exhibit primary breakage (79% of *Omomys*, 80% of other mammals, 73% of birds, 56% of other non-mammalian vertebrates), even if they were broken again later. This confirms that taphonomic agents resulting in fragmentation of most of the bones were in operation prior to deposition.

Small mammal bones tend to be completely destroyed by repeated trampling if exposed on the surface for long enough (Andrews, 1990). In experiments using large mammals and small bones from owl pellets (which included already broken bones), most major postcranial bones and smaller bones such as the vertebrae, foot bones, and ribs remained largely intact after six trampling events, while the jaws were unrecognizable by this stage, and all of the teeth had become isolated (Andrews, 1990). Soft sediments were found to be important, because even small, fragile bones sitting on soft sediment can be trampled down into it without being crushed. Trampling also causes dispersal of bones (Andrews, 1990). It is difficult to assess the possible effects of trampling on the *Omomys* Quarry assemblage. Isolated teeth are relatively numerous (Table IV), however this is also a breakage pattern produced by predators (Andrews, 1990). Most of the major postcranial bones are broken, and some of the most delicate and well-preserved specimens of *Omomys* include mandibles with almost complete dentitions (Figure 2C). Considering the fine-grained soft sediments in which the bones were deposited, it is unlikely that trampling would have affected only part of the quarry assemblage. The presence of well-preserved jaws, as well as the mostly horizontal orientation of long bones as previously discussed, suggests that trampling of the assemblage did not occur.

5.6. Predation

One of the major causes of mortality in small animals is predation (Andrews, 1990). Bone breakage, skeletal element frequencies, and amount of digestive corrosion are critical to diagnoses of predator accumulations, and have inspired a number of studies of modern predator assemblages for comparison with the fossil record (Andrews, 1990; Badgley, 1986; Behrensmeyer, 1975, 1978; Dodson and Wexlar, 1979; Fisher, 1981; Haynes, 1983; Hoffman, 1987; Korth, 1979; Kusmer, 1990; Mayhew, 1977; Mellet, 1974). In general, bone assemblages from carnivore scats are characterized by relatively high breakage with a high incidence of unidentifiable bone fragments, digestive corrosion, tooth marks, and a high variance in the relative representation of skeletal elements (Andrews, 1990). In most cases they are readily distinguishable from owl pellet assemblages. Pellets of Falconiformes contain less bone than owl pellets, and differences in behavior and physiology result in higher breakage and more complete digestion of bone and tooth enamel (Andrews 1990; Kusmer, 1990). In general, owls produce a high representation of bone types with a low variance in their relative representations, and a high number of identifiable specimens with little or no surface modification. Although owl pellet assemblages have been shown to be quantitatively different from other assemblages, Kusmer (1990) pointed out that due to a number of behavioral and physiological factors, owls can deposit remains with a higher degree of fragmentation and element loss than the small body of actualistic data indicate. For example, cases in which only the heads of prey were eaten have been observed, especially in periods of prey abundance (Raczynski and Ruprecht, 1974), resulting in the over-representation of cranial to postcranial specimens in pellets. Most owls are nocturnal but some also hunt by day, especially in the early morning and late evening, and pellet accumulations form on and below nesting platforms (Andrews, 1990), night time feeding stations, and daytime roosts (Glue, 1992).

Predators as agents of microvertebrate accumulations have been well documented (Andrews, 1990; Andrews and Evans, 1983; Dodson and Wexlar, 1979; Haynes, 1983; Hoffman, 1988; Kusmer, 1990; Maas, 1985; Mayhew, 1977; Mellett, 1974). Such accumulations can occur as the result of predators carrying prey items to a preferred feeding location, storing excess food items in caches, and from the accumulation and/or deposition of pellets and scats with indigestible remains contained within them (Andrews, 1990). Because predators always eat selectively, accumulations of their prey reflect the predator's tastes (Andrews, 1990) and/or a particularly available food source, rather than being a potentially accurate reflection of the living fauna from which they were derived. In contrast, "unmodified" death assemblages such as those formed as a result of catastrophic mortality, or attritional assemblages that result from a combination of predation, starvation, and disease tend to more accurately reflect the composition of a population or community (Behrensmeyer, 1978). In summary, predator assemblages exhibit biases in favor of certain species, age or other groupings of a population within a given species, or parts of individual animals.

The age bias in the *Omomys* population discussed earlier suggests that these remains were accumulated by a predator. Although purely speculative, it is interesting to consider conditions which would lead to the preferential selection of *Omomys* and within this species why adults would be apparently preferred or most available. Behavioral, ecological, social, and anatomical characteristics of modern prey species appear to influence the likelihood of their selection by modern monkey-eating eagles (Kerbis, 1997), and it is possible that adults and young adults of *Omomys* were selected at least in part for behavioral or social reasons. Most young adult and adult mammals exhibit higher activity patterns than juveniles and older adults, which makes mammals of these age groups relatively visible and therefore more available to certain predators than those which are less active, and in the case of smaller mammals such as *Omomys*, often hiding. Parental care behavior in modern prosimians is diverse and results in various degrees of protection for the young. In Dwarf Galagos for example, the mother hides her young while she forages at night, while in Golden Pottos the young cling to the belly of the mother for three to four months, after which time they are weaned and begin to follow the mother or ride dorsally (Van Horn and Eaton, 1979). It is possible that juvenile *Omomys* may not have been as available to predators as adults because they were hidden and less active, however juvenile bone is less calcified than adult bone, and is consequently more susceptible to weathering, abrasion, and damage by predators and scavengers (Badgley, 1986). This could have contributed towards the absence of juvenile remains in the *Omomys* sample, although one would still expect some juvenile teeth to be present. If the *Omomys* accumulation formed over a short time span, such as a single season, the absence of juvenile remains could be indicative of accumulation prior to the time of year in which young *Omomys* were born. Despite the sample population bias against juvenile *Omomys*, the small number of older adults probably reflects in part the likelihood that few individuals survived to old age.

Based on the taphonomic data, most *Omomys* remains were fragmented prior to deposition, and weathering, transport, and trampling do not appear to have contributed significantly to fragmentation. Furthermore, the data suggest that the *Omomys* remains are associated, and it is obvious that the *Omomys* accumulation is unusual in terms of the large number of specimens. In combination with the population structure of the *Omomys* death assemblage, which appears to be predominantly young adults and adults, these lines of evidence further suggest that the *Omomys* were accumulated by a predator.

Identification of the type of predator which could have accumulated *Omomys* based on bone breakage and element frequencies is difficult despite the relatively good preservation of the bones. Patterns of predator modification, including bone breakage and digestive corrosion, are specific to modern species, and may not be applicable to Eocene predators. Since skeletal morphology indicates that *Omomys* was an arboreal and nocturnal animal (Covert, 1995), potential predators would most likely be arboreal mammals, reptiles or birds. In addition, the large number of *Omomys* specimens present in the assemblage probably reflects a preference for and/or ready availability of this animal to an arboreal or avian predator which was possibly nocturnal. Because the bone accumulations of living mammalian carnivores exhibit a relatively high proportion of juvenile remains (Badgley, 1986), a non-mammalian predator is most likely given the absence of juveniles in the *Omomys* sample. Arboreal primates are most vulnerable to predation when they are most exposed in areas such as forest edges and tops of tree canopies (Isbell, 1994), as they could be when they are active. Such locations would also make them particularly vulnerable to avian predators. Predation on modern primates by owls and diurnal birds of prey is well documented (Andrews, 1990; Kerbis, 1997; Wright, 1994).

The preservation of almost complete mandibles, well preserved proximal and distal ends of long bones, lack of visible digestive corrosion, and lack of tooth marks, does seem to rule out mammalian carnivores and points to an avian predator. The oldest reliable occurrence of owls (Order Strigiformes) is from the Paleocene of Colorado, while Falconiformes are known from as far back as the early Eocene of England (Fedducia, 1996). According to Fedducia (1996), the earliest reliable North American occurrence of the Order Falconiformes is from the middle Oligocene, however specimens identified as falconiform were identified in the *Omomys* Quarry avian assemblage (this study), and have been identified from other UCM Bridger localities. The *Omomys* sample contains a relatively high number of elements and fragments similar to those found in modern pellets of both modern owls and some species of Falconiformes (Andrews, 1990; Kusmer, 1990). However, because owls in general produce very little or no digestive corrosion of tooth enamel (Andrews, 1990), the absence of enamel corrosion seems to be the best criteria for distinguishing owl-derived assemblages from those of Falconiformes (Kusmer, 1990). Based on the actualistic data, an owl (category 1 predator of Andrews, 1990) is the most likely predator for this *Omomys* sample since digestive corrosion of tooth enamel was not observed on any of the *Omomys* teeth.

General characteristics of the *Omomys* and avian fauna suggest that they were accumulated by different taphonomic processes. Avian long bones are the longest specimens in the quarry sample (up to 12 cm), with a much higher of representation of shafts than the *Omomys* sample, in which long bone specimens consist mostly of proximal and distal ends. In fact, the only complete long bones from the quarry are avian, and some are apparently associated, as is evidenced by their probable articulation at or shortly before burial, and occurrence with normally rare avian eggshell fragments. Like the *Omomys*, weathering, trampling, and hydraulic sorting do not appear to have significantly modified the avian assemblage. Unlike the *Omomys*, however, predation does not appear to have played a significant role in the accumulation of the avifaunal remains.

6. Conclusions

6.1. Depositional Environment

Lakes and their margins were important preservational environments during the Bridgerian, as evidenced by the relatively high numbers of fossils preserved in limestones and associated mudstones. Micrite, the dominant form of calcite found in fossiliferous upper Bridger limestones and calcareous mudstones, was formed as a bi-product of algal photosynthesis, and was originally precipitated as lime. These calcium carbonate-rich sediments accumulated on lake bottoms, and were incorporated into muddier shoreline sediments during 1) fluctuations in water level including floods, and 2) the retreat of lake waters associated with the terminal draining, drying up, or filling in of the lakes. Calcium carbonate-rich waters and sediments are associated with buffered (non-acidic to alkaline) conditions (Wetzel, 1983), and provide a favorable chemical environment for the preservation of bones, teeth, and molluscan shell. Unusually dense concentrations of these remains in water or sediments already rich in calcium carbonate provide a localized and additional carbonate source which could have possibly meant the difference between good preservation and total decay.

Terrestrial vertebrates were only rarely deposited in distal shoreface open water sediments of upper Bridger lakes. Behrensmeyer (1975) has shown that larger bones are comparable to coarse sand to pebble grain sizes in terms of hydraulic equivalence, although buoyancy can clearly have an important effect on the dispersal of bones that dry on land prior to entering water (Behrensmeyer, 1975; Dodson, 1973). The overall lack of sediments coarser than fine sand, and general absence of terrestrial vertebrates in the distal shoreface facies of upper Bridger lakes, implies a predominance of low hydraulic energy. Some Bridger lakes and ponds, including the one which deposited the HMLs, appear to have been sustained largely by groundwater and small localized streams. This is suggested by the absence of medium to coarse-grained sand that occurs in river channel deposits of the Bridger, and lack of deltaic facies in association with lacustrine deposits (Murphey, 2001).

In the upper Bridger, localized concentrations of mixed aquatic and terrestrial microvertebrates occur most often in thin calcareous mudstones overlying micritic limestones. These mudstones are postulated to represent "shoreline" deposits, and a simple, preliminary depositional model for their development is proposed (Figure 5). Shoreline deposits formed at the intersection of lake waters with muddy and marshy lake margins, and shoreline environments migrated with changes in water level. Upper Bridger lake margins were typically forested based on the presence of arboreal mammals, wood fragments, and ferns. Animal remains accumulated in shoreline areas largely through attrition. Because terrestrial vertebrates were not preserved in distal shoreface deposits, shoreline areas were where remains derived from both terrestrial and aquatic animals were deposited and mixed in muddy marginal sediments. This mixing probably occurred by various means, including washing or falling into water, incorporation into lake sediments during fluctuations in lake water level, and bioturbation. The initiation of the development of these deposits, or at least those portions of these deposits that were preserved, appears to coincide with the time of the maximum geographic extent of each lake. As lakes drained, dried up or filled in, shoreline deposits migrated basinward with the retreating waters.

6.2. Formation of the *Omomys* Quarry Accumulation

Accumulations as taxonomically unusual and with such a high diversity of well-preserved skeletal elements as the *Omomys* Quarry are not known to

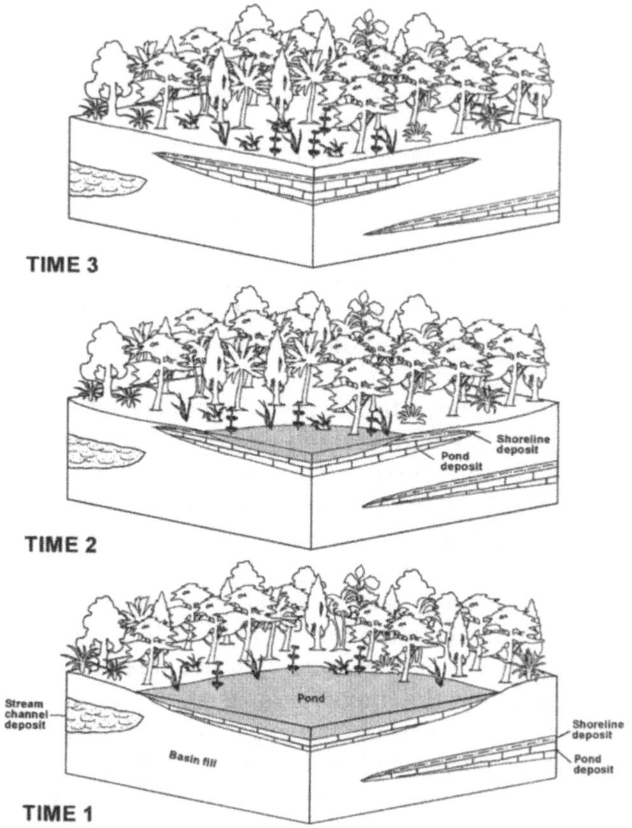

FIGURE 5. Depositional model for the development of "shoreline" deposits in relation to lateral and underlying lacustrine deposits in the upper Bridger Formation.

occur elsewhere in the upper Bridger Formation. In addition to the unusually high concentration of *Omomys*, birds, and eggshell, the various patterns of bone breakage, element frequencies, primary weathering, digestive corrosion, hydraulic sorting, and preservation, suggest that multiple taphonomic processes were involved in the formation of the accumulation. In this study, comparisons were made between subsets of the assemblage in order to study the different taphonomic processes involved, but the quantitative contributions of each of these processes may never be fully known. In addition to physical, chemical, or biological conditions which may have favored localized preservation, four types of accumulations with different taphonomic histories are postulated for the formation of the *Omomys* Quarry assemblage. The first represents an accumulation of predominantly aquatic vertebrates and invertebrates which formed by attrition in lake bottom sediments. These remains were heavily scavenged and include a low diversity of mostly highly fragmented remains of fish, amphibians, crocodilians, turtles, and gastropods. In addition to their occurrence at the *Omomys* Quarry, similar accumulations are locally preserved throughout the Hickey Mountain limestone.

The second type of accumulation is also attritional, but includes terrestrial and aquatic vertebrate and invertebrate remains which were deposited in lake shoreline sediments (as discussed above), and which are locally preserved in the mudstone overlying the HMLs. Referred to earlier in this paper as the typical Bridger, or "background" fauna, this accumulation approximates many other upper Bridger microvertebrate assemblages in terms of bone frequencies, preservation, and taxonomic composition. In general, these fossils exhibit a greater range of primary weathering, and some exhibit evidence of digestive corrosion. Elementally, this assemblage is numerically dominated by isolated teeth, dentaries, vertebrae, and phalanges, unlike the more elementally diverse *Omomys* and avian samples. Some of the avian remains which are not associated with the dominant type of eggshell, such as the hawk bones, could belong with this attritional accumulation.

Although also attritional, a third type of accumulation includes associated bird bones and eggshell fragments which were deposited in lake shoreline sediments and preserved in the mudstone overlying the HMLs. Because of the high concentration of eggshell, many of these remains were probably deposited in close proximity to a nesting area or rookery, and may have been buried relatively quickly during shoreline flooding. Some of these bones appear to have been articulated at or shortly before burial, and all of the eggs were broken prior to burial. The avian skeletal remains exhibit little primary weathering or modifications by trampling or transport, and are relatively complete and elementally diverse. Although the majority of eggshell fragments are avian, identification of the dominant type of avian eggshell was not possible due to poor preservation of ultrastructure. Owls and Falconiformes, however, are ruled out based on other eggshell characters. Since the avifauna of this shoreline deposit includes ground-nesting wading and shore-edge birds (flamingoes and rails), it is possible that much of the eggshell was accumulated by these types of birds.

It is hoped that further studies of the avian and eggshell samples will clarify this relationship.

The fourth type is interpreted as a predator accumulation of *Omomys*. Based on population and taphonomic data, mostly young adult and adult remains of these nocturnal animals were concentrated in owl pellets. The unusually dense concentration of remains suggests association and close proximity to the site of origin. Pellets may have been dropped from an owl's daytime roost or nighttime feeding station, and concentrated on the muddy shoreline substrate, or dropped directly into a flooded shoreline environment from emergent vegetation. In either case, the lack of primary weathering and hydraulic sorting associated with these remains indicates relatively rapid burial. Burial would have been expedited by even low energy flooding such as would result from episodic fluctuations of lake water level. In addition to the large number of relatively complete specimens, other taphonomic evidence for the owl hypothesis includes a low degree of digestive corrosion of bones, an absence of digestive corrosion of tooth enamel, and a high diversity of element types with a relatively low variance in their relative representation. The absence of owl bones in association with the *Omomys* is not surprising, because bones of predators are only rarely found in their prey accumulations (Andrews, 1990). Some other vertebrate specimens from the quarry are probably associated with the *Omomys* accumulation.

Unusual fossil accumulations provide tests for the applicability of actualistic taphonomic data to fossil assemblages. With the *Omomys* Quarry assemblage, the sedimentologic, faunal, and taphonomic data reveal a more complex taphonomic history than was initially hypothesized. As is the case with many other fossil microvertebrate accumulations, the processes of compaction, lithification and excavation added significantly to the difficulty of making taphonomic interpretations. As the body of actualistic data grow, particularly in avian taphonomy and eggshell studies, it is hoped that it will be possible to further interpret this assemblage.

ACKNOWLEDGMENTS. We thank Peter Robinson (UCM) for finding the *Omomys* Quarry locality, and students of the 1994 UCM Paleontological Field School for assistance with the excavation, stratigraphic sections, and mapping. The quarry excavation was planned by Peter Robinson after consultation with Paola Villa. We thank Steven Manchester (Florida Museum of Natural History) for identifications of fossil plants, the late Karl Hirsch (UCM) for preliminary investigations of fossil eggshell from the *Omomys* Quarry, Howard Hutchison (University of California Berkeley) for identifications of fossil turtles, Rosanne Humphrey (UCM) for assistance with preparation of figures, Bert Covert (UCM) for assistance with *Omomys* identifications, and John Drexler (University of Colorado Geol. Dept.) for use of the SEM. We also thank Judith Harris (UCM) and Paola Villa (Institute of Quaternary Research, Bordeaux) for informal discussions during the preparation of this manuscript, and Linda D. Chandler for comments on the fossil birds section. We thank Curators and Collection Managers at the following institutions for access to avian osteological specimens: David Willard (Field Museum of Natural History, Chicago), David Steadman and Tom Webber (Florida Museum of Natural History, Gainesville), and Elizabeth Wright (Museum of Natural History at the University of Georgia, Athens). We graciously acknowledge the Bureau of Land Management, and in particular Laurie Bryant at the Wyoming BLM State Office, for ongoing support of UCM's field studies in the Bridger Basin. Finally, we thank Gregg Gunnell and John Alexander for the invitation to participate in this interesting and informative symposium. Field work for this study was partially funded by grants from the University of Colorado Museum and the Sigma Xi Scientific Research Society.

References

Andrews, P., 1990, *Owls Caves and Fossils: Predation, Preservation and Accumulation of Small Mammal Bones in Caves, with Analysis of the Pleistocene Cave Faunas from Westbury-Sub-Mendip, Somerset, UK*, The Natural History Museum, UK.

Andrews, P., and Evans, E., 1983, Small mammal bone accumulations produced by mammalian carnivores, *Paleobiology*, **9**:289–307.

Austin, O. L., 1977, *Birds of the World*, Hamlyn Press, London.

Badgley, C., 1986, Counting individuals in mammalian fossil assemblages from fluvial environments, *Palaios* **1**:328–338.

Behrensmeyer, A. K., 1975, The taphonomy and paleoecology of Plio-Pleistocene vertebrate assemblages east of Lake Rudolf, Kenya, *Bull. Mus. Comp. Zool.* **146(10)**:473–578.

Behrensmeyer, A. K., 1978, Taphonomic and ecologic information on bone weathering, *Paleobiology*, **4**:150–162.

Bickart, K. J., 1981, A new thick-knee, *Burhinus*, from the Miocene of Nebraska, with comments on the habitat requirements of the Burhinidae (Aves: Charadriiformes), *Journal of Vertebrate Paleontology* **1(3–4)**:273–277.

Blob, R. W., and Fiorillo, A. R., 1996, The significance of vertebrate microfossil size and shape distributions for faunal abundance reconstructions: A late Cretaceous example, *Paleobiology* **22(3)**:422–435.

Bochenski, Z. M., and Tomek, T., 1997, Preservation of bird bones: erosion versus digestion by owls, *International Journal of Osteoarcheology* **7**:372–387.

Bradley, W. H., 1964, Geology of Green River Formation and associated Eocene rocks in southwestern Wyoming and adjacent parts of Colorado and Utah, *United States Geological Survey Professional Paper* **496-A**:1–86.

Bray, E. S., and Hirsch, K. F., (in press) Eggshell from the Upper Jurassic Morrison Formation, *Modern Geology Special Paper.*

Brodkorb, P., 1967, Catalogue of fossil birds, Part 3 (Ralliformes, Ichthyorniformes, Charadriiformes), *Bulletin Florida State Museum*, **11(3)**:99–220.

Cassiliano, M., 1997, Taphonomy of mammalian fossils across the Blancan-Irvingtonian boundary: Palm Spring Formation, Anza-Borrego desert of southern California, *Palaeogeography, Palaeoclimatology, Palaeoecology* **129**:81–108.

Chandler, R. M., 1997, New discoveries of *Titanis walleri* (Aves: Phorusrhacidae) and a new phylogenetic hypothesis for the phorusrhacids, *Journal of Vertebrate Paleontology* supplement to **17(3)**:36–37.

Cope, E. D., 1872, Descriptions of some new Vertebrata from the Bridger Group of the Eocene, *Proceedings of the American Philosphical Society* **12**:460–465.

Cope, E. D., 1873, On the extinct Vertebrata of the Eocene of Wyoming, observed by the expedition of 1872, with notes on the geology, in: *Sixth Annual Report of the United States Geological Survey of the Territories for 1872* (F. V. Hayden, ed.), pp. 545–649.

Covert, H. H., 1995, Locomotor adaptations of Eocene primates: adaptive diversity among the earliest prosimians, in: *Creatures of the Dark: The Nocturnal Prosimians* (L. Alterman, G. A. Doyle, and M. K. Izard, eds.), Plenum Press, New York, pp. 495–509.

Covert, H. H., and Murphey, P. C., 1994, New middle Eocene omomyine skeletal material, *American Journal of Physical Anthropology*, Supplement to **18**:72.

Cracraft, J., 1969, Systematics and evolution of the Gruiformes (Class, Aves). 1. The Eocene family Geranoididae and the early history of the Gruiformes, *American Museum Novitates*, **2388**.

Cracraft, J., 1973, Systematics and evolution of the Gruiformes (Class Aves). 3. Phylogeny of the Suborder Grues, *American Museum of Natural History Bulletin* **151(1)**.

D'Errico, F., and Villa, P., 1997, Holes and grooves: The contribution of microscopy and taphonomy to the problem of art origins, *Journal of Human Evolution* **33**:1–31.

Dodson, P., 1973, The significance of small bones in paleoecologic interpretation, *Contributions to Geology* **12(1)**:15–19.

Dodson, P., and Wexlar, D., 1979, Taphonomic investigations of owl pellets, *Paleobiology* **5(3)**:275–284.

Erben, H. K., 1970, Ultrastrukturen und mineralisation rezenter und fossiler eischalen bei vogeln und reptilien, *Biomineralisation* **1**:1–65.

Evanoff, E., Brand, L., and Murphey, P. C., 1998, Bridger Formation (middle Eocene) of southwest Wyoming: widespread marker units and subdivisions of the Bridger B through D, *Dakoterra* **5**:115–122.

Feduccia, A., 1980, A thick-knee (Aves: Burhinidae) from the Pleistocene of North America, and its bearing on ice age climates, *Contributions to Science, Natural History Museum of Los Angeles County* **330**:115–118.

Fedducia, A., 1996, *The Origin and Evolution of Birds*, Yale University Press, New Haven.

Fernandez-Jalvo, Y., 1995, Small mammal taphonomy at La Trinchera de Atapuerca (Burgos, Spain). A remarkable example of taphonomic criteria used for stratigraphic correlations and paleoenvironmental interpretations, *Palaeogeography, Palaeoclimatology, Palaeoecology* **114**:167–195.

Fernandez-Jalvo, Y., Denys, C., Andrews, P., Williams, T., Dauphin, Y., and Humphrey, L., 1998, Taphonomy and palaeoecology of Olduvai Bed-1 (Pleistocene, Tanzania), *Journal of Human Evolution* **34**:137–172.

Fisher, D. C., 1981, Crocodilian scatology, microvertebrate concentrations, and enamel-less teeth, *Paleobiology* **7**:262–275.

Gazin, C. L., 1976, Mammalian faunal zones of the Bridger middle Eocene, *Smithsonian Contributions to Paleobiology* **26**:1–25.

Gilinsky, N. L., and Bennington, J. B., 1994, Estimating numbers of whole individuals from collections of body parts: A taphonomic limitation of the paleontologic record, *Paleobiology* **20(2)**:245–258.

Glue, D., 1992, Owl pellets, in: *Owls of the World* (J. A. Burton, ed.), Peter Lowe Publishing, Eurobook Ltd., pp. 189–192.

Golubev, S. N., 1988, [Principal questions of biomineralization], in: *Sovremennaya Paleontologiya* (V. V. Menner, and V. P. Makridin, eds.) [In Russian], pp. 241–252.

Granger, W., 1908, A revision of the American Eocene Horses, *Bulletin of the American Museum of Natural History* **24(15)**:221–264.

Gunnell, G. F., 1997, Wasatchian-Bridgerian (Eocene) paleoecology of the western interior of North America: Changing paleoenvironments and taxonomic composition of omomyid (Tarsiiformes) primates, *Journal of Human Evolution* **32**:105–132.

Gunnell, G. F., and Bartels, W. S., 1994, Early Bridgerian (middle Eocene) vertebrate paleontology and paleoecology of the southern Green River Basin, Wyoming, *Contr. Geol., Univ. Wyoming* **30**:57–70.

Hanley, J. H., 1974, Systematics, paleoecology, and biostratigraphy of nonmarine mollusca from the Green River and Wasatch formations (Eocene), southwestern Wyoming and northwestern Colorado, *University of Wyoming Ph.D Dissertation*, 285 pp.

Haynes, G., 1983, A guide for differentiating mammalian carnivore taxa responsible for gnaw damage to herbivore limb bones, *Paleobiology* **9(2)**:164–172.

Hillson, S., 1986, *Teeth*, Cambridge University Press, UK.

Hirsch, K. F., 1985, Fossil crocodilian eggs from the Eocene of Colorado, *Journal of Paleontology* **59(3)**:531–542.

Hirsch, K. F., 1994, Upper Jurassic eggshells from the Western Interior of North America, in: *Dinosaur Eggs and Babies* (K. Carpenter, K. F. Hirsch, and J. R. Horner, eds.), Cambridge University Press, pp. 137–150.

Hirsch, K. F., and Kohring, R., 1992, Crocodilian eggs from the middle Eocene Bridger Formation, Wyoming, *Journal of Vertebrate Paleontology* **12(1)**:59–65.

Hirsch, K. F., and Quinn, B., 1990, Eggs and eggshell fragments from the Upper Cretaceous Two Medicine Formation of Montana, *Journal of Vertebrate Paleontology* **10(4)**:491–511.

Hoffman, R., 1988, The contribution of raptorial birds to patterning in small mammal assemblages, *Paleobiology* **14(1)**:81–90.

Holtzman, R. C., 1979, Maximum likelihood estimation of fossil assemblage composition, *Paleobiology* **5(2)**:77–89.

Isbell, L. A., 1994, Predation on primates: ecological patterns and evolutionary consequences, *Evolutionary Anthropology* **3(2)**:61–71.

Kerbis, J. C., 1997, Taphonomy of tropical forest environments: Monkey-eating eagles as agents of bone accumulation, *Journal of Vertebrate Paleontology* **17**(supplement to 3): p. 56a.

Khajuria, C. K., and Prasad, G. V. R., 1998, Taphonomy of a Late Cretaceous mammal-bearing microvertebrate assemblage from the Deccan inter-trappean beds of Naskal, peninsular India, *Palaeogeography, Palaeoclimatology, Palaeoecology* **137**:153–172.

Kidwell, S. M., 1985, Palaeobiological and sedimentological implications of fossil concentrations, *Nature* **318**:457–460.

Korth, W., 1979, Taphonomy of microvertebrate fossil assemblages, *Annals of the Carnegie Museum* **48**:235–285.

Krishtalka, L., West, R. M., Black, C. C., Dawson, M. R., Flynn, J. J., Turnbull, W. D., Stucky, R. K., McKenna, M. C., Bown, T. M., Golz, D. J., and Lillegraven, J. A., 1987, Eocene (Wasatchian through Duchesnean) biochronology of North America, in: *Cenozoic Mammals of North America: Geochronology and Biostratigraphy* (M. O. Woodburne, ed.), University of California Press, Berkeley, pp. 77–117.

Kusmer, K. D., 1990, Taphonomy of owl pellet deposition, *J. Paleont.* **64(4)**:629–637.

Leidy, J., 1869, Notice of some extinct vertebrates from Wyoming and Dakota, *Proceedings of the Academy of Natural Sciences of Philadelphia* (1869), pp. 63–67.

Leidy, J., 1871, Notice of some extinct rodents, *Proceedings of the Academy of Natural Sciences of Philadelphia* (1871), pp. 230–232.

Leidy, J., 1872, On the fossil vertebrates of the early Tertiary of Wyoming, in: *United States Geological Survey of Montana and Portions of Adjacent Territories, Fifth Annual Report* (F. V. Hayden, ed.), pp. 353–372.

Lyman, R. L., 1994, Relative abundances of skeletal specimens and taphonomic analysis of vertebrate remains, *Palaios* **9**:288–298.

Maas, M. C., 1985, Taphonomy of the late Eocene microvertebrate locality: Wind River Basin, Wyoming (USA), *Palaeogeography, Palaeoclimatology, Palaeoecology* **52**:123–142.

Marsh, O. C., 1871, Notice of some new fossil mammals from the Tertiary Formation: *American Journal of of Science and Arts* **2**:35–44.

Marsh, O. C., 1886, Dinocerata: A monograph of an extinct order of gigantic mammals, *Monographs of the United States Geological Survey* **10**:1–243.

Matthew, W. D., 1909, The Carnivora and Insectivora of the Bridger Basin, middle Eocene, *Memoirs of the American Museum of Natural History* **9(6)**:291–567.

Mayhew, D. F., 1977, Avian predators as accumulators of fossil mammal material, *Boreas* **6**:25–31.

Mellett, J. S., 1974, Scatological origin of microvertebrate remains, *Science* **185**:349–350.

Mikhailov, K. E., 1991, Classification of fossil eggshells of amniotic vertebrates, *Acta Paleontologica Polonica*, **36(2)**:193–238.

Mikhailov, K. E., 1992, The microstructure of avian and dinosaurian eggshell: phylogenetic implications, *Natural History Museum of Los Angeles County, Science Series, Papers in avian paleontology honoring Pierce Brodkorb* (K. E. Campbell, ed.), **36**:361–373.

Mikhailov, K. E., 1994a, Theropod and protoceratopsian dinosaur eggs from the Cretaceous of Mongolia and Kazakhstan, *Paleontological Journal* **28(2)**:101–120.

Mikhailov, K. E., 1994b, Eggs of sauropod and ornithopod dinosaurs from the Cretaceous deposits of Mongolia, *Paleontological Journal*, **28(3)**:141–159.

Mikhailov, K. E., 1995, Eggshell structure in the shoebill and pelecaniform birds: comparison with hammerkop, herons, ibises and storks, *Canadian Journal of Zoology* **73**:1754–1770.

Mikhailov, K. E., 1997a, Avian eggshells: an atlas of scanning electron micrographs, *British Ornithologists' Club Occasional Publications* **3**, Hertfordshire U.K.

Mikhailov, K. E., 1997b, Fossil and recent eggshell in amniotic vertebrates: fine structure, comparative morphology and classification, *Special Papers in Palaeontology, The Palaeontological Association*, London, **56**.

Mikhailov, K. E., Sabath, K., Kurzanov, S., 1994, Eggs and nests from the Cretaceous of Mongolia, in: *Dinosaur Eggs and Babies*, (K. Carpenter, K. F. Hirsch, and J. R. Horner, eds.) Cambridge University Press, pp. 88–115.

Mikhailov, K. E., Bray, E. S., and Hirsch, K. F., 1996, Parataxonomy of fossil egg remains (Veterovata)—principles and applications, *Journal of Vertebrate Paleontology* **16(4)**:763–769.

Murphey, P. C., 1995, Paleontology, sedimentology, and depositional history of a middle Eocene lacustrine deposit in the Bridger Formation, southwest Wyoming, *University of Colorado Masters Thesis*, 121 pp.

Murphey, P. C., 2001, Stratigraphy, fossil distribution, and depositional environments of the upper Bridger Formation (middle Eocene) of southwestern Wyoming, and the taphonomy of an unusual Bridger microfossil accumulation, *University of Colorado Doctoral Dissertation*, 344 pp.

Murphey, P. C., and Burger, B. J., 1996, Depositional setting and fauna of the *Omomys* Quarry, a possible owl site in the Bridger Formation (middle Eocene) of southwestern Wyoming, *Journal of Vertebrate Paleontology* **16**(supplement to 3):55A.

Murphey, P. C., Lester, A., Bohor, B., Robinson, P., Evanoff, E., and Larson, E., 1999, 40Ar/39Ar dating of volcanic ash deposits in the Bridger Formation (middle Eocene) of southwestern Wyoming, *Geological Society of America Abstracts with Programs* **31(7)**.

Olson, S. L., 1977, A synopsis of the fossil Rallidae, in: *Rails of the World: A Monograph on the Family Rallidae* (S. D. Ripley, ed.), Godine Press, Boston, pp. 509–525.

Olson, S. L., 1985, The fossil record of birds, in: *Avian Biology, Volume 8* (D. S. Farner, J. R. King, and K. C. Parks, eds.), Academic Press, New York, pp. 79–252.

Olson, S. L., and Feduccia, A., 1980, Relationships and evolution of flamingos (Aves: Phoenicopteridae), *Smithsonian Contributions to Zoology*, no. 316.

Osborn, H. F., 1929, The Titanotheres of ancient Wyoming, Dakota, and Nebraska, *United States Geological Survey Monograph* **1(55)**:1–701.

Packard, M. J., 1980, Ultrastructural morphology of the shell and shell membranes of eggs of common snapping turtles (*Chelydra serpentina*), *Journal of Morphology* **165**:187–204.

Pooley, A. S., 1979, Ultrastructural relationships of minerals and organic matter in avian eggshells, *Scanning Electron Microscopy* **II**:475–482.

Raczynski, J., and Ruprecht, A., 1974, The effect of digestion on the osteological composition of owl pellets, *Acta Ornithologica* **14**:1–12.

Roehler, H. W., 1992, Introduction to Greater Green River Basin geology, physiography, and history of investigations, *U. S. Geological Survey Professional Paper* **1506-A**:A1–A14.

Roehler, H. W., 1993, Eocene climates, deposition, and geography, Greater Green River Basin, Wyoming, Utah, and Colorado, *U. S. Geological Survey Professional Paper* **1506-F**:1–74.

Shotwell, J. A., 1955, An approach to the paleoecology of mammals, *Ecology* **36**:327–337.

Sinclair, W. J., 1906, Volcanic ash in the Bridger beds of Wyoming, *American Museum of Natural History Bulletin* **47**:1–91.

Tyler, C., 1964, Wilhelm von Nathusius (1821–1899) on avian eggshells, a translated and edited version of his work, Berkshire Printing Company, U.K.

Tyler, C., 1969a, Avian egg shells: Their structure and characteristics, in: International Review of General and Experimental Zoology (W. J. L. Felts, and R. J. Harrison, eds.), Academic Press, London, pp. 82–130.

Tyler, C., 1969b, A study of the egg shells of Gaviiformes, Procellariiformes, Podicipediformes and Pelecaniformes, *Journal of Zoology, London* **158**:395–412.

Tyler, C., and Simkiss, K., 1959, A study of the egg shells of ratite birds, *Proceedings of the Zoological Society of London* **133**:201–243.

Van Horn, R. N., and Eaton, G. G., 1979, Reproductive physiology and behavior in prosimians, in: *The Study of Prosimian Behavior* (G. A. Doyle, and R. D. Martin, eds.), Academic Press, New York, pp. 79–122.

Vianey-Liaud, M., Mallan, P., Buscail, O., and Montgelard, C., 1994, Review of French eggshells: morphology, structure, mineral and organic composition, in: *Dinosaur Eggs and Babies* (K. Carpenter, K. F. Hirsch, and J. R. Horner, eds.), Cambridge University Press, Cambridge, pp. 151–183.

Voorhies, M. R., 1969, Taphonomy and population dynamics of an early Pliocene vertebrate fauna, Knox County, Nebraska, *Contrib. Geol., Spec. Pap.*, **1**:1–69.

Weigelt, J., 1927, Rezente Wirbeltierleichen und ihre Paläobiologische Bedeutung, Verlag von Max Weg, Leipzig.

Wetzel, R. G., 1983, *Limnology*, 2nd ed., Harcourt Brace Jovanovich, Inc., Orlando.

Wetmore, A., 1931, Two primitive rails from the Eocene of Colorado and Wyoming, *Condor* **33**:107–109.

Winkler, D. A., 1983, Paleoecology of an early Eocene mammalian fauna from paleosols in the Clarks Fork Basin, northwestern Wyoming, *Palaeogeography, Palaeoclimatology, and Palaeoecology* **43**:261–298.

Wolff, R. G., 1973, Hydrodynamic sorting and ecology of a Pleistocene mammalian assemblage from California (USA), *Palaeogeography, Palaeoclimatology, and Palaeoecology* **13**:91–101.

Wright, P. C., 1994, Night watch on the Amazon, *Natural History*, **103(5)**:45–50.

Zhao, Z. K., 1975, The microstructure of dinosaurian eggshells of Nanhsiung Basin, Guangdong Province, *Vertebrata PalAsiatica* **13(2)**:105–117.

Zhao, Z. K., 1979, The advancement of research on the dinosaurian eggs in China, in: *Mesozoic and Cenozoic Red Beds in Southern China* (Institute of Vertebrate Paleontology, Paleoanthropology and Nanjing Institute of Paleontology, eds.), Science Press, Beijing, pp. 330–340.

Zhao, Z. K., and Jiang, Y. K., 1974, Microscopic studies on the dinosaurian eggshells from Laiyang, Shangtung Province, *Scientia Sinica* **17**:73–83.

Zhao, Z. K., and Li, R., 1993, First record of Late Cretaceous hypsilophodontid eggs from Bayan Manduhu, Inner Mongolia, *Vertebrata PalAsiatica* **31(2)**:77–84.

Chapter 16

Basin Margins, Biodiversity, Evolutionary Innovation, and the Origin of New Taxa

GREGG F. GUNNELL and WILLIAM S. BARTELS

GREGG F. GUNNELL • Museum of Paleontology, University of Michigan, Ann Arbor, MI 48109-1079. WILLIAM S. BARTELS • Department of Geological Sciences, Albion College, Albion, MI 49224.

Eocene Biodiversity: Unusual Occurrences and Rarely Sampled Habitats, edited by Gregg F. Gunnell, Kluwer Academic/Plenum Publishers, New York, 2001.

1. Introduction

A biochronologic framework for Eocene deposits in the Rocky Mountain Interior of North America has been developed over nearly one and a half centuries of study of these sequences and the biotic components contained within them. The definition of a series of North American Land Mammal Ages (NALMAs) for the Eocene (and other epochs) was based directly on these studies (Wood *et al.*, 1941) with subsequent refinement required as knowledge of faunas and geographic coverage broadened (Krishtalka *et al.*, 1987). Eocene NALMAs include from oldest to youngest: the Wasatchian, Bridgerian, Uintan, Duchesnean, and Chadronian (Prothero and Swisher, 1992). In every case, the faunal samples used to define each of these ages are derived from richly fossiliferous lowland areas usually found near the centers of intermontain basins where thick sequences of fluvial rocks were deposited. As such, faunal samples derived from a relatively limited number of habitats are those that represent the "type faunas" for each land mammal age and the stratigraphic ranges of contained taxa are based on first and last appearances as documented in these samples.

Black (1967), noting the presence of several late (now known to be middle) Eocene archaic mammalian taxa thought to have gone extinct in the earliest Eocene, suggested that known middle Eocene faunal samples were all derived from similar ecological circumstances and did not reflect the true diversity of mammals during the middle Eocene. Black (1967, page 62) further suggested that "faunas of different ecological aspect may be recovered from sediments along, and in, the mountain fronts of northwestern Wyoming."

Recent paleontological and geological fieldwork along the northeastern margin of the Green River Basin in southwestern Wyoming, in the area known as South Pass (Figure 1), has confirmed Black's suspicions and indicates that vertebrate faunal samples derived from basin margins differ in content and composition from time equivalent samples derived from basin center depositional environments (Bartels and Gunnell, 1997; Gunnell and Bartels, 1997, 1998). In the pages that follow, we will provide a definition of what we mean by basin margin settings, discuss why these areas are of interest and importance, describe geological and paleontological aspects that characterize basin margin assemblages, and finally, will offer an interpretive framework that we believe explains many of the puzzling aspects of the these faunal samples.

2. Geology of South Pass, Wyoming

The South Pass field area is adjacent to the Wind River Thrust and Continental (high angle) faults. Late Laramide movements continued on the faults until middle Eocene (Bridgerian) time providing sediments to the northeastern part of the Green River Basin (Steidtmann *et al.*, 1983).

Approximately 500 meters of late early to early middle Eocene strata are exposed in the vicinity of South Pass. The sequence consists of tongues of the

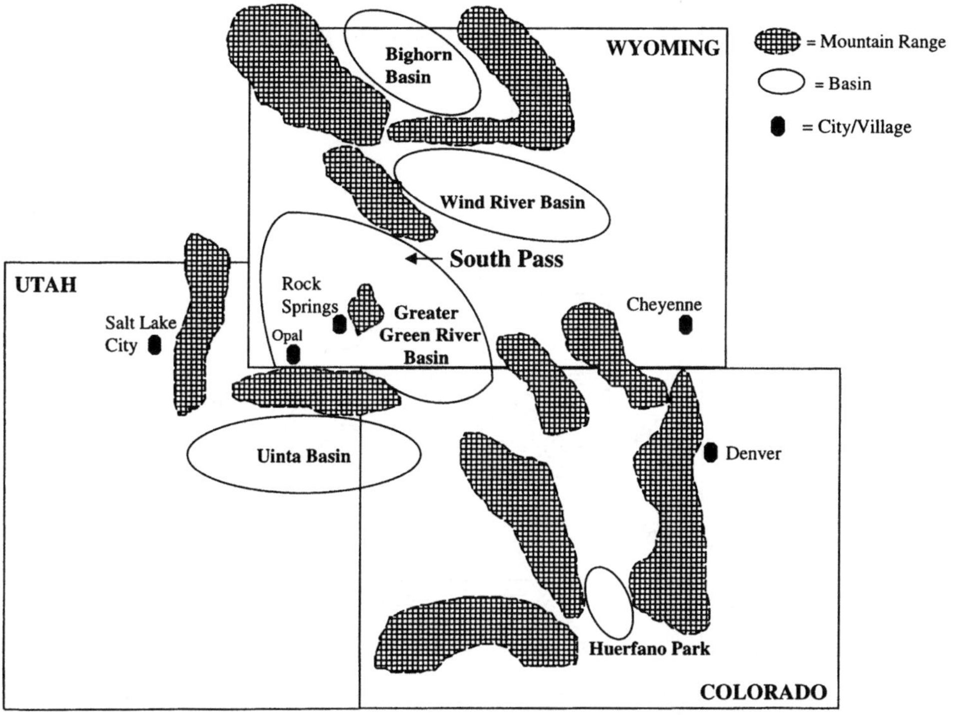

FIGURE 1. Map of Wyoming, Utah, and Colorado indicating the location of the South Pass field area with respect to major geographic features and cities. Also shown are the locations of Huerfano Park in Colorado and the Wind River Basin and Opal areas in Wyoming.

predominantly fluvial Wasatch and Bridger formations interfingering with various members of the predominantly lacustrine Green River Formation (Figure 2).

The Wasatch Formation consists of highly variegated (predominantly red) mudrocks, orange to buff sandstones, and dark gray conglomerates. The uppermost part of the lowest unit of the formation (Main Body) is exposed only at The Pinnacles in the southern portion of the field area. The upper unit, the coarser grained Cathedral Bluff Tongue, is well exposed along Honeycomb Buttes in the central and northern portions of the field area. The Cathedral Bluffs Tongue is bracketed by the Tipton and Laney shale members of the Green River Formation representing major incursions of Ancient Lake Gosiute. These units consist of thick brown paper shales interbedded with relatively pure carbonate, trona, and gypsum beds. The carbonates are often oolitic or stromatolitic and can be very laterally persistent, forming broad benches and plateaus between the badland exposures of the Wasatch and Bridger formations.

The Bridger Formation consists of pastel colored mudrocks interbedded with dark brown to red sandstones, thin carbonates, and tuffaceous beds exposed above Honeycomb Buttes along the flanks of Continental Peak and Oregon Buttes. The unit becomes redder and coarser upwards and represents

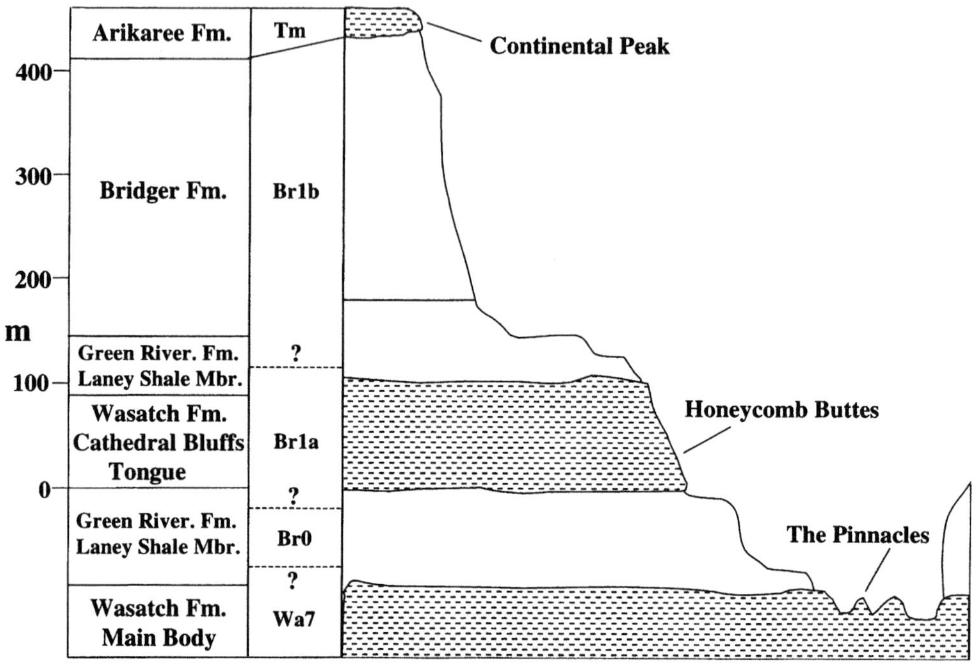

FIGURE 2. Generalized stratigraphy of the South Pass section. Measured section begins at base of Cathedral Bluffs Tongue of the Wasatch Formation and is measured in meters. Note that the earliest Gardnerbuttean (Biochron Br0) has not been located at South Pass and is probably contained within the lacustrine Laney Shale Member of the Green River Formation.

a shift from a primarily Wind River source to a mixed source including Absarokan volcanics.

Tetrapod remains are variably common in the Wasatch Formation, where they occur as channel lags, sheet wash concentrations, and along ancient lake "strand lines." Tetrapods have not been recovered from the Green River Formation at South Pass. Although large aquatic vertebrate remains are locally abundant in the Bridger Formation, terrestrial vertebrates are quite rare.

3. The Bridgerian Land Mammal Age

In a 1909 monograph on the southern Green River Basin, Matthew proposed a five-part division of the Bridger Formation, designating intervals A through E. The lowest 200 meters of the Bridger Formation represents interval A (Bridgerian Zone Br1b, Gunnell and Yarborough, 2000). Bridger A was viewed by Matthew (1909) as poorly fossiliferous, at least in terms of mammals. Interval B (Bridgerian Zone Br2, Gunnell and Bartels, 1994) represents the succeeding 150 meters of the Bridger Formation and is abundantly fossiliferous. Bridger B is separated from the C interval by the laterally extensive and relatively thick Sage Creek White Layer (calcareous limestone), while interval D is separated from Bridger C by the laterally extensive, calcareous Lone Tree White Layer. Bridger intervals C and D (Bridgerian Zone Br3, Gunnell and Bartels, 1994) together represent approximately 300 meters of section. Interval E represents approximately the upper 50 meters of the Bridger Formation in the Green River Basin.

Wood (1934) proposed grouping Matthew's intervals A and B into the Black's Fork "Member" of the Bridger Formation, based on faunal differences between this interval and those of Bridger C and D (termed the Twin Butte "Member" by Wood). The terms Blacksforkian and Twinbuttean have been used to designate land mammal subages of the Bridgerian Land Mammal Age (Krishtalka and others, 1987). The Bridgerian Land Mammal Age was named by the Wood Committee (1941) although, as has often been pointed out (Tedford, 1970; Woodburne, 1977, 1987; Prothero, 1998), the "Provincial Ages" proposed by this committee are not strictly conformable to the North American Code of Stratigraphic Nomenclature (1983) and as such, should be viewed as informal designations only. None-the-less, these "provincial ages" have proven useful in a general stratigraphic context.

In 1966, Robinson proposed an additional subage of the Bridgerian, the Gardnerbuttean, that was recognized as faunally transitional between the Wasatchian and Bridgerian Land Mammal Ages based on faunal assemblages from Huerfano Park, Colorado. Stucky (1984) named the *Palaeosyops borealis* Assemblage Zone for the stratigraphic sequence of sediments in the Wind River Basin denoted by the first appearance of the brontotheriid perissodactyl *P. borealis*. Stucky equated this with Robinson's (1966) Gardnerbuttean subage of the Bridgerian Land Mammal Age. Stucky (1984, p. 287) also noted the possibility that an additional biostratigraphic interval might be present in the Wind River Basin stratigraphically above the *Palaeosyops borealis* Assemblage Zone. Gunnell and Bartels (1994) designated the Gardnerbuttean as Br0.

Gunnell and Yarborough (2000) have confirmed Stucky's earlier suspicions. The Gardnerbuttean can be subdivided into an early portion (retained as Br0) represented by the first appearance of *Eotitanops borealis* (Stucky's *Palaeosyops borealis*) and a later portion represented by the first appearance of *Eotitanops minimus* and *Palaeosyops fontinalis*. The "*Palaeosyops fontinalis* Assemblage Zone" encompasses the later part of the Gardnerbuttean. It also encompasses the earliest part of the Blacksforkian (=Bridgerian A). Gunnell and Yarborough (2000) subdivided the Bridgerian Biostratigraphic Interval Br1 of Gunnell and Bartels (1994) into Br1a, representing the latest Gardnerbuttean and Br1b representing the earliest Blacksforkian or Bridgerian A. The mammalian faunas from the latest Gardnerbuttean (Br1a) and Bridgerian A (Br1b) are similar but there are differences that suggest that these two intervals are not contemporaneous (Gunnell, 1998). Figure 3 summarizes Bridgerian Land Mammal Age biochronology.

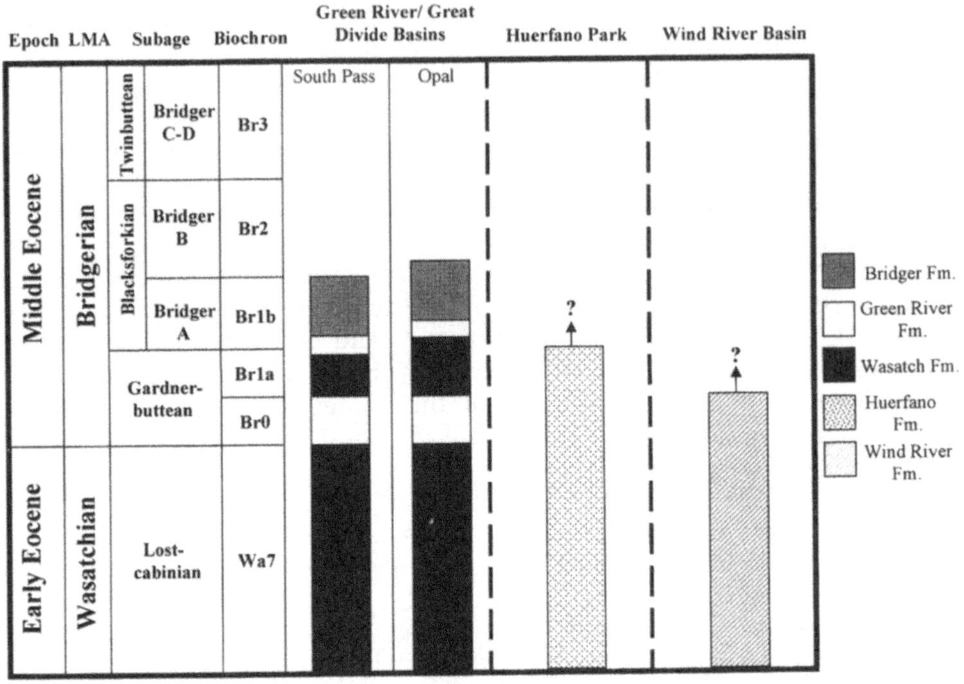

FIGURE 3. Comparison of general lithostratigraphy in four areas where the Wasatchian/Bridgerian transition is recorded. Note that the early Gardnerbuttean (Biochron Br0) is represented by fluvial sediments only at Huerfano and in the Wind River Basin. The late Gardnerbuttean (Biochron Br1a) is best represented at South Pass, Opal, and Huerfano, while the early Blacksforkian (Bridgerian A) is best known only from the Green River/Great Divide Basin.

4. Basin Margins

4.1. Basin Margins Defined

Our definition of basin margin settings contains geographic, physiographic, ecologic, and geologic components. In the broadest sense, basin margins can be thought of as comparable to upland areas. Geographically, basin margins are located adjacent to mountainous areas and include mountain fronts and immediate foothill areas. Physiographically, basin margins are characterized by much greater topographic relief than are basin centers and, because of this increased topographic complexity, are ecologically diverse with a much greater variety of habitats and microhabitats than are generally available in basin centers. Geologically, basin margins are much closer to source areas of basin fill than are basin centers and as such are characterized by higher energy depositional settings (alluvial fans and braided streams) with an abundance of coarser and more immature sediments. All of these features combine to offer vertebrates a more complex and more demanding environment than they generally encounter in the less topographically complex, less ecologically diverse, and generally lower energy environments of basin centers.

4.2. Importance of Basin Margin Areas

As noted in the introduction, areas other than those of basin centers have rarely been sampled in western North America. Thus, basin margin faunal samples are important in that they provide a picture of community structure that has never been adequately represented in previous studies of Eocene vertebrate evolution (Lillegraven and Tabrum, 1983; Bartels and Gunnell, 1997; Gunnell and Bartels, 1997, 1998). As ecologically diverse areas, basin margins can provide evidence to address questions of habitat utilization and habitat preference among Eocene vertebrates that is not possible by examining only the relatively homogenous habitats of basin center environments. Basin margin faunas can also be used to test biostratigraphic hypotheses by comparing first and last appearances of taxa between these areas and basin center faunal samples. If habitat preference is an important component of vertebrate spatial diversity (as is appears that it is based on modern ecological models), then this should have direct effects on when and where taxa first appear and last disappear in the fossil record. Finally, because of the increase in topographic relief and elevation in marginal areas, habitat zones become more restricted both horizontally and vertically in basin margin environments, leading to increased diversity in modern upland environments compared to lowland, more homogeneous habitats (DiMichele *et al.*, 1985; Horner *et al.*, 1992; Rosenzweig, 1995; Kerr and Packer, 1997). The same should be true for fossil vertebrate assemblages from basin margins.

5. Recognition of Ancient Basin Margin Environments

5.1. Geographic and Geologic Evidence

Ancient basin margin habitats can be recognized by a number of characteristic features. First, in terms of paleogeography, basin margins are defined by their proximity to upland source areas. Unless structural features suggest otherwise, areas that are on the margins of structural intracratonic basins today probably were in a similar position in the Tertiary. These areas include mountain fronts and adjacent foothills near these fronts. Sediments deposited in basin margins can be recognized by a suite of features characteristic of alluvial fan and braided stream depositional environments. At South Pass, these sedimentary rocks are typically immature, consisting of petromictite conglomerates, litharenite and sublitharenite sandstones and areneceous or conglomeratic mudrocks. Texturally, the sediments are also immature. Sorting is moderate to very poor, and grain shape is typically subangular. The coarser conglomerates tend to have a closed framework, but finer units are usually matrix supported indicating mudflow deposition. Stratigraphically, the sequences are thick and repetitive with abrupt and substantial horizontal and lateral grain size changes. Sediment geometries include coarse clastic wedges, abundant cut and fill, and blankets of finer sheet-wash deposits. The overall sequence and individual coarse units show rapid basinward thinning and fining.

5.2. Paleontological Evidence

Nearly 2,000 fossil vertebrate specimens have now been collected by field teams from The University of Michigan and Albion College from the basin margin sediments at South Pass (Table 1). These sediments comprise the Cathedral Bluffs Tongue of the Wasatch Formation and the lowest beds of the Bridger Formation in the vicinity of Oregon Buttes and Continental Peak (Figures 1 and 2). The faunal samples are of earliest Bridgerian age, spanning Bridgerian Biochronologic zones Br1a (late Gardnerbuttean subage) and Br1b (Bridgerian A) as defined by Gunnell and Yarborough (2000). Vertebrates found at South Pass include several species of fish, birds, and amphibians, a wide variety of reptiles including lizards, snakes, crocodylians, and turtles, and 57 species of mammals.

There are four general characteristics of South Pass vertebrate assemblages that can be viewed as indicators of basin margin environments. These are the presence of: 1) distinctive taxa; 2) unique taxa; 3) taxa of unique morphological form; and 4) anachronistic taxa. Distinctive taxa are those that are uncommon in basin center assemblages but are abundant in basin margin samples. Unique taxa are unknown from basin center assemblages and either represent new taxa only known at South Pass or taxa only known from South Pass and other basin margin samples. Unique morphological forms are represented by specimens that are morphologically distinct from the same taxa known from basin center assemblages, some of which may represent new taxa. Anachronistic taxa are those that occur in the basin margin during intervals when they are not represented in the basin center. Most importantly, these include examples of taxa that do not overlap biostratigraphically in basin center assemblages but co-occur in the same localities and horizons in basin margin samples, often representing presumed ancestor-descendent couplets.

6. Basin Margin Faunas

6.1. Distinctive Taxa

Some taxa that are rarely found in basin center sediments in the Green River Basin are relatively common from the basin margin sediments at South Pass. Cataloged amphibian specimens are approximately four times more numerous from South Pass, while birds are about twice as common at South

Table 1. Faunal list of tetrapods from South Pass by biochronological interval (Wa-7 to Brlb).

Wa7 (Latest Wasatchian, Lostcabinian)

- Testudines
  - Trionychidae
    - Gen. et sp. indet.
  - Emydidae
    - Gen. et sp. indet.
- Lacertilia
  - Anguidae
    - *Glyptosaurus sylvestris*
    - Large glyptosaurine sp.
    - *Xestops vagans*
  - Xenosauridae
    - *Restes rugosus*
  - Xantusiidae
    - *Palaeoxantusia* sp.
- Crocodylia
  - Crocodylidae
    - Gen. et sp. indet.
  - Crocodylidae (?)
    - *Pristichampsus vorax*
  - Caimanidae (?)
    - "*Diplocynodon*" sp.
  - Alligatoridae
    - *Procaimanoidea* sp.
- Mammalia
  - Didelphomorpha
    - Didelphidae
      - *Peradectes* sp.
      - *Peratherium* sp.
  - Palaeanodonta
    - Epoicotheriidae
      - Epoicotheriid sp. indet.
  - Condylarthra
    - Hyopsodontidae
      - *Hyopsodus* sp.
  - Cimolesta
    - Pantolestidae
      - Pantolestid sp. indet.
    - Apatemyidae
      - *Apatemys* sp.
  - Creodonta
    - Hyaenodontidae
      - *Prototomus* sp.
      - *Prolimnocyon* sp.
  - Mesonychia
    - Hapalodectidae
      - *Hapalodectes* sp.
  - Carnivora
    - Miacidae
      - *Vulpavus* sp., cf. *V. australis*
    - Viverravidae
      - *Viverravus* sp.
  - Erinaceomorpha
    - Sespedectidae
      - *Scenopagus* sp.
  - Plesiadapiformes
    - Microsyopidae
      - *Microsyops knightensis*
- Primates
  - Notharctidae
    - *Notharctus venticolis*
    - *Cantins nunienus*
  - Omomyidae
    - *Absarokius* sp.
    - *Loveina* sp.
- Perissodactyla
  - Equidae
    - *Hyracotherium* sp.
  - Brontotheriidae (?)
    - *Lambdotherium popoagicum*
- Artiodactyla
  - Dichobunidae
    - *Bunophorus* sp., cf. *B. sinclairi*
    - *Diacodexis* sp.
  - Homacodontidae
    - *Hexacodus* sp.
- Rodentia
  - Paramyidae
    - *Paramys* sp.
    - *Microparamys* sp.
  - Sciuravidae
    - Sciuravid sp. indet.

Br1a (Late Gardnerbuttean)

- Testudines
  - Baenidae
    - Gen. et sp. indet.
  - Dermatemydidae
    - *Baptemys wyomingensis*
  - Carettochelyidae
    - *Anosteira ornata*
  - Trionychidae
    - *Amyda aequa*
    - *Aspideretes* sp.
    - Gen. et sp. nov.
  - Emydidae
    - *Echmatemys wyomingensis*
    - *Echmatemys septaria*
    - Gen. et sp. nov.
- Lacertilia
  - Anguidae
    - *Glyptosaurus sylvestris*
    - Large glyptosaurine sp.
    - *Xestops vagans*
    - *Machaerosaurus torrejonensis*
  - Agamidae
    - *Tinosaurus* sp.
  - Varanidae
    - *Saniwa ensidens*
  - Xenosauridae
    - *Restes rugosus*
  - Iguanidae
    - *Parasauromalus olseni*
  - Xantusiidae
    - *Palaeoxantusia* sp.

(*continued*)

Table 1. (Continued)

- Amphisbaenia
  - Amphisbaenidae
    - Gen. et sp. indet.
- Ophidia
  - Boidae
    - *Boavus occidentalis*
    - *Dunnophis microechinus*
    - *Calamagras primus*
    - *Coniophis* sp.
- Crocodylia
  - Crocodylidae
    - "*Crocodylus*" *affinis*
  - Crocodylidae (?)
    - *Pristichampsus vorax*
  - Caimanidae (?)
    - "*Diplocynodon*" sp.
  - Alligatoridae
    - *Procaimanoidea* sp.
- Mammalia
  - Didelphomorpha
    - Didelphidae
      - *Peradectes* sp.
      - *Peratherium* sp.
  - Palaeanodonta
    - Epoicotheriidae
      - Epoicotheriid sp.
    - Metacheiromyidae
      - Metacheiromyid sp.
      - *Brachianodon westorum*
  - Dinocerata
    - Uintatheriidae
      - *Bathyopsis* sp.
  - Taeniodonta
    - Stylinodontidae
      - Stylinodont sp.
  - Tillodontia
    - Esthonychidae
      - *Esthonyx* sp.
      - *Trogosus* sp.
  - Condylarthra
    - Hyopsodontidae
      - *Hyopsodus* sp.
  - Cimolesta
    - Pantolestidae
      - Pantolestid sp.
    - Apatemyidae
      - *Apatemys* sp.
  - Leptictida
    - Leptictidae
      - Leptictid sp.
  - Creodonta
    - Hyaenodontidae
      - *Prolimnocyon* sp.
      - *Sinopa* sp.
      - *Thinocyon* sp.
      - *Tritemnodon* sp.
    - Oxyaenidae
      - *Patriofelis* sp.
  - Mesonychia
    - Mesonychidae
      - *Mesonyx*? sp.
  - Carnivora
    - Miacidae
      - *Miacis* sp.
      - *Vulpavus* sp., cf. *V. australis*
      - *Vulpavus* sp., cf. *V. canavus*
      - *Uintacyon* sp.
      - *Oödectes* sp.
    - Viverravidae
      - *Didymictis altidens*
      - *Viverravus gracilis*
      - *Viverravus sicarius*
      - *Viverravus minutus*
  - Erinaceomorpha
    - Sespedectidae
      - *Scenopagus* sp.
  - Plesiadapiformes
    - Microsyopidae
      - *Microsyops knightensis*
      - *Microsyops* sp., cf. *M. scottianus*
  - Primates
    - Notharctidae
      - *Notharctus venticolis*
      - *Cantins nunienus*
      - *Notharctus robinsoni*
      - *Smilodectes* sp. nov.
    - Omomyidae
      - *Absarokius witteri*
      - *Absarokius nocerai*
      - *Washakius izetti*
      - *Washakius insignis*
      - *Utahia* nov.
      - *Trogolemur amplior*
      - *Uintanius rutherfurdi*
      - *Omomys carteri*
      - *Omomys lloydi*
  - Perissodactyla
    - Equidae
      - *Hyracotherium* sp.
      - *Orohippus* sp.
    - Helaletidae
      - *Helaletes* sp.
      - *Hyrachyus* sp.
    - Isectolophidae
      - Isectolophid sp.
    - Brontotheriidae
      - *Eotitanops minimus*
      - *Palaeosyops fontinalis*
    - Artiodactyla
      - Dichobunidae
        - *Bunophorus* sp.
      - Homacodontidae
        - *Antiacodon* sp. small
        - *Antiacodon* sp. large
        - *Hexacodus* sp. nov.
        - *Microsus* sp. nov.
    - Rodentia
      - Cylindrodontidae
        - Cylindrodontid sp.
      - Paramyidae
        - *Microparamys* sp.
        - Paramyid sp. (3-4 species)
        - Sciuravidae
          - Sciuravid sp.

(continued)

Table 1. (Continued)

<u>Br1b (Early Blacksforkian, Bridgerian A)</u>

- Testudines
  - Baenidae
    - *Baena arenosa*
  - Dermatemydidae
    - *Baptemys wyomingensis*
  - Carettochelyidae
    - *Anosteira ornata*
  - Trionychidae
    - Gen. et sp. indet.
  - Emydidae
    - Gen. et sp. indet.
- Lacertilia
  - Anguidae
    - *Glyptosaurus* sp.
    - *Xestops vagans*
  - Varanidae
    - *Saniwa ensidens*
  - Xantusiidae
    - *Palaeoxantusia* sp.
- Ophidia
  - Boidae
    - *Boavus occidentalis*
- Crocodylia
  - Crocodylidae
    - "*Crocodylus*" *affinis*
    - "*Crocodylus*" *acer*
  - Caimanidae
    - "*Diplocynodon*" sp.
  - Alligatoridae
    - Gen. et sp. indet.
- Mammalia
  - Condylarthra
    - Hyopsodontidae
      - *Hyopsodus* sp.
  - Carnivora
    - Viverravidae
      - *Viverravus minutus*
  - Primates
    - Notharctidae
      - *Notharctus robinsoni*
      - *Smilodectes mcgrewi*
    - Omomyidae
      - *Anaptomorphus westi*
  - Perissodactyla
    - Helaletidae
      - *Helaletes* sp.
      - *Hyrachyus* sp.
    - Brontotheriidae
      - *Palaeosyops fontinalis*
  - Rodentia
    - Sciuravidae
      - Sciuravid sp. indet.

Pass as they are from basin center deposits. Among squamates, snakes and varanid and anguid lizards are all more common at South Pass with snakes and varanids being nearly absent from basin center assemblages (Figure 4). In addition, the burrowing amphisbaenians are common at South Pass but are absent from earliest Bridgerian sediments in the central part of the Green River Basin. In a negative sense, the dermatemydid turtle *Baptemys*, which is very abundant in the basin center, is represented by a single specimen at South Pass. Among crocodylians, alligatorids, and crocodylids are more common in basin center assemblages, but pristichampsines, which are exceedingly rare in the basin center, are abundant in basin margin assemblages (Figure 5). Alligators are present at South Pass but not common, while crocodylids are relatively common in both places.

Two mammalian taxa that are rarely found in basin center assemblages are relatively common at South Pass. Both the hyaenodontid creodont *Tritemnodon* and the esthonychid tillodont *Trogosus* are commonly found in the basin margin sediments at South Pass whereas *Tritemnodon* is absent from University of Michigan (UM) collections from basin center sediments (it does occur rarely in other collections) and *Trogosus* is represented by only three specimens from basin center collections at UM.

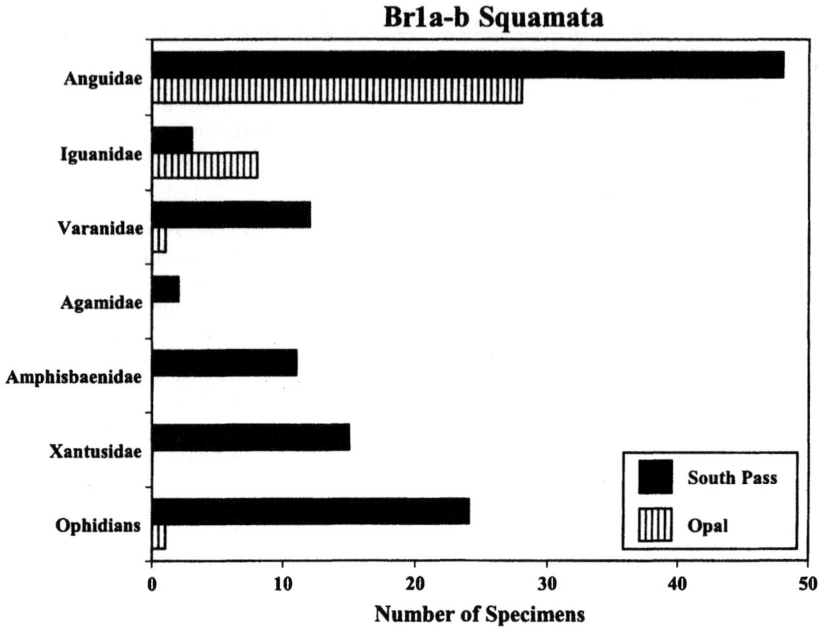

FIGURE 4. Number of specimens representing Squamata in UM collections from South Pass and Opal.

6.2. Unique Taxa

Most reptilian groups from South Pass are represented by previously known taxa, the only exceptions being probable new species of trionychid and emydid turtles. The new trionychid exhibits two character states not seen in basin center specimens, a scalloping of marginals between rib ends and the presence of fontanelles between the nuchal and first neural. The new emydid is a relatively small form similar in some regards to *Echmatemys wyomingensis* (a common basin center taxon), but distinguished by an extremely thin and wide gutter and short epiplastral lobe. Among mammals there are as many as eight new species represented in the South Pass assemblage. There are three new species of artiodactyls including small species of *Hexacodus* and *Antiacodon* and a larger species of *Microsus*. Artiodactyls in general are more diverse at South Pass than from any other earliest Bridgerian assemblage. New species of the limnocyonid creodonts *Prolimnocyon* and *Thinocyon* are also present, the former representing the latest known occurrence of *Prolimnocyon* and the latter representing the earliest known occurrence of *Thinocyon*.

There is a new taxon of mesonychian present at South Pass. It is difficult to determine whether it represents *Mesonyx* or *Synoplotherium* but whichever genus it belongs to it does represent a new species. A new, small isectolophid perissodactyl is also present in the South Pass basin margin assemblage. It shares resemblances with *Homogalax*, *Heptodon*, and *Isectolophus* but in all probability represents a new genus. Finally, there is a new species of omomyiform primate in the South Pass fauna representing the genus *Utahia*. Other, poorly represented taxa, such as *Palaeanodon* and *Esthonyx* may also be represented by new species but better specimens will be needed in order to make that determination.

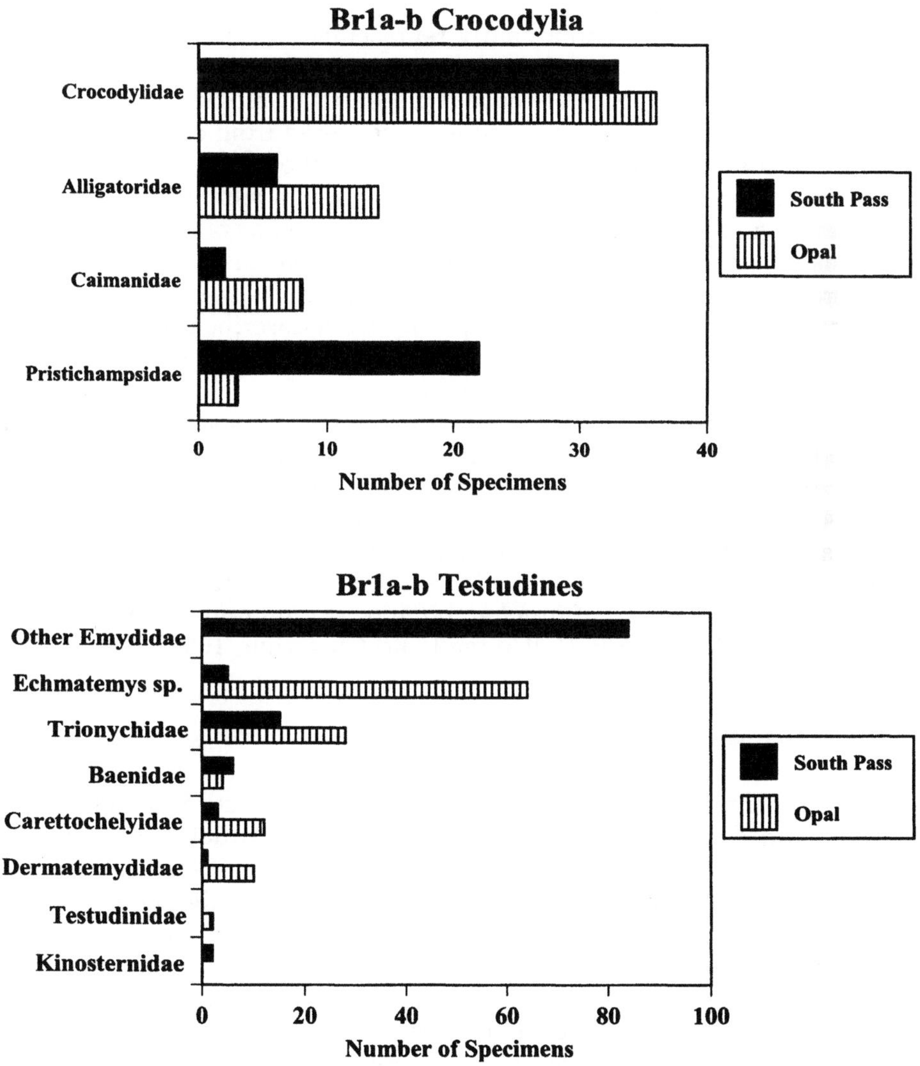

FIGURE 5. Number of specimens representing Testudines (bottom) and Crocodylia (top) in UM collections from South Pass and Opal.

6.3. Unique Morphological Forms

Several South Pass basin margin taxa differ in morphological detail from their counterparts in basin center assemblages but may not necessarily represent new species. There is reason to believe that populations occurring in marginal areas may exhibit higher variability than populations from more homogeneous habitats (Smith *et al.*, 1997; Kunin, 1998) and that some of the unique morphology shown by these taxa may be reflecting this increased variation. In some cases these forms appear as intermediates between earlier occurring and later occurring taxa.

Emydid turtles are predominantly represented at South Pass by the new taxon discussed above, but the common basin center taxa, *Echmatemys wyomingensis* and *E. septaria* are each represented by single specimens. These individuals are larger than any specimens recovered from coeval basin center deposits and are also both characterized by morphological details that differ from basin center specimens. Anguid lizards show a similar size variance in the basin margin where specimens indicate the presence of individuals (taxa?) approximately twice the size of the largest coeval basin center examples. Among crocodylians, the alligatorid *Procaimanoidea* (represented by five fairly complete skulls) is distinct morphologically from described basin center taxa (*P. utahensis* Gilmore, 1942 and *P. kayi* Mook, 1941) and from undescribed specimens form the Green River Formation.

Unique morphological forms among mammals consist almost entirely of taxa with intermediate morphologies between those seen in the preceding Lostcabinian subage and in the succeeding Blacksforkian subage. These taxa include *Antiacodon* sp., cf. *A. huerfanensis, Didymictis* sp., cf. *D. altidens, Prototomus* sp., cf. *P. multicuspis,* and *Microsyops* sp., cf. *M. scottianus.* The one exception to this may be *Brachianodon* sp., cf. *B. westorum* which appears to be a morphological variant of *Brachianodon westorum* previously only known from Bridgerian Biochron Br1b (Gunnell and Gingerich, 1993).

6.4. Anachronistic Taxa

The most interesting occurrences documented at South Pass are those that we have termed anachronistic taxa, taxa with substantial range extensions which may include presumed ancestor-descendant pairs (Bartels and Gunnell, 1997; Gunnell and Bartels, 1997, 1998). Based on previously understood, biostratigraphically ordered, occurrences of basin center taxa, certain ancestor-descendant lineages have been hypothesized. Anachronistic taxa are recognized by the co-occurrence in the same localities and horizons of these supposed ancestor-descendant taxa representing either precocious first appearances of descendants or delayed last occurrences of ancestors.

The snakes *Calamagras primus* (normally restricted to Br1a) co-occurs at South Pass with *Boavus occidentalis* (first appearing in Br1b in the basin center). Among lizards, *Palaeoxantusia* first appears at South Pass but is not found until Br1b in the basin center and both *Restes rugosus* and *Machaerosaurus torrejonensis* appear at South Pass well before their Br2 debut in the basin center (Walker, 1999). Both of these taxa are also known from considerably older (Paleocene and early Eocene) basin center deposits.

Anachronistic mammalian taxa present at South Pass include the homacodontid artiodactyl *Hexacodus* and its descendants *Microsus* and *Antiacodon*, the limnocyonid creodont *Prolimnocyon* and its descendant *Thinocyon*, the equid perissodactyl *Hyracotherium* and its descendant *Orohippus*, the brontotheriid perissodactyl *Eotitanops* and its descendant *Palaeosyops*, the esthonychid tillodont *Esthonyx* and its descendant *Trogosus*, and the metacheiromyid palaeanodont *Palaeanodon* and its descendant *Brachianodon*.

7. Anachronism and Cladogensis

7.1. Interpretations of Anachronism

The general occurrence of taxa in the basin margin prior to and or subsequent to their occupation of basin center suggests that these uplands could serve as both "incubation zones" and refugia for taxa that were not as yet or no longer competitive in the basin center environment. There are several possible explanations for the existence of ancestor-descendant anachronistic taxa at South Pass. The anachronistic pattern might be a simple sampling artifact, it could be the result of time averaging of older and younger samples, it could represent immigration from outside the Green River Basin proper, it could be the result of a collapsing or vertical compression of elevational habitat zones, or anachronism could be interpreted as direct evidence of speciation in the fossil record.

7.1.1. Sampling

Generally speaking for fossil assemblages, the larger the sample size the larger the diversity represented. That is to say, the probability of sampling rare taxa increases with larger and larger sample sizes, while the probability of "missing" a rare taxon increases if samples sizes are small (Buzas *et al.*, 1982; Koch, 1987; Badgley and Gingerich, 1988; Badgley, 1990; Smith, 1994). Just the opposite seems to be true at South Pass with reference to anachronistic taxa. Despite having a relatively small sample size (1,947 vertebrate specimens) the number of anachronistic taxa is quite high (Figure 6). There are literally thousands of specimens of fossil mammals from basin center areas in the greater Green River Basin yet never have the co-occurrences documented above for South Pass been found in these much larger samples. For example, it would be expected that, if *Prolimnocyon* and *Thinocyon* did co-exist, it would be much more likely to have been documented with the large sample from basin center sediments rather than the comparatively paltry sample from the basin

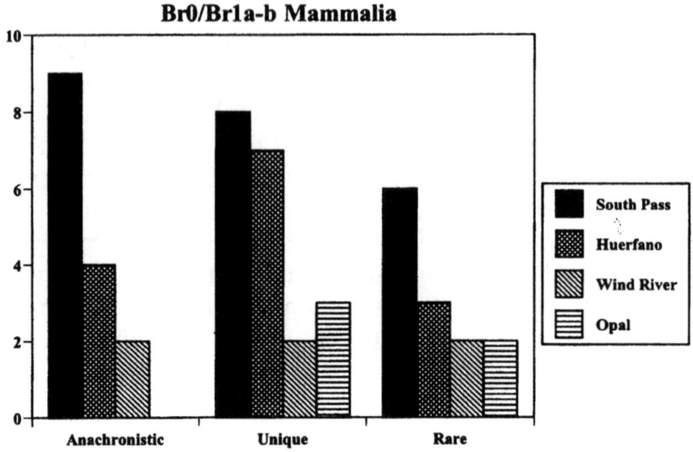

FIGURE 6. Number of rare, unique, and anachronistic mammalian taxa represented from South Pass, Huerfano, Wind River Basin, and Opal.

margin. This suggests that these co-occurrences are not just the result of sampling effects.

7.1.2. Time Averaging

It is possible that the co-occurrence of ancestor and descendant taxa at South Pass represents a faunal assemblages derived from a mixing of older and younger sediments. There is some evidence to suggest that the stratigraphic sequence at South Pass does represent a relatively long period of time. Paleosol horizons developed in the Wasatch Formation are relatively mature with relatively thick B horizons, indicating relatively long exposure times (Bown and Kraus, 1981a, b, 1993). However, other evidence argues against time averaging.

Preliminary paleomagnetic analysis (Clyde, pers. comm., 1999) at South Pass indicates a nearly identical pattern of facies shifts and ordering of reversals to that documented at Opal in the basin center (Clyde *et al.*, 1997). This does not support time averaging as an explanation of anachronism as one would expect much different sediment and reversal patterns if substantially different facies and time spans were being compared.

Reworking of older deposits is unlikely since the South Pass section was deposited directly adjacent to its continuously elevated "erosive" Precambrian source area. In addition, there is no taphonomic evidence to suggest reworking or even differential transportation of various elements of the assemblage.

There is evidence for anachronism from other basin margin assemblages such as Huerfano Park in Colorado (co-occurrences of *Esthonyx* and *Trogosus*, see McKenna, 1976; *Eotitanops* and *Palaeosyops*, see Robinson, 1966) and the Owl Creek Range in Wyoming (co-occurrence of the omomyid primates *Shoshonius* and *Absarokius* and their descendants *Washakius* and *Aycrossia*, respectively, see Bown, 1982; Honey, 1990). This indicates that anachronism is not just a local phenomenon at South Pass. Finally, the nature of the anachronisms themselves are such that they seem difficult to reconcile with time averaging. This will be discussed further below.

7.1.3. Immigration and Compression of Vertical Habitat Zones

Both of these explanations require further evidence to either support or reject them. Immigration into the basin margin area at South Pass from outside the greater Green River Basin is always a possible explanation for these co-occurrences but it is difficult to imagine where such immigrants might have come from as there is no evidence from adjacent basins to suggest the presence of the unique South Pass anachronistic sister taxon (one of the sister taxa in each anachronistic pair is nearly always a taxon unique to South Pass—see below for further discussion). Compression of vertical habitat zones is difficult to evaluate but remains a plausible possibility that requires further investigation including a much clearer understanding of modern vertical habitat zones and the interaction of vertebrate populations living within these zones. There is no evidence for long distance transport of most South Pass specimens so unless diverse habitats were directly adjacent to one another (a possibility in these marginal areas) it may be unlikely that faunal mixing from radically different habitats happened regularly during deposition (see Kunin, 1998 for a discussion of spatial "mass effects" in plant communities).

7.1.4. Cladogensis

The most intriguing possible explanation for the occurrence of anachronistic taxa is that they represent the products of cladogenesis and that speciation via cladogenesis is being documented in the basin margin area at South Pass. The recognition of ancestor-descendant taxa in the fossil record remains a contentious issue (Engelmann and Wiley, 1977; Paul, 1992; Fisher, 1994; Smith, 1994; Alroy, 1995). Foote (1996) has demonstrated that the probability of ancestor-descendant pairs co-occurring is quite high, especially under certain models of speciation. If so, the existence of anachronistic taxa at South Pass may provide additional fossil evidence to support this suggestion.

A case of simple cladogenesis is presented in Figure 7. In this case taxon A gives rise to taxa B and C via cladogenesis. This is an example of bifurcation (Foote, 1996) in which two new taxa are produced from a single ancestral taxon. Another variation of this is that taxon A may produce only a single new taxon (say taxon C) while continuing to exist itself (in this case the term budding is applied, see Foote, 1996; Pearson, 1998). One possible outcome of cladogenesis in the fossil record is presented at the bottom of Figure 7. Taxon A undergoes cladogenesis and produces new taxa B and C where B is a new species of genus A while C is a new genus, recognized as such because it shares derived character states with a previously known genus from a younger stratigraphic horizon (that is to say, it is then the ancestral taxon to the younger

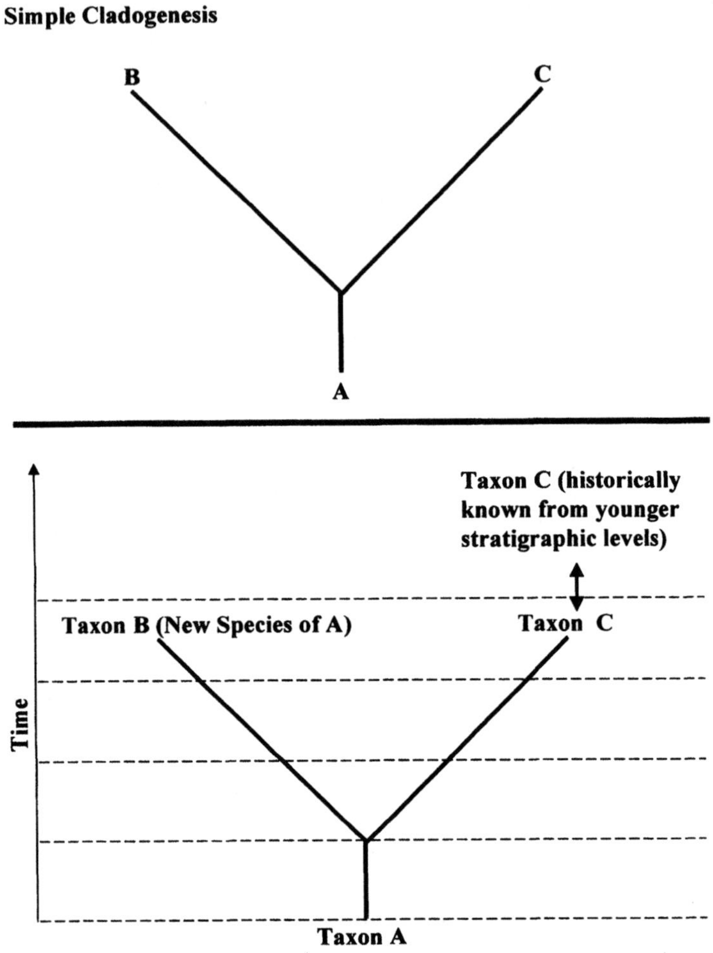

FIGURE 7. Top represents a simple example of cladogenesis where taxon A gives rise to taxa B and C. Bottom represents one possible representation of this cladogenic event as viewed in the fossil record where time is now added to the vertical axis. In this example, taxon A gives rise to two new taxa, B and C, where B represents a new species of the taxon A genus while C represents a new genus distinct from the genus represented by taxa A and B. In this case, taxon C is recognized as a new genus (or a distinct genus) because it shares derived character states with a previously known genus from a younger stratigraphic interval.

occurring genus). Taxon C may or may not be the same species as that represented by the later occurring genus. If the example presented in Figure 7 were instead a case of budding, then taxon B would simply be the same genus and species as taxon A.

This gets us back to the nature of anachronism at South Pass as mentioned above in section 7.1.2. In almost every case at South Pass, taxon B represents a new species of taxon A. It is difficult to reconcile this with simple time averaging as an explanation because one would then expect taxon C and taxon A to co-occur without the presence of taxon B. The fact that taxon B exists and represents a new species of taxon A suggests that mixing of time horizons is not a plausible explanation for the occurrence of anachronistic taxa, at least in the South Pass assemblage.

7.2. Examples of Anachronism

If we examine the co-occurrence of presumed ancestors and descendant at South Pass more closely it becomes apparent that, in most cases, bifurcate cladogenesis is the implied pattern of speciation. Figure 8 shows the pattern of homacodontid artiodactyl evolution for the early middle Eocene. In the top

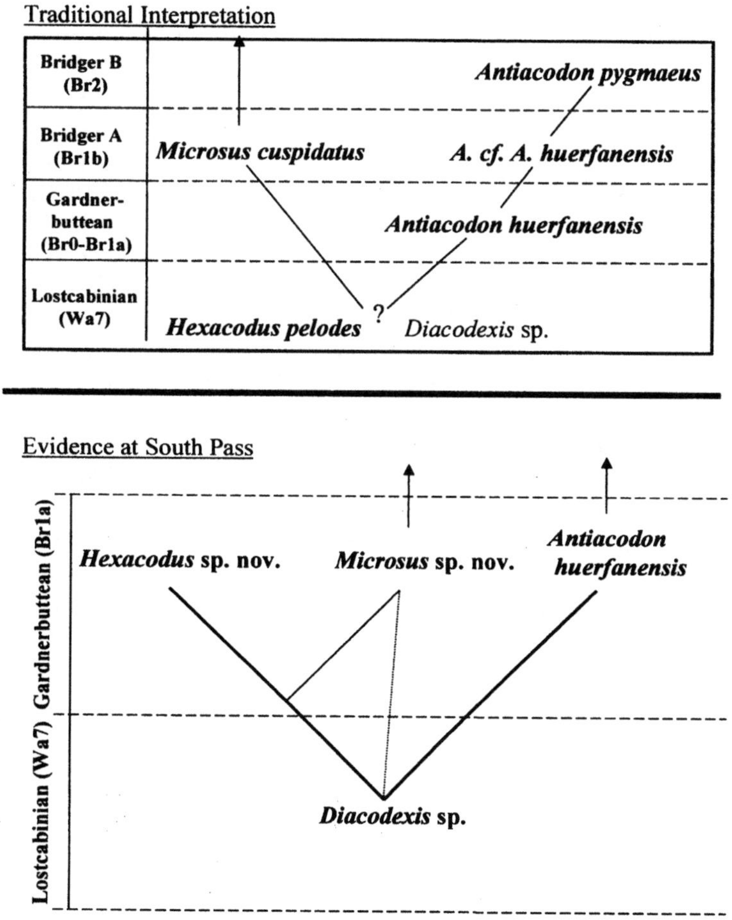

FIGURE 8. Top represents traditional interpretation of Wasatchian/Bridgerian homacodontid artiodactyl phylogeny where *Hexacodus* gives rise to *Microsus* and succeeding species of *Antiacodon*. Bottom represents situation at South Pass where *Hexacodus, Microsus*, and *Antiacodon* all co-occur.

figure is the traditional interpretation with *Hexacodus pelodes* giving rise to *Antiacodon huerfanensis* and *Microsus cuspidatus* via cladogenesis with no overlap of ancestor and descendant taxa. The *Antiacodon* lineage then undergoes phyletic transformation (Foote, 1996) via pseudoextinction to produce *Antiacodon pygmaeus* from *Antiacodon huerfanensis*.

The bottom figure shows the evolutionary pattern documented at South Pass. *Hexacodus* sp. (probably *H. pelodes*) occurs in the Lostcabinian and gives rise to three co-occurring taxa in the Gardnerbuttean, a new species of *Hexacodus*, a new species of *Microsus*, and *Antiacodon huerfanensis*. The new species of *Hexacodus* and *Microsus* are recognized as representing those genera because they share derived character states with a previously occurring form in the case of *Hexacodus* and a later occurring form in the case of *Microsus*, while *Antiacodon huerfanensis* is recognized as such because it shares derived character states with that species from a later occurring horizon. The new species of *Hexacodus* and *Microsus* differ from *H. pelodes* and *M. cuspidatus*, respectively, in size and morphological detail. There may be an additional species of *Antiacodon* present at South Pass further complicating interpretation of the pattern of artiodactyl evolution documented there.

A similar pattern is documented among brontotheriid perissodactyls both from South Pass and Huerfano Park, another basin margin or upland area in southern Colorado. The traditional pattern (Figure 9) suggests that *Eotitanops borealis* gave rise to *Palaeosyops fontinalis* and then the rest of the *Palaeosyops* lineage via phyletic transformation and/or cladogenesis, possibly from a *Lambdotherium*-like ancestry. At both South Pass and Huerfano, *Palaeosyops fontinalis* co-occurs with a new species of *Eotitanops* (or at least a different species of *Eotitanops*), *E. minimus*. In each of these cases, the pattern is consistent with cladogenesis via bifurcation.

Other examples of bifurcation exist at South Pass including co-occurrences of *Prolimnocyon* and *Thinocyon*, *Utahia* and *Washakius*, and possibly *Echmatemys* and the new emydid turtle.

There are also two cases of possible budding at South Pass wherein the ancestral taxon co-occurs with its descendant without significant modification of the ancestor. One of these occurs in metacheiromyid palaeanodonts where *Palaeanodon* sp. and its presumed descendant *Brachianodon* sp., cf. *B. westorum* co-occur. It is difficult to be certain of the specific identity of the South Pass *Palaeanodon* specimens so it remains to be seen if this clade does represent budding cladogenesis or not.

The other case involves equid perissodactyls where *Orohippus* sp. co-occurs with two distinct species of *Hyracotherium*, *H. vasacciense* and *H. index*, both of which are known from the previous Lostcabinian interval. There are several complicating factors involved in the interpretation of early Bridgerian equid evolution. First, it is not clear which of the two *Hyracotherium* species might be more closely related to *Orohippus*. The two genera differ in only a few diagnostic characters mostly involving the upper premolars such that incomplete specimens are difficult to place taxonomically. Secondly *Orohippus* may well occur in the Lostcabinian (Korth and Evander, 1982). I

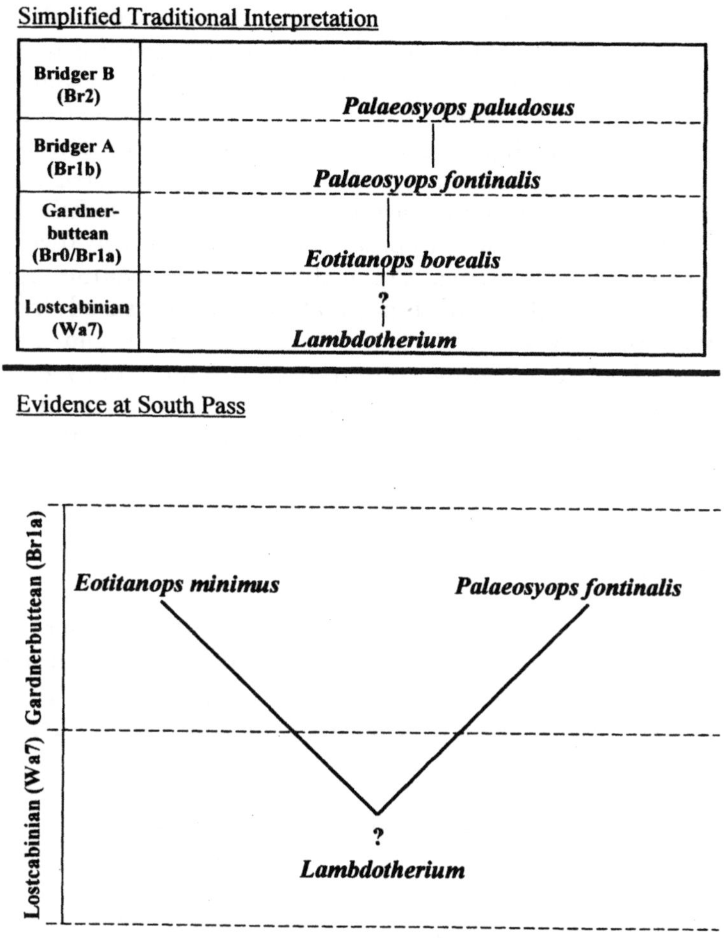

FIGURE 9. Top represents traditional interpretation of Wasatchian/Bridgerian brontotheriid perissodactyl phylogeny where *Lambdotherium* gives rise to *Eotitanops* and then succeeding species of *Palaeosyops*. Bottom represents situation at South Pass where *Eotitanops* and *Palaeosyops* co-occur.

so, then the pattern at South Pass may simply reflect a continuation of co-occurring lineages present in the Lostcabinian.

8. Speciation

Thus far we have presented an argument suggesting that, based on the pattern of co-occurrence of taxa documented at South Pass, several mammalian lineages are undergoing cladogenesis during this time and that the patterns are consistent with a bifurcating speciation model for the most part (Foote, 1996). What other lines of evidence can be gathered that might indicate that speciation actually is being documented in the fossil record at South Pass? Vertebrate populations in basin margin environments should undergo higher levels of stress from competing populations as they are more closely spaced and more restricted in their ranges. Under these conditions of isolation (from main basin center populations) and higher populational stress (assumptions of the peripatric speciation model, see Mayr, 1974), it could be expected that evolutionary innovation might have a greater chance of being successfully incorporated into the genetic structure of vertebrate populations and in turn increase the chances of speciation occurring. Evidence of such innovation might be manifest as increased variability within populations and increased diversity within higher taxonomic categories (Williamson, 1981; Kerr and Packer, 1997). It should also lead to differing community structures between upland and lowland faunas with a higher incidence of unique taxa in upland or basin margin communities (reflected as increased diversity and differing taxonomic composition).

8.1. Variation

Figure 10 shows the coefficient of variation (CV) for three different taxa of mammals from South Pass compared with the same genera (not necessarily the same species, however) from basin center samples (Wind River Basin for *Notharctus* and Opal/Big Piney for *Microsyops* and *Hyopsodus*). The CV's are

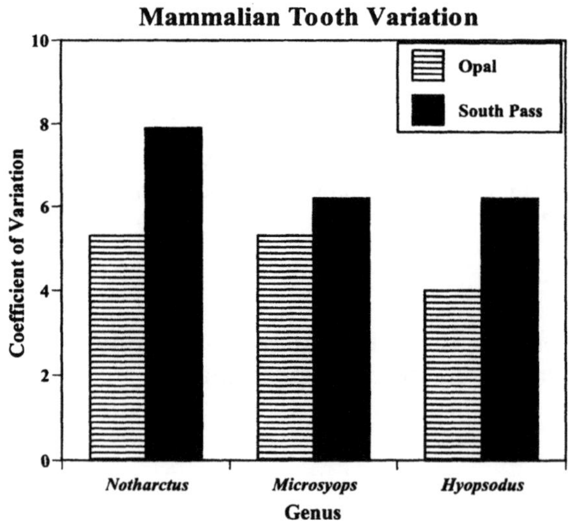

FIGURE 10. Mammalian tooth variability as expressed by coefficient of variation (CV) of tooth area for three different mammalian orders, Primates (*Notharctus*), Plesiadapiformes (*Microsyops*), and Condylarthra (*Hyopsodus*). Note that in each case, tooth variability is higher in the South Pass samples than those from Opal. All CV's are based on at least ten individual specimens. CV = Standard Deviation/Mean × 100.

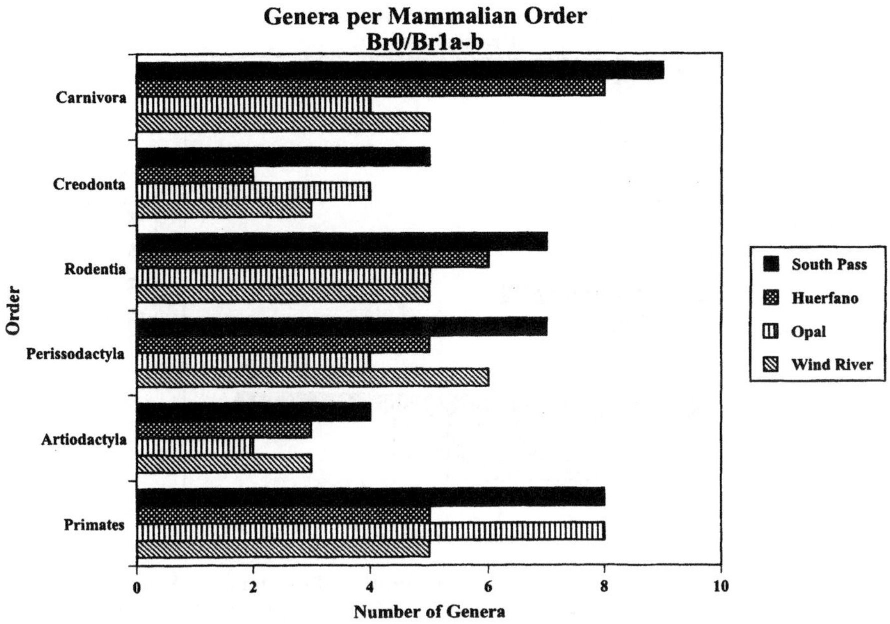

FIGURE 11. Number of genera representing six orders of mammals in Wasatchian/Bridgerian transitional samples from the Wind River Basin, Huerfano Park, South Pass, and Opal.

based on upper first molars for *Notharctus*, lower second molars for *Microsyops*, and lower first molars for *Hyopsodus*. These teeth were chosen because they represent the largest available sample for each taxon from South Pass (N = 10 for all South Pass species and greater than 10 for all basin center samples). In each case, the variation is higher in the South Pass sample than it is in the basin center sample. While not necessarily significant, these differences do suggest that mammalian variability as expressed by within species tooth size is greater in basin margin habitats.

As mentioned above trionychid and emydid turtles from South Pass exhibit greater variability than is seen in their counterparts from basin center habitats. South Pass specimens of the alligatorid *Procaimanoidea* are also highly variable. Among the four complete skulls of *Procaimanoidea* known from South Pass, each is more different from the others in morphological detail than are the five previously described basin center *Procaimanoidea* specimens in total.

8.2. Diversity

Figures 11 and 12 show mammalian diversity expressed as genera per order and species per family, respectively. In both figures, South Pass is compared with two basin center assemblages, the Wind River Basin in

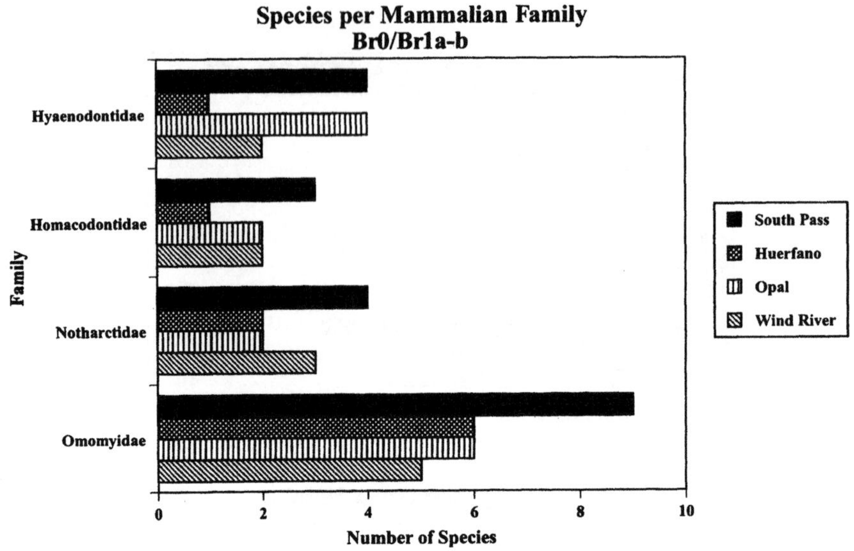

FIGURE 12. Number of species representing five families of mammals in Wasatchian/Bridgerian transitional samples from the Wind River Basin, Huerfano Park, South Pass, and Opal.

southcentral Wyoming and the Opal/Big Island Badlands in southwestern Wyoming and a basin margin or upland assemblage from Huerfano Park in Colorado. The Wind River assemblage is from the early Gardnerbuttean (Bridgerian Biochron Br0), the Opal/Big Island assemblage is from the earliest Blacksforkian (Br1b), and Huerfano and South Pass represent the late Gardnerbuttean (Br1a). In each case, the number of genera representing five of the more common orders of mammals from each area is as high or higher at South Pass than it is in all other areas. The same pattern holds for the number of species per family with South Pass having an equal or greater number of species for any given family than does any other assemblage. Interestingly, the Huerfano Park upland assemblage is not as diverse as that of South Pass, and in some cases, the Wind River and Opal/Big Island assemblages (probably partially due to the relatively small sample sizes from Huerfano).

8.3. Vertebrate Community Structure

The composition of the vertebrate community at South Pass differs substantially from communities based on basin center assemblages reflecting both differing habitat utilization and preferences and vertebrate diversity. As noted above, South Pass crocodylians are dominated by crocodylids and pristichampsids while crocodylids, caimanids, and alligatorids are more common in basin center assemblages. Basin margin squamates are dominated by anguid and varanid lizards and snakes while iguanids and anguid lizards are the dominate squamates from basin center assemblages (see Figures 4 and 5). Among turtles, emydids other than *Echmatemys* are the most common turtles at South Pass while *Echmatemys*, trionychids, and dermatemydids are more common from basin center areas.

The most common orders of mammals at South Pass in terms of total numbers of specimens are rodents, perissodactyls, primates, condylarths, and plesiadapiforms. Except for a larger number of plesiadapiforms at Huerfano and a much larger number of primates from Opal/Big Island, the same orders dominate each of these assemblages. It is within orders that taxonomic composition differs between these areas.

Within Perissodactyla (Figure 13) helaletids are by far the most common family at South Pass, while equids, brontotheriids, and isectolophids make up smaller percentages of perissodactyls. This pattern is similar to the basin center assemblage from Opal/Big Island but differs from the upland Huerfano assemblage and the basin center Wind River assemblage, both of which are dominated by brontotheriids. Helaletids are less common at Huerfano and Wind River while equids are about the same as at South Pass and Opal/Big Island. Only the South Pass and Wind River assemblages contain isectolophids.

A closer examination of perissodactyls (Figure 14) reveals other interesting patterns. Within brontotheriids for example, the Wind River assemblage contains only *Eotitanops* while the basin margin samples from Huerfano and South Pass contain co-occurring *Eotitanops* and *Palaeosyops*. At Opal/Big Island only *Palaeosyops* occurs. Within equids, *Hyracotherium* and *Orohippus*

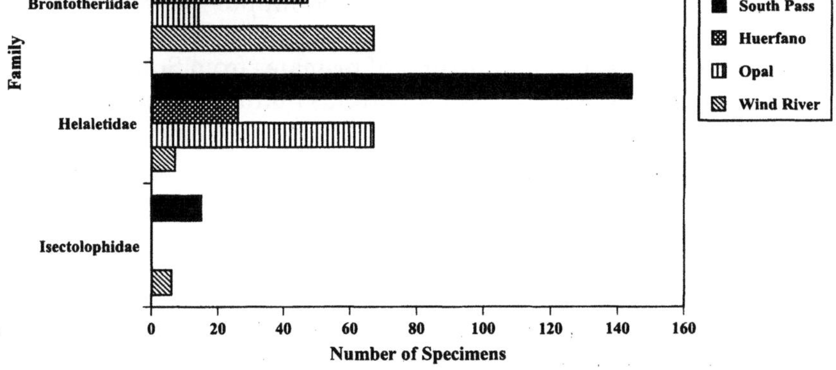

FIGURE 13. Percentage of the number of specimens representing four different perissodactyl families in Wasatchian/Bridgerian transitional samples from the Wind River Basin, Huerfano Park, South Pass, and Opal.

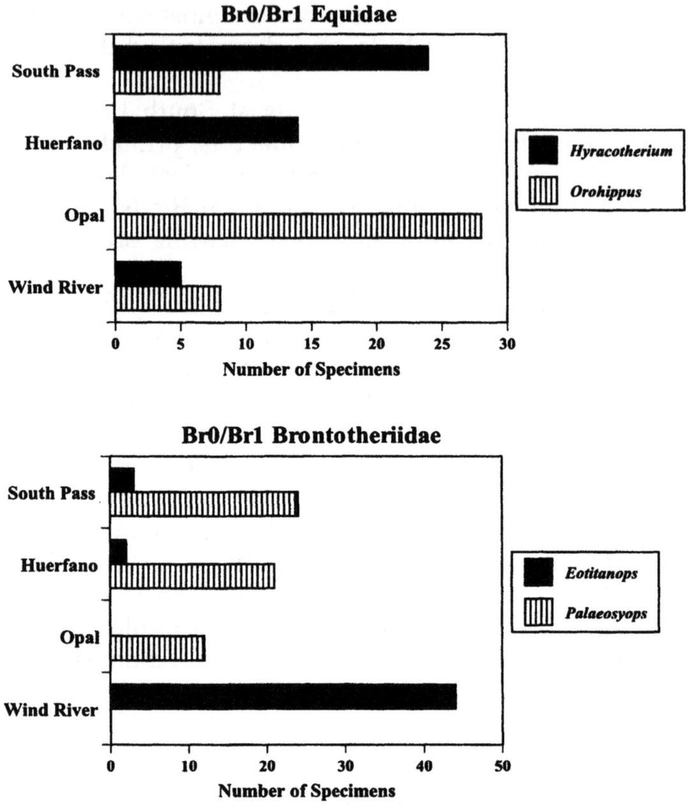

FIGURE 14. Number of specimens representing equid perissodactyls (top) and brontotheriid perissodactyls (bottom) in Wasatchian/Bridgerian transitional samples from the Wind River Basin, Huerfano Park, South Pass, and Opal.

both occur in the assemblages at Wind River and South Pass but only *Hyracotherium* is present at Huerfano and only *Orohippus* is present at Opal/Big Island.

Figure 15 examines the composition of primates from South Pass, Opal/Big Island, and Huerfano. At South Pass notharctids outnumber omomyids by more than two to one, while at Huerfano and Opal/Big Island both families are about equally represented.

9. Conclusions

Faunal assemblages from the basin margin at South Pass are different in content and composition from time equivalent basin center assemblages from the greater Green River Basin. The available evidence excludes sampling effects, time averaging, immigration, and mixing of faunal samples from adjacent differing habitats as explanations for this phenomenon.

The preponderance of evidence from higher diversity, higher variability, differing community structure, and the presence of anachronistic taxa all strongly suggest that allopatric speciation via bifurcate cladogenesis was occurring in these basin margin habitats and that the faunal assemblages at South Pass reflect the outcome of these speciation events. We believe that marginal areas provided places where speciation could and did occur because of the increased likelihood of successful evolutionary innovation in these more heterogeneous habitats and because of increased populational stresses experienced by these margin-dwelling vertebrate communities due to restriction of habitat ranges and because of genetic isolation of these populations from main populations living in basin center areas.

As such, marginal areas may have acted as "species pumps" by providing new species and increasing genetic variability as populations experimented with different morphological adaptations. The "sudden" appearance of rare new species in more static basin center assemblages can be explained by invasion of these areas by marginal populations. It is clear that biochronologies based on traditional faunal samples from basin center sediments under-represent biodiversity and can lead to under-estimation of the temporal range of vertebrate taxa. Such under-estimation will directly effect NALMA biochronology because it is based on first and last appearance data of index taxa. For this reason, sampling across wide ranging geographic areas and multiple habitats is crucial to the construction of more rigorous biochronological sequences.

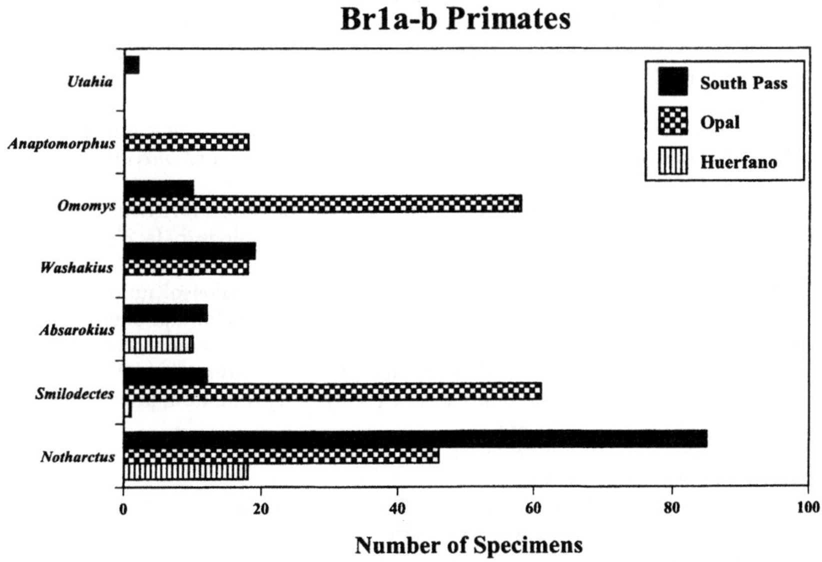

FIGURE 15. Number of specimens representing seven different primate genera in Wasatchian/Bridgerian transitional samples from Huerfano Park, South Pass, and Opal.

ACKNOWLEDGMENTS. We wish to thank all of our colleagues who have worked with us at South Pass. Without their dedication little could have been accomplished. Special thanks go to George H. Junne, Jr., Conduff G. Childress, M. Elizabeth Strasser, Rachael J. Walker, John-Paul Zonneveld, William C. Clyde, Charles Coupar, James A. Pope, Jonathan I. Bloch, Darrell L. Robbins, Stephanie B. Gaswirth, Joshua A. Trapani, Magdalena K. Hutna, and Kathleen M. Muldoon. The authors have benefited greatly from discussions with Daniel C. Fisher, Philip D. Gingerich, Catherine E. Badgley, Jonathan I. Bloch, David L. Fox, Mark D. Uhen, Robyn J. Burnham, Linda C. Ivany, Ellen R. Miller, Amy Chew, and William J. Sanders. We also thank the staff at the Bureau of Land Management, Wyoming, especially Laurie Bryant, who have facilitated our work in numerous ways. Field work has been supported by the National Science Foundation (EAR-9619908 to WSB and SBR-9705264 to GFG), the Wenner-Gren Foundation, the Carnegie-Mellon Fund at Albion College, and the Office of the Vice President for Research and the Museum of Paleontology at the University of Michigan.

References

Alroy, J., 1995, Continuous Track Analysis: A new phylogenetic and biogeographic method, *Systematic Biology* **44:**152–178.

Badgley, C., 1990, A statistical assessment of last appearances in the Eocene record of mammals, in: *Dawn of the Age of Mammals in the northern part of the Rocky Mountain Interior, North America,* Special Paper 243 (T. M. Bown and K. D. Rose, eds.), Geological Society of America, pp. 153–167.

Badgley, C., and Gingerich, P. D., 1988, Sampling and faunal turnover in early Eocene mammals, *Palaeogeography, Palaeoclimatology, Palaeoecology* **63:**141–157.

Bartels, W. S., and Gunnell, G. F., 1997, Basin margin faunas and the origin of North American Land Mammal Age faunal turnover, *Journal of Vertebrate Paleontology* **17(3):**31A.

Black, C. C., 1967, Middle and late Eocene mammal communities: A major discrepancy, *Science* **156:**62–64.

Bown, T. M., 1982, Geology, paleontology, and correlation of Eocene volcaniclastic rocks, southeast Absaroka Range, Hot Springs County, Wyoming, *Geological Survey Professional Paper* **1201-A:**A1–A75.

Bown, T. M., and Kraus, M. J., 1981a, Lower Eocene alluvial paleosols (Willwood Formation, northwestern Wyoming, U.S.A.) and their significance for paleoecology, paleoclimatology, and basin analysis, *Palaeogeography, Palaeoclimatology, Palaeoecology* **34:**1–30.

Bown, T. M., and Kraus, M. J., 1981b, Vertebrate fossil-bearing paleosol units (Willwood Formation, lower Eocene, northwest Wyoming, U.S.A.): Implications for taphonomy, biostratigraphy, and assemblage analysis, *Palaeogeography, Palaeoclimatology, Palaeoecology* **34:**31–56.

Bown, T. M., and Kraus, M. J., 1993, Time-stratigraphic reconstruction and integration of paleopedologic, sedimentologic, and biotic events (Willwood Formation, lower Eocene, northwest Wyoming, U.S.A.), *Palaios* **8:**68–80.

Buzas, M. A., Koch, C. F., Culver, S. J., and Sohl, N. F., 1982, On the distribution of species occurrence, *Paleobiology* **8:**142–150.

Clyde, W. C., Zonneveld, J.-P., Stamatakos, J., Gunnell, G. F., and Bartels, W. S., 1997, Magnetostratigraphy across the Wasatchian/Bridgerian NALMA Boundary (Early to Middle Eocene) in the western Green River Basin, Wyoming, *Journal of Geology* **105:**657–669.

DiMichele, W. A., Phillips, T. L., and Peppers, R. A., 1985, The influence of climate and depositional environment on the distribution and evolution of Pennsylvanian coal-swamp plants, in: *Geological factors and the evolution of plants* (B. Tiffney, ed.), Yale University Press, New Haven, pp. 223–256.

Engelmann, G. F., and Wiley, E. O., 1977, The place of ancestor-descendant relationships in phylogeny, *Systematic Zoology* **26**:1–11.

Fisher, D. C., 1994, Stratocladistics: morphological and temporal patterns and their relation to phylogenetic process, in: *Interpreting the hierarchy of nature: from systematic patterns to evolutionary process theories*, (L. Grande and O. Rieppel, eds.), Academic Press, San Diego, pp. 133–171.

Foote, M., 1996, On the probability of ancestors in the fossil record, *Paleobiology* **22**:141–151.

Gilmore, C. W., 1942, A new crocodilian from the Eocene of Utah, *Journal of Paleontology* **20**:62–67.

Gunnell, G. F., and Bartels, W. S., 1994, Early Bridgerian (middle Eocene) vertebrate paleontology and paleoecology of the southern Green River Basin, Wyoming, *Contributions to Geology, University of Wyoming* **30**:57–70.

Gunnell, G. F., and Bartels, W. S., 1997, Basin-margin mammalian assemblages from the Wasatch Formation (Bridgerian) of the northeastern Green River Basin, Wyoming—Anachronistic taxa and the origin of new genera, *Journal of Vertebrate Paleontology* **17**:51A.

Gunnell, G. F., and Bartels, W. S., 1998, Basin margins and morphologic divergence: Paleontologic documentation of cladogenesis and evolutionary innovation, *Journal of Vertebrate Paleontology* **18**:47A.

Gunnell, G. F., and Gingerich, P. D., 1993, Skeleton of *Brachianodon westorum*, a new middle Eocene metacheiromyid (Mammalia, Palaeanodonta) from the early Bridgerian (Bridger A) of the southern Green River Basin, Wyoming. *Contributions from the Museum of Paleontology, University of Michigan* **28**:365–392.

Gunnell, G. F., and Yarborough, V. L., 2000, Brontotheriidae (Perissodactyla) from the late early and middle Eocene (Bridgerian), Wasatch and Bridger Formations, southern Green River Basin, southwestern Wyoming, *Journal of Vertebrate Paleontology* **20**:349–368.

Honey, J. G., 1990, New washakin primates (Omomyidae) from the Eocene of Wyoming and Colorado, and comments on the evolution of the Washakiini, *Journal of Vertebrate Paleontology* **10**:206–221.

Horner, J. R., Varricchio, D. J., and Goodwin, M. B., 1992, Marine transgressions and the evolution of Cretaceous dinosaurs, *Nature* **358**:59–61.

Kerr, J. T., and Packer, L., 1997, Habitat heterogeneity as a determinant of mammal species richness in high-energy regions, *Nature* **385**:252–254.

Koch, C. F., 1987, Prediction of sample size effects on the measured temporal and geographic distribution patterns of species, *Paleobiology* **13**:100–107.

Korth, W. W., and Evander, R. L., 1982, A new species of *Orohippus* (Perissodactyla, Equidae) from the early Eocene of Wyoming, *Journal of Vertebrate Paleontology* **2**:167–171.

Krishtalka, L., West, R. M., Black, C. C., Dawson, M. R., Flynn, J. J., Turnbull, W. D., Stucky, R. K., McKenna, M. C., Bown, T. M., Golz, D. J., and Lillegraven, J. A., 1987, Eocene (Wasatchian through Duchesnean) biochronology of North America, in: *Cenozoic mammals of North America—Geochronology and biostratigraphy*, (M. O. Woodburne, ed.), University of California Press, Berkeley, pp. 77–117.

Kunin, W. E., 1998, Biodiversity at the edge: A test of the importance of spatial "mass effects" in the Rothamsted Park grass experiments, *Proceedings of the National Academy of Science, USA* **95**:207–212.

Lillegraven, J. A., and Tabrum, A. R., 1983, A new species of *Centetodon* (Mammalia, Insectivora, Geolabididae) from southwestern Montana and its biogeographical implications, *Contributions to Geology, University of Wyoming* **22**:57–73.

Matthew, W. D., 1909, The Carnivora and Insectivora of the Bridger Basin, middle Eocene, *Memoirs of the American Museum of Natural History* **9**:291–567.

Mayr, E., 1974, Cladistic analysis or cladistic classification, *Zeitschrift für Zoologische Systematik und Evolutionsforschung* **12**:94–128.

McKenna, M. C., 1976, *Esthonyx* in the upper faunal assemblage, Huerfano Formation, Eocene of Colorado, *Journal of Paleontology* **50**:354–355.

Mook, C. C., 1941, A new crocodilian, *Hassiacosuchus kayi*, from the Bridger Eocene beds of Wyoming, *Annals of the Carnegie Museum* **28**:207–221.

North American Commission on Stratigraphic Nomenclature, 1983, North American Stratigraphic Code, *The American Association of Petroleum Geologists Bulletin* **67**:841–875.

Paul, C. R. C., 1992, The recognition of ancestors, *Historical Biology* **6**:239–250.

Pearson, P. N., 1998, Speciation and extinction asymmetries in paleontological phylogenies: evidence for evolutionary progress?, *Paleobiology* **24**:305–335.

Prothero, D. R., 1998, The chronological, climatic, and paleogeographic background to North American mammalian evolution, in: *Evolution of Tertiary Mammals of North America—Volume 1: Terrestrial Carnivores, Ungulates, and Ungulatelike Mammals*, (C. M. Janis, K. M. Scott, and L. L. Jacobs, eds.), Cambridge University Press, Cambridge, pp. 9–36.

Prothero, D. R., and Swisher III, C. C., 1992, Magnetostratigraphy and geochronology of the terrestrial Eocene-Oligocene transition in North America, in: *Eocene-Oligocene climatic and biotic evolution*, (D. R. Prothero and W. A. Berggren, eds.), Princeton University Press, Princeton, pp. 46–73.

Robinson, P., 1966, Fossil Mammalia of the Huerfano Formation, Eocene of Colorado, *Bulletin of the Peabody Museum of Natural History, Yale University* **21**:1–95.

Rosenzweig, M. L., 1995, *Species diversity in space and time*, Cambridge University Press, Cambridge.

Smith, A. B., 1994, *Systematics and the fossil record documenting evolutionary patterns*, Blackwell Scientific Publications, Oxford.

Smith, T. B., Wayne, R. K., Girman, D. J., and Bruford, M. W., 1997, A role for ecotones in generating rainforest biodiversity, *Science* **276**:1855–1857.

Steidtmann, J. R., McGee, L. C., and Middleton, L. T., 1983, Laramide sedimentation, folding, and faulting in the southern Wind River Range, Wyoming, in: *Rocky Mountain foreland basins and uplifts*, (J. D. Lowell and R. Gries, eds.), Rocky Mountain Association of Geologists, Denver, pp. 161–168.

Stucky, R. K., 1984, Revision of the Wind River faunas, early Eocene of central Wyoming. Part 5. Geology and biostratigraphy of the upper part of the Wind River Formation, northeastern Wind River Basin, *Annals of the Carnegie Museum* **53**:231–294.

Tedford, R. H., 1970, Principles and practices of mammalian geochronology in North America, *Proceedings of the North American Paleontological Convention* **2F**:666–703.

Walker, R. J., 1999, An analysis of the herpetofauna and paleoenvironment of the Wasatch and Bridger formations (middle Eocene), at South Pass, Wyoming, Unpublished Ph.D. Dissertation, Michigan State University, East Lansing, Michigan, 1–536.

Williamson, P. G., 1981, Paleontological documentation of speciation in Cenozoic molluscs from Turkana Basin, *Nature* **293**:437–443.

Wood II, H. E., 1934, Revision of the Hyrachyidae, *Bulletin of the American Museum of Natural History* **67**:181–295.

Wood II, H. E., Chaney, R. W., Clark, J., Colbert, E. H., Jepsen, G. L., Reeside Jr., J. B., and Stock, C., 1941, Nomenclature and correlation of the North American continental Tertiary, *Geological Society of America Bulletin* **52**:1–48.

Woodburne, M. O., 1977, Definition and characterization in mammalian chronostratigraphy, *Journal of Paleontology* **51**:220–234.

Woodburne, M. O., (ed.), 1987, *Cenozoic Mammals of North America—Geochronology and Biostratigraphy*, University of California Press, Berkeley, 336 pages.

Index

GPSR Compliance

The European Union's (EU) General Product Safety Regulation (GPSR) is a set of rules that requires consumer products to be safe and our obligations to ensure this.

If you have any concerns about our products, you can contact us on ProductSafety@springernature.com

In case Publisher is established outside the EU, the EU authorized representative is:

Springer Nature Customer Service Center GmbH
Europaplatz 3
69115 Heidelberg, Germany

Batch number: 10755329

Printed by Printforce, the Netherlands